# Cambridge IGCSE™

# International Mathematics

## Third edition

Ric Pimentel
Frankie Pimentel
Terry Wall

Endorsement indicates that a resource has passed Cambridge International's rigorous quality-assurance process and is suitable to support the delivery of a Cambridge International syllabus. However, endorsed resources are not the only suitable materials available to support teaching and learning, and are not essential to be used to achieve the qualification. Resource lists found on the Cambridge International website will include this resource and other endorsed resources. Any example answers to questions taken from past question papers, practice questions, accompanying marks and mark schemes included in this resource have been written by the authors and are for guidance only. They do not replicate examination papers. In examinations the way marks are awarded may be different. Any references to assessment and/or assessment preparation are the publisher's interpretation of the syllabus requirements. Examiners will not use endorsed resources as a source of material for any assessment set by Cambridge International. While the publishers have made every attempt to ensure that advice on the qualification and its assessment is accurate, the official syllabus specimen assessment materials and any associated assessment guidance materials produced by the awarding body are the only authoritative source of information and should always be referred to for definitive guidance. Cambridge International recommends that teachers consider using a range of teaching and learning resources based on their own professional judgement of their students' needs. Cambridge International has not paid for the production of this resource, nor does Cambridge International receive any royalties from its sale. For more information about the endorsement process, please visit www.cambridgeinternational.org/endorsed-resources.

Cambridge International copyright material in this publication is reproduced under licence and remains the intellectual property of Cambridge Assessment International Education.

**Photo credits**

**p.3** © Aleksandra Antic/Shutterstock; **p.4** tl © Natthapon Ngamnithiporn/123RF; **p.4** tr © Pavel Lipskiy/123RF; **p.4** b © William Rossin/123RF; **pp.22–3** © Aleksandra Antic/Shutterstock; **p.23** © Dinodia Photos/Alamy Stock Photo; **pp.114–5** © katjen/Shutterstock; **p.115** © Eduard Kim/Shutterstock; **pp.206–7** © Viktar Malyshchyts/123RF; **p.207** Image from *The Nine Chapters on the Mathematical Art*, published in 1820 via Wikimedia Commons (https://en.wikipedia.org/wiki/Public_domain); **pp.274–5** © Halfpoint/Shutterstock; **p.275** © Georgios Kollidas/123RF; **pp.302–3** © ESB Professional/Shutterstock; **p.303** © Classic Image/Alamy Stock Photo; **pp.364–5** © WitR/Shutterstock; **p.365** © Print Collector/HIP/TopFoto; **pp.414–5** © 3Dsculptor/Shutterstock; **p.415** © Granger, NYC/TopFoto; **pp.484–5** © Anton Petrus/Shutterstock; **p.485** © Science History Images/Alamy Stock Photo; **pp.522–3** © Harvepino/Shutterstock; **p.523** © Bernard 63/Fotolia; **pp.554–5** © Shutterstock; **p.555** © Caifas/stock.adobe.com

Every effort has been made to trace all copyright holders, but if any have been inadvertently overlooked, the Publishers will be pleased to make the necessary arrangements at the first opportunity. All exam-style questions and sample answers in this title were written by the authors.

Although every effort has been made to ensure that website addresses are correct at time of going to press, Hodder Education cannot be held responsible for the content of any website mentioned in this book. It is sometimes possible to find a relocated web page by typing in the address of the home page for a website in the URL window of your browser.

Hachette UK's policy is to use papers that are natural, renewable and recyclable products and made from wood grown in well-managed forests and other controlled sources. The logging and manufacturing processes are expected to conform to the environmental regulations of the country of origin.

Orders: please contact Hachette UK Distribution, Hely Hutchinson Centre, Milton Road, Didcot, Oxfordshire, OX11 7HH. Telephone: +44 (0)1235 827827. Email education@hachette.co.uk Lines are open from 9 a.m. to 5 p.m., Monday to Friday. You can also order through our website: www.hoddereducation.com

© Ric Pimentel, Terry Wall and Frankie Pimentel 2023

First published in 2011

Second edition published in 2018

This edition published in 2023 by
Hodder Education, an Hachette UK Company, Carmelite House, 50 Victoria Embankment, London EC4Y 0DZ

www.hoddereducation.com

| Impression number | 10 9 8 7 6 5 4 3 2 1 |
|---|---|
| Year | 2027 2026 2025 2024 2023 |

All rights reserved. Apart from any use permitted under UK copyright law, no part of this publication may be reproduced or transmitted in any form or by any means, electronic or mechanical, including photocopying and recording, or held within any information storage and retrieval system, without permission in writing from the publisher or under licence from the Copyright Licensing Agency Limited. Further details of such licences (for reprographic reproduction) may be obtained from the Copyright Licensing Agency Limited, www.cla.co.uk

Cover photo © xamtiw – stock.adobe.com

Illustrations by Pantek Arts Ltd and Integra Software Services

Typeset in Times Ten LT Std Roman 10/12 by Integra Software Services Pvt. Ltd., Pondicherry, India

Printed in Italy

A catalogue record for this title is available from the British Library.

ISBN: 978 1 3983 7394 5

# Contents

| | | Introduction | vi |
|---|---|---|---|
| **TOPIC 0** | | Introduction to the graphic display calculator | 3 |
| | Section 1 | The history of the calculator | 4 |
| | Section 2 | The graphic display calculator | 4 |
| | Section 3 | Plotting graphs | 9 |
| | Section 4 | Tables of values | 15 |
| | Section 5 | Lists | 17 |
| **TOPIC 1** | | Number | 22 |
| | Chapter 1 | Vocabulary for sets of numbers | 24 |
| | Chapter 2 | Calculations and order of operations | 31 |
| | Chapter 3 | Fractions, decimals and percentages | 34 |
| | Chapter 4 | Further percentages | 47 |
| | Chapter 5 | Ratio and proportion | 59 |
| | Chapter 6 | Approximation and rounding | 65 |
| | Chapter 7 | Laws of indices | 70 |
| | Chapter 8 | Standard form | 74 |
| | Chapter 9 | Surds | 78 |
| | Chapter 10 | Speed, distance and time | 82 |
| | Chapter 11 | Money and finance | 91 |
| | Chapter 12 | Set notation and Venn diagrams | 95 |
| | | Investigations, modelling and ICT 1 | 103 |
| | | Student assessments 1 | 108 |
| **TOPIC 2** | | Algebra | 114 |
| | Chapter 13 | Algebraic representation and manipulation | 116 |
| | Chapter 14 | Further algebraic representation and manipulation | 121 |
| | Chapter 15 | Linear and simultaneous equations | 132 |
| | Chapter 16 | Solving quadratic equations | 150 |
| | Chapter 17 | Using a graphic display calculator to solve equations | 155 |
| | Chapter 18 | Inequalities | 162 |
| | Chapter 19 | Indices and algebra | 174 |
| | Chapter 20 | Sequences | 177 |
| | Chapter 21 | Direct and inverse variation | 189 |
| | | Investigations, modelling and ICT 2 | 195 |
| | | Student assessments 2 | 198 |
| **TOPIC 3** | | Functions | 206 |
| | Chapter 22 | Function notation | 208 |
| | Chapter 23 | Recognising graphs of common functions | 211 |
| | Chapter 24 | Transformation of graphs | 221 |

# CONTENTS

| | | |
|---|---|---|
| Chapter 25 | Using a graphic display calculator to sketch and analyse functions | 226 |
| Chapter 26 | Finding a quadratic function from key values | 234 |
| Chapter 27 | Finding the equation of other common functions | 247 |
| Chapter 28 | Composite functions | 253 |
| Chapter 29 | Inverse functions | 257 |
| Chapter 30 | Logarithmic functions | 259 |
| | Investigations, modelling and ICT 3 | 266 |
| | Student assessments 3 | 270 |

## TOPIC 4 Coordinate geometry — 274

| | | |
|---|---|---|
| Chapter 31 | Coordinates | 276 |
| Chapter 32 | Line segments | 280 |
| Chapter 33 | Equation of a straight line | 285 |
| | Investigations, modelling and ICT 4 | 296 |
| | Student assessment 4 | 301 |

## TOPIC 5 Geometry — 302

| | | |
|---|---|---|
| Chapter 34 | Geometrical vocabulary | 304 |
| Chapter 35 | Symmetry | 310 |
| Chapter 36 | Measuring and drawing angles and bearings | 316 |
| Chapter 37 | Angle properties | 320 |
| Chapter 38 | Similarity | 334 |
| Chapter 39 | Properties of circles | 341 |
| | Investigations, modelling and ICT 5 | 352 |
| | Student assessments 5 | 357 |

## TOPIC 6 Mensuration — 364

| | | |
|---|---|---|
| Chapter 40 | Measures | 366 |
| Chapter 41 | Perimeter and area of simple plane shapes | 370 |
| Chapter 42 | Circumference and area of a circle | 373 |
| Chapter 43 | Arc length and area of a sector | 377 |
| Chapter 44 | Area and volume of further plane shapes and prisms | 382 |
| Chapter 45 | Surface area and volume of other solids | 390 |
| | Investigations, modelling and ICT 6 | 404 |
| | Student assessments 6 | 407 |

## TOPIC 7 Trigonometry — 414

| | | |
|---|---|---|
| Chapter 46 | Pythagoras' theorem | 416 |
| Chapter 47 | Sine, cosine and tangent ratios | 423 |
| Chapter 48 | Special angles and their trigonometric ratios | 432 |
| Chapter 49 | The sine and cosine rules | 441 |
| Chapter 50 | Applications of trigonometry | 451 |

| | | |
|---|---|---|
| **Chapter 51** | Trigonometric graphs, properties and transformations | 464 |
| | Investigations, modelling and ICT 7 | 473 |
| | Student assessments 7 | 477 |

## TOPIC 8  Vectors and transformations — 484

| | | |
|---|---|---|
| **Chapter 52** | Simple vectors | 486 |
| **Chapter 53** | Magnitude of a vector | 492 |
| **Chapter 54** | Transformations | 497 |
| **Chapter 55** | Further transformations | 510 |
| | Investigations, modelling and ICT 8 | 515 |
| | Student assessments 8 | 518 |

## TOPIC 9  Probability — 522

| | | |
|---|---|---|
| **Chapter 56** | Theoretical probability | 524 |
| **Chapter 57** | Tree diagrams | 529 |
| **Chapter 58** | Use of Venn diagrams in probability | 535 |
| **Chapter 59** | Laws of probability | 537 |
| **Chapter 60** | Relative frequency | 543 |
| | Investigations, modelling and ICT 9 | 546 |
| | Student assessments 9 | 551 |

## TOPIC 10  Statistics — 554

| | | |
|---|---|---|
| **Chapter 61** | Basic graphs and charts | 556 |
| **Chapter 62** | Stem-and-leaf diagrams | 567 |
| **Chapter 63** | Averages and ranges | 570 |
| **Chapter 64** | Cumulative frequency | 579 |
| **Chapter 65** | Scatter diagrams, correlation and lines of best fit | 585 |
| | Investigations, modelling and ICT 10 | 594 |
| | Student assessments 10 | 597 |

| | |
|---|---|
| **Glossary** | 601 |
| **Index** | 613 |

Answers are available at www.hoddereducation.com/cambridgeextras

Explore the book cover: how are pineapples mathematical?

The Fibonacci sequence begins 0, 1, 1, 2, 3, 5, 8, …. Each number in the sequence is equal to the sum of the preceding two numbers. You can see the Fibonacci sequence if you count the number of sections in a pineapple travelling in a spiral from bottom to top. Look up 'Fibonacci in pineapples' to discover more.

# Introduction

This book has been written for all students of the Cambridge IGCSE™ International Mathematics syllabus (0607) and supports the full syllabus. It provides the detail and guidance that are needed to support you throughout your course and help you to prepare for your examinations.

# How to use this book

To make your study of mathematics as rewarding and successful as possible, this Cambridge endorsed textbook offers the following important features:

## Learning objectives

Each topic starts with an outline of the subject material and syllabus topics to be covered. To differentiate between the Core and Extended curriculum, please use the colour coding system used in the chapters. Black refers to Core material, purple – to Extended material. Further notes and examples are given for some topics within the syllabus document available at www.cambridgeinternational.org. You should always check the appropriate syllabus document for the year of examination to confirm details and for more information.

## Organisation

Topics follow the order of the syllabus and are divided into chapters. Within each chapter there is a blend of teaching, worked examples and exercises to help you build confidence and develop the skills and knowledge you need. At the end of each topic there are comprehensive Student assessments. You will also find sets of questions linked to the **Boost eBook** (boost-learning.com), which offer practice in topic areas that students often find difficult.

## Investigations, modelling and ICT

The syllabus specifically refers to 'Applying mathematical techniques to solve problems', and this is fully integrated into the exercises and assessments in the book. There are also sections called 'Investigations, modelling and ICT', which include problem-solving questions and ICT activities. It is not possible to provide answers to many of these problems, where students are invited to provide their own proof or complete their own investigation, because there are multiple possibilities. In these cases, students are advised to check with their teacher. In the **Boost eBook** there is a selection of videos which offer support in problem-solving strategies and encourage reflective practice.

## Key terms and glossary

It is important to understand and use mathematical terms; therefore, all key terms are highlighted in bold and explained in the glossary.

*How to use this book*

## Calculator and non-calculator questions

All exercise questions that should be attempted without a calculator are indicated by ✗. You should do as many calculations as possible without using a calculator. This will help to build understanding and confidence. Some areas of mathematics, such as those using powers and roots, π, trigonometry and calculations with decimals, are more likely to require a calculator.

## Worked example

The worked examples cover important techniques and question styles. They are designed to reinforce the explanations, and give you step-by-step help for solving problems.

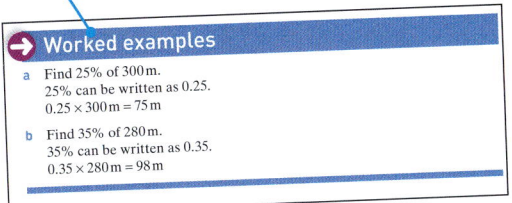

## Student assessment

End-of-topic questions are included to test your understanding of the key concepts and help prepare you for your exam.

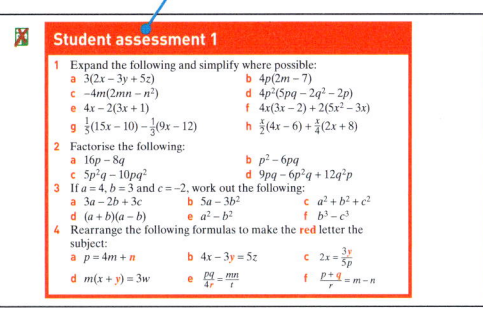

## Exercise

These appear throughout the text, and allow you to apply what you have learned. There are plenty of routine questions covering important examination techniques.

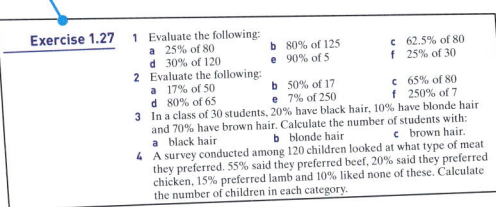

## Callouts

These commentaries provide additional explanations and encourage full understanding of mathematical principles.

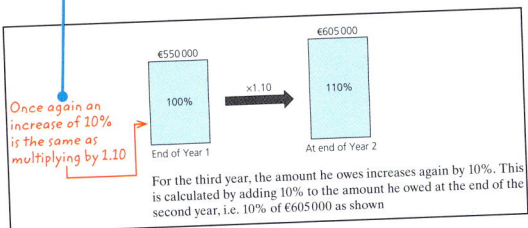

## Tables

Tables show you how to solve problems using your Casio or Texas calculator.

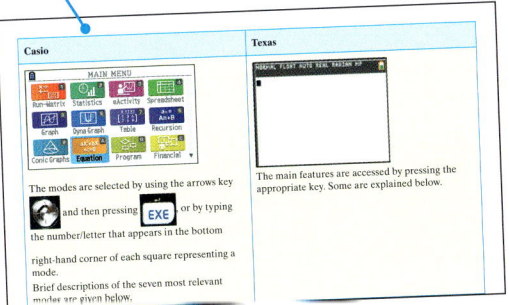

## Investigations, modelling and ICT

More real-world problem-solving activities are provided at the end of each section to put what you've learned into practice.

vii

# Assessment

The information in this section is taken from the Cambridge International syllabus. You should always refer to the appropriate syllabus document for the year of examination to confirm the details and for more information. The syllabus document is available on the Cambridge International website at www.cambridgeinternational.org. For Cambridge IGCSE International Mathematics you will take three papers. Students studying the Core syllabus will take Paper 1, Paper 3 and Paper 5. Students studying the Extended syllabus will take Paper 2, Paper 4 and Paper 6.

|  | Length | Type of question | Calculator? |
| --- | --- | --- | --- |
| Paper 1 (Core) | 1 hour 15 minutes | Structured and unstructured questions | No |
| Paper 2 (Extended) | 1 hour 30 minutes | Structured and unstructured questions | No |
| Paper 3 (Core) | 1 hour 15 minutes | Structured and unstructured questions | Yes |
| Paper 4 (Extended) | 1 hour 30 minutes | Structured and unstructured questions | Yes |
| Paper 5 (Core) | 1 hour 15 minutes | One investigation | Yes |
| Paper 6 (Extended) | 1 hour 30 minutes | One investigation section and one modelling section | Yes |

## Command words

The command words that may appear in your question papers are listed below. The command word used will relate to the context of the question.

| Command word | What it means |
| --- | --- |
| Calculate | work out from given facts, figures or information |
| Comment | give an informed opinion |
| Compare | identify/comment on similarities and/or differences |
| Determine | establish with certainty |
| Describe | state the points of a topic / give characteristics and main features |
| Explain | set out purposes or reasons / make the relationships between things clear / say why and/or how and support with relevant evidence |
| Give | produce an answer from a given source or recall/memory |
| Plot | mark point(s) on a graph |
| Revise | change to reflect further given information |
| Show (that) | provide structured evidence that leads to a given result |
| Sketch | make a simple freehand drawing showing the key features |
| State | express in clear terms |
| Work out | calculate from given facts, figures or information with or without the use of a calculator |
| Write | give an answer in a specific form |
| Write down | give an answer without significant working |

### Examination techniques

Make sure you check the instructions on the question paper, the length of the paper and the number of questions you have to answer. In the case of Cambridge IGCSE Mathematics examinations you will have to answer every question as there will be no choice.

Allocate your time sensibly between each question. Students may let themselves down by spending too long on some questions and too little time (or no time at all) on others.

Make sure you show your working to show how you've reached your answer.

# From the authors

*Mathematics* comes from the Greek word meaning *knowledge* or *learning*. Galileo Galilei (1564–1642) wrote 'the universe cannot be read until we learn the language in which it is written. It is written in mathematical language'. Mathematics is used in science, engineering, medicine, art, finance etc., but mathematicians have always studied the subject for pleasure. They look for patterns in nature, for fun, as a game or a puzzle. A mathematician may find that his or her puzzle solving helps to solve 'real life' problems. But trigonometry was developed without a 'real life' application in mind; it happened to be applied to navigation and many other things afterwards. Similarly, the algebra of curves was not 'invented' to send a rocket to the moon. The study of mathematics exists in all lands and cultures. A mathematician in Kenya may be working with another in Japan to extend work done by a Brazilian in the USA.

People in all cultures have tried to understand the world around them, and mathematics has been a common way of furthering that understanding. Each Topic in this textbook has an introduction which tries to show how, over thousands of years, mathematical ideas have been passed from one culture to another. So when you are studying from this textbook, remember that you are following in the footsteps of earlier mathematicians who were excited by the discoveries they had made. These discoveries have shaped our current world.

You may find some of the questions in this book difficult. Rather than immediately asking your teacher for help, remember that mathematics is intended to stretch the mind. If you are trying to get physically fit you do not stop as soon as things get hard. It is the same with mental fitness. Think logically. Try harder. In the end you are responsible for your own learning. Teachers and textbooks can only guide you. Be confident that you can solve that difficult problem.

Ric Pimentel<br>
Terry Wall<br>
Frankie Pimentel

# Introduction to the graphic display calculator

## Contents
1. The history of the calculator
2. The graphic display calculator
3. Plotting graphs
4. Tables of values
5. Lists

**INTRODUCTION TO THE GRAPHIC DISPLAY CALCULATOR**

## SECTION 1 The history of the calculator

There are many different types of calculators available today. These include basic calculators, scientific calculators and the latest graphic display calculators. The history of the calculator is a long one.

The abacus was invented between 2300BC and 500BC. It was used mainly for addition and subtraction and is still widely used in parts of Southeast Asia.

The slide rule was invented in 1621. It was able to do more complex operations than the abacus and continued to be widely used into the early 1970s.

The first mechanical calculator was invented by Blaise Pascal in 1642. It used a system of gears.

The first handheld calculator appeared in 1967 as a result of the development of the integrated circuit.

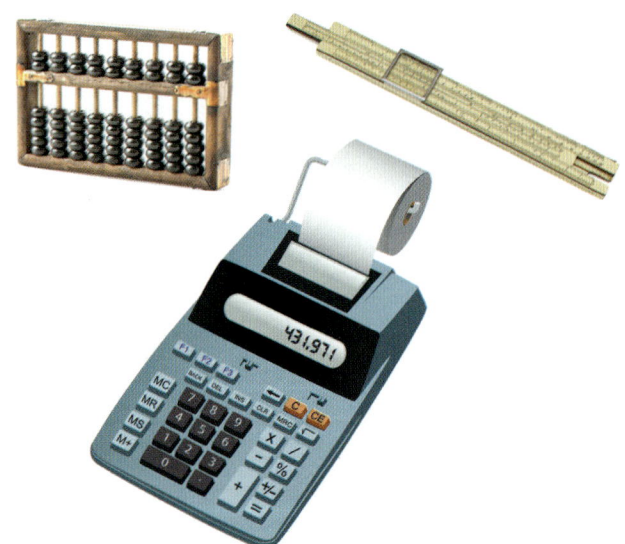

## SECTION 2 The graphic display calculator

Graphic display calculators are powerful tools used for the study of mathematics in the modern classroom. However, as with all tools, their effectiveness is only apparent when used properly. This section will look at some of the key features of the graphic display calculator, so that you start to understand some of its potential. More detailed

SECTION 2 The graphic display calculator

exploration of its capabilities is integrated into the relevant sections throughout this book. The two models used are the Casio *fx*-CG50 and the Texas TI-84 Plus CE-T. Many graphic display calculators have similar capabilities to the calculators shown. However, if your calculator is different, it is important that you take the time to familiarise yourself with it.

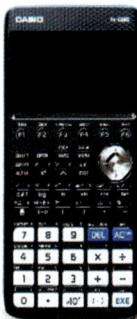 

Here is the home screen (menu/applications) for both calculators.

| Casio | Texas |
|---|---|
| *[Main Menu screen showing modes: Run-Matrix, Statistics, eActivity, Spreadsheet, Graph, Dyna Graph, Table, Recursion, Conic Graphs, Equation, Program, Financial]* | *[Blank home screen showing: NORMAL FLOAT AUTO REAL RADIAN MP]* |
| 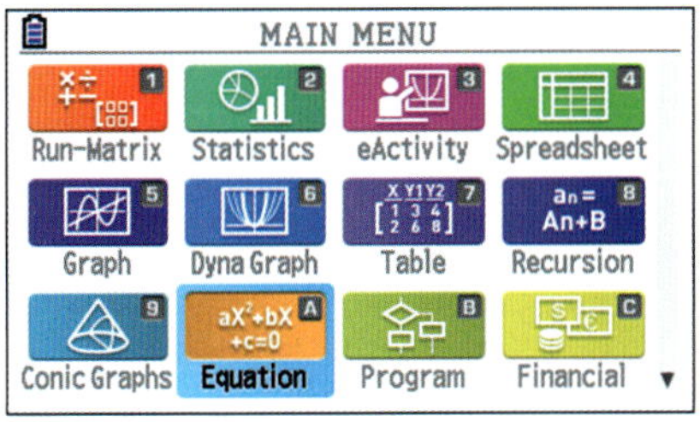 The modes are selected by using the arrows key  ![arrow key] and then pressing  ![EXE], or by typing the number/letter that appears in the top right-hand corner of each square representing a mode.<br><br>Brief descriptions of the six most relevant modes are given below. | 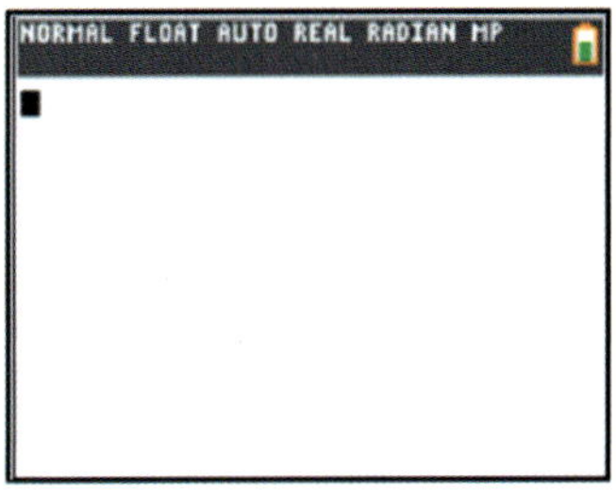 The main features are accessed by pressing the appropriate key. Some are explained below. |
| **1 Run-Matrix** is used for arithmetic calculations. | Arithmetic calculations can be typed directly from the home screen. |
| **2 Statistics** is used for statistical calculations and for drawing graphs. |  ![test/math key] is used to access numerical operations. |

5

# INTRODUCTION TO THE GRAPHIC DISPLAY CALCULATOR

| | |
|---|---|
| **4 Spreadsheet** is a spreadsheet and can be used for calculations and graphs. | **list stat** is used for statistical calculations and for drawing graphs of the data entered. |
| **5 Graph** is used for entering the equations of graphs and plotting them. | **stat plot f1 y=** is used for entering the equations of graphs. |
| **7 Table** is used to generate a table of results from an equation. | **table f5 graph** is used for graphing functions. |
| **A Equation** is used to solve different types of equations. | |

*Exercises may be done as a class activity.*

## Basic calculations

The aim of the following exercise is to familiarise you with some of the buttons dealing with basic mathematical operations on your calculator. It is assumed that you will already be familiar with the mathematical content.

**Exercise 1** Using your graphic display calculator, find the value of the following:

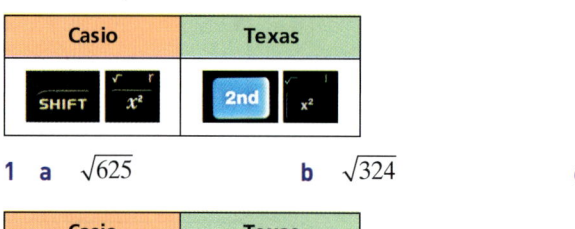

1  a  $\sqrt{625}$  b  $\sqrt{324}$  c  $2\sqrt{8} \times \sqrt[5]{2}$

| Casio | Texas |
|---|---|
| SHIFT ^ | math 5 |

2  a  $\sqrt[3]{1728}$  b  $\sqrt[4]{1296}$  c  $\sqrt[5]{3125}$

| Casio | Texas |
|---|---|
| $x^2$ | $x^2$ |

3  a  $13^2$  b  $8^2 \div 4^2$  c  $\sqrt{5^2 + 12^2}$

## SECTION 2 The graphic display calculator

### Exercise 1 (cont)

| Casio | Texas |
|---|---|
| $\sqrt[x]{\phantom{x}}$ ⌃ | π H ⌃ |

**4 a** $6^3$   **b** $9^4 \div 27^2$   **c** $\sqrt[4]{\dfrac{4^3 \times 2^8}{8^2}}$

| Casio | Texas |
|---|---|
| ×10$^x$ | 2nd EE , |

**5 a** $(2.3 \times 10^3) + (12.1 \times 10^2)$
 **b** $(4.03 \times 10^3) + (15.6 \times 10^4) - (1.05 \times 10^4)$
 **c** $\dfrac{13.95 \times 10^6}{15.5 \times 10^3} - (9 \times 10^2)$

Graphic display calculators also have a large number of memory channels. Use these to store answers which are needed for subsequent calculations. This will minimise rounding errors.

| Casio | Texas |
|---|---|
| ALPHA followed by a letter of the alphabet and EXE | alpha followed by a letter of the alphabet and enter |

### Exercise 2

**1** In the following expressions, $a = 5$, $b = 4$ and $c = 2$.
Enter each of these values in memory channels A, B and C, respectively, of your calculator and work out the following:

 **a** $a + b + c$   **b** $a - (b + c)$   **c** $(a + b)^2 - c$

 **d** $\dfrac{2(b+c)^3}{(a-c)}$   **e** $\dfrac{\sqrt[4]{a^2 - b^2}}{c}$   **f** $\dfrac{(ac)^2 + ba^2}{a + b + c}$

**2** Circles A, B, C and D have radii 10 cm, 6 cm, 4 cm and 1 cm respectively.

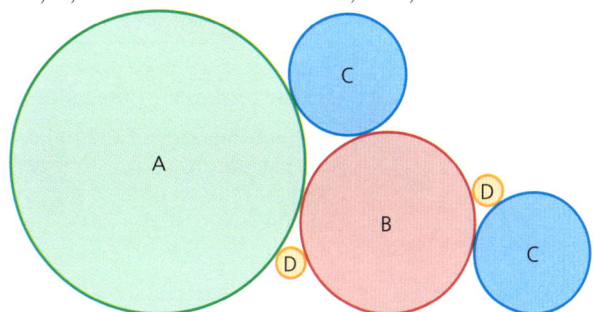

# INTRODUCTION TO THE GRAPHIC DISPLAY CALCULATOR

  a  Calculate the area of circle A and store your answer in memory channel A.
  b  Calculate the area of circle B and store your answer in memory channel B.
  c  Calculate the area of each of the circles C and D, storing the answers in memory channels C and D respectively.
  d  Using your calculator, find the value of A + B + 2C + 2D.
  e  What does the answer to Q.2d represent?

3 The diagram shows a child's shape-sorting toy. The top consists of a rectangular piece of wood of dimension 30 cm × 12 cm. Four shapes W, X, Y and Z are cut out of it.

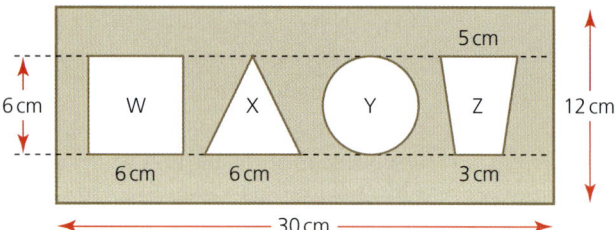

  a  Calculate the area of the triangle X. Store the answer in your calculator's memory.
  b  Calculate the area of the trapezium Z. Store the answer in your calculator's memory.
  c  Calculate the total area of the shapes W, X, Y and Z.
  d  Calculate the area of the rectangular piece of wood left once the shapes have been cut out.

4 Three balls just fit inside a cylindrical tube as shown. The radius ($r$) of each ball is 5 cm.

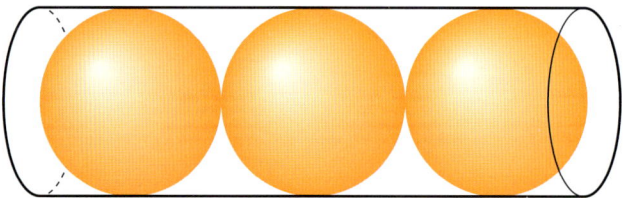

  a  Use the formula for the volume of a sphere, $V = \frac{4}{3}\pi r^3$, to calculate the volume of one of the balls. Store the answer in the memory of your calculator.
  b  Calculate the volume of the cylinder.
  c  Calculate the volume of the cylinder **not** occupied by the three balls.

# SECTION 3 Plotting graphs

One of a graphic display calculator's principal features is to plot graphs of functions. This helps to visualise what the function looks like and, later on, will help solve a number of different types of problems. This section aims to show how to graph a variety of different functions.

For example, to plot a graph of the function $y = 2x + 3$, use the following buttons on your calculator:

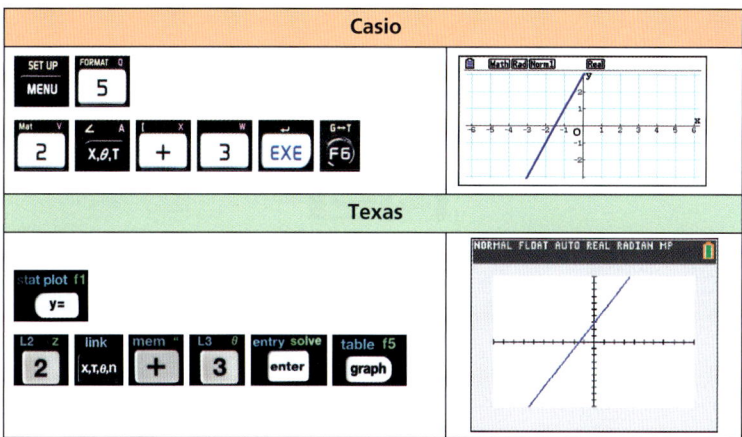

It may be necessary to change the scale on the axes in order to change how much of the graph, or what part of the graph, can be seen. This can be done in several ways, two of which are described here:

» by using the zoom facility

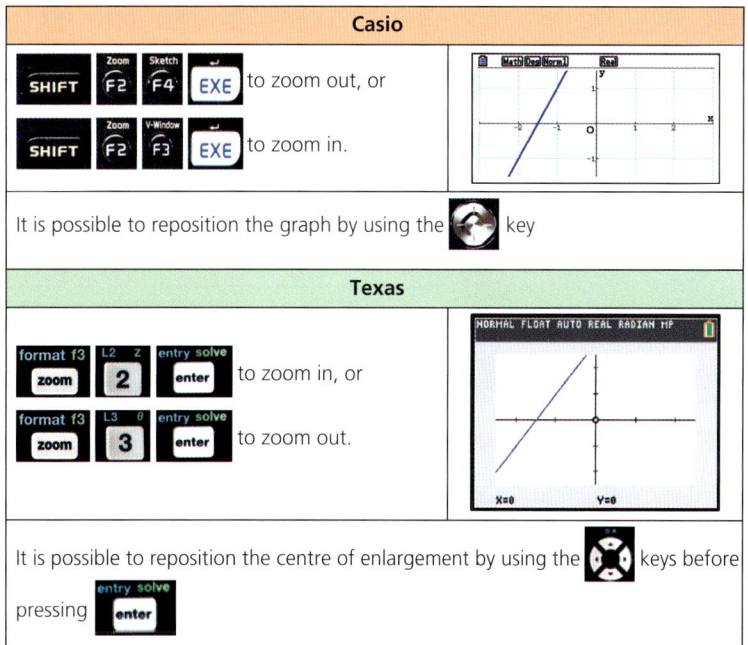

# INTRODUCTION TO THE GRAPHIC DISPLAY CALCULATOR

» by changing the scale manually.

| Casio |
|---|
| 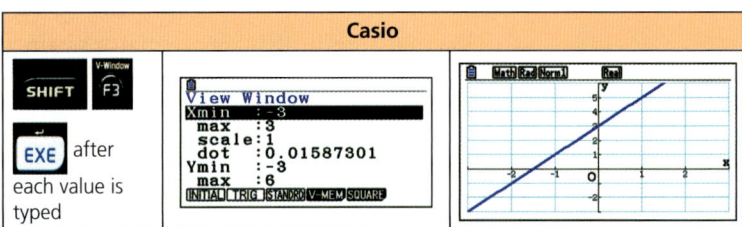 |
| Xmin: minimum value on the *x*-axis, Xmax: maximum value on the *x*-axis, Xscale: distance between the tick marks on the *x*-axis, Xdot: value that relates to one *x*-axis dot (this is set automatically). |
| Texas |
| 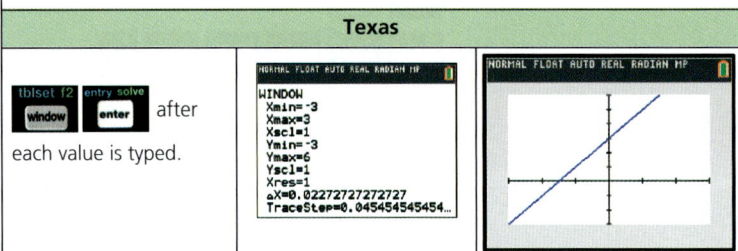 |
| Xmin: minimum value on the *x*-axis, Xmax: maximum value on the *x*-axis, Xscl: distance between the tick marks on the *x*-axis. |

**Exercise 3**

In Q.1–4 the axes have been set to their default settings, i.e.
Xmin = –10, Xmax = 10, Xscale = 1, Ymin = –10, Ymax = 10, Yscale = 1.

i) 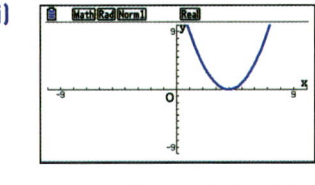  ii)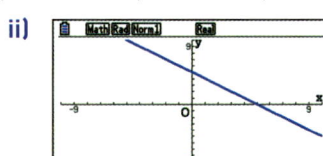

iii) 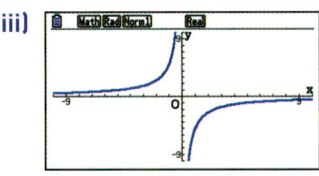  iv)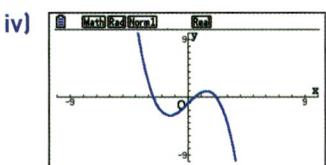

v) 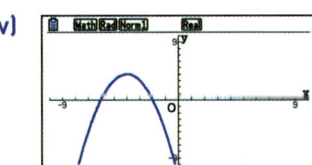  vi)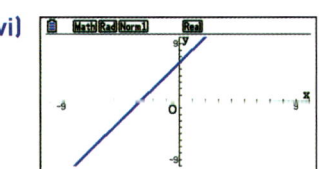

# Exercise 3 (cont)

vii)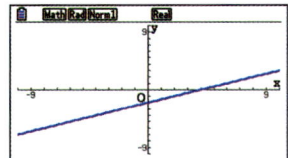

viii)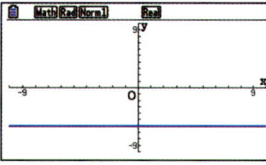

ix)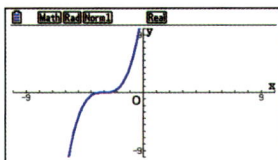

x)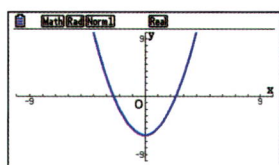

**1** Use your graphic display calculator to graph the following functions and match each of them to the correct graph.

**a** $y = 2x + 6$      **b** $y = \frac{1}{2}x - 2$

**c** $y = -x + 5$      **d** $y = -\frac{5}{x}$

**e** $y = x^2 - 6$      **f** $y = (x - 4)^2$

**g** $y = -(x + 4)^2 + 4$      **h** $y = \frac{1}{2}(x + 3)^3$

**i** $y = -\frac{1}{3}x^3 + 2x - 1$      **j** $y = -6$

**2** In each of the following, a function and its graph is given. Using your graphic display calculator, enter a function that produces a reflection of the original function in the $x$-axis.

**a** $y = x + 5$      **b** $y = -2x + 4$

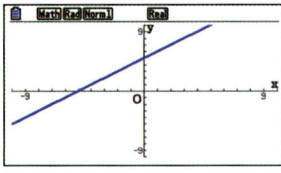 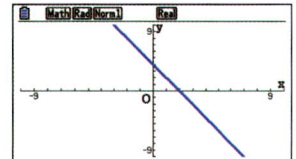

**c** $y = (x + 5)^2$      **d** $y = (x - 5)^2 + 3$

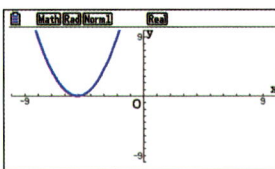 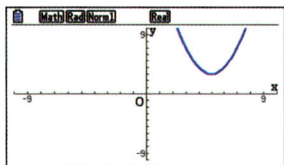

**3** Using your graphic display calculator, enter a function that produces a reflection in the $y$-axis of each of the original functions in Q.2.

# INTRODUCTION TO THE GRAPHIC DISPLAY CALCULATOR

4  By entering appropriate functions into your graphic display calculator:
   i   make your calculator screen look like the screens shown (note: don't worry about the colours)
   ii  write down the functions you used.

a 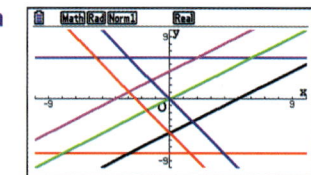   b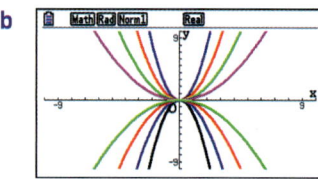

c 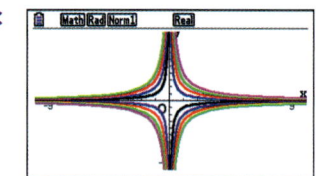   d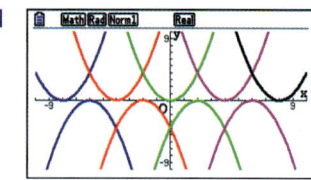

## Intersections

When graphing a function it is often necessary to find where it intersects one or both of the axes. If more than one function is graphed on the same axes, it may also be necessary to find where the graphs intersect each other. Graphic display calculators can be used to find the coordinates of any points of intersection.

### → Worked example

Find where the graph of $y = \frac{1}{5}(x+3)^3 + 2$ intersects both the $x$- and $y$-axes.

The graph shows that $y = \frac{1}{5}(x+3)^3 + 2$ intersects each of the axes once.

To find the coordinates of the points of intersection:

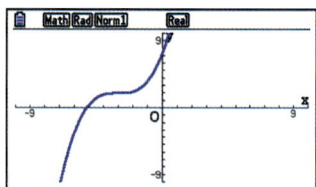

# SECTION 3 Plotting graphs

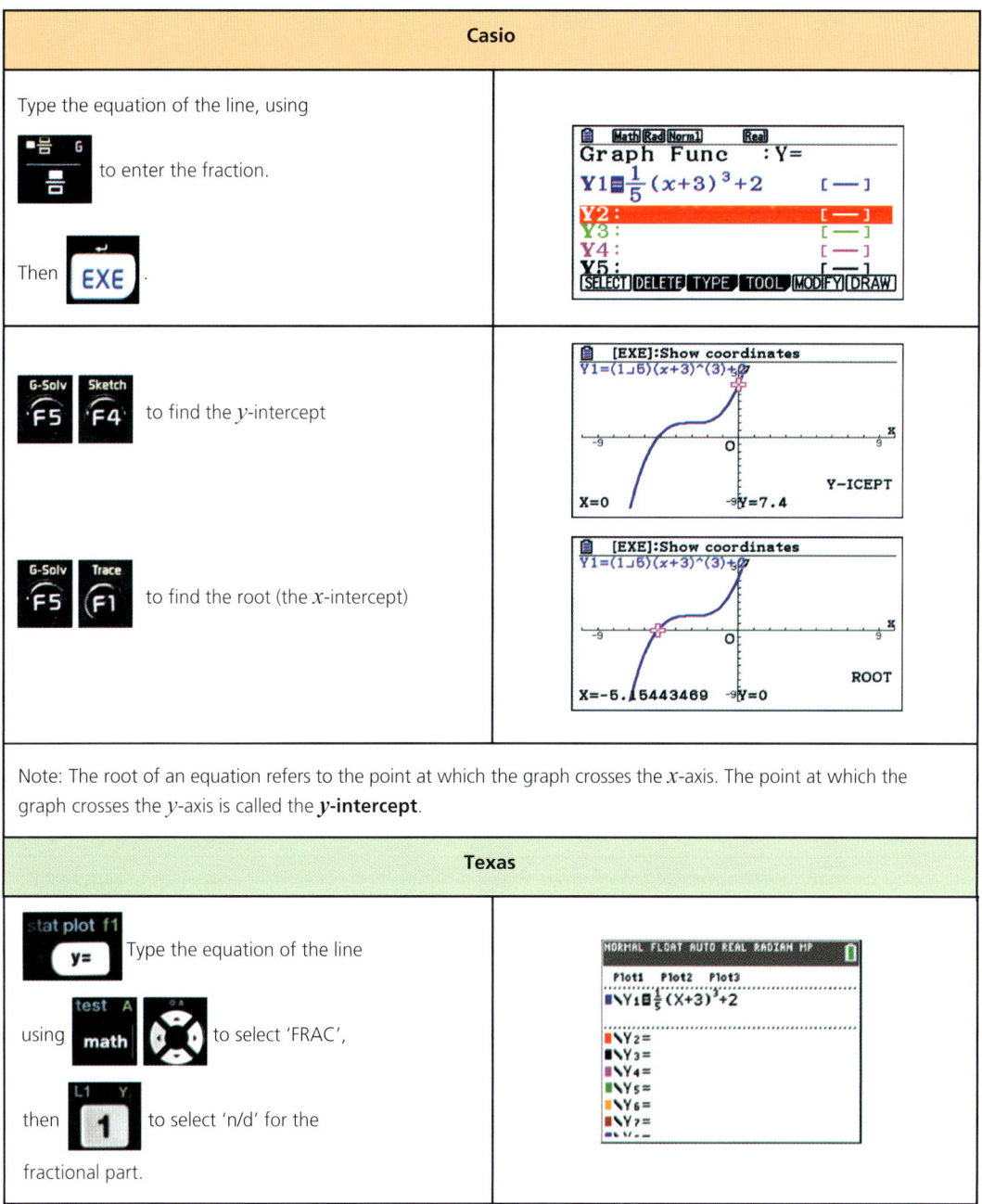

| Casio |
|---|
| Type the equation of the line, using ![frac key] to enter the fraction. Then ![EXE]. |
| ![F5 F4] to find the $y$-intercept |
| ![F5 F1] to find the root (the $x$-intercept) |

Note: The root of an equation refers to the point at which the graph crosses the $x$-axis. The point at which the graph crosses the $y$-axis is called the **$y$-intercept**.

| Texas |
|---|
| ![y=] Type the equation of the line using ![math] ![arrow] to select 'FRAC', then ![1] to select 'n/d' for the fractional part. |

# INTRODUCTION TO THE GRAPHIC DISPLAY CALCULATOR

 enter $x = 0$ to find the $y$-intercept.

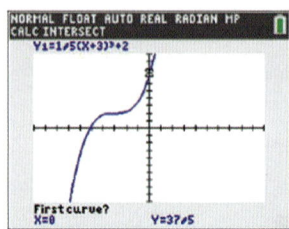

Use ⊕ to move the cursor on the graph to a point to the left of the intersection with the $x$-axis, then `enter`

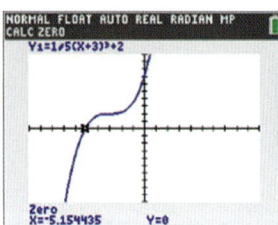

Use ⊕ to move the cursor on the graph to a point to the right of the intersection with the $x$-axis, then `enter` and `enter` again.

Note: The TI-84 Plus CE-T prompts the user to identify a point to the left of the intersection with the $x$-axis (left bound) and then a point to the right of the intersection (right bound).

---

**Exercise 4**

1  Find an approximate solution to where the following graphs intersect both the $x$- and $y$-axes using your graphic display calculator.
   a  $y = x^2 - 3$
   b  $y = (x + 3)^2 + 2$
   c  $y = \frac{1}{2}x^3 - 2x^2 + x + 1$
   d  $y = \frac{-5}{x + 2} + 6$

2  Find the coordinates of the point(s) of intersection of each of the following pairs of equations using your graphic display calculator.
   a  $y = x + 3$ and $y = -2x - 2$
   b  $y = -x + 1$ and $y = \frac{1}{2}(x^2 - 3)$
   c  $y = -x^2 + 1$ and $y = \frac{1}{2}(x^2 - 3)$
   d  $y = -\frac{1}{4}x^3 + 2x^2 - 3$ and $y = \frac{1}{2}x^2 - 2$

# SECTION 4 Tables of values

A function such as $y = \frac{3}{x} + 2$ implies that there is a relationship between $y$ and $x$.

To plot the graph manually, the coordinates of several points on the graph need to be calculated and then plotted. Graphic display calculators have the facility to produce a table of values giving the coordinates of some of the points on the graph.

## ➜ Worked example

For the function $y = \frac{3}{x} + 2$, complete the following table of values using the table facility of your graphic display calculator:

| $x$ | −3 | −2 | −1 | 0 | 1 | 2 | 3 |
|---|---|---|---|---|---|---|---|
| $y$ | | | | | | | |

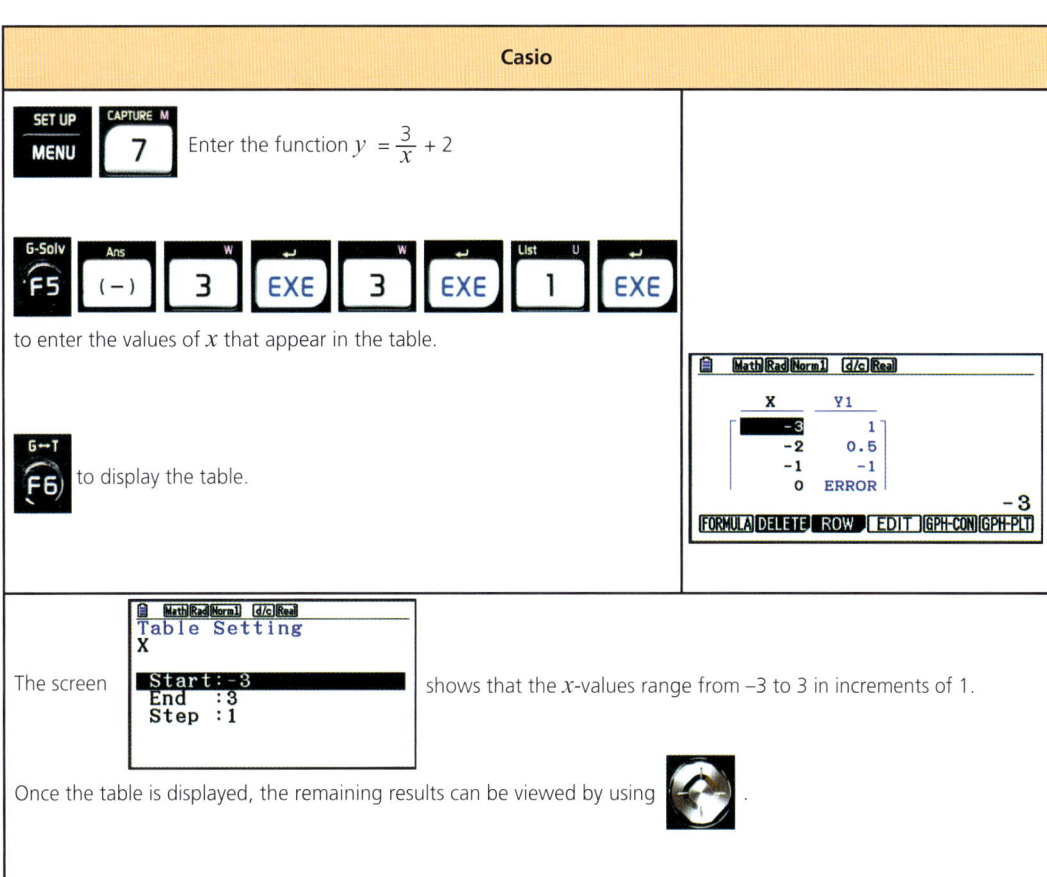

# INTRODUCTION TO THE GRAPHIC DISPLAY CALCULATOR

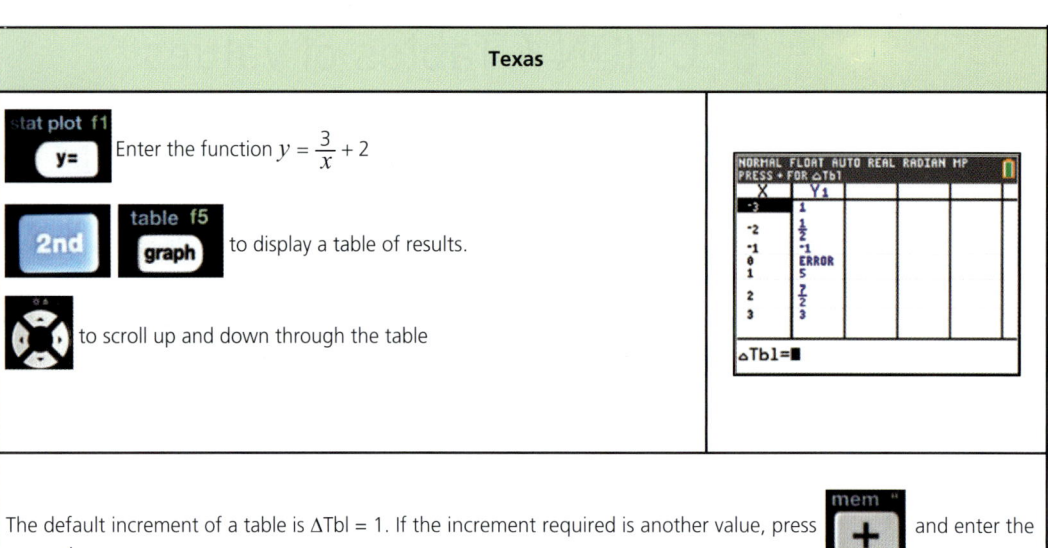

The default increment of a table is ΔTbl = 1. If the increment required is another value, press [+] and enter the new value.

## Exercise 5

1  Copy and complete the tables of values for the following functions using the table facility of your graphic display calculator:

a  $y = x^2 + x - 4$

| $x$ | −3 | −2 | −1 | 0 | 1 | 2 | 3 |
|---|---|---|---|---|---|---|---|
| $y$ | | | | | | | |

b  $y = x^3 + x^2 - 10$

| $x$ | −3 | −2 | −1 | 0 | 1 | 2 | 3 |
|---|---|---|---|---|---|---|---|
| $y$ | | | | | | | |

c  $y = \dfrac{4}{x}$

| $x$ | 0 | 0.5 | 1 | 1.5 | 2 | 2.5 | 3 |
|---|---|---|---|---|---|---|---|
| $y$ | | | | | | | |

## Exercise 5 (cont)

d  $y = \sqrt{(x+1)}$

| $x$ | −1 | −0.5 | 0 | 0.5 | 1 | 1.5 | 2 | 2.5 | 3 |
|---|---|---|---|---|---|---|---|---|---|
| $y$ | | | | | | | | | |

2  A car accelerates from rest. Its speed, $y$ m/s, $x$ seconds after starting, is given by the equation $y = 1.8x$.
   a  Using the table facility of your graphic display calculator, calculate the speed of the car every 2 seconds for the first 20 seconds.
   b  How fast is the car travelling after 10 seconds?

3  A ball is thrown vertically upwards. Its height, $y$ metres, $x$ seconds after launch, is given by the equation $y = 15x - 5x^2$.
   a  Using the table facility of your graphic display calculator, calculate the height of the ball each $\frac{1}{2}$ second during the first 4 seconds.
   b  What is the greatest height reached by the ball?
   c  How many seconds after its launch did the ball reach its highest point?
   d  After how many seconds did the ball hit the ground?
   e  In the context of this question, why can the values for $x = 3.5$ and 4 be ignored?

# SECTION 5 Lists

Data is often collected and then analysed so that observations and conclusions can be made. Graphic display calculators have the facility to store data as lists. Once stored as a list, many different types of calculations can be carried out. This section will explain how to enter data as a list and then how to carry out some simple calculations.

## ➡ Worked examples

a  An athlete records her time (seconds) in ten races for running 100 m. These times are shown below:

   12.4   12.7   12.6   12.9   12.4   12.3   12.7   12.4   12.5   13.1

Calculate the mean, median and mode for this set of data using the list facility of your graphic display calculator.

# INTRODUCTION TO THE GRAPHIC DISPLAY CALCULATOR

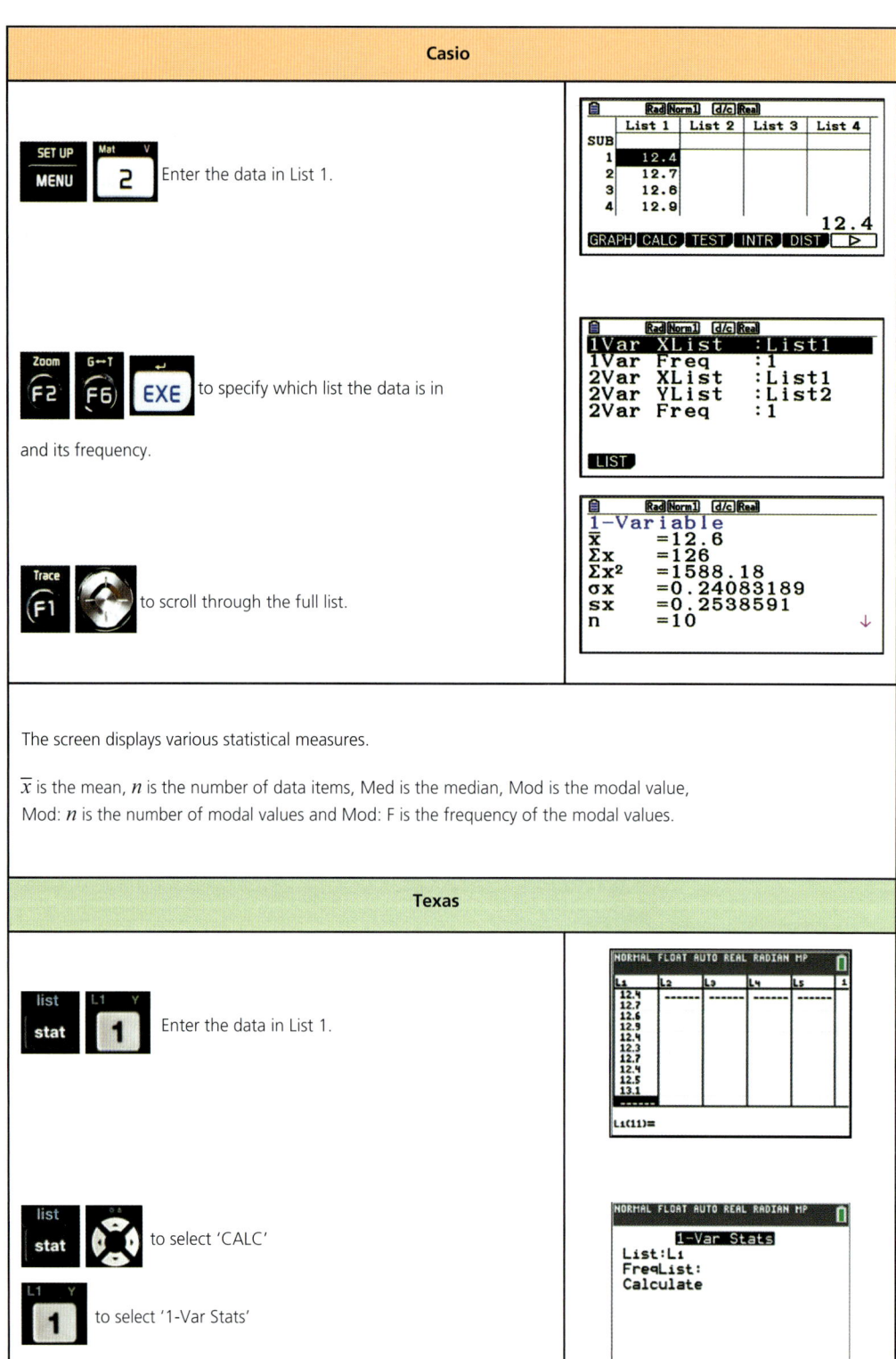

The screen displays various statistical measures.

$\bar{x}$ is the mean, $n$ is the number of data items, Med is the median, Mod is the modal value, Mod: $n$ is the number of modal values and Mod: F is the frequency of the modal values.

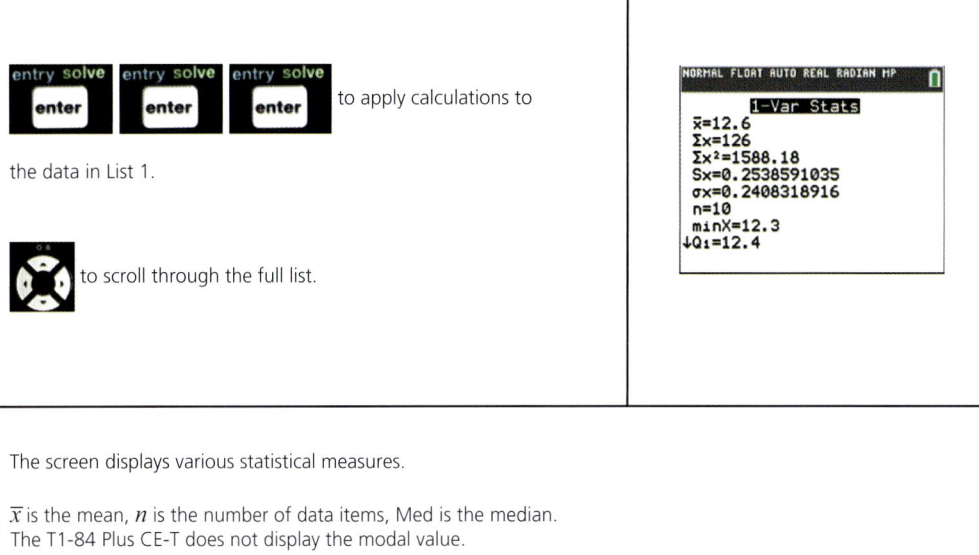

entry solve enter, entry solve enter, entry solve enter to apply calculations to the data in List 1.

⊕ to scroll through the full list.

The screen displays various statistical measures.

$\bar{x}$ is the mean, $n$ is the number of data items, Med is the median.
The T1-84 Plus CE-T does not display the modal value.

If a lot of data is collected, it is often presented in a frequency table.

**b** The numbers of students in 30 Maths classes are shown in the frequency table below:

| Number of students | Frequency |
|---|---|
| 27 | 4 |
| 28 | 6 |
| 29 | 9 |
| 30 | 7 |
| 31 | 3 |
| 32 | 1 |

Calculate the mean, median and mode for this set of data using the list facility of your graphic display calculator.

# INTRODUCTION TO THE GRAPHIC DISPLAY CALCULATOR

## Casio

  Enter the number of students in List 1 and the frequency in List 2.

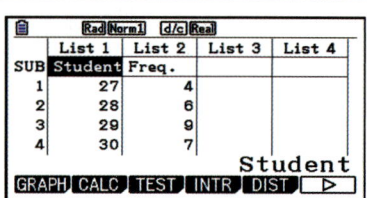

 to select 'CALC'

 to specify which lists the data and the frequency are in.

 to select '1 VAR'

 to scroll through the full list.

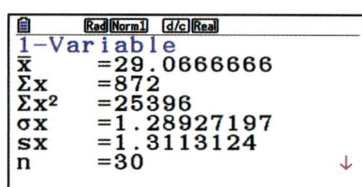

## Texas

  Enter the number of students in List 1 and the frequency in List 2.

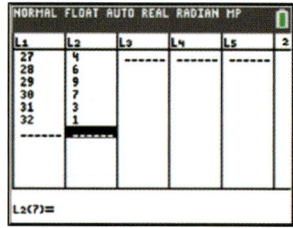

  to select 'CALC'

 to select '1-Var Stats'

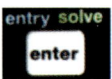   to specify that the data is in $L_1$ and the frequency in $L_2$.

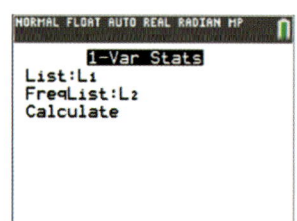

SECTION 5 Lists

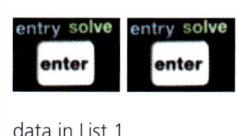 to apply the calculations to the data in List 1

 to scroll through the full set of results.

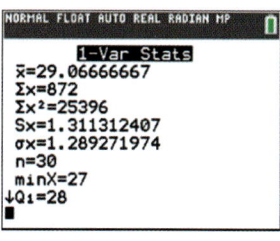

## Exercise 6

1  Find the mean, the median and, if possible, the mode of these sets of numbers using the list facility of your graphic display calculator.
   a  3, 6, 10, 8, 9, 10, 12, 4, 6, 10, 9, 4
   b  12.5, 13.6, 12.2, 14.4, 17.1, 14.8, 20.9, 12.2

2  During a board game a player makes a note of the numbers he rolls on the dice. These are shown in the frequency table below:

| Number on dice | 1 | 2 | 3 | 4 | 5 | 6 |
|---|---|---|---|---|---|---|
| Frequency | 3 | 8 | 5 | 2 | 5 | 7 |

Find the mean, the median and, if possible, the modal dice roll using the list facility of your graphic display calculator.

3  A class of 30 students sat two Maths tests. Their scores out of 10 are recorded in the frequency tables below:

| Test A | | | | | | | | | | |
|---|---|---|---|---|---|---|---|---|---|---|
| Score | 1 | 2 | 3 | 4 | 5 | 6 | 7 | 8 | 9 | 10 |
| Frequency | 3 | 2 | 4 | 3 | 1 | 8 | 3 | 1 | 3 | 2 |

| Test B | | | | | | | | | | |
|---|---|---|---|---|---|---|---|---|---|---|
| Score | 1 | 2 | 3 | 4 | 5 | 6 | 7 | 8 | 9 | 10 |
| Frequency | 4 | 1 | 0 | 0 | 0 | 24 | 0 | 0 | 0 | 1 |

a  Find the mean, the median and, if possible, the mode for each test using the list facility of your graphic display calculator.
b  Comment on any similarities or differences in your answers to Q.3a.
c  Which test did the class find easiest? Give reasons for your answer.

# TOPIC 1

# Number

## Contents

Chapter 1   Vocabulary for sets of numbers (C1.1, C1.3, E1.1, E1.3)
Chapter 2   Calculations and order of operations (C1.6, E1.6)
Chapter 3   Fractions, decimals and percentages (C1.4, C1.6, E1.4, E1.6)
Chapter 4   Further percentages (C1.12, C1.15, E1.12, E1.15)
Chapter 5   Ratio and proportion (C1.10, E1.10)
Chapter 6   Approximation and rounding (C1.9, E1.9)
Chapter 7   Laws of indices (C1.7, E1.7)
Chapter 8   Standard form (C1.8, E1.8)
Chapter 9   Surds (E1.17)
Chapter 10  Speed, distance and time (C1.11, C1.13, C1.14, E1.11, E1.13, E1.14)
Chapter 11  Money and finance (C1.11, C1.12, C1.15, E1.11, E1.12, E1.15)
Chapter 12  Set notation and Venn diagrams (C1.2, E1.2)

---

### Learning objectives

**C1.1        E1.1**

Identify and use:
- natural numbers (0, 1, 2, …)
- integers (positive, zero and negative)
- prime numbers
- square numbers
- cube numbers
- triangle numbers
- common factors
- common multiples
- rational and irrational numbers
- reciprocals

**C1.2        E1.2**

Understand and use set language, notation and Venn diagrams to describe sets **and represent relationships between sets**

Definition of sets:

e.g.
$A = \{x \mid x \text{ is a natural number}\}$
$B = \{(x, y) \mid y = mx + c\}$
$C = \{x \mid a \leq x \leq b\}$
$D = \{a, b, c...\}$

**C1.3      E1.3**
Calculate with the following:
- squares
- square roots
- cubes
- cube roots of numbers
- other powers and roots

**C1.4      E1.4**
1. Use the language and notation of the following in appropriate contexts
   - proper and improper fractions
   - mixed numbers
   - decimals
   - percentages
2. Recognise equivalence and convert between these forms

**C1.6      E1.6**
Use the four operations for calculations with integers, fractions and decimals, including correct ordering of operations and use of brackets

**C1.7      E1.7**
1. Understand and use indices (positive, zero, negative, **fractional**)
2. Use the rules of indices

**C1.8      E1.8**
1. Use the standard form $A \times 10^n$ where $n$ is a positive or negative integer, and $1 \leqslant A < 10$
2. Convert numbers into and out of standard form
3. Calculate with values in standard form

**C1.9      E1.9**
1. Round values to a specified degree of accuracy
2. Make estimates for calculations involving numbers, quantities and lengths
3. Round answers to a reasonable degree of accuracy in the context of a given problem

**C1.10     E1.10**
Understand and use ratio and proportion, to:
- give ratios in simplest form
- divide a quantity in a given ratio
- use proportional reasoning and ratios in context

**C1.11     E1.11**
1. Use common measures of rate
2. Solve problems involving average speed

**C1.12     E1.12**
1. Calculate a given percentage of a quantity
2. Express one quantity as a percentage of another
3. Calculate percentage increase or decrease
4. Calculate with simple and compound interest
5. **Calculations using reverse percentages**

**C1.13     E1.13**
1. Use a calculator efficiently
2. Enter values appropriately on a calculator
3. Interpret the calculator display appropriately

**C1.14     E1.14**
1. Calculations involving time: seconds (s), minutes (min), hours (h), days, weeks, months, years, including the relationship between units
2. Calculate times in terms of the 24-hour and 12-hour clock
3. Read clocks and timetables

**C1.15     E1.15**
1. Calculate with money
2. Convert from one currency to another

**E1.17**
1. **Understand and use surds (radicals), including simplifying expressions**
2. **Rationalise the denominator**

Elements in purple refer to the Extended curriculum only.

# Hindu mathematicians

*Aryabatta (476–550)*

In 1300BC a Hindu teacher named Laghada used geometry and trigonometry for his astronomical calculations. At around this time, other Indian mathematicians solved quadratic and simultaneous equations.

Much later, in about AD500, another Indian teacher, Aryabatta, worked on approximations for pi and on the trigonometry of the sphere. He realised that not only did the planets go round the Sun, but that their paths were elliptic.

Brahmagupta, a Hindu, was the first to treat zero as a number in its own right. This helped to develop the decimal system of numbers.

One of the greatest mathematicians of all time was Bhascara, who, in the twelfth century, worked in algebra and trigonometry. He discovered that:

$$\sin(A + B) = \sin A \cos B + \cos A \sin B$$

His work was taken to Arabia and later to Europe.

Still alive today is the Indian mathematician Raman Parimala (born in 1948). Her work is famous in the fields of algebra and its connections with algebraic geometry and number theory.

# 1 Vocabulary for sets of numbers

## Natural numbers

A child learns to count 'one, two, three, four, …'. These are sometimes called the counting numbers or natural numbers.

The child will say 'I am three', or 'I live at number 73'.

If you include the number 0, then you have the set of numbers called the **natural numbers**.

The set of natural numbers = {0, 1, 2, 3, 4, …}.

## Integers

On a cold day, the temperature may be 4°C at 10 p.m. If the temperature drops by a further 6°C, then the temperature is 'below zero'; it is −2°C.

If you are overdrawn at the bank by $200, this might be shown as −$200.

The set of **integers** = {…, −3, −2, −1, 0, 1, 2, 3, …}.

{1, 2, 3, 4, ...} are the **positive integers**. {..., −4, −3, −2, −1} are the **negative integers**.

The integers are therefore an extension of the natural numbers. Every natural number is an integer.

> *Another way of describing an integer is a whole number.*

## Rational numbers

A child may say 'I am three'; she may also say 'I am three and a half', or even 'three and a quarter'. $3\frac{1}{2}$ and $3\frac{1}{4}$ are **rational numbers**. All rational numbers can be written as a fraction whose denominator is not zero. All terminating and recurring decimals are rational numbers as they can also be written as fractions, e.g.

$$0.2 = \frac{1}{5} \quad 0.3 = \frac{3}{10} \quad 7 = \frac{7}{1} \quad 1.53 = \frac{153}{100} \quad 0.\dot{2} = \frac{2}{9}$$

The set of rational numbers is an extension of the set of integers.

> *The dot above the 2 implies that the 2 is repeated. It is known as a recurring decimal.*

## Irrational numbers

Numbers which cannot be written as a fraction are not rational numbers; they are **irrational numbers**.

*Cube numbers*

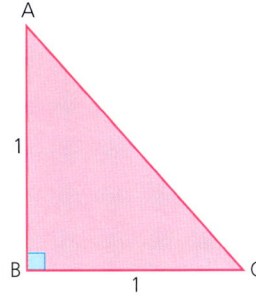

Using Pythagoras' theorem in the diagram to the left, the length of the hypotenuse AC is found as:

$(AC)^2 = 1^2 + 1^2$
$(AC)^2 = 2$
$AC = \sqrt{2}$

$\sqrt{2} = 1.41421356\ldots$ . The digits in this number do not recur or repeat. This is a property of all irrational numbers. Another example of an irrational number you will come across is $\pi$ (pi). It is the ratio of the circumference of a circle to the length of its diameter. Although it is often rounded to 3.142, the digits continue indefinitely, never repeating themselves.

## Prime numbers

A **prime number** is one whose only factors are 1 and itself. (Note that 1 is not a prime number.)

 **Exercise 1.1**

1  In a 10 by 10 square, write the numbers 1 to 100.
   Cross out number 1.
   Cross out all the even numbers after 2 (these have 2 as a factor).
   Cross out every third number after 3 (these have 3 as a factor).
   a  Continue with 5, 7, 11 and 13 and list all the numbers that are left.
   b  What do you notice about the numbers that are left?

## Square numbers

 **Exercise 1.2**

1  In a 10 by 10 square write the numbers 1 to 100.
   Shade in 1 and then $2 \times 2$, $3 \times 3$, $4 \times 4$, $5 \times 5$ etc.
   These are the **square numbers**.
   $3 \times 3$ can be written $3^2$ (pronounced three squared)
   $7 \times 7$ can be written $7^2$ (the 2 is called an index; plural **indices**)

## Cube numbers

$3 \times 3 \times 3$ can be written $3^3$ (pronounced three cubed)
$5 \times 5 \times 5$ can be written $5^3$
$2 \times 2 \times 2 \times 5 \times 5$ can be written $2^3 \times 5^2$

**Exercise 1.3**

1  Write the following using indices:
   a  $9 \times 9$                    b  $12 \times 12$
   c  $8 \times 8$                    d  $7 \times 7 \times 7$
   e  $4 \times 4 \times 4$           f  $3 \times 3 \times 2 \times 2 \times 2$
   g  $5 \times 5 \times 5 \times 2 \times 2$   h  $4 \times 4 \times 3 \times 3 \times 2 \times 2$

# 1 VOCABULARY FOR SETS OF NUMBERS

## Triangle numbers

Consider the pattern of dots below:

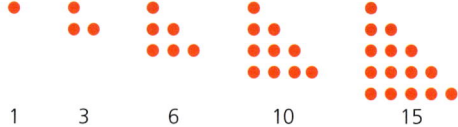

1    3    6    10    15

The number of dots in each triangular pattern is a **triangle number**. The next two triangle numbers in this sequence are therefore 21 and 28.

## Factors

The factors of 12 are all the numbers which will divide exactly into 12, i.e. 1, 2, 3, 4, 6 and 12.

The **factors** of a number are therefore all the numbers that go into it exactly.

### Exercise 1.4

1  List all the factors of the following numbers:
   a  6       b  9       c  7       d  15      e  24
   f  36      g  35      h  25      i  42      j  100

## Prime factors

The factors of 12 are 1, 2, 3, 4, 6 and 12.

Of these, 2 and 3 are prime numbers, so 2 and 3 are the **prime factors** of 12.

### Exercise 1.5

1  List the prime factors of the following numbers:
   a  15      b  18      c  24      d  16      e  20
   f  13      g  33      h  35      i  70      j  56

An easy way to find prime factors is to divide by the prime numbers in order, starting with the smallest first.

### ➡ Worked examples

a  Find the prime factors of 18 and write it as a product of prime numbers:

|   | 18 |
|---|----|
| 2 | 9  |
| 3 | 3  |
| 3 | 1  |

$18 = 2 \times 3 \times 3$ or $2 \times 3^2$

# Highest common factor

**b** Find the prime factors of 24 and write it as a product of prime numbers:

|   | 24 |
|---|----|
| 2 | 12 |
| 2 | 6  |
| 2 | 3  |
| 3 | 1  |

$24 = 2 \times 2 \times 2 \times 3$ or $2^3 \times 3$

**c** Find the prime factors of 75 and write it as a product of prime numbers:

|   | 75 |
|---|----|
| 3 | 25 |
| 5 | 5  |
| 5 | 1  |

$75 = 3 \times 5 \times 5$ or $3 \times 5^2$

### Exercise 1.6

1 Find the prime factors of the following numbers and write them as a product of prime numbers:
- **a** 12
- **b** 32
- **c** 36
- **d** 40
- **e** 44
- **f** 56
- **g** 45
- **h** 39
- **i** 231
- **j** 63

## Highest common factor

The factors of 12 are 1, 2, 3, 4, 6, 12.

The factors of 18 are 1, 2, 3, 6, 9, 18.

The **common factors** of 12 and 18 are 1, 2, 3 and 6.

So you can see that the **highest common factor** (HCF) is 6.

### Exercise 1.7

1 Find the HCF of the following numbers:
- **a** 8, 12
- **b** 10, 25
- **c** 22, 110
- **d** 39, 52
- **e** 60, 144
- **f** 12, 18, 24
- **g** 15, 21, 27
- **h** 36, 63, 108
- **i** 32, 56, 72
- **j** 34, 51, 68

# 1 VOCABULARY FOR SETS OF NUMBERS

## Multiples

The **multiples** of a number are the numbers that can be divided exactly by that number.

The multiples of 5 are 5, 10, 15, 20, 25, 30 ...

The multiples of 2 are 2, 4, 6, 8, ...

The multiples of 3 are 3, 6, 9, 12, 15, ...

The multiples of 6 are 6, 12, 18, 24, 30 ...

The **common multiples** of 2 and 3 are 6, 12, 18, 24, ...

The **lowest common multiple** (LCM) of 2 and 3 is 6, since 6 is the smallest number divisible by 2 and 3.

The LCM of 3 and 5 is 15.

The LCM of 5 and 6 is 30.

### Exercise 1.8

1  Find the LCM of the following:
   a  3, 5      b  4, 6       c  2, 7       d  4, 7
   e  4, 8      f  6, 14      g  4, 15      h  3, 4, 6
   i  3, 4, 5   j  3, 5, 12

2  Find the LCM of the following:
   a  2, 3, 5      b  2, 3, 4       c  2, 7, 10     d  3, 9, 10
   e  6, 8, 20     f  3, 5, 7       g  4, 5, 10     h  3, 7, 11
   i  6, 10, 16    j  25, 40, 100

## Square roots

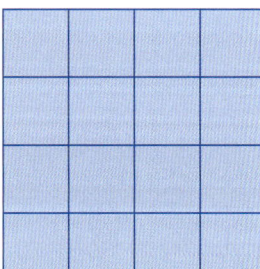

The square shown contains 16 squares. It has sides of length 4 units. So the square root of 16 is 4.

This can be written as $\sqrt{16} = 4$.

Note that $4 \times 4$ is 16 so 4 is the square root of 16.

However, $-4 \times -4$ is also 16 so $-4$ is also the square root of 16.

By convention, $\sqrt{16}$ means 'the positive square root of 16' so $\sqrt{16} = 4$ but the square root of 16 is $\pm 4$, i.e. $+4$ or $-4$.

Note: $-16$ has no square root since any integer squared is positive.

### Exercise 1.9

1  Find the following:
   a  $\sqrt{25}$      b  $\sqrt{9}$       c  $\sqrt{49}$      d  $\sqrt{100}$
   e  $\sqrt{121}$     f  $\sqrt{169}$     g  $\sqrt{0.01}$    h  $\sqrt{0.04}$
   i  $\sqrt{0.09}$    j  $\sqrt{0.25}$

# Further powers and roots

2. Use the √ key on your calculator to check your answers to question 1.
3. Calculate the following:

   a $\sqrt{\frac{1}{9}}$  b $\sqrt{\frac{1}{16}}$  c $\sqrt{\frac{1}{25}}$  d $\sqrt{\frac{1}{49}}$

   e $\sqrt{\frac{1}{100}}$  f $\sqrt{\frac{4}{9}}$  g $\sqrt{\frac{9}{100}}$  h $\sqrt{\frac{49}{81}}$

   i $\sqrt{2\frac{7}{9}}$  j $\sqrt{6\frac{1}{4}}$

## Using a graph

**Exercise 1.10**

1. Copy and complete the table below for the equation $y = \sqrt{x}$.

| x | 0 | 1 | 4 | 9 | 16 | 25 | 36 | 49 | 64 | 81 | 100 |
|---|---|---|---|---|----|----|----|----|----|----|-----|
| y |   |   |   |   |    |    |    |    |    |    |     |

Plot the graph of $y = \sqrt{x}$.

Use your graph to find approximate values of the following:

a $\sqrt{35}$  b $\sqrt{45}$  c $\sqrt{55}$  d $\sqrt{60}$  e $\sqrt{2}$

2. Check your answers to question 1 using the √ key on your calculator.

## Cube roots

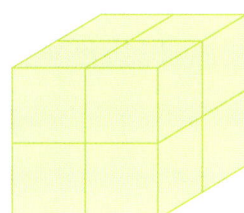

The cube shown has sides of 2 units and occupies 8 cubic units of space.

(That is, 2 × 2 × 2.)

So, the cube root of 8 is 2.

This can be written as $\sqrt[3]{8} = 2$.

$\sqrt[3]{\phantom{x}}$ is read as 'the cube root of ....'.

$\sqrt[3]{64}$ is 4, since 4 × 4 × 4 = 64.

Note that $\sqrt[3]{64}$ cannot also be −4 since −4 × −4 × −4 = −64, therefore $\sqrt[3]{-64}$ is −4.

**Exercise 1.11**

1. Find the following cube roots:

   a $\sqrt[3]{8}$  b $\sqrt[3]{125}$  c $\sqrt[3]{27}$  d $\sqrt[3]{0.001}$

   e $\sqrt[3]{0.027}$  f $\sqrt[3]{216}$  g $\sqrt[3]{1000}$  h $\sqrt[3]{1\,000\,000}$

   i $\sqrt[3]{-8}$  j $\sqrt[3]{-27}$  k $\sqrt[3]{-1000}$  l $\sqrt[3]{-1}$

## Further powers and roots

You have seen that the square of a number is the same as raising that number to the power of 2, for example the square of 5 is written as $5^2$ and means 5 × 5. Similarly, the cube of a number is the same as raising that number to the power of 3, for example, the cube of 5 is written as $5^3$ and means 5 × 5 × 5.

# 1 VOCABULARY FOR SETS OF NUMBERS

Numbers can be raised to other powers too. Therefore, 5 raised to the power of 6 can be written as $5^6$ and means $5 \times 5 \times 5 \times 5 \times 5 \times 5$.

You will find a button on your calculator to help you to do this. On most calculators, it will look like $y^x$.

You have also seen that the square root of a number can be written using the symbol $\sqrt{\phantom{x}}$.

Therefore the square root of 16 is written as $\sqrt{16}$ and is equal to 4.

The cube root of a number can be written using the $\sqrt[3]{\phantom{x}}$ symbol. Therefore, the cube root of 27 is written as $\sqrt[3]{27}$ and is 3 because $3 \times 3 \times 3 = 27$.

Numbers can be rooted by other values as well. The fourth root of a number can be written using the symbol $\sqrt[4]{\phantom{x}}$. Therefore the fourth root of 625 can be expressed as $\sqrt[4]{625}$. And is ±5 because both $5 \times 5 \times 5 \times 5 = 625$ and $(-5) \times (-5) \times (-5) \times (-5) = 625$.

You will find a button on your calculator to help you to calculate with roots. On most calculators, it will look like $\sqrt[x]{y}$.

### Exercise 1.12

1 Work out the following:
- a $6^4$
- b $3^5 + 2^4$
- c $(3^4)^2$
- d $0.1^6 \div 0.01^4$
- e $\sqrt[4]{2401}$
- f $\sqrt[8]{256}$
- g $\left(\sqrt[5]{243}\right)^3$
- h $\left(\sqrt[9]{36}\right)^9$
- i $2^7 \times \sqrt{\frac{1}{4}}$
- j $\sqrt[6]{\frac{1}{64}} \times 2^7$
- k $\sqrt[4]{5^4}$
- l $\left(\sqrt[10]{59049}\right)^2$

## Reciprocals

The **reciprocal** of a number is 1 divided by that number.

e.g. The reciprocal of 2 is $\frac{1}{2}$

The reciprocal of 10 is $\frac{1}{10}$

*Reciprocals of fractions and their use are covered later in this topic.*

The reciprocal of $\frac{1}{3}$ is $\frac{1}{\frac{1}{3}} = 3$

The reciprocal of $\frac{2}{5}$ is $\frac{1}{\frac{2}{5}} = \frac{5}{2}$

# 2 Calculations and order of operations

When doing calculations, take care that you do them in the correct order.

> ## Worked examples

a Use a calculator to work out the answer to the following:

$2 + 3 \times 4$  $= 14$

b Use a calculator to work out the answer to the following:

$(2 + 3) \times 4$

 $= 20$

The reason why different answers are obtained is because, by convention, the operations have different priorities. These are as follows:

1 brackets
2 powers
3 multiplication/division
4 addition/subtraction

Therefore in **Worked example a**, $3 \times 4$ is evaluated first, and then the 2 is added, while in **Worked example b**, $(2 + 3)$ is evaluated first, followed by multiplication by 4.

c Work out the answer to the following:

$-4 \times (8 + -3)$

The $(8 + -3)$ is evaluated first as it is in the brackets; the answer 5 is then multiplied by $-4$ to give an answer of $-20$.

d Work out the answer to the following:

$-4 \times 8 + -3$

The $-4 \times 8$ is evaluated first as it is a multiplication; the answer $-32$ then has $-3$ added to it giving an answer of $-35$.

## 2 CALCULATIONS AND ORDER OF OPERATIONS

**Exercise 1.13** In the following questions, work out the answers:
i in your head,
ii using a calculator.

1. a $8 \times 3 + 2$    b $4 \div 2 + 8$
   c $12 \times 4 - 6$    d $4 + 6 \times 2$
   e $10 - 6 \div 3$    f $6 - 3 \times 4$

2. a $7 \times 2 + 3 \times 2$    b $12 \div 3 + 6 \times 5$
   c $9 + 3 \times 8 - 1$    d $36 - 9 \div 3 - 2$
   e $-14 \times 2 - 16 \div 2$    f $4 + 3 \times 7 - 6 \div 3$

3. a $(4 + 5) \times 3$    b $8 \times (12 - 4)$
   c $3 \times (-8 + -3) - 3$    d $(4 + 11) \div (7 - 2)$
   e $4 \times 3 \times (7 + 5)$    f $24 \div 3 \div (10 - 5)$

**Exercise 1.14** In each of the following questions:
i Copy the calculation and put in any brackets which are needed to make it correct.
ii Check your answer using a calculator.

1. a $6 \times 2 + 1 = 18$    b $1 + 3 \times 5 = 16$
   c $8 + 6 \div 2 = 7$    d $9 + 2 \times 4 = 44$
   e $9 \div 3 \times 4 + 1 = 13$    f $3 + 2 \times 4 - 1 = 15$

2. a $12 \div 4 - 2 + 6 = 7$    b $12 \div 4 - 2 + 6 = 12$
   c $12 \div 4 - 2 + 6 = -5$    d $12 \div 4 - 2 + 6 = 1.5$
   e $4 + 5 \times 6 - 1 = 33$    f $4 + 5 \times 6 - 1 = 29$
   g $4 + 5 \times 6 - 1 = 53$    h $4 + 5 \times 6 - 1 = 45$

For calculations involving fractions, use the fraction button on the calculator. Make sure that you do the calculations in the correct order.

### ➡ Worked example

Find the value of the following using a calculator:

$$\frac{90 + 38}{4^3}$$

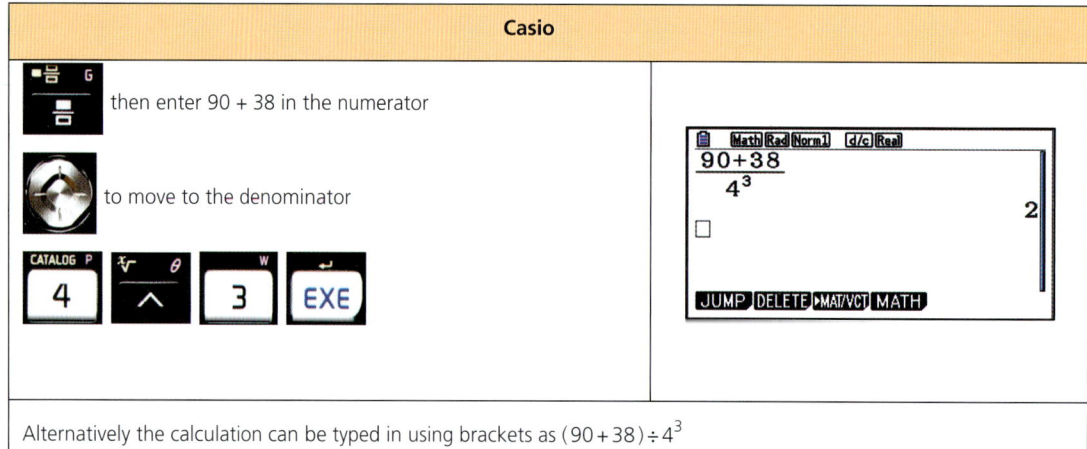

Alternatively the calculation can be typed in using brackets as $(90 + 38) \div 4^3$

# Calculations and order of operations

| Texas |
|---|
| [math] to select 'FRAC', then  [1] to select 'n/d' for the fractional part.  90 + 38 in the numerator  [↓] to move to the denominator.  [4] [^] [3] [enter]  ... NORMAL FLOAT AUTO REAL RADIAN MP  $\frac{90+38}{4^3}$  2. |
| Alternatively the calculation can be typed in using brackets as $(90+38) \div 4^3$ |

**Exercise 1.15** Using a calculator, work out the following:

1. a  $\dfrac{9+3}{6}$     b  $\dfrac{30-6}{5+3}$     c  $\dfrac{40+9}{12-5}$

    d  $\dfrac{15 \times 2}{7+8} + 2$     e  $\dfrac{100+21}{11} + 4 \times 3$     f  $\dfrac{7+2 \times 4}{7-2} - 3$

2. a  $\dfrac{4^2 - 6}{2+8}$     b  $\dfrac{3^2 + 4^2}{5}$     c  $\dfrac{6^3 - 4^2}{4 \times 25}$

    d  $\dfrac{3^3 \times 4^4}{12^2} + 2$     e  $\dfrac{3+3^3}{5} + \dfrac{4^2 - 2^3}{8}$     f  $\dfrac{(6+3) \times 4}{2^3} - 2 \times 3$

# 3 Fractions, decimals and percentages

## Fractions

A single unit can be broken into equal parts called fractions, e.g. $\frac{1}{2}, \frac{1}{3}, \frac{1}{6}$.

If, for example, the unit is broken into ten equal parts and three parts are then taken, the fraction is written as $\frac{3}{10}$. That is, three parts out of ten parts.

In the fraction $\frac{3}{10}$:

The three is called the **numerator**.

The ten is called the **denominator**.

In a **proper fraction**, the numerator is less than the denominator, e.g. $\frac{3}{4}$.

In an **improper fraction**, the numerator is more than the denominator, e.g. $\frac{9}{2}$.

A **mixed number** is made up of a natural number and a proper fraction, e.g. $4\frac{1}{5}$.

## A fraction of an amount

> ### Worked examples
> 
> a  Find $\frac{1}{5}$ of 35.
> 
>    This means 'divide 35 into 5 equal parts'.
> 
>    $\frac{1}{5}$ of 35 is 7.
> 
> b  Find $\frac{3}{5}$ of 35.
> 
>    Since $\frac{1}{5}$ of 35 is 7, $\frac{3}{5}$ of 35 is $3 \times 7$.
> 
>    That is, 21.

*A fraction of an amount*

**Exercise 1.16**

1  Work out the following:

   a  $\frac{1}{5}$ of 40      b  $\frac{3}{5}$ of 40      c  $\frac{1}{9}$ of 36

   d  $\frac{5}{9}$ of 36      e  $\frac{1}{8}$ of 72      f  $\frac{7}{8}$ of 72

   g  $\frac{1}{4}$ of 60      h  $\frac{5}{12}$ of 60     i  $\frac{1}{4}$ of 8

   j  $\frac{3}{4}$ of 8

2  Work out the following:

   a  $\frac{3}{4}$ of 12      b  $\frac{4}{5}$ of 20      c  $\frac{4}{9}$ of 45

   d  $\frac{5}{8}$ of 64      e  $\frac{3}{11}$ of 66     f  $\frac{9}{10}$ of 80

   g  $\frac{5}{7}$ of 42      h  $\frac{8}{9}$ of 54      i  $\frac{7}{8}$ of 240

   j  $\frac{4}{5}$ of 65

# Changing a mixed number to an improper fraction

> **Worked examples**

a  Change $2\frac{3}{4}$ to an improper fraction.

$1 = \frac{4}{4}$

$2 = \frac{8}{4}$

$2\frac{3}{4} = \frac{8}{4} + \frac{3}{4}$

$= \frac{11}{4}$

b  Change $3\frac{5}{8}$ to an improper fraction.

$3\frac{5}{8} = \frac{24}{8} + \frac{5}{8}$

$= \frac{24 + 5}{8}$

$= \frac{29}{8}$

35

## 3 FRACTIONS, DECIMALS AND PERCENTAGES

 **Exercise 1.17**   Change the following mixed numbers to improper fractions:

a  $4\frac{2}{3}$    b  $3\frac{3}{5}$    c  $5\frac{7}{8}$    d  $2\frac{5}{6}$

e  $8\frac{1}{2}$    f  $9\frac{5}{7}$    g  $6\frac{4}{9}$    h  $4\frac{1}{4}$

i  $5\frac{4}{11}$   j  $7\frac{6}{7}$    k  $4\frac{3}{10}$   l  $11\frac{3}{13}$

## Changing an improper fraction to a mixed number

 **Worked example**

Change $\frac{27}{4}$ to a mixed number.

*24 is the largest multiple of 4 below 27.*

$\frac{27}{4} = \frac{24+3}{4}$

$= \frac{24}{4} + \frac{3}{4}$

$= 6\frac{3}{4}$

 **Exercise 1.18**   1  Change the following improper fractions to mixed numbers:

a  $\frac{29}{4}$   b  $\frac{33}{5}$   c  $\frac{41}{6}$   d  $\frac{53}{8}$   e  $\frac{49}{9}$

f  $\frac{17}{12}$  g  $\frac{66}{7}$   h  $\frac{33}{10}$  i  $\frac{19}{2}$   j  $\frac{73}{12}$

## Decimals

| H | T | U | . | $\frac{1}{10}$ | $\frac{1}{100}$ | $\frac{1}{1000}$ |
|---|---|---|---|---|---|---|
|   |   | 3 | . | 2 | 7 |   |
|   |   | 0 | . | 0 | 3 | 8 |

3.27 is 3 units, 2 tenths and 7 hundredths

3.27 = $3 + \frac{2}{10} + \frac{7}{100}$

0.038 is 3 hundredths and 8 thousandths

0.038 = $\frac{3}{100} + \frac{8}{1000}$

## Percentages

Note that 2 tenths and 7 hundredths is equivalent to 27 hundredths

$$\frac{2}{10} + \frac{7}{100} = \frac{27}{100}$$

and that 3 hundredths and 8 thousandths is equivalent to 38 thousandths

$$\frac{3}{100} + \frac{8}{1000} = \frac{38}{1000}$$

### Exercise 1.19

1  Make a table similar to the one above. List the digits in the following numbers in their correct position:
   a  6.023            b  5.94            c  18.3
   d  0.071            e  2.001           f  3.56

2  Work out the following without using a calculator:
   a  2.7 + 0.35 + 16.09        b  1.44 + 0.072 + 82.3
   c  23.8 − 17.2               d  16.9 − 5.74
   e  121.3 − 85.49             f  6.03 + 0.5 − 1.21
   g  72.5 − 9.08 + 3.72        h  100 − 32.74 − 61.2
   i  16.0 − 9.24 − 5.36        j  1.1 − 0.92 − 0.005

# Percentages

A fraction whose denominator is 100 can be written as a percentage.

$\frac{29}{100}$ can be written as 29%

$\frac{45}{100}$ can be written as 45%

## Changing a fraction to a percentage

By using equivalent fractions to change the denominator to 100, other fractions can be written as percentages.

### ➡ Worked example

Change $\frac{3}{5}$ to a percentage.

$$\frac{3}{5} = \frac{3}{5} \times \frac{20}{20} = \frac{60}{100}$$

$\frac{60}{100}$ can be written as 60%

# 3 FRACTIONS, DECIMALS AND PERCENTAGES

 **Exercise 1.20**

1. Write each of the following as a fraction with denominator 100, then write them as percentages:

   a $\frac{29}{50}$   b $\frac{17}{25}$   c $\frac{11}{20}$   d $\frac{3}{10}$

   e $\frac{23}{25}$   f $\frac{19}{50}$   g $\frac{3}{4}$   h $\frac{2}{5}$

2. Copy and complete the table of equivalents below.

| Fraction | Decimal | Percentage |
|---|---|---|
| $\frac{1}{10}$ | | |
| | 0.2 | |
| | | 30% |
| $\frac{4}{10}$ | | |
| | 0.5 | |
| | | 60% |
| | 0.7 | |
| $\frac{4}{5}$ | | |
| | 0.9 | |
| $\frac{1}{4}$ | | |
| | | 75% |

## Equivalent fractions

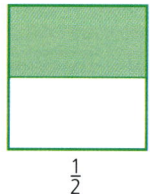
$\frac{1}{2}$

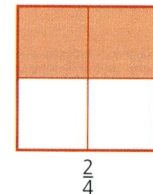
$\frac{2}{4}$

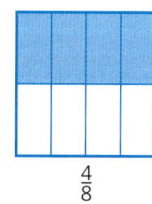
$\frac{4}{8}$

It should be apparent that $\frac{1}{2}, \frac{2}{4}$ and $\frac{4}{8}$ are equivalent fractions.

Similarly, $\frac{1}{3}, \frac{2}{6}, \frac{3}{9}$ and $\frac{4}{12}$ are equivalent, as are $\frac{1}{5}, \frac{10}{50}$ and $\frac{20}{100}$.

Equivalent fractions are mathematically the same as each other. In the diagrams above $\frac{1}{2}$ is mathematically the same as $\frac{4}{8}$. However, $\frac{1}{2}$ is a simplified form of $\frac{4}{8}$.

## Equivalent fractions

When carrying out calculations involving fractions it is usual to give your answer in its **simplest form**.

> ## Worked examples

a Write $\frac{4}{22}$ in its simplest form.

Divide both the numerator and the denominator by their highest common factor.
The highest common factor of 4 and 22 is 2.
Dividing both 4 and 22 by 2 gives $\frac{2}{11}$.
Therefore $\frac{2}{11}$ is $\frac{4}{22}$ written in its simplest form.

b Write $\frac{12}{40}$ in its simplest form.

Divide both the numerator and the denominator by their highest common factor.
The highest common factor of 12 and 40 is 4.
Dividing both 12 and 40 by 4 gives $\frac{3}{10}$.
Therefore $\frac{3}{10}$ is $\frac{12}{40}$ written in its simplest form.

 **Exercise 1.21**

1 Copy the following sets of equivalent fractions and fill in the blanks:

a $\frac{2}{5} = \frac{4}{\_\_} = \frac{\_\_}{20} = \frac{\_\_}{50} = \frac{16}{\_\_}$

b $\frac{3}{8} = \frac{6}{\_\_} = \frac{\_\_}{24} = \frac{15}{\_\_} = \frac{\_\_}{72}$

c $\frac{\_\_}{7} = \frac{8}{14} = \frac{12}{\_\_} = \frac{\_\_}{56} = \frac{36}{\_\_}$

d $\frac{5}{\_\_} = \frac{\_\_}{27} = \frac{20}{36} = \frac{\_\_}{90} = \frac{55}{\_\_}$

2 Write the following fractions in their simplest form:

a $\frac{5}{10}$

b $\frac{7}{21}$

c $\frac{8}{12}$

d $\frac{16}{36}$

e $\frac{75}{100}$

f $\frac{81}{90}$

# 3 FRACTIONS, DECIMALS AND PERCENTAGES

## Addition and subtraction of fractions

To add or subtract fractions, the denominators need to be the same.

### → Worked examples

a $\frac{3}{11} + \frac{5}{11} = \frac{8}{11}$

b $\frac{7}{8} + \frac{5}{8} = \frac{12}{8} = 1\frac{1}{2}$

c $\frac{1}{2} + \frac{1}{3} = \frac{3}{6} + \frac{2}{6} = \frac{5}{6}$

d $\frac{4}{5} - \frac{1}{3} = \frac{12}{15} - \frac{5}{15} = \frac{7}{15}$

When dealing with calculations involving mixed numbers, it is sometimes easier to change them to improper fractions first.

### → Worked examples

a $5\frac{3}{4} - 2\frac{5}{8}$

$= \frac{23}{4} - \frac{21}{8}$

$= \frac{46}{8} - \frac{21}{8}$

$= \frac{25}{8} = 3\frac{1}{8}$

b $1\frac{4}{7} + 3\frac{3}{4}$

$= \frac{11}{7} + \frac{15}{4}$

$= \frac{44}{28} + \frac{105}{28}$

$= \frac{149}{28} = 5\frac{9}{28}$

## Exercise 1.22

Work out each of the following and write the answer as a fraction in its simplest form:

1. a $\frac{3}{7} - \frac{2}{7}$
   b $\frac{4}{5} - \frac{7}{10}$
   c $\frac{8}{9} - \frac{1}{3}$
   d $\frac{7}{12} - \frac{1}{2}$
   e $\frac{5}{8} - \frac{2}{5}$
   f $\frac{3}{4} - \frac{2}{5} + \frac{7}{10}$

2. a $\frac{3}{4} + \frac{1}{5} - \frac{2}{3}$
   b $\frac{3}{8} + \frac{7}{11} - \frac{1}{2}$
   c $\frac{4}{5} - \frac{3}{10} + \frac{7}{20}$
   d $\frac{9}{13} + \frac{1}{3} - \frac{4}{5}$
   e $\frac{9}{10} - \frac{1}{5} - \frac{1}{4}$
   f $\frac{8}{9} - \frac{1}{3} - \frac{1}{2}$

3. a $2\frac{1}{2} + 3\frac{1}{4}$
   b $3\frac{3}{5} + 1\frac{7}{10}$
   c $6\frac{1}{2} - 3\frac{2}{5}$
   d $8\frac{5}{8} - 2\frac{1}{3}$
   e $5\frac{7}{8} - 4\frac{3}{4}$
   f $3\frac{1}{4} - 2\frac{5}{9}$

# Equivalent fractions

4  a  $2\frac{1}{2} + 1\frac{1}{4} + 1\frac{3}{8}$      b  $2\frac{4}{5} + 3\frac{1}{8} + 1\frac{3}{10}$

   c  $4\frac{1}{2} - 1\frac{1}{4} - 3\frac{5}{8}$      d  $6\frac{1}{2} - 2\frac{3}{4} - 3\frac{2}{5}$

   e  $2\frac{4}{7} - 3\frac{1}{4} - 1\frac{3}{5}$      f  $4\frac{7}{20} - 5\frac{1}{2} + 2\frac{2}{5}$

## Multiplication and division of fractions

To multiply fractions simply multiply the numerators together to give the numerator of the answer and multiply the denominators together to give the denominator of the answer.

### → Worked examples

a  $\frac{3}{4} \times \frac{2}{3}$

    $= \frac{6}{12}$

    $= \frac{1}{2}$

This can be visualised as follows:

$\frac{3}{4} \times \frac{2}{3}$ can be read as $\frac{3}{4}$ of $\frac{2}{3}$

The lowest common multiple of 3 and 4 is 12, therefore a rectangle split into 12 parts can be used.

Shade $\frac{2}{3}$ of the rectangle:

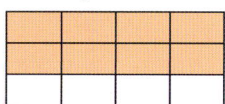

Then shade $\frac{3}{4}$ of the $\frac{2}{3}$:

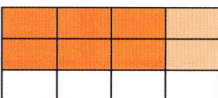

$\frac{6}{12}$ of the original rectangle now has the darker shading and represents $\frac{3}{4}$ of $\frac{2}{3}$

Therefore $\frac{3}{4} \times \frac{2}{3} = \frac{6}{12}$, which can be obtained simply by multiplying the numerators together and the denominators together.

$\frac{6}{12}$ is then simplified to $\frac{1}{2}$

b  $3\frac{1}{2} \times 4\frac{4}{7}$

    $= \frac{7}{2} \times \frac{32}{7}$

    $= \frac{224}{14}$

    $= 16$

# 3 FRACTIONS, DECIMALS AND PERCENTAGES

This can be visualised as follows:

$3\frac{1}{2} \times 4\frac{4}{7}$ can be read as $3\frac{1}{2}$ of $4\frac{4}{7}$

The lowest common multiple of 2 and 7 is 14, therefore a rectangle split into 14 parts can be used.

There are $4\frac{4}{7}$ of these rectangles:

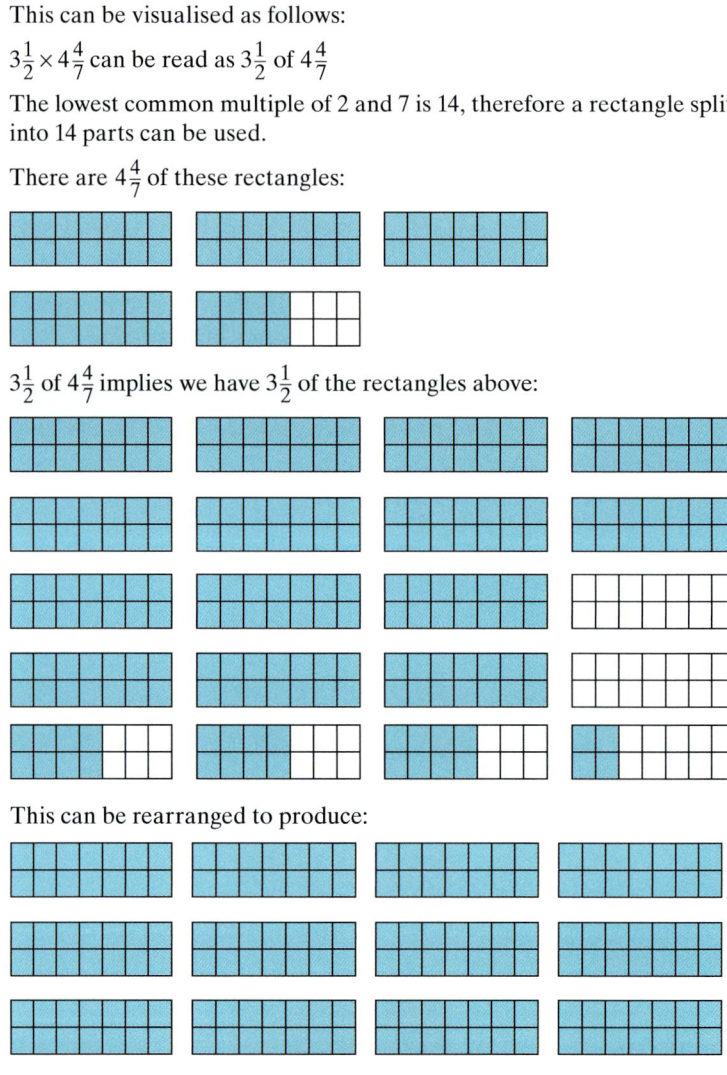

$3\frac{1}{2}$ of $4\frac{4}{7}$ implies we have $3\frac{1}{2}$ of the rectangles above:

This can be rearranged to produce:

16 of the original rectangles are now shaded, therefore $3\frac{1}{2} \times 4\frac{4}{7} = 16$

Or $\frac{7}{2} \times \frac{32}{7} = 16$, which can simply be obtained by multiplying the numerators together and the denominators together, and then simplifying the answer.

# Equivalent fractions

The **reciprocal** of a number is obtained when 1 is divided by that number. The reciprocal of 5 is $\frac{1}{5}$, the reciprocal of $\frac{2}{5}$ is $\frac{5}{2}$, and so on.

## → Worked examples

Dividing fractions is the same as multiplying by the reciprocal.

a  $\frac{3}{8} \div \frac{3}{4}$

$= \frac{3}{8} \times \frac{4}{3}$

$= \frac{12}{24}$

$= \frac{1}{2}$

$\frac{3}{8} \div \frac{3}{4}$ can be read as "How many $\frac{3}{4}$ go into $\frac{3}{8}$?"

The lowest common multiple of 8 and 4 is 8, therefore a rectangle split into 8 parts can be used.

$\frac{3}{8}$ can be shaded and compared with $\frac{3}{4}$

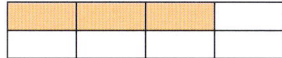 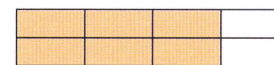

Only half of $\frac{3}{4}$ can "go into" $\frac{3}{8}$ i.e. $\frac{3}{8} \div \frac{3}{4} = \frac{1}{2}$

This can simply be simply calculated by multiplying $\frac{3}{8}$ by the reciprocal of $\frac{3}{4}$:

$\frac{3}{8} \times \frac{4}{3} = \frac{12}{24} = \frac{1}{2}$

b  $5\frac{1}{2} \div 3\frac{2}{3}$

$= \frac{11}{2} \div \frac{11}{3}$

$= \frac{11}{2} \times \frac{3}{11}$

$= \frac{3}{2}$

$5\frac{1}{2} \div 3\frac{2}{3}$ can be read as "How many $3\frac{2}{3}$ go into $5\frac{1}{2}$?"

The lowest common multiple of 3 and 2 is 6, therefore a rectangle split into 6 parts can be used.

Shade $5\frac{1}{2}$ of these rectangles:

 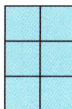 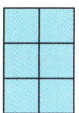 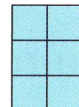 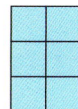 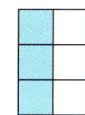

# 3 FRACTIONS, DECIMALS AND PERCENTAGES

Shade $3\frac{2}{3}$ of these:

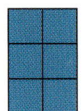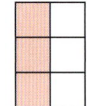

Only $1\frac{1}{2}$ of the $3\frac{2}{3}$ go into $5\frac{1}{2}$

Therefore $5\frac{1}{2} \div 3\frac{2}{3} = 1\frac{1}{2}$

Or $\frac{11}{2} \div \frac{11}{3} = \frac{3}{2}$

This can simply be calculated by multiplying $\frac{11}{2}$ by the reciprocal of $\frac{11}{3}$:

$\frac{11}{2} \times \frac{3}{11} = \frac{33}{22} = \frac{3}{2}$

## Exercise 1.23

1. Write the reciprocal of each of the following:
   a $\frac{3}{4}$
   b $\frac{5}{9}$
   c $7$
   d $\frac{1}{9}$
   e $2\frac{3}{4}$
   f $4\frac{5}{8}$

2. Find the value of the following:
   a $\frac{3}{8} \times \frac{4}{9}$
   b $\frac{2}{3} \times \frac{9}{10}$
   c $\frac{5}{7} \times \frac{4}{15}$
   d $\frac{3}{4}$ of $\frac{8}{9}$
   e $\frac{5}{6}$ of $\frac{3}{10}$
   f $\frac{7}{8}$ of $\frac{2}{5}$

3. Find the value of the following:
   a $\frac{5}{8} \div \frac{3}{4}$
   b $\frac{5}{6} \div \frac{1}{3}$
   c $\frac{4}{5} \div \frac{7}{10}$
   d $1\frac{2}{3} \div \frac{2}{5}$
   e $\frac{3}{7} \div 2\frac{1}{7}$
   f $1\frac{1}{4} \div 1\frac{7}{8}$

4. Find the value of the following:
   a $(\frac{3}{8} \times \frac{4}{5}) + (\frac{1}{2}$ of $\frac{3}{5})$
   b $(1\frac{1}{2} \times 3\frac{3}{4}) - (2\frac{3}{5} \div 1\frac{1}{2})$
   c $(\frac{3}{5}$ of $\frac{4}{9}) + (\frac{4}{9}$ of $\frac{3}{5})$
   d $(1\frac{1}{3} \times 2\frac{5}{8})^2$

5. Using the correct order of operations, work out the following:
   a $\frac{1}{4} + \frac{2}{7} \times \frac{3}{4}$
   b $\frac{1}{2} + \frac{3}{8} \div \frac{1}{4} - 2\frac{1}{2}$
   c $\left(\frac{1}{6}\right)^2 - \frac{1}{8} \times \frac{5}{9}$
   d $\left(5\frac{2}{3} - \frac{14}{3}\right)^2 \div \frac{2}{7} + 3\frac{1}{2}$

# Changing a fraction to a decimal

To change a fraction to a decimal, divide the numerator by the denominator.

## Changing a decimal to a fraction

> **Worked examples**

a  Change $\frac{5}{8}$ to a decimal.

Divide 5 by 8 as follows:

$$8\overline{)5.0^20^40} \quad 0.6\ 2\ 5$$

Therefore $\frac{5}{8} = 0.625$

b  Change $2\frac{3}{5}$ to a decimal.

This can be represented as $2 + \frac{3}{5}$

First divide 3 by 5 as shown

$$5\overline{)3.0} \quad 0.6$$

Therefore $2\frac{3}{5} = 2 + 0.6 = 2.6$

 **Exercise 1.24**

1  Change the following fractions to decimals:

   a  $\frac{3}{4}$     b  $\frac{4}{5}$     c  $\frac{9}{20}$     d  $\frac{17}{50}$

   e  $\frac{1}{3}$     f  $\frac{3}{8}$     g  $\frac{7}{16}$     h  $\frac{2}{9}$

2  Change the following mixed numbers to decimals:

   a  $2\frac{3}{4}$     b  $3\frac{3}{5}$     c  $4\frac{7}{20}$     d  $6\frac{11}{50}$

   e  $5\frac{2}{3}$     f  $6\frac{7}{8}$     g  $5\frac{9}{16}$     h  $4\frac{2}{9}$

## Changing a decimal to a fraction

Changing a decimal to a fraction is done by knowing the 'value' of each of the digits in any decimal.

> **Worked examples**

a  Change 0.45 from a decimal to a fraction.

| units | . | tenths | hundredths |
|---|---|---|---|
| 0 | . | 4 | 5 |

0.45 is therefore equivalent to 4 tenths and 5 hundredths, which in turn is the same as 45 hundredths.

Therefore $0.45 = \frac{45}{100} = \frac{9}{20}$

# 3 FRACTIONS, DECIMALS AND PERCENTAGES

**b** Change 2.325 from a decimal to a fraction.

| units | . | tenths | hundredths | thousandths |
|---|---|---|---|---|
| 2 | . | 3 | 2 | 5 |

Therefore $2.325 = 2\frac{325}{1000} = 2\frac{13}{40}$

## Exercise 1.25

1 Change the following decimals to fractions:
- **a** 0.5
- **b** 0.7
- **c** 0.6
- **d** 0.75
- **e** 0.825
- **f** 0.05
- **g** 0.050
- **h** 0.402
- **i** 0.0002

2 Change the following decimals to mixed numbers:
- **a** 2.4
- **b** 6.5
- **c** 8.2
- **d** 3.75
- **e** 10.55
- **f** 9.204
- **g** 15.455
- **h** 30.001
- **i** 1.0205

# 4 Further percentages

You will already be familiar with the percentage equivalent of simple fractions and decimals as outlined in the table below:

| Fraction | Decimal | Percentage |
|---|---|---|
| $\frac{1}{2}$ | 0.5 | 50% |
| $\frac{1}{4}$ | 0.25 | 25% |
| $\frac{3}{4}$ | 0.75 | 75% |
| $\frac{1}{8}$ | 0.125 | 12.5% |
| $\frac{3}{8}$ | 0.375 | 37.5% |
| $\frac{5}{8}$ | 0.625 | 62.5% |
| $\frac{7}{8}$ | 0.875 | 87.5% |
| $\frac{1}{10}$ | 0.1 | 10% |
| $\frac{2}{10}$ or $\frac{1}{5}$ | 0.2 | 20% |
| $\frac{3}{10}$ | 0.3 | 30% |
| $\frac{4}{10}$ or $\frac{2}{5}$ | 0.4 | 40% |
| $\frac{6}{10}$ or $\frac{3}{5}$ | 0.6 | 60% |
| $\frac{7}{10}$ | 0.7 | 70% |
| $\frac{8}{10}$ or $\frac{4}{5}$ | 0.8 | 80% |
| $\frac{9}{10}$ | 0.9 | 90% |

## Simple percentages

### ➜ Worked examples

a   Of 100 sheep in a field, 88 are ewes.

   i   What percentage of the sheep are ewes?
       88 out of 100 are ewes
       = 88%

# 4 FURTHER PERCENTAGES

  ii What percentage are not ewes?
  
  12 out of 100 are not ewes
  
  = 12%

b A gymnast scored marks out of 10 from five judges.

  The marks awarded were: 8.0, 8.2, 7.9, 8.3, 7.6.
  
  Write these marks as percentages.
  
  $\frac{8.0}{10} = \frac{80}{100} = 80\%$ $\qquad$ $\frac{8.2}{10} = \frac{82}{100} = 82\%$ $\qquad$ $\frac{7.9}{10} = \frac{79}{100} = 79\%$
  
  $\frac{8.3}{10} = \frac{83}{100} = 83\%$ $\qquad$ $\frac{7.6}{10} = \frac{76}{100} = 76\%$

c Convert the following percentages into fractions and decimals:
  
  i 27% $\qquad\qquad$ ii 5%
  
  $\frac{27}{100} = 0.27$ $\qquad\qquad$ $\frac{5}{100} = 0.05$

---

 **Exercise 1.26**

1 In a survey of 100 cars, 47 were white, 23 were blue and 30 were red. Write each of these numbers as a percentage of the total.

2 $\frac{7}{10}$ of the surface of the Earth is water. Write this as a percentage.

3 There are 200 birds in a flock. 120 of them are female. What percentage of the flock are:
  a female?  b male?

4 Write these percentages as fractions of 100:
  a 73%  b 28%
  c 10%  d 25%

5 Write these fractions as percentages:
  a $\frac{27}{100}$  b $\frac{3}{10}$
  c $\frac{7}{50}$  d $\frac{1}{4}$

6 Convert the following percentages to decimals:
  a 39%  b 47%  c 83%
  d 7%   e 2%   f 20%

7 Convert the following decimals to percentages:
  a 0.31  b 0.67  c 0.09
  d 0.05  e 0.2   f 0.75

## Calculating a percentage of a quantity

### → Worked examples

a  Find 25% of 300 m.
   25% can be written as 0.25.
   0.25 × 300 m = 75 m

b  Find 35% of 280 m.
   35% can be written as 0.35.
   0.35 × 280 m = 98 m

 **Exercise 1.27**

1  Work out the following:
   a  25% of 80      b  80% of 125     c  62.5% of 80
   d  30% of 120     e  90% of 5       f  25% of 30
2  Work out the following:
   a  17% of 50      b  50% of 17      c  65% of 80
   d  80% of 65      e  7% of 250      f  250% of 7
3  In a class of 30 students, 20% have black hair, 10% have blonde hair and 70% have brown hair. Calculate the number of students with:
   a  black hair    b  blonde hair    c  brown hair.
4  A survey conducted among 120 children looked at what type of meat they preferred. 55% said they preferred beef, 20% said they preferred chicken, 15% preferred lamb and 10% liked none of these. Calculate the number of children in each category.
5  A survey was carried out in a school to see what nationality its students were. Of the 220 students in the school, 65% were English, 20% were Pakistani, 5% were Greek and 10% belonged to other nationalities. Calculate the number of students of each nationality.
6  A shopkeeper keeps a record of the number of items he sells in one day. Of the 150 items he sold, 46% were newspapers, 24% were pens, 12% were books while the remaining 18% were other assorted items. Calculate the number of each item he sold.

## Expressing one quantity as a percentage of another

To express one quantity as a percentage of another, write the first quantity as a fraction of the second and then multiply by 100.

# 4 FURTHER PERCENTAGES

 **Worked example**

In an examination a girl obtains 69 marks out of 75. Write this result as a percentage.

$\frac{69}{75} \times 100\% = 92\%$

### Exercise 1.28

1. Write the first quantity as a percentage of the second.
   - **a** 24 out of 50
   - **b** 46 out of 125
   - **c** 7 out of 20
   - **d** 45 out of 90
   - **e** 9 out of 20
   - **f** 16 out of 40
   - **g** 13 out of 39
   - **h** 20 out of 35

2. A hockey team plays 42 matches. It wins 21, draws 14 and loses the rest. Write each of these results as a percentage of the total number of games played.

3. Four candidates stood in an election:
   - A received 24 500 votes
   - B received 18 200 votes
   - C received 16 300 votes
   - D received 12 000 votes
   
   Write each of these as a percentage of the total votes cast.

4. A car manufacturer produces 155 000 cars a year. The cars are available for sale in six different colours. The numbers sold of each colour were:
   - Red 55 000
   - Blue 48 000
   - White 27 500
   - Silver 10 200
   - Green 9300
   - Black 5000
   
   Write each of these as a percentage of the total number of cars produced. Give your answers to 3 s.f.

## Percentage increases and decreases

 **Worked examples**

**a** A garage increases the price of a truck by 12%. If the original price was $14 500, calculate its new price.

Note: the original price represents 100%, therefore the increased price can be represented as 112%.

New price = 112% of $14 500
= 1.12 × $14 500
= $16 240

*Percentage increases and decreases*

**b** A Saudi doctor has a salary of 16 000 Saudi riyals per month. If his salary increases by 8%, calculate:

  **i** the amount extra he receives a month

  **ii** his new monthly salary.

  **i** Increase = 8% of 16 000 riyals = 0.08 × 16 000 riyals = 1280 riyals

  **ii** New salary = old salary + increase = 16 000 + 1280 riyals per month = 17 280 riyals per month

**c** A shop is having a sale. It sells a set of tools costing $130 at a 15% discount. Calculate the sale price of the tools.

Note: The old price represents 100%, therefore the new price can be represented as:

$(100 - 15)\% = 85\%$.

85% of $130 = 0.85 × $130 = $110.50

*Unless the answer is a whole number, for questions involving money, you should always give answers to → 2 decimal places.*

### Exercise 1.29

**1** Increase the following by the given percentage:
  **a** 150 by 25%  **b** 230 by 40%  **c** 7000 by 2%
  **d** 70 by 250%  **e** 80 by 12.5%  **f** 75 by 62%

**2** Decrease the following by the given percentage:
  **a** 120 by 25%  **b** 40 by 5%  **c** 90 by 90%
  **d** 1000 by 10%  **e** 80 by 37.5%  **f** 75 by 42%

**3** In each part below, the first number is increased to become the second number. Calculate the percentage increase in each case.
  **a** 50 → 60  **b** 75 → 135  **c** 40 → 84
  **d** 30 → 31.5  **e** 18 → 33.3  **f** 4 → 13

**4** In each part below, the first number is decreased to become the second number. Calculate the percentage decrease in each case.
  **a** 50 → 25  **b** 80 → 56  **c** 150 → 142.5
  **d** 3 → 0  **e** 550 → 352  **f** 20 → 19

**5** A farmer increases the yield on his farm by 15%. If his previous yield was 6500 tonnes, what is his current yield?

**6** The cost of a computer in a Brazilian computer store is reduced by 12.5% in a sale. If the computer was priced at 7800 Brazilian real (BRL), what is its price in the sale?

**7** A winter coat is priced at £100. In the sale, its price is reduced by 25%.
  **a** Calculate the sale price of the coat.
  **b** After the sale its price is increased by 25% again. Calculate the price of the coat after the sale.

**8** A farmer takes 250 chickens to be sold at a market. In the first hour she sells 8% of her chickens. In the second hour she sells 10% of those that were left.
  **a** How many chickens has she sold in total?
  **b** What percentage of the original number did she sell in the two hours?

**9** The number of fish on a fish farm increases by approximately 10% each month. If there were originally 350 fish, calculate to the nearest 100 how many fish there would be after 12 months.

# 4 FURTHER PERCENTAGES

## Simple interest

**Interest** is money added by a bank to sums deposited by customers, or money charged by a bank to customers for borrowing. The money deposited or borrowed is called the **principal** (sometimes called the capital). The **percentage interest** is the given rate and the money is usually left or borrowed for a fixed period of time.
The following formula can be used to calculate **simple interest**:

*You will need to know this formula.*

$$I = \frac{Prn}{100}$$

where  $I$ = the simple interest paid
$P$ = the principal (sometimes called the capital) (the amount borrowed or lent)
$n$ = number of time periods (usually years)
$r$ = percentage rate

To understand this formula and why it works, you can look at using percentages as shown in the worked example below.

To work out 15% of $300, simply calculate $\frac{15}{100} \times 300$.

If this is repeated 4 times the calculation becomes $\frac{15}{100} \times 300 \times 4$.

This can also be written as $\frac{15 \times 300 \times 4}{100}$.

Therefore to work out $r$% of $P$, the calculation is $\frac{r}{100} \times P$

which can be written as $\frac{Pr}{100}$. If this is repeated $n$ times, the calculation is $\frac{Prn}{100}$.

### → Worked examples

a  Find the simple interest earned on $250 deposited for six years at 8% per year

$$I = \frac{Prn}{100}$$

$$= \frac{250 \times 8 \times 6}{100}$$

$$= 120$$

The interest paid is $120.

b  How long will it take for a sum of €250 invested at 8% per year to earn interest of €80?

$$I = \frac{Prn}{100}$$

$$80 = \frac{250 \times 8 \times n}{100}$$

$$8000 = 2000n$$

$$n = 4$$

It will take 4 years.

## Simple interest

**c** What rate per year must be paid for a principal of £750 to earn interest of £180 in four years?

$$I = \frac{Prn}{100}$$

$$180 = \frac{750 \times r \times 4}{100}$$

$$180 = 30r$$

$$r = 6$$

A rate of 6% is required.

The total amount, $A$, after simple interest is added is given by the formula:
$$A = P + \frac{Prn}{100}$$
This is an example of a linear sequence. These are covered in more detail in Topic 2.

### Exercise 1.30

All rates of interest are annual rates.

1 Find the simple interest paid in each of the following cases:

| | Principal | Rate | Time period |
|---|---|---|---|
| a | NZ$300 | 6% | 4 years |
| b | £750 | 8% | 7 years |
| c | ¥425 | 6% | 4 years |
| d | 2800 baht | 4.5% | 2 years |
| e | HK$880 | 6% | 7 years |

2 How long will it take for the following amounts of interest to be earned?

| | Principal | Rate | Interest |
|---|---|---|---|
| a | R$500 | 4% | R$150 |
| b | ¥5800 | 4% | ¥96 |
| c | A$4000 | 7.5% | A$1500 |
| d | £2800 | 8.5% | £1904 |
| e | €900 | 4.5% | €243 |
| f | 400 Ft | 9% | 252 Ft |

3 Calculate the rate of interest per year that will earn the given amount of interest in the stated time period:

| | Principal | Time period | Interest |
|---|---|---|---|
| a | €400 | 4 years | €1120 |
| b | US$800 | 7 years | US$224 |
| c | ₹2000 | 3 years | ₹210 |
| d | £1500 | 6 years | £675 |
| e | €850 | 5 years | €340 |
| f | A$1250 | 2 years | A$275 |

# 4 FURTHER PERCENTAGES

**Exercise 1.30 (cont)**

4  Calculate the principal that will earn the interest stated, in the number of years and at the given rate, in each of the following cases:

|   | Interest | Time period | Rate |
|---|---|---|---|
| a | 80 Ft | 4 years | 5% |
| b | NZ$36 | 3 years | 6% |
| c | €340 | 5 years | 8% |
| d | 540 baht | 6 years | 7.5% |
| e | €540 | 3 years | 4.5% |
| f | US$348 | 4 years | 7.25% |

5  What rate of interest is paid on a deposit of £2000 that earns £400 interest in five years?

6  How long will it take a principal of €350 to earn €56 interest at 8% per year?

7  A principal of 480 Ft earns 108 Ft interest in five years. What rate of interest was being paid?

8  A principal of €750 becomes a total of €1320 in eight years. What rate of interest was being paid?

9  A$1500 is invested for six years at 3.5% per year. What is the interest earned?

10  500 baht is invested for 11 years and becomes 830 baht in total. What rate of interest was being paid?

## Compound interest

**Compound interest** means that interest is paid not only on the principal amount, but also on the interest itself: it is compounded (or added to). This sounds complicated but the example below will make it clearer.

e.g. A builder is going to build six houses on a plot of land in Spain. He borrows €500 000 at 10% compound interest and will pay the loan off in full after three years.

10% of €500 000 is €50 000, therefore at the end of the first year he will owe a total of €550 000 as shown below.

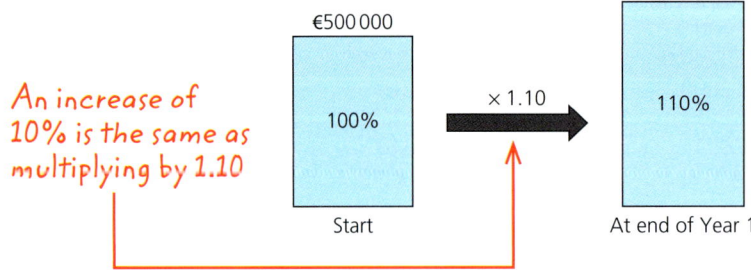

An increase of 10% is the same as multiplying by 1.10

*Compound interest*

For the second year, the amount he owes increases again by 10%, but this is calculated by adding 10% to the amount he owed at the end of the first year, i.e. 10% of €550 000. This can be represented using the diagram below.

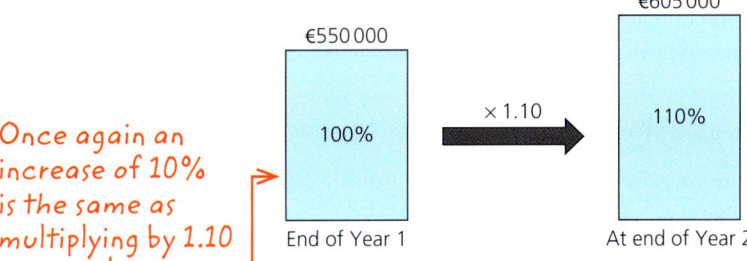

Once again an increase of 10% is the same as multiplying by 1.10

For the third year, the amount he owes increases again by 10%. This is calculated by adding 10% to the amount he owed at the end of the second year, i.e. 10% of €605 000 as shown.

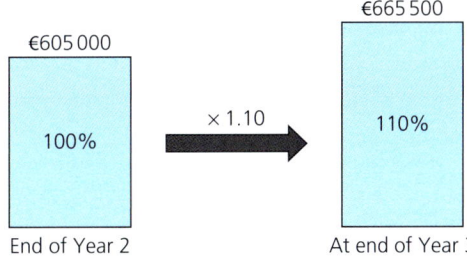

Therefore, the compound interest he has to pay at the end of three years is €665 500 − €500 000 = €165 500.

By looking at the diagrams above it can be seen that the principal amount has in effect been multiplied by 1.10 three times (this is the same as multiplying by $1.10^3$), i.e. €500 000 × $1.10^3$ = €665 500.

The time taken for a debt to grow at compound interest can be calculated as shown in the worked example below.

## → Worked example

How long will it take for a debt to double at a compound interest rate of 27% per annum?

An interest rate of 27% implies a multiplier of 1.27.
The effect of applying this multiplier to a principal amount $P$ is shown in the table:

## 4 FURTHER PERCENTAGES

| Time (years) | 0 | 1 | 2 | 3 |
|---|---|---|---|---|
| Debt | $P$ | $1.27P$ | $1.27^2 P = 1.61P$ | $1.27^3 P = 2.05P$ |

$\times 1.27 \quad \times 1.27 \quad \times 1.27$

The debt will have more than doubled after 3 years.

*Exponential growth is relevant to the Extended syllabus only.*

Compound interest is an example of an exponential sequence and **exponential growth**. This is covered in more detail in Topic 2 and Topic 3.

The interest is usually calculated annually, but there can be other time periods. Compound interest can be charged yearly, half-yearly, quarterly, monthly or daily. (In theory any time period can be chosen.)

### → Worked examples

**a** Find the compound interest paid on a loan of $600 for 3 years at an annual percentage rate (APR) of 5%.

An increase of 5% is equivalent to a multiplier of 1.05.

Therefore 3 years at 5% is calculated as $600 \times 1.05^3 = 694.58$ (2 d.p).

The total payment is $694.58, so the interest paid is $694.58 – $600 = $94.58.

*This is known as quarterly payments as 3 months are a quarter of a year. Note this content is relevant to the Extended syllabus only.*

**b** Find the compound interest paid on a loan of $600 for 3 years at an APR of 5%, if the interest is calculated every 3 months.

> **Note**
>
> As 3 months is a quarter of a year, the interest for each time period of 3 months is $5 \div 4 = 1.25\%$

An increase of 1.25% is equivalent to a multiplier of 1.0125.

There will be 12 time periods of 3 months each over the three-year period.

Therefore, the total amount to pay back is $600 \times 1.0125^{12} = 696.45$

The interest paid is therefore $696.45 – $600 = $96.45

In the worked example part a above, the compound interest was calculated annually and the total interest to pay back after three years was $94.58.

In the worked example part b above, the compound interest was calculated quarterly (every 3 months) and the total interest to pay back after three years was $96.45.

# Compound interest

*Why might lenders prefer quarterly calculations compared with annual calculations?*

The time period for which the compound interest is calculated therefore affects the total amount of interest to be paid.

You can calculate the amount paid using the formula for compound interest:

$$A = P\left(1 + \frac{r}{100n}\right)^{nt}$$

where  $A$ is the total amount paid

$P$ is the principal amount (i.e. the amount borrowed or lent)

$r$ is the percentage rate

*If the compound interest is calculated annually then $n = 1$*

$n$ is the number of interest payments made in one year

$t$ represents the overall time period in years

Therefore, applying this formula to the worked example part b above gives:

$$A = 600\left(1 + \frac{5}{100 \times 4}\right)^{12} = 696.45$$

If the total amount of interest paid ($I$) needs to be calculated, then the principal amount needs to be subtracted from the total.

*The formula for calculating compound interest will **not** be given on the formula sheet.*

i.e.  $$I = P\left(1 + \frac{r}{100n}\right)^{nt} - P$$

### Exercise 1.31

1  A shipping company borrows $70 million at 5% annual compound interest to build a new cruise ship. If it repays the debt after 3 years, how much interest will the company pay?
2  A woman borrows $100 000 for home improvements. The interest rate is 15% and she repays it in full after 3 years. How much interest does she pay?
3  A man owes $5000 on his credit cards. The APR is 20%. If he doesn't repay any of the debt, how much will he owe after 4 years?
4  A school increases its intake by 10% each year. If it starts with 1000 students, how many will it have at the beginning of the fourth year of expansion?
5  8 million tonnes of fish were caught in the North Sea in 2012. If the catch is reduced by 20% each year for 4 years, what weight is caught at the end of this time?
6  How many years will it take for a debt to double at 42% annual compound interest?
7  How many years will it take for a debt to double at 15% annual compound interest?
8  A car loses value at a rate of 27% each year. How long will it take for its value to halve?

# 4 FURTHER PERCENTAGES

## Reverse percentages

### → Worked examples

a  In a test, Ahmed answered 92% of the questions correctly. If he answered 23 questions correctly, how many did he get wrong?

92% of the marks is equivalent to 23 questions.

1% of the marks therefore is equivalent to $\frac{23}{92}$ questions.

So 100% is equivalent to $\frac{23}{92} \times 100 = 25$ questions.

Ahmed got 2 questions wrong.

b  A boat is sold for £15 360. This represents a profit of 28% to the seller. What did the boat originally cost the seller?

The selling price is 128% of the original cost to the seller. 128% of the original cost is £15 360.

1% of the original cost is $\frac{£15360}{128}$.

100% of the original cost is $\frac{£15360}{128} \times 100$, i.e. £12 000.

### Exercise 1.32

1  Calculate the value of $X$ in each of the following:
   a  40% of $X$ is 240
   b  24% of $X$ is 84
   c  85% of $X$ is 765
   d  4% of $X$ is 10
   e  15% of $X$ is 18.75
   f  7% of $X$ is 0.105

2  Calculate the value of $Y$ in each of the following:
   a  125% of $Y$ is 70
   b  140% of $Y$ is 91
   c  210% of $Y$ is 189
   d  340% of $Y$ is 68
   e  150% of $Y$ is 0.375
   f  144% of $Y$ is −54.72

3  In a Geography textbook, 35% of the pages are coloured. If there are 98 coloured pages, how many pages are there in the whole book?

4  A town has 3500 families who own a car. If this represents 28% of the families in the town, how many families are there in total?

5  In a test, Isabel scored 88%. If she got three questions incorrect, how many did she get correct?

6  Water expands when it freezes. Ice is less dense than water so it floats. If the increase in volume is 4%, what volume of water will make an iceberg of 12 700 000 m³? Give your answer to three significant figures.

# 5 Ratio and proportion

## Direct proportion

Workers in a pottery factory are paid according to how many plates they produce. The wage paid to them is said to be in **direct proportion** to the number of plates made. As the number of plates made increases so does their wage. Other workers are paid for the number of hours worked. For them the wage paid is in **direct proportion** to the number of hours worked. There are two main methods for solving problems involving direct proportion: the ratio method and the unitary method.

### ➡ Worked example

A bottling machine fills 500 bottles in 15 minutes. How many bottles will it fill in $1\frac{1}{2}$ hours?

Note: The time units must be the same, so for either method the $1\frac{1}{2}$ hours must be changed to 90 minutes.

### The ratio method

Let $x$ be the number of bottles filled. Then:

$$\frac{x}{90} = \frac{500}{15}$$

so $x = \frac{500 \times 90}{15} = 3000$

3000 bottles are filled in $1\frac{1}{2}$ hours.

### The unitary method

In 15 minutes 500 bottles are filled.

Therefore in 1 minute $\frac{500}{15}$ bottles are filled.

So in 90 minutes $90 \times \frac{500}{15}$ bottles are filled.

In $1\frac{1}{2}$ hours, 3000 bottles are filled.

**Exercise 1.33**

Use either the ratio method or the unitary method to solve the problems below.
1 A machine prints four books in 10 minutes. How many will it print in 2 hours?
2 A farmer plants five apple trees in 25 minutes. If he continues to work at a constant rate, how long will it take him to plant 200 trees?

# 5 RATIO AND PROPORTION

**Exercise 1.33 (cont)**

3. A television set uses 3 units of electricity in 2 hours. How many units will it use in 7 hours? Give your answer to the nearest unit.
4. A bricklayer lays 1500 bricks in an 8-hour day. Assuming she continues to work at the same rate, calculate:
   a. how many bricks she would expect to lay in a five-day week
   b. how long to the nearest hour it would take her to lay 10 000 bricks.
5. A machine used to paint white lines on a road uses 250 litres of paint for each 8 km of road marked. Calculate:
   a. how many litres of paint would be needed for 200 km of road
   b. what length of road could be marked with 4000 litres of paint.
6. An aircraft is cruising at 720 km/h and covers 1000 km. How far would it travel in the same period of time if the speed increased to 800 km/h?
7. A production line travelling at 2 m/s labels 150 tins. In the same period of time how many will it label at:
   a. 6 m/s    b. 1 m/s    c. 1.6 m/s?
8. A car travels at an average speed of 80 km/h for 6 hours.
   a. How far will it travel in the 6 hours?
   b. What average speed will it need to travel at in order to cover the same distance in 5 hours?

If the information is given in the form of a ratio, the method of solution is the same.

## → Worked example

Tin and copper are mixed in the ratio 8:3. How much tin is needed to mix with 36 g of copper?

### The ratio method

Let $x$ grams be the mass of tin needed.

$$\frac{x}{36} = \frac{8}{3}$$

Therefore $x = \frac{8 \times 36}{3}$

$x = 96$

So 96 g of tin is needed.

### The unitary method

3 g of copper mixes with 8 g of tin.

1 g of copper mixes with $\frac{8}{3}$ g of tin.

So 36 g of copper mixes with $36 \times \frac{8}{3}$ g of tin.

Therefore 36 g of copper mixes with 96 g of tin.

## Exercise 1.34

1. Sand and gravel are mixed in the ratio 5:3 to form ballast.
   a. How much gravel is mixed with 750 kg of sand?
   b. How much sand is mixed with 750 kg of gravel?
2. A recipe uses 150 g butter, 500 g flour, 50 g sugar and 100 g currants to make 18 small cakes.
   a. Write the ratio of the amount of butter : flour : sugar : currants in its simplest form.
   b. How much of each ingredient will be needed to make 72 cakes?
   c. How many whole cakes could be made with 1 kg of butter?
3. A map is drawn to a scale of 1:5000. The distance between two historic monuments is measured on the map to be 15.5 cm. Calculate the actual distance between the monuments, giving your answer in metres.
4. A breakfast cereal is available in packets of two sizes, 350 g and 800 g. The prices of the two packets are $2.25 and $4.95, respectively. Which packet provides better value for money? Justify your answer clearly.
5. A small car's petrol tank can hold 22 litres of petrol. It can travel on average 530 km on a full tank. A large car's petrol tank can hold 47 litres of petrol. It can travel on average 972 km on a full tank.
   a. On average how many kilometres can the small car travel on 1 litre of petrol?
   b. Which car has the most efficient fuel consumption? Show your working.

# Divide a quantity in a given ratio

### ➡ Worked examples

a. Divide 20 m in the ratio 3:2.

### The ratio method

3:2 gives 5 parts.

$\frac{3}{5} \times 20\,\text{m} = 12\,\text{m}$

$\frac{2}{5} \times 20\,\text{m} = 8\,\text{m}$

20 m divided in the ratio 3:2 is 12 m : 8 m.

### The unitary method

3:2 gives 5 parts.
5 parts is equivalent to 20 m.
1 part is equivalent to $\frac{20}{5}$ m.

Therefore 3 parts is $3 \times \frac{20}{5}$ m; that is 12 m.

Therefore 2 parts is $2 \times \frac{20}{5}$ m; that is 8 m.

b. A factory produces cars in red, blue, white and green in the ratio 7:5:3:1. Out of a production of 48 000 cars, how many are white?

7 + 5 + 3 + 1 gives a total of 16 parts.

Therefore the total number of white cars = $\frac{3}{16} \times 48\,000 = 9000$.

# 5 RATIO AND PROPORTION

**Exercise 1.35**

1. a  Divide 150 in the ratio 2:3.
   b  Divide 72 in the ratio 2:3:4.
   c  Divide 5 kg in the ratio 13:7.
   d  Divide 45 min in the ratio 2:3.
   e  Divide 1 hour in the ratio 1:5.

2. $\frac{7}{8}$ of a can of drink is water, the rest is syrup. What is the ratio of water to syrup?

3. $\frac{5}{9}$ of a litre carton of orange is pure orange juice, the rest is water. How many millilitres of each are in the carton?

4. 55% of students in a school are boys.
   a  What is the ratio of boys to girls?
   b  How many boys and how many girls are there if the school has 800 students?

5. a  A piece of wood is cut in the ratio 2:3. What fraction of the length is the longer piece?
   b  If the piece of wood is 80 cm long, how long is the shorter piece?

6. A gas pipe is 7 km long. A valve is positioned in such a way that it divides the length of the pipe in the ratio 4:3. Calculate the distance of the valve from each end of the pipe.

7. The size of the angles of a quadrilateral are in the ratio 1:2:3:3. Calculate the size of each angle.

8. The angles of a triangle are in the ratio 3:5:4. Calculate the size of each angle.

9. A millionaire leaves 1.4 million dollars in his will to be shared between his three children in the ratio of their ages. If they are 24, 28 and 32 years old, calculate to the nearest dollar the amount they will each receive.

10. A small company makes a profit of $8000. This is divided between the directors in the ratio of their initial investments. If Alex put $20 000 into the firm, Maria $35 000 and Ahmet $25 000, calculate the amount of the profit they will each receive.

11. A paint mix uses red and white paint in a ratio of 1:12.
    a  How much white paint will be needed to mix with 1.4 litres of red paint?
    b  If a total of 15.5 litres of paint is mixed, calculate the amount of white paint and the amount of red paint used. Give your answers to the nearest 0.1 litre.

12. A tulip farmer sells sacks of mixed bulbs to local people. The bulbs develop into two different colours of tulips, red and yellow. The colours are packaged in a ratio of 8:5 respectively.
    a  If a sack contains 200 red bulbs, calculate the number of yellow bulbs.
    b  If a sack contains 351 bulbs in total, how many of each colour would you expect to find?
    c  One sack is packaged with a bulb mixture in the ratio 7:5 by mistake. If the sack contains 624 bulbs, how many more yellow bulbs would you expect to have compared with a normal sack of 624 bulbs?

# Inverse proportion

**13** A pure fruit juice is made by mixing the juices of oranges and mangoes in the ratio of 9:2.
  **a** If 189 litres of orange juice are used, calculate the number of litres of mango juice needed.
  **b** If 605 litres of the juice are made, calculate the number of litres of orange juice and mango juice used.

## Inverse proportion

Sometimes an increase in one quantity causes a decrease in another quantity. For example, if fruit is to be picked by hand, the more people there are picking the fruit, the less time it will take.

### → Worked examples

**a** If 8 people can pick the apples from the trees in 6 days, how long will it take 12 people?

8 people take 6 days.

1 person will take $6 \times 8$ days.

Therefore 12 people will take $\frac{6 \times 8}{12}$ days, i.e. 4 days.

**b** A cyclist averages a speed of 27 km/h for 4 hours. At what average speed would she need to cycle to cover the same distance in 3 hours?

Completing it in 1 hour would require cycling at $27 \times 4$ km/h.

Completing it in 3 hours requires cycling at

$\frac{27 \times 4}{3}$ km/h; that is 36 km/h.

### Exercise 1.36

**1** A teacher shares sweets among 8 students so that they get 6 each. How many sweets would each student get if there were 12 students?

**2** The table below represents the relationship between the speed and the time taken for a train to travel between two stations.

| Speed (km/h) | 60 |   |   | 120 | 90 | 50 | 10 |
|---|---|---|---|---|---|---|---|
| Time (h) |   | 2 | 3 | 4 |   |   |   |

Copy and complete the table.

**3** Six people can dig a trench in 8 hours.
  **a** How long would it take:
    **i** 4 people        **ii** 12 people        **iii** 1 person?
  **b** How many people would it take to dig the trench in:
    **i** 3 hours        **ii** 16 hours        **iii** 1 hour?

# 5 RATIO AND PROPORTION

**Exercise 1.36 (cont)**

4 Chairs in a hall are arranged in 35 rows of 18.
   a How many rows would there be with 21 chairs to a row?
   b How many chairs would there be in each row if there were 15 rows?
5 A train travelling at 100 km/h takes 4 hours for a journey. How long would it take a train travelling at 60 km/h?
6 A worker in a sugar factory packs 24 cardboard boxes with 15 bags of sugar in each. If he had boxes which held 18 bags of sugar each, how many fewer boxes would be needed?
7 A swimming pool is filled in 30 hours by two identical pumps. How much quicker would it be filled if five similar pumps were used instead?

# 6 Approximation and rounding

In many instances, exact numbers are not necessary or even desirable. In those circumstances, approximations are given. The approximations can take several forms. The common types of approximations are outlined below.

## Rounding

If 28 617 people attend a gymnastics competition, this figure can be reported to various levels of accuracy.

To the nearest 10 000 this figure would be rounded up to 30 000.

To the nearest 1000 the figure would be rounded up to 29 000.

To the nearest 100 the figure would be rounded down to 28 600.

In this type of situation it is unlikely that the exact number would be reported.

**Exercise 1.37**

1. Round the following numbers to the nearest 1000:
   - a  68 786
   - b  74 245
   - c  89 000
   - d  4020
   - e  99 500
   - f  999 999

2. Round the following numbers to the nearest 100:
   - a  78 540
   - b  6858
   - c  14 099
   - d  8084
   - e  950
   - f  2984

3. Round the following numbers to the nearest 10:
   - a  485
   - b  692
   - c  8847
   - d  83
   - e  4
   - f  997

## Decimal places

A number can also be approximated to a given number of decimal places (d.p.). This refers to the number of digits written after a decimal point.

 **Worked examples**

a  Write 7.864 to 1 d.p.

The answer needs to be written with one digit after the decimal point. However, to do this, the second digit after the decimal point also needs to be considered. If it is 5 or more then the first digit is rounded up.

i.e. 7.864 is written as 7.9 to 1 d.p.

## 6 APPROXIMATION AND ROUNDING

**b** Write 5.574 to 2 d.p.

The answer here is to be given with two digits after the decimal point. In this case the third digit after the decimal point needs to be considered. As the third digit after the decimal point is less than 5, the second digit is not rounded up.
i.e. 5.574 is written as 5.57 to 2 d.p.

### Exercise 1.38

1  Round the following to 1 d.p.:
   a  5.58   b  0.73   c  11.86   d  157.39   e  4.04
   f  15.045   g  2.95   h  0.98   i  12.049

2  Round the following to 2 d.p.:
   a  6.473   b  9.587   c  16.476   d  0.088   e  0.014
   f  9.3048   g  99.996   h  0.0048   i  3.0037

## Significant figures

Numbers can also be approximated to a given number of significant figures (s.f.). In the number 43.25, the 4 is the most significant figure as it has a value of 40. In contrast, the 5 is the least significant as it only has a value of 5 hundredths.

### Worked examples

**a** Write 43.25 to 3 s.f.

Although only the three most significant digits are written, the fourth digit needs to be considered to see whether the third digit is to be rounded up or not.
i.e. 43.25 is written as 43.3 to 3 s.f.

**b** Write 0.0043 to 1 s.f.

In this example only two digits have any significance, the 4 and the 3. The 4 is the most significant and therefore is the only one of the two to be written in the answer.
i.e. 0.0043 is written as 0.004 to 1 s.f.

### Exercise 1.39

1  Write the following to the number of significant figures written in brackets:
   a  48 599 (1 s.f.)   b  48 599 (3 s.f.)   c  6841 (1 s.f.)
   d  7538 (2 s.f.)   e  483.7 (1 s.f.)   f  2.5728 (3 s.f.)
   g  990 (1 s.f.)   h  2045 (2 s.f.)   i  14.952 (3 s.f.)

2  Write the following to the number of significant figures written in brackets:
   a  0.08562 (1 s.f.)   b  0.5932 (1 s.f.)   c  0.942 (2 s.f.)
   d  0.954 (1 s.f.)   e  0.954 (2 s.f.)   f  0.00305 (1 s.f.)
   g  0.00305 (2 s.f.)   h  0.00973 (2 s.f.)   i  0.00973 (1 s.f.)

# Appropriate accuracy

In many instances, calculations carried out using a calculator produce answers which are not natural numbers. A calculator will give the answer to as many decimal places as will fit on its screen. In most cases this degree of accuracy is neither desirable nor necessary. Unless another degree of accuracy is stated, answers involving lengths should be given to three significant figures and angles to one decimal place.

###  Worked example

Calculate 4.64 ÷ 2.3 giving your answer to an appropriate degree of accuracy.

The calculator will give the answer to 4.64 ÷ 2.3 as 2.017 391 3. However, the answer given to 3 s.f. is sufficient.

Therefore 4.64 ÷ 2.3 = 2.02 (3 s.f.).

# Estimating answers to calculations

Even though many calculations can be done quickly and effectively on a calculator, often an estimate for an answer can be a useful check. This is done by rounding each of the numbers in such a way that the calculation becomes relatively straightforward.

###  Worked examples

a  Estimate the answer to 57 × 246.
   Here are two possibilities:
   i   60 × 200 = 12 000
   ii  50 × 250 = 12 500
b  Estimate the answer to 6386 ÷ 27.
   6000 ÷ 30 = 200

**Exercise 1.40**

1  Calculate the following, giving your answer to an appropriate degree of accuracy:
   a  23.456 × 17.89     b  0.4 × 12.62     c  18 × 9.24
   d  76.24 ÷ 3.2        e  $7.6^2$         f  $16.42^3$
   g  $\frac{2.3 \times 3.37}{4}$   h  $\frac{8.31}{2.02}$   i  $9.2 \div 4^2$

 2  Without using a calculator, estimate the answers to the following:
   a  78.45 + 51.02      b  168.3 − 87.09   c  2.93 × 3.14

# 6 APPROXIMATION AND ROUNDING

**Exercise 1.40 (cont)**

3 Without using a calculator, estimate the answers to the following:
   a  $62 \times 19$      b  $270 \times 12$      c  $55 \times 60$
   d  $4950 \times 28$      e  $0.8 \times 0.95$      f  $0.184 \times 475$

4 Without using a calculator, estimate the answers to the following:
   a  $3946 \div 18$      b  $8287 \div 42$      c  $906 \div 27$
   d  $5520 \div 13$      e  $48 \div 0.12$      f  $610 \div 0.22$

5 Estimate the shaded areas of the following shapes. Do *not* work out an exact answer.

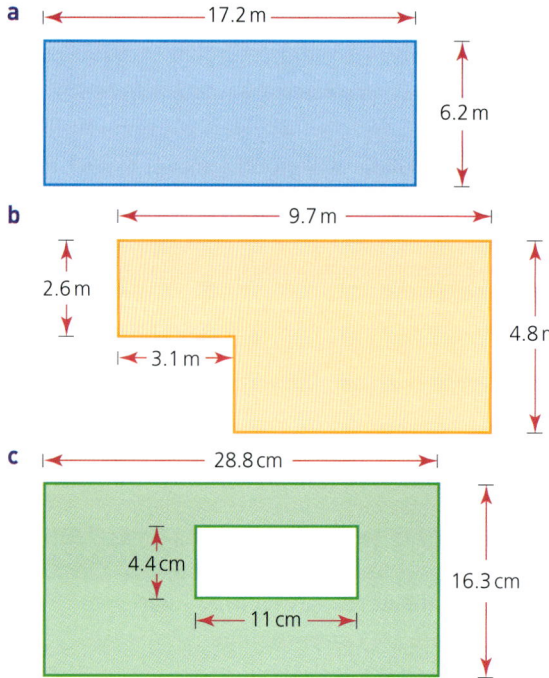

6 Estimate the volume of each of the solids below. Do *not* work out an exact answer.

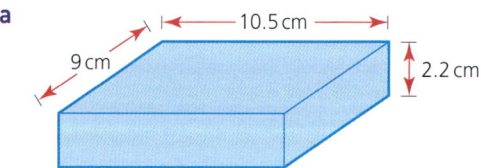

*Estimating answers to calculations*

**b**

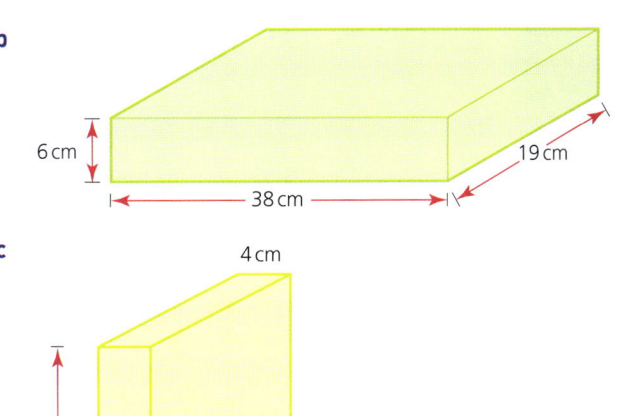

**c**

# 7 Laws of indices

The index refers to the power to which a number is raised. In the example $5^3$, the number 5 is raised to the power 3. The 3 is known as the **index**. Indices is the plural of index.

### → Worked examples

a  $5^3 = 5 \times 5 \times 5$
   $= 125$

b  $7^4 = 7 \times 7 \times 7 \times 7$
   $= 2401$

c  $3^1 = 3$

## Laws of indices

When working with numbers involving indices, there are three basic laws which can be applied. These are:

1  $a^m \times a^n = a^{m+n}$
2  $a^m \div a^n$ or $\dfrac{a^m}{a^n} = a^{m-n}$
3  $(a^m)^n = a^{mn}$

## Positive indices

### → Worked examples

a  Simplify $4^3 \times 4^2$

   $4^3 \times 4^2 = 4^{(3+2)}$

   $= 4^5$

b  Simplify $2^5 \div 2^3$

   $2^5 \div 2^3 = 2^{(5-3)}$

   $= 2^2$

c  Find the value of $3^3 \times 3^4$

   $3^3 \times 3^4 = 3^{(3+4)}$

   $= 3^7$

   $= 2187$

d  Find the value of $(4^2)^3$

   $(4^2)^3 = 4^{(2 \times 3)}$

   $= 4^6$

   $= 4096$

*The zero index*

 **Exercise 1.41**

1. Using indices, simplify the following expressions:
   a. $3 \times 3 \times 3$
   b. $2 \times 2 \times 2 \times 2 \times 2$
   c. $4 \times 4$
   d. $6 \times 6 \times 6 \times 6$
   e. $8 \times 8 \times 8 \times 8 \times 8 \times 8$
   f. $5$

2. Simplify the following using indices:
   a. $2 \times 2 \times 2 \times 3 \times 3$
   b. $4 \times 4 \times 4 \times 4 \times 4 \times 5 \times 5$
   c. $3 \times 3 \times 4 \times 4 \times 4 \times 5 \times 5$
   d. $2 \times 7 \times 7 \times 7 \times 7$
   e. $1 \times 1 \times 6 \times 6$
   f. $3 \times 3 \times 3 \times 4 \times 4 \times 6 \times 6 \times 6 \times 6 \times 6$

3. Write out the following in full:
   a. $4^2$
   b. $5^7$
   c. $3^5$
   d. $4^3 \times 6^3$
   e. $7^2 \times 2^7$
   f. $3^2 \times 4^3 \times 2^4$

4. Without a calculator, work out the value of the following:
   a. $2^5$
   b. $3^4$
   c. $8^2$
   d. $6^3$
   e. $10^6$
   f. $4^4$
   g. $2^3 \times 3^2$
   h. $10^3 \times 5^3$

 **Exercise 1.42**

1. Simplify the following using indices:
   a. $3^2 \times 3^4$
   b. $8^5 \times 8^2$
   c. $5^2 \times 5^4 \times 5^3$
   d. $4^3 \times 4^5 \times 4^2$
   e. $2^1 \times 2^3$
   f. $6^2 \times 3^2 \times 3^3 \times 6^4$
   g. $4^5 \times 4^3 \times 5^5 \times 5^4 \times 6^2$
   h. $2^4 \times 5^7 \times 5^3 \times 6^2 \times 6^6$

2. Simplify the following:
   a. $4^6 \div 4^2$
   b. $5^7 \div 5^4$
   c. $2^5 \div 2^4$
   d. $6^5 \div 6^2$
   e. $\dfrac{6^5}{6^2}$
   f. $\dfrac{8^6}{8^5}$
   g. $\dfrac{4^8}{4^5}$
   h. $\dfrac{3^9}{3^2}$

3. Simplify the following:
   a. $(5^2)^2$
   b. $(4^3)^4$
   c. $(10^2)^5$
   d. $(3^3)^5$
   e. $(6^2)^4$
   f. $(8^2)^3$

4. Simplify the following:
   a. $\dfrac{2^2 \times 2^4}{2^3}$
   b. $\dfrac{3^4 \times 3^2}{3^5}$
   c. $\dfrac{5^6 \times 5^7}{5^2 \times 5^8}$
   d. $\dfrac{(4^2)^5 \times 4^2}{4^7}$
   e. $\dfrac{4^4 \times 2^5 \times 4^2}{4^3 \times 2^3}$
   f. $\dfrac{6^3 \times 6^3 \times 8^5 \times 8^6}{8^6 \times 6^2}$
   g. $\dfrac{(5^5)^2 \times (4^4)^3}{5^8 \times 4^9}$
   h. $\dfrac{(6^3)^4 \times 6^3 \times 4^9}{6^8 \times (4^2)^4}$

# The zero index

The zero index indicates that a number is raised to the power 0. A number raised to the power 0 is equal to 1. This can be explained by applying the laws of indices:

$a^m \div a^n = a^{m-n}$    therefore    $\dfrac{a^m}{a^m} = a^{m-m}$

$\phantom{a^m \div a^n = a^{m-n}  therefore  \dfrac{a^m}{a^m}} = a^0$

However,    $\dfrac{a^m}{a^m} = 1$

therefore    $a^0 = 1$

# 7 LAWS OF INDICES

## Negative indices

A negative index indicates that a number is being raised to a negative power, e.g. $4^{-3}$

Another law of indices states that $a^{-m} = \frac{1}{a^m}$. It can be proved as follows:

$a^{-m} = a^{0-m}$

$\qquad = \frac{a^0}{a^m}$ (from the second law of indices)

$\qquad = \frac{1}{a^m}$

therefore $a^{-m} = \frac{1}{a^m}$

 **Exercise 1.43** Without using a calculator, work out the following:

1. 
   a. $2^3 \times 2^0$
   b. $5^2 \div 6^0$
   c. $5^2 \times 5^{-2}$
   d. $6^3 \times 6^{-3}$
   e. $(4^0)^2$
   f. $4^0 \div 2^2$

2. 
   a. $4^{-1}$
   b. $3^{-2}$
   c. $6 \times 10^{-2}$
   d. $5 \times 10^{-3}$
   e. $100 \times 10^{-2}$
   f. $10^{-3}$

3. 
   a. $9 \times 3^{-2}$
   b. $16 \times 2^{-3}$
   c. $64 \times 2^{-4}$
   d. $4 \times 2^{-3}$
   e. $36 \times 6^{-3}$
   f. $100 \times 10^{-1}$

4. 
   a. $\frac{3}{2^{-2}}$
   b. $\frac{4}{2^{-3}}$
   c. $\frac{9}{5^{-2}}$
   d. $\frac{5}{4^{-2}}$
   e. $\frac{7^{-3}}{7^{-4}}$
   f. $\frac{8^{-6}}{8^{-8}}$

## Fractional indices

$16^{\frac{1}{2}}$ can be written as $(4^2)^{\frac{1}{2}}$.

$(4^2)^{\frac{1}{2}} = 4^{(2 \times \frac{1}{2})}$

$\qquad = 4^1$

$\qquad = 4$

Therefore $16^{\frac{1}{2}} = 4$

But $\sqrt{16} = 4$

Therefore $16^{\frac{1}{2}} = \sqrt{16}$

Similarly:

$27^{\frac{1}{3}}$ can be written as $(3^3)^{\frac{1}{3}}$

$(3^3)^{\frac{1}{3}} = 3^{(3 \times \frac{1}{3})}$

$\qquad = 3^1$

$\qquad = 3$

Therefore $27^{\frac{1}{3}} = 3$

But $\qquad \sqrt[3]{27} = 3$

Therefore $27^{\frac{1}{3}} = \sqrt[3]{27}$

## Fractional indices

In general:
$$a^{\frac{1}{n}} = \sqrt[n]{a}$$

$$a^{\frac{m}{n}} = (a^m)^{\frac{1}{n}} = \sqrt[n]{a^m} \text{ or } a^{\frac{m}{n}} = \left(a^{\frac{1}{n}}\right)^m = \left(\sqrt[n]{a}\right)^m$$

### → Worked examples

**a** Find the value of $16^{\frac{1}{4}}$.

$16^{\frac{1}{4}} = \sqrt[4]{16}$          Alternatively: $16^{\frac{1}{4}} = (2^4)^{\frac{1}{4}}$
$= \sqrt[4]{(2^4)}$                              $= 2^1$
$= 2$                                              $= 2$

**b** Find the value of $25^{\frac{3}{2}}$.

$25^{\frac{3}{2}} = \left(25^{\frac{1}{2}}\right)^3$          Alternatively: $25^{\frac{3}{2}} = (5^2)^{\frac{3}{2}}$
$= \left(\sqrt{25}\right)^3$                              $= 5^3$
$= 5^3$                                              $= 125$
$= 125$

### Exercise 1.44

Work out all parts:

**1** a $16^{\frac{1}{2}}$   b $25^{\frac{1}{2}}$   c $100^{\frac{1}{2}}$
   d $27^{\frac{1}{3}}$   e $81^{\frac{1}{2}}$   f $1000^{\frac{1}{3}}$

**2** a $16^{\frac{1}{4}}$   b $81^{\frac{1}{4}}$   c $32^{\frac{1}{5}}$
   d $64^{\frac{1}{6}}$   e $216^{\frac{1}{3}}$   f $256^{\frac{1}{4}}$

**3** a $4^{\frac{3}{2}}$   b $4^{\frac{5}{2}}$   c $9^{\frac{3}{2}}$
   d $16^{\frac{3}{2}}$   e $1^{\frac{5}{2}}$   f $27^{\frac{2}{3}}$

**4** a $125^{\frac{2}{3}}$   b $32^{\frac{3}{5}}$   c $64^{\frac{5}{6}}$
   d $1000^{\frac{2}{3}}$   e $16^{\frac{5}{4}}$   f $81^{\frac{3}{4}}$

### Exercise 1.45

Find the value of the following:

**1** a $\dfrac{27^{\frac{2}{3}}}{3^2}$   b $\dfrac{7^{\frac{3}{2}}}{\sqrt{7}}$   c $\dfrac{4^{\frac{5}{2}}}{4^2}$
   d $\dfrac{16^{\frac{3}{2}}}{2^6}$   e $\dfrac{27^{\frac{5}{3}}}{\sqrt{9}}$   f $\dfrac{6^{\frac{4}{3}}}{6^{\frac{1}{3}}}$

**2** a $5^{\frac{2}{3}} \times 5^{\frac{4}{3}}$   b $4^{\frac{1}{4}} \times 4^{\frac{1}{4}}$   c $8 \times 2^{-2}$
   d $3^{\frac{4}{3}} \times 3^{\frac{5}{3}}$   e $2^{-2} \times 16$   f $8^{\frac{5}{3}} \times 8^{-\frac{4}{3}}$

**3** a $\dfrac{2^{\frac{1}{2}} \times 2^{\frac{5}{2}}}{2}$   b $\dfrac{4^{\frac{5}{6}} \times 4^{\frac{1}{6}}}{4^{\frac{1}{2}}}$   c $\dfrac{2^3 \times 8^{\frac{3}{2}}}{\sqrt{8}}$
   d $\dfrac{(3^2)^{\frac{3}{2}} \times 3^{-\frac{1}{2}}}{3^{\frac{1}{2}}}$   e $\dfrac{8^{\frac{1}{3}} \times 7}{27^{\frac{1}{3}}}$   f $\dfrac{9^{\frac{1}{2}} \times 3^{\frac{5}{2}}}{3^{\frac{2}{3}} \times 3^{-\frac{1}{6}}}$

# 8 Standard form

In 1610, Galileo and a German astronomer, Marius, independently discovered Jupiter's four largest moons, Io, Europa, Ganymede and Callisto. At that time it was believed that the Sun revolved around the Earth. Galileo was one of the few people who believed that the Earth revolved around the Sun. As a result of this, the Church declared that he was a heretic and imprisoned him. It took the Church a further 350 years to officially accept that he was correct; he was only pardoned in 1992.

Facts about Jupiter:

It has a mass of 1 900 000 000 000 000 000 000 000 000 kg.

It has a diameter of 142 800 000 m.

It has a mean distance from the Sun of 778 000 000 km.

*As these numbers have large numbers of zeros it can be difficult to grasp how big they are.*

Standard form is also known as standard index form or sometimes as scientific notation. It involves writing large numbers or very small numbers in terms of powers of 10.

## A positive index

$100 = 1 \times 10^2$

$1000 = 1 \times 10^3$

$10\,000 = 1 \times 10^4$

$3000 = 3 \times 10^3$

For a number to be in standard form, it must take the form $a \times 10^n$ where the index $n$ is a positive or negative integer and $a$ must lie in the range $1 \leq a < 10$.

For example, 3100 can be written in many different ways:

$3.1 \times 10^3$   $31 \times 10^2$   $0.31 \times 10^4$   etc.

However, only $3.1 \times 10^3$ agrees with the above conditions and is therefore the only one which is written in standard form.

### ➡ Worked examples

a Write 72 000 in standard form.
$7.2 \times 10^4$

b Of the numbers below, ring those which are written in standard form:

($4.2 \times 10^3$)   $0.35 \times 10^2$   $18 \times 10^5$   ($6 \times 10^3$)   $0.01 \times 10^1$

*A positive index*

c  Write 5 thousand million in standard form.

$$5 \text{ thousand million} = 5\,000\,000\,000$$
$$= 5 \times 10^9$$

d  Multiply the following and write your answer in standard form:

$$600 \times 4000$$
$$= 2\,400\,000$$
$$= 2.4 \times 10^6$$

*Core students will be allowed to use calculators for any calculations involving standard form.* →

e  Multiply the following and write your answer in standard form:

$$(2.4 \times 10^4) \times (5 \times 10^7) = 12 \times 10^{11} = 1.2 \times 10^{12}$$ when written in standard form

f  Divide the following and write your answer in standard form:

$$(6.4 \times 10^7) \div (1.6 \times 10^3) = 4 \times 10^4$$

g  Add the following and write your answer in standard form:

$$(3.8 \times 10^6) + (8.7 \times 10^4)$$

Changing the indices to the same value gives the sum:

$$(380 \times 10^4) + (8.7 \times 10^4)$$
$$= 388.7 \times 10^4$$
$$= 3.887 \times 10^6 \text{ when written in standard form}$$

h  Subtract the following and write your answer in standard form:

$$(6.5 \times 10^7) - (9.2 \times 10^5)$$

Changing the indices to the same value gives the sum:

$$(650 \times 10^5) - (9.2 \times 10^5)$$
$$= 640.8 \times 10^5$$
$$= 6.408 \times 10^7 \text{ when written in standard form}$$

Your calculators have a standard form button and will also give answers in standard form if the answer is very large. For example, to enter the number $8 \times 10^4$ into the calculator, use the following keys on your calculator:

| Casio | Texas |
|---|---|
| 8  ×10^x  4 | 8  2nd  ,  4 |
| Note: A number such as 1 000 000 000 000 000 would appear on the screen as $1 \times 10^{15}$ | Note: A number such as 1 000 000 000 000 000 would appear on the screen as 1 E 15 |

# 8 STANDARD FORM

## Exercise 1.46

1. Which of the following are not in standard form?
   a  $6.2 \times 10^5$
   b  $7.834 \times 10^{16}$
   c  $8.0 \times 10^5$
   d  $0.46 \times 10^7$
   e  $82.3 \times 10^6$
   f  $6.75 \times 10^1$

2. Write the following numbers in standard form:
   a  600 000
   b  48 000 000
   c  784 000 000 000
   d  534 000
   e  7 million
   f  8.5 million

3. Write the following in standard form:
   a  $68 \times 10^5$
   b  $720 \times 10^6$
   c  $8 \times 10^5$
   d  $0.75 \times 10^8$
   e  $0.4 \times 10^{10}$
   f  $50 \times 10^6$

4. Multiply the following and write your answers in standard form:
   a  $200 \times 3000$
   b  $6000 \times 4000$
   c  7 million $\times$ 20
   d  $500 \times 6$ million
   e  3 million $\times$ 4 million
   f  $4500 \times 4000$

5. Light from the Sun takes approximately 8 minutes to reach Earth. If light travels at a speed of $3 \times 10^8$ m/s, calculate to three significant figures (s.f.) the distance from the Sun to the Earth.

*Students studying the Extended syllabus should attempt these calculations without a calculator where possible.* →

6. Find the value of the following and write your answers in standard form:
   a  $(4.4 \times 10^3) \times (2 \times 10^5)$
   b  $(6.8 \times 10^7) \times (3 \times 10^3)$
   c  $(4 \times 10^5) \times (8.3 \times 10^5)$
   d  $(5 \times 10^9) \times (8.4 \times 10^{12})$
   e  $(8.5 \times 10^6) \times (6 \times 10^{15})$
   f  $(5.0 \times 10^{12})^2$

7. Find the value of the following and write your answers in standard form:
   a  $(3.8 \times 10^8) \div (1.9 \times 10^6)$
   b  $(6.75 \times 10^9) \div (2.25 \times 10^4)$
   c  $(9.6 \times 10^{11}) \div (2.4 \times 10^5)$
   d  $(1.8 \times 10^{12}) \div (9.0 \times 10^7)$
   e  $(2.3 \times 10^{11}) \div (9.2 \times 10^4)$
   f  $(2.4 \times 10^8) \div (6.0 \times 10^3)$

8. Find the value of the following and write your answers in standard form:
   a  $(3.8 \times 10^5) + (4.6 \times 10^4)$
   b  $(7.9 \times 10^9) + (5.8 \times 10^8)$
   c  $(6.3 \times 10^7) + (8.8 \times 10^5)$
   d  $(3.15 \times 10^9) + (7.0 \times 10^6)$
   e  $(5.3 \times 10^8) - (8.0 \times 10^7)$
   f  $(6.5 \times 10^7) - (4.9 \times 10^6)$
   g  $(8.93 \times 10^{10}) - (7.8 \times 10^9)$
   h  $(4.07 \times 10^7) - (5.1 \times 10^6)$

9. The following list shows the average distance of the planets of the Solar System from the Sun.
   Jupiter     778 million km
   Mercury     58 million km
   Mars        228 million km
   Uranus      2870 million km
   Venus       108 million km
   Neptune     4500 million km
   Earth       150 million km
   Saturn      1430 million km

   Write each of the distances in standard form and then arrange them in order of magnitude, starting with the distance of the planet closest to the Sun.

## A negative index

A negative index is used when writing a number between 0 and 1 in standard form.

e.g. 
$100 = 1 \times 10^2$
$10 = 1 \times 10^1$
$1 = 1 \times 10^0$
$0.1 = 1 \times 10^{-1}$
$0.01 = 1 \times 10^{-2}$
$0.001 = 1 \times 10^{-3}$
$0.0001 = 1 \times 10^{-4}$

Note that $a$ must still lie within the range $1 \leqslant a < 10$.

### Worked examples

a  Write 0.0032 in standard form.
   $3.2 \times 10^{-3}$

b  Write the following numbers in order of magnitude, starting with the largest:
   $3.6 \times 10^{-3}$   $5.2 \times 10^{-5}$   $1 \times 10^{-2}$   $8.35 \times 10^{-2}$   $6.08 \times 10^{-8}$

   $8.35 \times 10^{-2}, 1 \times 10^{-2}, 3.6 \times 10^{-3}, 5.2 \times 10^{-5}, 6.08 \times 10^{-8}$

### Exercise 1.47

1  Write the following numbers in standard form:
   a  0.0006   b  0.000053   c  0.000864
   d  0.000000088   e  0.0000007   f  0.0004145

2  Write the following numbers in standard form:
   a  $68 \times 10^{-5}$   b  $750 \times 10^{-9}$   c  $42 \times 10^{-11}$
   d  $0.08 \times 10^{-7}$   e  $0.057 \times 10^{-9}$   f  $0.4 \times 10^{-10}$

3  Find the value of $n$ in each of the following cases:
   a  $0.00025 = 2.5 \times 10^n$   b  $0.00357 = 3.57 \times 10^n$
   c  $0.00000006 = 6 \times 10^n$   d  $0.004^2 = 1.6 \times 10^n$
   e  $0.00065^2 = 4.225 \times 10^n$   f  $0.0002^n = 8 \times 10^{-12}$

4  Write these numbers in order of magnitude, starting with the largest:
   $3.2 \times 10^{-4}$   $6.8 \times 10^5$   $5.57 \times 10^{-9}$   $6.2 \times 10^3$
   $5.8 \times 10^{-7}$   $8.414 \times 10^2$   $6.741 \times 10^{-4}$

# 9 Surds

The roots of some numbers produce rational answers, for example:
$$\sqrt{16} = 4 \qquad \sqrt[5]{32} = 2 \qquad \sqrt{\frac{4}{9}} = \frac{2}{3}$$

If roots cannot be written as rational numbers, they are known as **surds** (or **radicals**). Surds are therefore irrational numbers. Examples of surds include $\sqrt{2}$, $\sqrt{3}$ and $\sqrt[4]{10}$.

If an answer to a question is a surd, then leaving the answer in surd form is exact. Using a calculator to calculate a decimal equivalent is only an approximation.

Consider the right-angled triangle below. Two of its sides are 1 cm in length. The length of the longer side (the hypotenuse) can be calculated, using Pythagoras' theorem, as having a length of $\sqrt{2}$ cm.

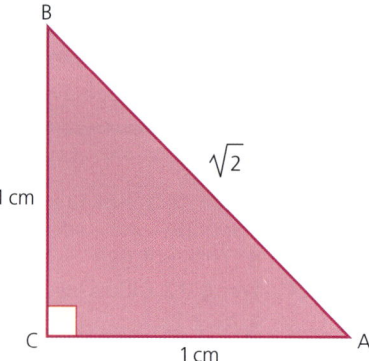

$\sqrt{2}$ is the exact length. A calculator will state that $\sqrt{2} = 1.414\,213\,562$, but this is only an approximation correct to nine decimal places.

You should always leave answers in exact form unless you are asked to give your answer to a certain number of decimal places.

## Simplification of surds

If $\sqrt{x}$ cannot be simplified further then it is in basic form. 17 and 43 are prime numbers so $\sqrt{17}$ and $\sqrt{43}$ cannot be simplified further. The square root of some numbers which are not prime such as $\sqrt{20}$, $\sqrt{63}$ and $\sqrt{363}$ can be simplified:

$$\sqrt{20} = \sqrt{4 \times 5} = \sqrt{4} \times \sqrt{5} = 2\sqrt{5}$$
$$\sqrt{63} = \sqrt{9 \times 7} = \sqrt{9} \times \sqrt{7} = 3\sqrt{7}$$
$$\sqrt{363} = \sqrt{121 \times 3} = \sqrt{121} \times \sqrt{3} = 11\sqrt{3}$$

Note: Each time, the original number is written as the product of two numbers, one of which is square.

## Simplification of surds

Surds can be manipulated and simplified according to a number of rules. These are:

| Rule | Example |
|---|---|
| $\sqrt{a} \times \sqrt{a} = a$ | $\sqrt{3} \times \sqrt{3} = 3$ |
| $\sqrt{a} \times \sqrt{b} = \sqrt{ab}$ | $\sqrt{3} \times \sqrt{5} = \sqrt{15}$ |
| $\dfrac{\sqrt{a}}{\sqrt{b}} = \sqrt{\dfrac{a}{b}}$ | $\dfrac{\sqrt{8}}{\sqrt{2}} = \sqrt{\dfrac{8}{2}} = \sqrt{4} = 2$ |
| $a\sqrt{b} \times \sqrt{c} = a\sqrt{bc}$ | $3\sqrt{5} \times \sqrt{6} = 3\sqrt{30}$ |
| $\sqrt{a} + \sqrt{b} \neq \sqrt{a+b}$ | $\sqrt{4} + \sqrt{9} \neq \sqrt{13}$ as $2 + 3 \neq \sqrt{13}$ |

### ➔ Worked examples

**a** Simplify $\sqrt{3} + \sqrt{12}$.

In order to add surds, they must both be multiples of the same surd.
$\sqrt{12} = \sqrt{4} \times \sqrt{3} = 2\sqrt{3}$

Therefore $\sqrt{3} + \sqrt{12}$ can be written as $\sqrt{3} + 2\sqrt{3}$
$= 3\sqrt{3}$.

**b** Expand and simplify $(2 + \sqrt{3})(3 - \sqrt{3})$.

Multiplying both terms in the first bracket by both terms in the second bracket gives: $2 \times 3 - 2\sqrt{3} + 3\sqrt{3} - \sqrt{3} \times \sqrt{3}$
$\Rightarrow 6 + \sqrt{3} - 3$
$\Rightarrow 3 + \sqrt{3}$

Your calculator is able to work with surds and the Casio will give answers in surd form too. The table below shows the screens of both calculators when working out **worked example b** above.

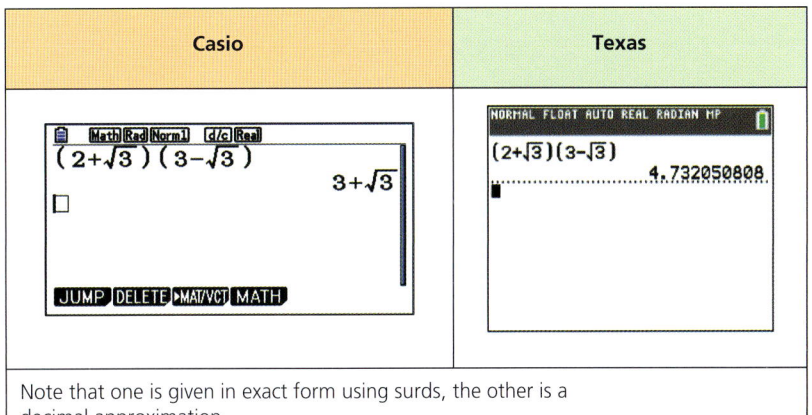

| Casio | Texas |
|---|---|

Note that one is given in exact form using surds, the other is a decimal approximation.

# 9 SURDS

 **Exercise 1.48**

1. Simplify the following surds:
   - a $\sqrt{24}$
   - b $\sqrt{48}$
   - c $\sqrt{75}$
   - d $\sqrt{700}$
   - e $\sqrt{162}$

2. Simplify the following where possible:
   - a $\sqrt{47}$
   - b $\sqrt{98}$
   - c $\sqrt{8}$
   - d $\sqrt{51}$
   - e $\sqrt{432}$

3. Simplify these expressions:
   - a $\sqrt{3} \times \sqrt{3}$
   - b $\sqrt{5} \times \sqrt{5}$
   - c $\sqrt{3} + \sqrt{3}$
   - d $\sqrt{2} + \sqrt{2}$
   - e $3\sqrt{5} - \sqrt{5}$
   - f $4\sqrt{7} + 3\sqrt{7}$

4. Simplify these expressions:
   - a $\sqrt{2} + \sqrt{8}$
   - b $\sqrt{7} + \sqrt{63}$
   - c $\sqrt{20} + \sqrt{45}$
   - d $3\sqrt{2} - 4\sqrt{8}$
   - e $5\sqrt{10} - \sqrt{40}$
   - f $\sqrt{28} - \sqrt{7}$

5. Expand the following expressions and simplify as far as possible:
   - a $(3 + \sqrt{2})(1 + \sqrt{2})$
   - b $(2 - \sqrt{2})(3 + \sqrt{2})$
   - c $(5 + \sqrt{5})(3 - \sqrt{5})$
   - d $(1 + 2\sqrt{3})(4 - 3\sqrt{3})$
   - e $(3 + 3\sqrt{2})(5 - 2\sqrt{2})$
   - f $(3 - 2\sqrt{5})(4 - 3\sqrt{5})$

## Rationalising the denominator

$\frac{1}{\sqrt{2}}$ is a fraction with a surd in the denominator.

It is considered mathematically neater if fractions are written without surds in their denominator. Removing surds from the denominator of a fraction is known as rationalising the denominator.

To rationalise, the fraction must be multiplied by a fraction that is equivalent to 1 but which eliminates the surd.

### → Worked example

Rationalise $\frac{1}{\sqrt{2}}$

Multiplying the fraction by $\frac{\sqrt{2}}{\sqrt{2}}$ gives:

$$\frac{1}{\sqrt{2}} \times \frac{\sqrt{2}}{\sqrt{2}} = \frac{\sqrt{2}}{2}$$

Note that $\frac{\sqrt{2}}{\sqrt{2}} = 1$ therefore $\frac{1}{\sqrt{2}}$ is unchanged when multiplied by 1.

In general, to rationalise a fraction of the form $\frac{a}{\sqrt{b}}$, multiply by $\frac{\sqrt{b}}{\sqrt{b}}$ to give $\frac{a\sqrt{b}}{b}$

## Rationalising the denominator

 **Exercise 1.49**

1  Rationalise the denominator in the following fractions, simplifying your answers where possible:

a  $\dfrac{1}{\sqrt{5}}$     b  $\dfrac{2}{\sqrt{7}}$     c  $\dfrac{2}{\sqrt{2}}$     d  $\dfrac{3}{\sqrt{3}}$

e  $\dfrac{4}{\sqrt{7}}$     f  $\dfrac{4}{\sqrt{8}}$     g  $\dfrac{5}{\sqrt{5}}$     h  $\dfrac{6}{\sqrt{3}}$

i  $\dfrac{5}{\sqrt{15}}$    j  $\dfrac{7}{\sqrt{3}}$    k  $\dfrac{12}{\sqrt{2}}$

2  Work out the following, leaving your answers in simplified and rationalised form:

a  $\dfrac{3}{\sqrt{2}} + \dfrac{1}{2}$     b  $\dfrac{1}{\sqrt{3}} + \dfrac{2}{\sqrt{6}}$     c  $\dfrac{2}{\sqrt{5}} + \dfrac{\sqrt{5}}{2}$

The denominator is not, however, always just a single term,

e.g. $\dfrac{1}{3+\sqrt{2}}$.

Rationalising the denominator in this type of fraction is not just a case of multiplying by $\dfrac{3+\sqrt{2}}{3+\sqrt{2}}$ as this will not eliminate the surd in the denominator,

i.e. $\dfrac{1}{3+\sqrt{2}} \times \dfrac{3+\sqrt{2}}{3+\sqrt{2}} = \dfrac{3+\sqrt{2}}{11+6\sqrt{2}}$.

To rationalise the denominator in this type of fraction requires an understanding of the difference of two squares, i.e. that $(a-b)(a+b) = a^2 - b^2$. This demonstrates that if either $a$ or $b$ is a surd, the result involving $a^2$ and $b^2$ will be rational.

Therefore, to rationalise $\dfrac{1}{3+\sqrt{2}}$:

$\dfrac{1}{3+\sqrt{2}} \times \dfrac{3-\sqrt{2}}{3-\sqrt{2}} = \dfrac{3-\sqrt{2}}{9-3\sqrt{2}+3\sqrt{2}-2} = \dfrac{3-\sqrt{2}}{7}$

###  Worked example

Rationalise the denominator of the fraction $\dfrac{1}{3-2\sqrt{2}}$.

$\dfrac{1}{3-2\sqrt{2}} \times \dfrac{3+2\sqrt{2}}{3+2\sqrt{2}} = \dfrac{3+2\sqrt{2}}{9+6\sqrt{2}-6\sqrt{2}-4\sqrt{2}\sqrt{2}}$

$\Rightarrow \dfrac{3+2\sqrt{2}}{9-8} = 3+2\sqrt{2}$

 **Exercise 1.50**  Rationalise the denominator in the following fractions. Where possible, leave your answers in simplified form.

1  $\dfrac{1}{\sqrt{2}+1}$     2  $\dfrac{1}{\sqrt{3}-1}$     3  $\dfrac{3}{2-\sqrt{3}}$     4  $\dfrac{2}{\sqrt{2}-1}$

5  $\dfrac{5}{2+\sqrt{5}}$    6  $\dfrac{7}{1+\sqrt{7}}$    7  $\dfrac{1}{3-\sqrt{3}}$    8  $\dfrac{2}{2-\sqrt{3}}$

9  $\dfrac{1}{\sqrt{6}-\sqrt{5}}$    10  $\dfrac{1}{\sqrt{2}+1} + \dfrac{1}{\sqrt{2}-1}$

# 10 Speed, distance and time

## Time

Times may be given in terms of the 12-hour clock, e.g. 'I get up at seven o'clock in the morning, play football at half past two in the afternoon, and go to bed before eleven o'clock'.

These times can be written as 7 a.m., 2.30 p.m. and 11 p.m.

In order to save confusion, most timetables are written using the 24-hour clock.

7 a.m. is written as 07 00

2.30 p.m. is written as 14 30

11 p.m. is written as 23 00

To change p.m. times to 24-hour clock times, add 12 hours.

To change 24-hour clock times later than 12.00 noon to 12-hour clock times, subtract 12 hours.

**Exercise 1.51**

1   The clocks below show times in the morning. Write down the times using both the 12-hour and the 24-hour clock.

   a    b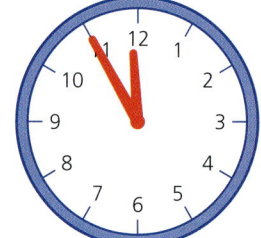

2   The clocks below show times in the afternoon. Write down the times using both the 12-hour and the 24-hour clock.

   a    b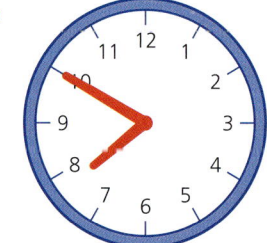

**3** Rewrite these times using the 24-hour clock:
   **a** 2.30 p.m.    **b** 9 p.m.    **c** 8.45 a.m.    **d** 6 a.m.
   **e** midday    **f** 10.55 p.m.    **g** 7.30 a.m.    **h** 7.30 p.m.
   **i** 1 a.m.    **j** midnight

**4** Rewrite these times using the 24-hour clock:
   **a** A quarter past seven in the morning
   **b** Eight o'clock at night
   **c** Ten past nine in the morning
   **d** A quarter to nine in the morning
   **e** A quarter to three in the afternoon
   **f** Twenty to eight in the evening

**5** These times are written for the 24-hour clock. Rewrite them using a.m. and p.m.
   **a** 0720    **b** 0900    **c** 1430    **d** 1825
   **e** 2340    **f** 0115    **g** 0005    **h** 1135
   **i** 1750    **j** 2359    **k** 0410    **l** 0545

**6 a** A journey to work takes a woman three quarters of an hour. If she catches the bus at the following times, when does she arrive?
     **i** 0720    **ii** 0755    **iii** 0820    **iv** 0845
  **b** The same woman catches buses home at the times shown below. The journey takes 55 minutes. If she catches the bus at the following times, when does she arrive?
     **i** 1725    **ii** 1750    **iii** 1805    **iv** 1820

**7** A boy cycles to school each day. His journey takes 70 minutes. When will he arrive if he leaves home at:
   **a** 0715    **b** 0825    **c** 0840    **d** 0855?

**8 a** The train into the city from a village takes 1 hour and 40 minutes. Copy and complete the following train timetable.

| Depart | Arrive |
|---|---|
| 0615 | |
| | 0810 |
| 0925 | |
| | 1200 |
| 1318 | |
| | 1628 |
| 1854 | |
| | 2105 |

## 10 SPEED, DISTANCE AND TIME

**Exercise 1.51 (cont)**

b The same journey by bus takes 2 hours and 5 minutes. Copy and complete the following bus timetable.

| Depart | Arrive |
|---|---|
| 06 00 | |
| | 08 50 |
| 08 55 | |
| | 11 14 |
| 13 48 | |
| | 16 22 |
| 21 25 | |
| | 00 10 |

9 A coach runs from Cambridge to the airports at Stansted, Gatwick and Heathrow. The time taken for the journey remains constant. Copy and complete the following timetables for the outward and return journeys.

| Cambridge | 04 00 | 08 35 | 12 50 | 19 45 | 21 10 |
|---|---|---|---|---|---|
| Stansted | 05 15 | | | | |
| Gatwick | 06 50 | | | | |
| Heathrow | 07 35 | | | | |

| Heathrow | 06 25 | 09 40 | 14 35 | 18 10 | 22 15 |
|---|---|---|---|---|---|
| Gatwick | 08 12 | | | | |
| Stansted | 10 03 | | | | |
| Cambridge | 11 00 | | | | |

10 The flight time from London to Johannesburg is 11 hours and 20 minutes. Copy and complete the following timetable.

| | London | Jo'burg | London | Jo'burg |
|---|---|---|---|---|
| Sunday | 06 15 | | 14 20 | |
| Monday | | 18 45 | | 05 25 |
| Tuesday | 07 20 | | 15 13 | |
| Wednesday | | 19 12 | | 07 30 |
| Thursday | 06 10 | | 16 27 | |
| Friday | | 17 25 | | 08 15 |
| Saturday | 09 55 | | 18 50 | |

*Speed, distance and time*

11  The flight time from Amsterdam to Kuala Lumpur is 13 hours and 45 minutes. Copy and complete the following timetable.

|  | Amsterdam | Kuala Lumpur | Amsterdam | Kuala Lumpur | Amsterdam | Kuala Lumpur |
|---|---|---|---|---|---|---|
| Sunday | 0828 |  | 1400 |  | 1830 |  |
| Monday |  | 2200 |  | 0315 |  | 0950 |
| Tuesday | 0915 |  | 1525 |  | 1755 |  |
| Wednesday |  | 2135 |  | 0400 |  | 0822 |
| Thursday | 0700 |  | 1345 |  | 1840 |  |
| Friday |  | 0010 |  | 0445 |  | 0738 |
| Saturday | 1012 |  | 1420 |  | 1908 |  |

12  A man sets off from Paris on an 'Around the world' bicycle ride for charity. He sets off on 1st January 2017. The number of days taken to reach certain cities is given in the following table.

| City | Number of days from the start |
|---|---|
| Rome | 8 |
| Moscow | 71 |
| Cairo | 148 |
| Mumbai | 210 |
| Singapore | 262 |
| Tokyo | 341 |
| Los Angeles | 387 |
| Brasilia | 412 |
| Paris | 500 |

Work out the dates at which he arrives at each of the cities.

# Speed, distance and time

You need to be aware of the following formulas:

  distance = speed × time

Rearranging the formula gives:

  $\text{speed} = \frac{\text{distance}}{\text{time}}$

Where the speed is not constant:

  $\text{average speed} = \frac{\text{total distance}}{\text{total time}}$

*In physics, you will see the term **velocity (v)**. Velocity is speed in a given direction. It is a vector. See Topic 8.*

# 10 SPEED, DISTANCE AND TIME

> **Worked example**
>
> A train covers the 480 km journey from Paris to Lyon at an average speed of 100 km/h. If the train leaves Paris at 08:35, when does it arrive in Lyon?
>
> Time taken = $\dfrac{\text{distance}}{\text{speed}}$
>
> Paris to Lyon = $\dfrac{480}{100}$ hours = 4.8 hours
>
> 4.8 hours is 4 hours and (0.8 × 60 minutes), which is 4 hours and 48 minutes.
>
> Departure time is 08:35; arrival time is 08:35 + 4 hours 48 minutes.
>
> Arrival time is 13:23.

Note that 4.80 h does not represent 4 h and 80 min. This is because time is not a decimal system which has 10 as its base number. Time is a **sexagesimal** number system with 60 as its base number, i.e. there are 60 minutes in an hour and 60 seconds in a minute.

As shown above, converting 0.8 h into minutes can be done by multiplying by sixty.

i.e. 0.8 h is equivalent to 0.8 × 60 = 48 min

*Degrees, minutes and seconds will look similar to* °′″

Your graphic display calculator will have a sexagesimal function and it will convert time given as a decimal into time in degrees, minutes and seconds.

The instructions below show how to do this:

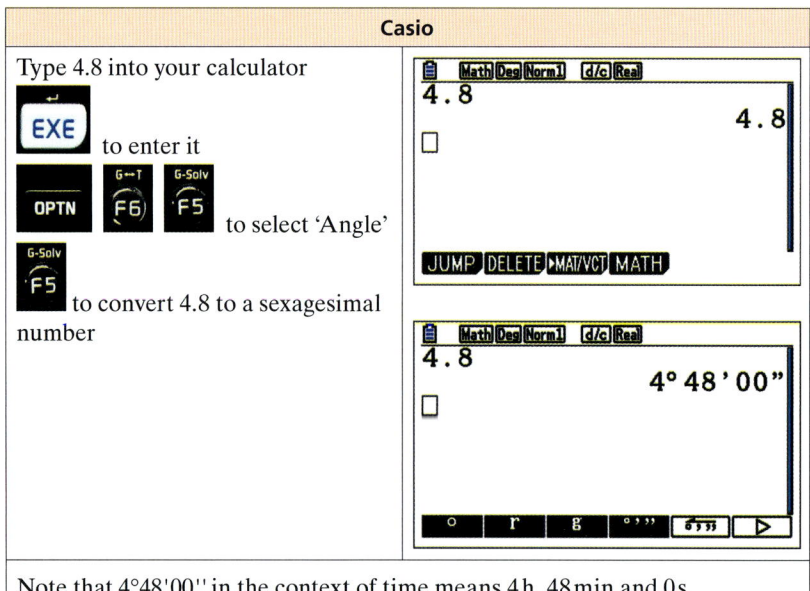

Note that 4°48′00″ in the context of time means 4 h, 48 min and 0 s

*Speed, distance and time*

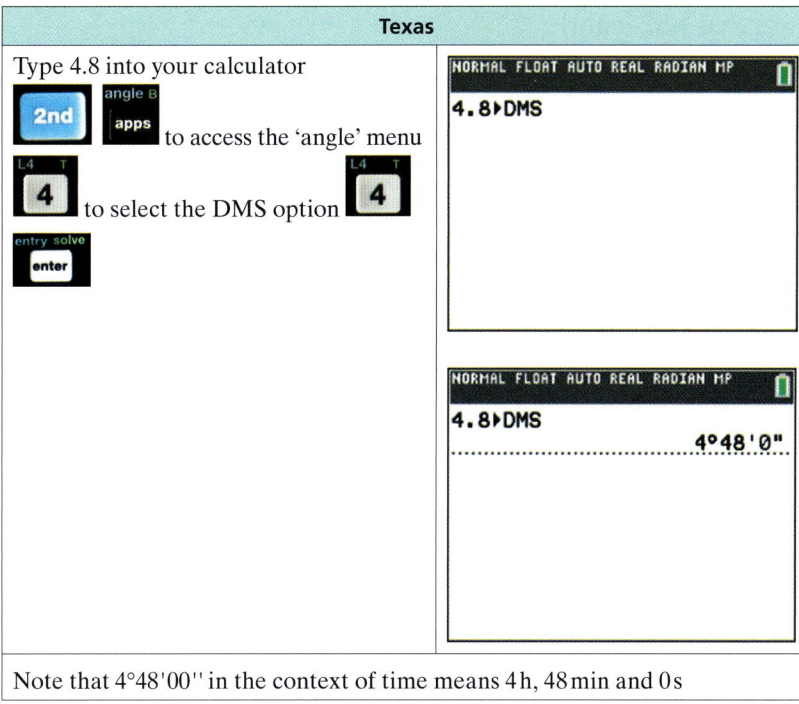

Note that 4°48'00" in the context of time means 4h, 48min and 0s

**Exercise 1.52**

1  Using your calculator, convert the following times written as decimals into time using hours, minutes and seconds.
   a  0.25 h          b  3.765 h          c  0.22 h

2  A train leaves a station at 06 24. The journey has 4 stops. Calculate the time the train arrives at each stop if the time taken from one stop to the next is as follows:
   Start to stop 1 takes 0.35 h
   Stop 1 to stop 2 takes 1.30 h
   Stop 2 to stop 3 takes 1.65 h
   Stop 3 to final stop takes 2.91 h

3  Find the average speed of an object moving:
   a  30 m in 5 s                b  48 m in 12 s
   c  78 km in 2 h               d  50 km in 2.5 h
   e  400 km in 2 h 30 min       f  110 km in 2 h 12 min

4  How far will an object travel during:
   a  10 s at 40 m/s             b  7 s at 26 m/s
   c  3 hours at 70 km/h         d  4 h 15 min at 60 km/h
   e  10 min at 60 km/h          f  1 h 6 min at 20 m/s?

5  How long will it take to travel:
   a  50 m at 10 m/s             b  1 km at 20 m/s
   c  2 km at 30 km/h            d  5 km at 70 m/s
   e  200 cm at 0.4 m/s          f  1 km at 15 km/h?

6  A train travels a distance of 420 km. The journey takes $3\frac{1}{2}$ hours and includes two stops, each of 15 minutes. Calculate the average speed of the train:
   a  for the whole journey      b  when it is moving.

# 10 SPEED, DISTANCE AND TIME

**Exercise 1.52 (cont)**

7  A plane flies from Boston, USA, to London, a distance of 5600 km. It leaves at 8 p.m. Boston local time and arrives at 8 a.m. local time in London. If the time in London is 5 hours ahead of the time in Boston, calculate the average speed of the plane.

8  How long does it take a plane to fly from New Delhi to Sydney, a distance of 10 420 km, if the plane flies at an average speed of 760 km/h? Give your answer:
   a  to two decimal places
   b  to the nearest minute.

9  A train leaves Paris at 8 p.m. Monday and travels to Istanbul, a distance of 4200 km. If the train travels at an average speed of 70 km/h and the time in Istanbul is two hours ahead of the time in Paris, give the day and time at which the train arrives in Istanbul.

The graph of an object travelling at a constant speed is a straight line as shown.

Gradient = $\frac{d}{t}$

The units of the gradient are m/s, and so the gradient of a distance–time graph represents the speed at which the object is travelling.

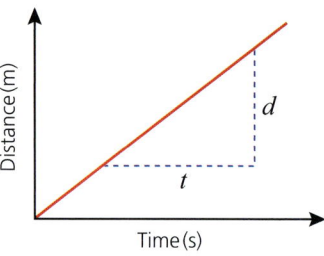

## → Worked examples

The graph shown represents an object travelling at constant speed.

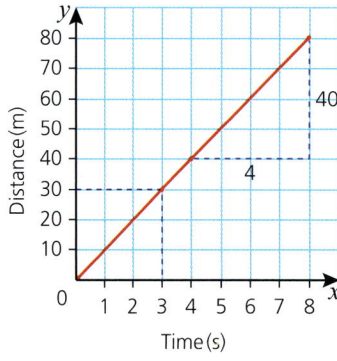

a  From the graph, calculate how long it took to cover a distance of 30 m.

   The time taken to travel 30 m is 3 seconds.

b  Calculate the gradient of the graph.

   Taking two points on the line, gradient = $\frac{40}{4}$ = 10.

c  Calculate the speed at which the object was travelling.

   Gradient of a distance–time graph = speed.

   Therefore the speed is 10 m/s.

## Speed, distance and time

**Exercise 1.53**

1  Draw a distance–time graph for the first 10 seconds of an object travelling at 6 m/s.
2  Draw a distance–time graph for the first 10 seconds of an object travelling at 5 m/s. Use your graph to estimate:
   a  the time taken to travel 25 m
   b  how far the object travels in 3.5 seconds.
3  Two objects A and B set off from the same point and move in the same straight line. B sets off first, while A sets off 2 seconds later.

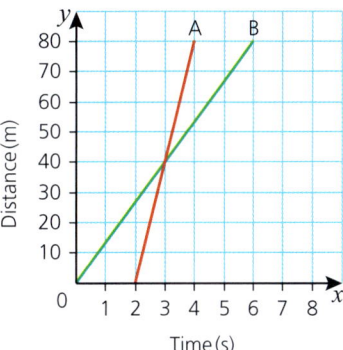

   Use the distance–time graph shown to estimate:
   a  the speed of each of the objects
   b  how far apart the objects will be 20 seconds after the start.
4  Three objects A, B and C move in the same straight line away from a point X.
   Both A and C change their speed during the journey, while B travels at the same constant speed throughout.

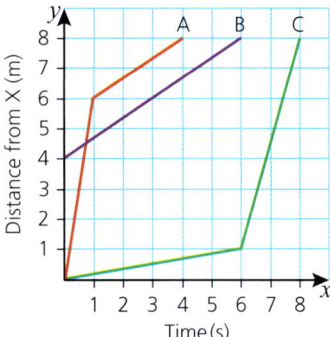

   From the distance–time graph estimate:
   a  the speed of object B
   b  the two speeds of object A
   c  the average speed of object C
   d  how far object C is from X 3 seconds from the start
   e  how far apart objects A and C are 4 seconds from the start.

# 10 SPEED, DISTANCE AND TIME

## Speed, distance and time graphs

The graphs of two or more journeys can be shown on the same axes. The shape of the graph gives a clear picture of the movement of each of the objects.

### Worked examples

Car X and Car Y both reach point B 100 km from A at 11 a.m.

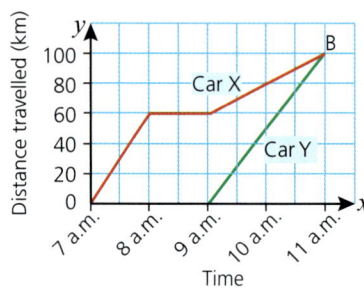

a Calculate the speed of Car X between 7 a.m. and 8 a.m.

speed = $\frac{\text{distance}}{\text{time}}$

= $\frac{60}{1}$ km/h

= 60 km/h

b Calculate the speed of Car Y between 9 a.m. and 11 a.m.

speed = $\frac{100}{2}$ km/h

= 50 km/h

c Explain what is happening to Car X between 8 a.m. and 9 a.m.

No distance has been travelled, therefore Car X is stationary.

### Exercise 1.54

1 Two friends Paul and Helena arrange to meet for lunch at noon. They live 50 km apart and the restaurant is 30 km from Paul's home. The travel graph below illustrates their journeys.

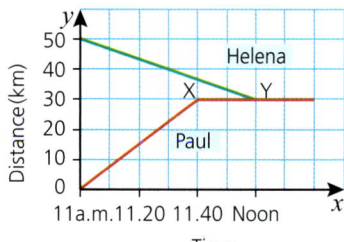

  a What is Paul's average speed between 11 a.m. and 11.40 a.m.?
  b What is Helena's average speed between 11 a.m. and noon?
  c What does the line XY represent?

2 A car travels at a speed of 60 km/h for 1 hour. It stops for 30 minutes and then continues at a constant speed of 80 km/h for a further 1.5 hours. Draw a distance–time graph for this journey.

3 A girl cycles for 1.5 hours at 10 km/h. She stops for an hour and then travels for a further 15 km in 1 hour. Draw a distance–time graph of the girl's journey.

# 11 Money and finance

## Currency conversions

In 2022, 1 euro (€) could be exchanged for 1.50 Australian dollars (A$).

> ### Worked examples
> a  How many Australian dollars can be bought for €400?
>
> €1 buys A$1.50.
>
> €400 buys 1.50 × 400 = A$600.
>
> b  How much does it cost to buy A$940?
>
> A$1.50 costs €1.
>
> A$940 costs $\frac{1 \times 940}{1.5}$ = €626.67

**Exercise 1.55**  The table shows the exchange rates for €1 into various currencies.

| Australia | 1.50 Australian dollars (A$) |
|---|---|
| India | 75 rupees |
| Zimbabwe | 412 800 Zimbabwean dollars (ZIM$) |
| South Africa | 15 rand |
| Turkey | 4.0 Turkish lira (L) |
| Japan | 130 yen |
| Kuwait | 0.35 dinar |
| USA | 1.15 US dollars (US$) |

1  Convert the following.
   a  €25 into Australian dollars     b  €50 into rupees
   c  €20 into Zimbabwean dollars     d  €300 into rand
   e  €130 into Turkish lira     f  €40 into yen
   g  €400 into dinar     h  €150 into US dollars

2  How many euros does it cost to buy the following?
   a  A$500     b  200 rupees
   c  ZIM$10 000     d  500 rand
   e  750 Turkish lira     f  1200 yen
   g  50 dinar     h  US$150

# 11 MONEY AND FINANCE

# Earnings

**Net pay** is what is left after deductions such as tax, insurance and pension contributions are taken from **gross earnings**.

That is, net pay = gross pay – deductions

A **bonus** is an extra payment that is sometimes added to an employee's **basic pay**.

In many companies, there is a fixed number of hours that an employee is expected to work. Any work done in excess of this **basic week** is paid at a higher rate, referred to as **overtime**. The overtime rate may be 1.5 times the basic pay, called **time and a half**, or it may be twice the basic pay, which is called **double time**.

**Piece work** is another method of payment. Employees are paid for the number of items made, not for the time taken.

## Exercise 1.56

1. Mr Ahmet's gross pay is $188.25. Deductions amount to $33.43. What is his net pay?
2. Miss Said's basic pay is $128. She earns $36 overtime and receives a bonus of $18. What is her net pay?
3. Mrs Hafar's gross pay is $203. She pays $54 tax and $18 towards her pension. What is her net pay?
4. Mr Wong works for 35 hours a week at an hourly rate of $8.30. What is his basic pay?
5. a  Miss Martinez works for 38 hours a week at an hourly rate of $4.15. In addition, she works 6 hours overtime at time and a half. What is her total gross pay?
   b  Deductions amount to 32% of her total gross pay. What is her net pay?
6. Pepe is paid $5.50 for each basket of grapes he picks. One week, he picks 25 baskets. How much is he paid?
7. Maria is paid €5 for every 12 plates that she makes. This is her record for one week.

   | Mon   | 240 |
   |-------|-----|
   | Tues  | 360 |
   | Wed   | 288 |
   | Thurs | 192 |
   | Fri   | 180 |

   How much is she paid?
8. Neo works at home making clothes. The patterns and materials are provided by the company. The table shows the rates she is paid and the number of items she makes in one week.

| Item | Rate | Number made |
|---|---|---|
| Jacket | 25 rand | 3 |
| Trousers | 11 rand | 12 |
| Shirt | 13 rand | 7 |
| Dress | 12 rand | 0 |

**a** What are her gross earnings?
**b** Deductions amount to 15% of gross earnings. What is her net pay?

# Profit and loss

Foodstuffs and manufactured goods are produced at a cost, known as the **cost price**, and sold at the **selling price**. If the selling price is greater than the cost price, then a profit is made.

### ➡ Worked example

A market trader buys oranges in boxes of 12 dozen for $14.40 per box. He buys three boxes and sells all the oranges for 12c each. What is his profit or **loss**?

Cost price: 3 × $14.40 = $43.20

Selling price: 3 × 144 × 12c = $51.84

In this case, he makes a profit of $51.84 – $43.20.

His profit is $8.64.

A second way of solving this problem would be:

$14.40 for a box of 144 oranges is 10c for each orange.

So, cost price of each orange is 10c and the selling price of each orange is 12c. The profit is 2c per orange.

So, 3 boxes would give a profit of 3 × 144 × 2c.

That is, $8.64.

Sometimes, particularly during sales or promotions, the selling price is reduced; this is known as a **discount**.

### ➡ Worked example

In a sale, a skirt that usually costs $35 is sold at a 15% discount. How much is the discount?

15% of $35 = 0.15 × $35 = $5.25

The discount is $5.25.

# 11 MONEY AND FINANCE

 **Exercise 1.57**

1. A market trader buys peaches in boxes of 120. He buys 4 boxes at a cost price of $13.20 per box. He sells 425 peaches at 12c each. The rest are ruined. How much profit or loss does he make?
2. A shopkeeper buys 72 bars of chocolate for $5.76. What is his profit if he sells them all for 12c each?
3. A holiday company charters an aircraft to fly to Malta at a cost of $22 000. It then sells 150 seats on the plane at $185 each and a further 35 seats at a 20% discount. Calculate the profit made per seat if the plane has 200 seats.
4. A car is priced at $7200. The car dealer allows a customer to pay a one-third deposit and 12 payments of $420 per month. How much extra does it cost the customer?
5. At an auction, a company sells 150 television sets for an **average** of $65 each. The production cost was $10 000. How much loss did the company make?

## Percentage profit and loss

Most profits or losses are expressed as a percentage.

Profit or loss, divided by cost price, multiplied by 100 = % profit or loss.

 **Worked example**

A woman buys a car for $7500 and sells it two years later for $4500. Calculate her loss over the two years as a percentage of the cost price.

Cost price = $7500    Selling price = $4500    Loss = $3000

% loss = $\frac{3000}{7500} \times 100 = 40\%$

Her loss is 40%.

When something becomes worth less over a period of time it is said to **depreciate** in value.

**Exercise 1.58**

1. Find the depreciation of the following cars as a percentage of the cost price. (C.P. = cost price, S.P. = selling price)
   a  VW      C.P. $4500     S.P. $4005
   b  Rover   C.P. $9200     S.P. $6900
2. A company manufactures electrical equipment for the kitchen. Find the percentage profit on each of the following items.
   a  Fridge    C.P. $50     S.P. $65
   b  Freezer   C.P. $80     S.P. $96
3. A developer builds a number of different types of houses. Which type of house gives the developer the largest percentage profit?
   Type A    C.P. $40 000    S.P. $52 000
   Type B    C.P. $65 000    S.P. $75 000
   Type C    C.P. $81 000    S.P. $108 000
4. Students in a school organise a disco. The disco company charges £350 hire charge. The students sell 280 tickets at £2.25 each. What is the percentage profit?

# 12 Set notation and Venn diagrams

## Sets

A **set** is a well-defined group of objects or symbols. The objects or symbols are called the **elements** of the set.

If an element $e$ belongs to a set $S$, this is represented as $e \in S$. If $e$ does not belong to set $S$ this is represented as $e \notin S$.

### Worked examples

a  A particular set consists of the following elements:
   {South Africa, Namibia, Egypt, Angola, ...}
   i  Describe the set.
      The elements of the set are countries of Africa.
   ii Add another two elements to the set.
      e.g. Zimbabwe, Ghana
   iii Is the set finite or infinite?
      Finite

b  Consider the set $A = \{x : x \text{ is a natural number}\}$
   i  Describe the set.
      The elements of the set are the natural numbers.
   ii Write down two elements of the set.
      e.g. 3 and 15

c  Consider the set $B = \{(x, y) : y = 2x - 4\}$
   i  Describe the set.
      The elements of the set are the coordinates of points found on the straight line with equation $y = 2x - 4$.
   ii Write down two elements of the set.
      e.g. $(0, -4)$ and $(10, 16)$

d  Consider the set $C = \{x : 2 \leqslant x \leqslant 8\}$
   i  Describe the set.
      The elements of the set include any number between 2 and 8 inclusive.
   ii Write down two elements of the set.
      e.g. 5 and 6.3

### Exercise 1.59

1  In the following questions:
   i  describe the set in words
   ii write down another two elements of the set.
   a  {Asia, Africa, Europe, ...}
   b  {2, 4, 6, 8, ...}
   c  {Sunday, Monday, Tuesday, ...}
   d  {1, 3, 6, 10, ...}
   e  {January, March, July, ...}

# 12 SET NOTATION AND VENN DIAGRAMS

**Exercise 1.59 (cont)**

    f  {Mehmet, Michael, Mustapha, Matthew, …}
    g  {11, 13, 17, 19, …}
    h  {a, e, i, …}
    i  {Earth, Mars, Venus, …}
    j  $A = \{x : 3 \leqslant x \leqslant 12\}$
    k  $S = \{y : -5 \leqslant y \leqslant 5\}$

2  The number of elements in a set $A$ is written as $n(A)$.
Give the value of $n(A)$ for the finite sets in Q.1a–k above.

## Subsets

If all the elements of one set $X$ are also elements of another set $Y$, then $X$ is said to be a **subset** of $Y$.

This is written as $X \subseteq Y$.

If a set $A$ is empty (i.e. it has no elements in it), then this is called the **empty set** and it is represented by the symbol $\varnothing$. The empty set is a subset of all sets.

e.g. Three girls, Winnie, Natalie and Emma, form a set $A$.

$A$ = {Winnie, Natalie, Emma}

All the possible subsets of $A$ are given below:

$B$ = {Winnie, Natalie, Emma}

$C$ = {Winnie, Natalie}

$D$ = {Winnie, Emma}

$E$ = {Natalie, Emma}

$F$ = {Winnie}

$G$ = {Natalie}

$H$ = {Emma}

$I = \varnothing$

Note that the sets $B$ and $I$ above are considered as subsets of $A$.

i.e. $B \subseteq A$ and $I \subseteq A$

Similarly, $G \nsubseteq H$ implies that $G$ is not a subset of $H$.

### ➡ Worked examples

$A$ = {1, 2, 3, 4, 5, 6, 7, 8, 9, 10}

  a  List subset $B$ {even numbers}.
     $B$ = {2, 4, 6, 8, 10}

  b  List subset $C$ {prime numbers}.
     $C$ = {2, 3, 5, 7}

*Set notation and Venn diagrams*

**Exercise 1.60**

1. $P = \{$natural numbers less than $30\}$
   a. List the subset $Q$ {even numbers}.
   b. List the subset $R$ {odd numbers}.
   c. List the subset $S$ {prime numbers}.
   d. List the subset $T$ {square numbers}.
   e. List the subset $U$ {triangle numbers}.

2. $A = \{$natural numbers between 50 and $70\}$
   a. List the subset $B$ {multiples of 5}.
   b. List the subset $C$ {multiples of 3}.
   c. List the subset $D$ {square numbers}.

3. State whether each of the following statements is true or false:
   a. {Algeria, Mozambique} $\subseteq$ {countries in Africa}
   b. {mango, banana} $\subseteq$ {fruit}
   c. $\{1, 2, 3, 4\} \subseteq \{1, 2, 3, 4\}$
   d. {volleyball, basketball} $\not\subseteq$ {team sports}
   e. {potatoes, carrots} $\subseteq$ {vegetables}

## The universal set

The **universal set** ($\cup$) for any particular problem is the set that contains all the possible elements for that problem.

The **complement** of a set $A$ is the set of elements that are in $\cup$ but not in $A$. The complement of $A$ is identified as $A'$.

### ➡ Worked examples

a. If $\cup = \{1, 2, 3, 4, 5, 6, 7, 8, 9, 10\}$ and $A = \{1, 2, 3, 4, 5\}$, what set is represented by $A'$?
   $A'$ consists of those elements in $\cup$ that are not in $A$.
   Therefore $A' = \{6, 7, 8, 9, 10\}$

b. If $\cup = \{$all 3D shapes$\}$ and $P = \{$**prisms**$\}$, what set is represented by $P'$?
   $P' = \{$all 3D shapes except prisms$\}$

# Set notation and Venn diagrams

**Venn diagrams** are the principal way of showing sets diagrammatically.

The method consists primarily of entering the elements of a set into a circle or circles.

Some examples of the uses of Venn diagrams are shown.

# 12 SET NOTATION AND VENN DIAGRAMS

$A = \{2, 4, 6, 8, 10\}$ can be represented as:

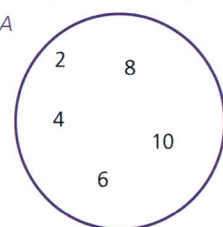

Elements that are in more than one set can also be represented using Venn diagrams.

$P = \{3, 6, 9, 12, 15, 18\}$ and $Q = \{2, 4, 6, 8, 10, 12\}$ can be represented as:

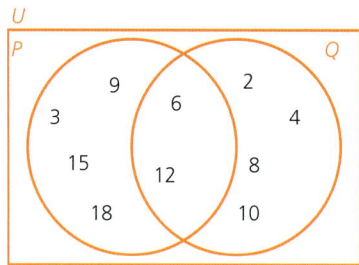

In the diagram above, it can be seen that those elements that belong to both sets are placed in the region of overlap of the two circles.

When two sets $P$ and $Q$ overlap as they do above, the notation $P \cap Q$ is used to denote the set of elements in the **intersection**, i.e. $P \cap Q = \{6, 12\}$.

Note that $6 \in P \cap Q$; $8 \notin P \cap Q$.

$X = \{1, 3, 6, 7, 14\}$ and $Y = \{3, 9, 13, 14, 18\}$ are represented as:

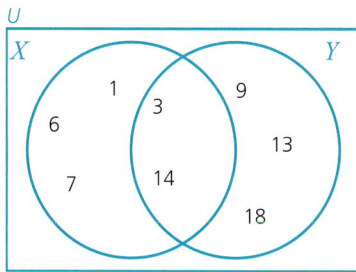

The **union** of two sets is everything that belongs to either or both of the sets and is represented by the symbol $\cup$.

Therefore, in the example above, $X \cup Y = \{1, 3, 6, 7, 9, 13, 14, 18\}$.

# Set notation and Venn diagrams

$J = \{10, 20, 30, 40, 50, 60, 70, 80, 90, 100\}$ and $K = \{60, 70, 80\}$; as discussed earlier, $K$ is a subset of $J$.

$K \subseteq J$ can be represented as shown below:

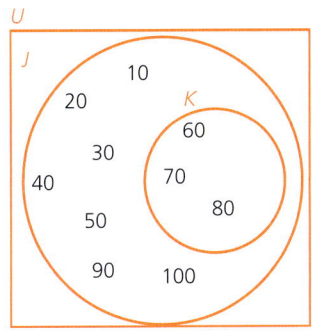

## Exercise 1.61

1  Using the Venn diagram, indicate whether the following statements are true or false. $\in$ means 'is an element of' and $\notin$ means 'is not an element of'.

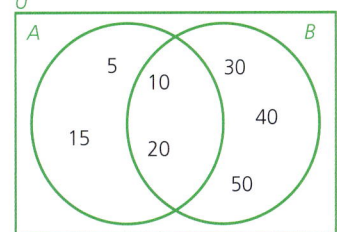

  a  $5 \in A$
  b  $20 \in B$
  c  $20 \notin A$
  d  $50 \in A$
  e  $50 \notin B$
  f  $A \cap B = \{10, 20\}$

2  Complete the statement $A \cap B = \{...\}$ for each of these Venn diagrams.

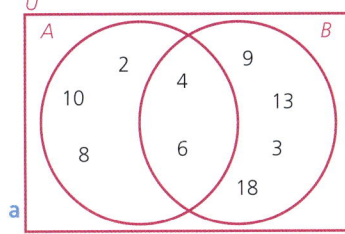
a

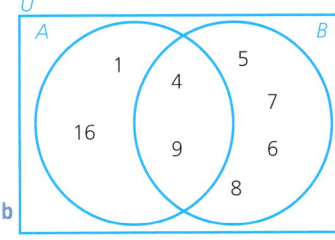
b

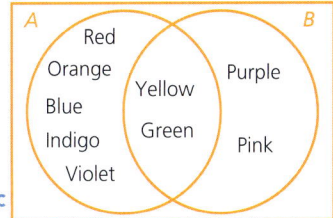
c

3  Complete the statement $A \cup B = \{...\}$ for each of the Venn diagrams in question 2.

4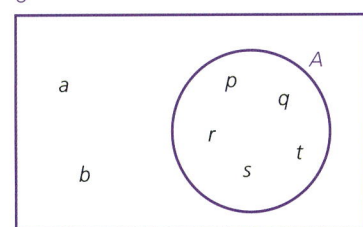

Copy and complete the following statements.
  a  $\cup = \{...\}$
  b  $A' = \{...\}$

# 12 SET NOTATION AND VENN DIAGRAMS

**Exercise 1.61 (cont)**

5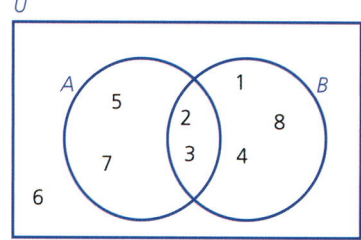

Copy and complete the following statements.
a  U = {...}
b  A' = {...}
c  A ∩ B = {...}
d  A ∪ B = {...}
e  (A ∩ B)' = {...}
f  A ∩ B' = {...}

6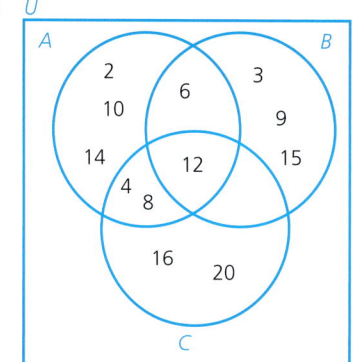

a  Describe in words the elements of:
   i   set A
   ii  set B
   iii set C
b  Copy and complete the following statements.
   i   A ∩ B = {...}
   ii  A ∩ C = {...}
   iii B ∩ C = {...}
   iv  A ∩ B ∩ C = {...}
   v   A ∪ B = {...}
   vi  C ∪ B = {...}

7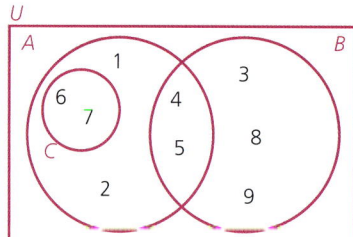

a  Copy and complete the following statements.
   i   A = {...}
   ii  B = {...}
   iii C' = {...}
   iv  A ∩ B = {...}
   v   A ∪ B = {...}
   vi  (A ∩ B)' = {...}
b  State, using set notation, the relationship between C and A.

# Problems involving sets

**Exercise 1.62**

1. $A = \{$Egypt, Libya, Morocco, Chad$\}$
   $B = \{$Iran, Iraq, Turkey, Egypt$\}$
   a  Draw a Venn diagram to represent the above information.
   b  Copy and complete the following statements.
      i   $A \cap B = \{...\}$
      ii  $A \cup B = \{...\}$

2. $P = \{2, 3, 5, 7, 11, 13, 17\}$
   $Q = \{11, 13, 15, 17, 19\}$
   a  Draw a Venn diagram to represent the above information.
   b  Copy and complete the following statements.
      i   $P \cap Q = \{...\}$
      ii  $P \cup Q = \{...\}$

3. $B = \{2, 4, 6, 8, 10\}$
   $A \cup B = \{1, 2, 3, 4, 6, 8, 10\}$
   $A \cap B = \{2, 4\}$
   Represent the above information on a Venn diagram.

4. $X = \{$a, c, d, e, f, g, l$\}$
   $Y = \{$b, c, d, e, h, i, k, l, m$\}$
   $Z = \{$c, f, i, j, m$\}$
   Represent the above information on a Venn diagram.

5. $P = \{1, 4, 7, 9, 11, 15\}$
   $Q = \{5, 10, 15\}$
   $R = \{1, 4, 9\}$
   Represent the above information on a Venn diagram.

## Problems involving sets

### → Worked example

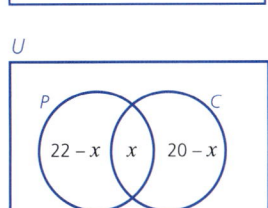

In a class of 31 students, some study physics and some study chemistry. If 22 study physics, 20 study chemistry and 5 study neither, calculate the number of students who take both subjects.

The information given above can be entered into a Venn diagram in stages.

The students taking neither physics nor chemistry can be put in first (as shown on the left).

This leaves 26 students to be entered into the set circles.

If $x$ students take both subjects, then

$n(P) = 22 - x + x$
$n(C) = 20 - x + x$
$P \cup C = 31 - 5 = 26$
Therefore $22 - x + x + 20 - x = 26$
$\qquad\qquad\qquad\qquad 42 - x = 26$
$\qquad\qquad\qquad\qquad\qquad x = 16$

Substituting the value of $x$ into the Venn diagram gives

## 12 SET NOTATION AND VENN DIAGRAMS

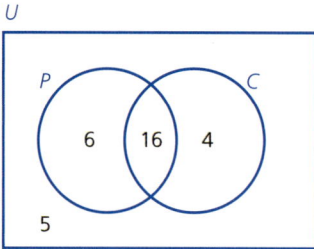

Therefore, the number of students taking both physics and chemistry is 16.

 **Exercise 1.63**

1. In a class of 35 students, 19 take Spanish, 18 take French and 3 take neither. Calculate how many take:
   a both French and Spanish
   b just Spanish
   c just French.
2. In a year group of 108 students, 60 liked football, 53 liked tennis and 10 liked neither. Calculate the number of students who liked football but not tennis.
3. In a year group of 133 students, 60 liked hockey, 45 liked rugby and 18 liked neither. Calculate the number of students who:
   a liked both hockey and rugby
   b liked only hockey.
4. One year, 37 students sat an examination in physics, 48 sat an examination in chemistry and 45 sat an examination in biology. 15 students sat physics and chemistry, 13 sat chemistry and biology, 7 sat physics and biology and 5 students sat all three subjects.
   a Draw a Venn diagram to represent this information.
   b Calculate $n(P \cup C \cup B)$.

# Investigations, modelling and ICT 1

The syllabus covered by this book examines investigations and modelling in Papers 5 and 6.

It can seem difficult to know how to begin an investigation. The suggestions below may help.

1. Read the question carefully and start with simple cases.
2. Draw simple diagrams to help.
3. Put the results from simple cases in a table.
4. Look for a pattern in your results.
5. Try to find a general rule in words.
6. Write your rule algebraically.
7. Test the rule for a new example.
8. Check that the original question has been answered.

## → Worked example

A mystic rose is created by placing a number of points evenly spaced on the circumference of a circle. Straight lines are then drawn from each point to every other point. The diagram shows a mystic rose with 20 points.

  i   How many straight lines are there?

  ii  How many straight lines would there be on a mystic rose with 100 points?

To answer these questions, you are not expected to draw either of the shapes and count the number of lines.

### 1/2 Try simple cases:
By drawing some simple cases and counting the lines, some results can be found:

a
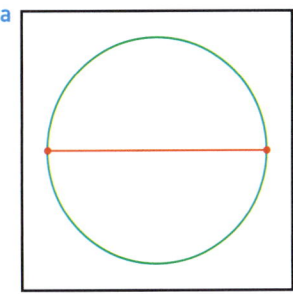
Mystic rose with 2 points
Number of lines = 1

b
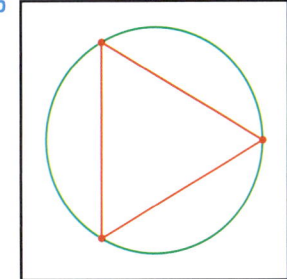
Mystic rose with 3 points
Number of lines = 3

# INVESTIGATIONS, MODELLING AND ICT 1

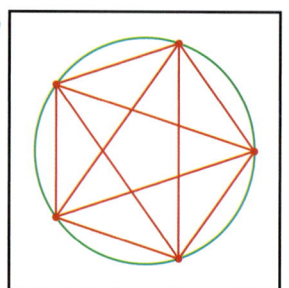

Mystic rose with 4 points
Number of lines = 6

Mystic rose with 5 points
Number of lines = 10

**3 Enter the results in an ordered table:**

| Number of points | 2 | 3 | 4 | 5 |
|---|---|---|---|---|
| Number of lines | 1 | 3 | 6 | 10 |

*You will find out more about recognising patterns in sequences in Chapter 20.*

**4/5 Look for a pattern in the results:**
There are two patterns.
The first shows how the values change.

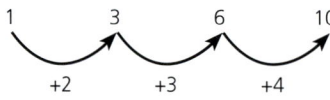

It can be seen that the difference between successive terms is increasing by one each time.
The problem with this pattern is that to find the 20th and 100th terms, it would be necessary to continue this pattern and find all the terms leading up to the 20th and 100th terms.
The second pattern is the relationship between the number of points and the number of lines.

| Number of points | 2 | 3 | 4 | 5 |
|---|---|---|---|---|
| Number of lines | 1 | 3 | 6 | 10 |

It is important to find a relationship that works for all values; for example, subtracting 1 from the number of points gives the number of lines in the first example only, so this is not useful. However, halving the number of points and multiplying this by 1 less than the number of points works each time; i.e.
Number of lines = (half the number of points) × (one less than the number of points).

*You will learn how to write algebraic expressions and formulas in Chapter 13.*

**6 Write the rule algebraically:**
The rule written in words above can be written more elegantly using algebra.
Let the number of lines be $l$ and the number of points be $p$.

$$l = \tfrac{1}{2} p(p - 1)$$

Note: Any letters can be used to represent the number of lines and the number of points, not just $l$ and $p$.

## 7 Test the rule:
The rule was derived from the original results. It can be tested by generating a further result.
If the number of points $p = 6$, then the number of lines $l$ is:

$$l = \tfrac{1}{2} \times 6\,(6-1)$$
$$= 3 \times 5$$
$$= 15$$

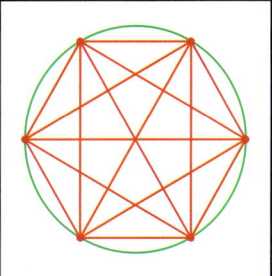

From the diagram, the number of lines can also be counted as 15.

## 8 Check that the original questions have been answered:
Using the formula, the number of lines in a mystic rose with 20 points is:

$$l = \tfrac{1}{2} \times 20\,(20-1)$$
$$= 10 \times 19$$
$$= 190$$

The number of lines in a mystic rose with 100 points is:

$$l = \tfrac{1}{2} \times 100\,(100-1)$$
$$= 50 \times 99$$
$$= 4950$$

# Investigation: Primes and squares

13, 41 and 73 are prime numbers.

Two different square numbers can be added together to make these prime numbers, e.g. $3^2 + 8^2 = 73$.

1. Find the two square numbers that can be added to make 13 and 41.
2. List the prime numbers less than 100.
3. Which of the prime numbers less than 100 can be shown to be the sum of two different square numbers?
4. Is there a rule to the numbers in Q.3?
5. Your rule is a predictive rule, not a formula. Discuss the difference.

# INVESTIGATIONS, MODELLING AND ICT 1

## Investigation: Football league

There are 18 teams in a football league.

1. If each team plays the other teams twice, once at home and once away, how many matches are played in a season?
2. If there are $t$ teams in a league, how many matches are played in a season?

## ICT: The school trip

A school is organising a trip to an amusement park for its students.

In order to work out how much to charge each student, the trip organiser decides to model the trip using a spreadsheet.

The information he has is as follows.
- There must be at least 1 teacher for every 10 students, but the minimum number of teachers on the trip is 2
- Coach hire costs $500 for a 30 seat coach
- Entrance to the amusement park costs $30 per student and $20 per accompanying teacher
- For every teacher that goes on the trip, the school must hire a replacement teacher for the day, which costs $200 per day.

1. Design a spreadsheet to calculate the cost of the trip.

   Suggested column headings are given below:

   |   | A | B | C | D | E | F | G | H | I |
   |---|---|---|---|---|---|---|---|---|---|
   | 1 | Number of Students | Number of Teachers | Number of Coaches | Cost of Teachers | Cost of Coaches | Student Entry Cost | Teacher Entry Cost | Total Cost | Cost per Student |
   | 2 | 1 | 2 | 1 | 400 | 500 | 30 | 40 | | |
   | 3 | | | | | | | | | |
   | 4 | | | | | | | | | |
   | 5 | | | | | | | | | |

2. What would be the total cost of the trip if:
   a. only one student goes on the trip?
   b. 10 students go on the trip?
   c. 75 students go on the trip?
3. In order to work out how much to charge each student, the total cost of the trip is divided by the number of students going on the trip. What is the cost per student if:
   a. only one student goes on the trip?
   b. 10 students go on the trip?
   c. 75 students go on the trip?

4   The school decides that the trip can only go ahead if the price per student is definitely not greater than $80.

What is the minimum number of students needed to ensure that the trip can go ahead?

# ICT Activity

In this activity you will be using a spreadsheet to track the price of a company's shares over a period of time.

1   a   Using the internet or a newspaper as a resource, find the value of a particular company's shares.
    b   Over a period of a month (or week), record the value of the company's shares. This should be carried out on a daily basis.
2   When you have collected all the results, enter them into a spreadsheet similar to the one shown below:

|   | A | B | C |
|---|---|---|---|
| 1 |   | Company Name | |
| 2 | Day | Share Price | Percentage Value |
| 3 | 1 | 3.26 | 100 |
| 4 | 2 | 3.29 | |
| 5 | 3 | 4.11 | |
| 6 | 4 | | |
| 7 | 5 | | |
| 8 | | | |
| 9 | | | |
| 10 | | | |
| 11 | etc | etc | |

3   In column C, enter formulas that will calculate the value of the shares as a percentage of their value on day 1.
4   When the spreadsheet is complete, produce a graph showing how the percentage value of the share price changed over time.
5   Write a short report explaining the performance of the company's shares during that time.

# Student assessments 1

## Student assessment 1

1. Write the reciprocal of the following numbers:
   a  8
   b  $\frac{1}{5}$
   c  $\frac{4}{7}$

2. Simplify the following expressions:
   a  $\sqrt{3} \times \sqrt{6}$
   b  $\sqrt{5} \times \sqrt{7}$
   c  $\sqrt{2} \times \sqrt{12}$

3. Simplify the following expressions:
   a  $\sqrt{3} + \sqrt{27}$
   b  $\sqrt{24} + \sqrt{54}$
   c  $3\sqrt{8} - \sqrt{32}$

4. Expand the following expressions and simplify as far as possible:
   a  $(1 - \sqrt{2})(3 + \sqrt{2})$
   b  $(3\sqrt{5} - 2)^2$

5. Rationalise the denominator for the following fractions:
   a  $\frac{3}{\sqrt{5}}$
   b  $\frac{5}{\sqrt{10}}$
   c  $\frac{4}{\sqrt{2} - 1}$
   d  $\frac{2 - \sqrt{3}}{2 + \sqrt{3}}$

6. a  A rod has a length of $\sqrt{3}$ cm. If three of these rods are placed end to end, decide whether the total length is a rational or irrational number. Give reasons for your answer.
   b  A square has side length $3\sqrt{5}$ cm. Decide whether the area of the square is rational or irrational. Give reasons for your answer.

7. Copy the calculations below and put in any brackets which are needed to make it correct.
   a  $2 + 3^2 \times 4 = 100$
   b  $20 + 4 \div 2^2 = 6$

8. Find the value of the following:
   a  $\frac{4^2 - 6}{2} + 5$
   b  $\frac{(5+7)^2 - 23}{11}$

9. Find the value of the following:
   a  $4^3$
   b  $15^2$
   c  $10^3$
   d  $\sqrt{196}$
   e  $\sqrt[3]{125}$
   f  $\sqrt[3]{1}$

10. Find the value of the following:
    a  $5^{-1}$
    b  $2^{-3}$
    c  $(2^3)^2$
    d  $81^{\frac{1}{2}}$
    e  $8^{-\frac{1}{3}}$
    f  $\sqrt[5]{32}$

## Student assessment 2

**1** Copy the table below and fill in the missing values:

| | Fraction | Decimal | Percentage |
|---|---|---|---|
| a | | 0.25 | |
| b | $\frac{3}{5}$ | | |

| | Fraction | Decimal | Percentage |
|---|---|---|---|
| c | | | $62\frac{1}{2}$ % |
| d | $2\frac{1}{4}$ | | |

**2** Find 30% of 2500 m.

**3** In a sale, a shop reduces its prices by 12.5%. What is the sale price of a desk previously costing €600?

**4** In the last six years, the value of a house has increased by 35%. If it cost £72 000 six years ago, what is its value now?

**5** Write the first quantity as a percentage of the second:
    **a** 35 minutes, 2 hours     **b** 650 g, 3 kg     **c** 5 m, 4 m
    **d** 15 s, 3 minutes     **e** 600 kg, 3 tonnes     **f** 35 cl, 3.5 l

**6** Shares in a company are bought for $600. After a year, the same shares are sold for $550. Calculate the percentage depreciation.

**7** In a sale, the price of a jacket originally costing 17 000 Japanese yen (¥) is reduced by ¥4000. Any item not sold by the last day of the sale is reduced by a further 50%. If the jacket is sold on the last day of the sale, calculate:
    **a** the price it is finally sold for
    **b** the overall percentage reduction in price.

**8** The population of a type of insect increases by approximately 10% each day. How many days will it take for the population to double?

**9** Find the compound interest on €5 million for 3 years at 6% interest per year

**10** A boat loses 15% of its value each year. How long will it take for its value to halve?

## Student assessment 3

1. Change the following fractions to mixed numbers:
   a $\frac{36}{7}$  b $\frac{65}{9}$  c $\frac{71}{5}$

2. Calculate the original price in each of the following:

   | Selling price | Profit |
   |---|---|
   | $224 | 12% |
   | $62.50 | 150% |
   | $660.24 | 26% |
   | $38.50 | 285% |

3. Calculate the original price in each of the following:

   | Selling price | Loss |
   |---|---|
   | $392.70 | 15% |
   | $2480 | 38% |
   | $3937.50 | 12.5% |
   | $4675 | 15% |

4. In an examination, Sarah obtained 87.5% by gaining 105 marks. How many marks did she lose?

5. At the end of a year, a factory has produced 38 500 television sets. If this represents a 10% increase in productivity on last year, calculate the number of sets that were made last year.

6. A computer manufacturer is expected to have produced 24 000 units by the end of this year. If this represents a 4% decrease on last year's output, calculate the number of units produced last year.

7. A farmer increased his yield by 5% each year over the last five years. If he produced 600 tonnes this year, calculate to the nearest tonne his yield five years ago.

8. A machine for making chocolate bars can produce 4000 bars of chocolate in 8 hours. How many chocolate bars can three of these machines make in 8 minutes?

9. Red and white paints are mixed in the ratio of 60 : 24. 130 ml of white paint is used for the mixture.
   a Express the ratio in its simplest form.
   b Calculate the total volume of paint in the mixture.

## Student assessment 4

1  Round the following numbers to the degree of accuracy shown in brackets:
   a  6472 (nearest 10)  b  88465 (nearest 100)
   c  64785 (nearest 1000)  d  6.7 (nearest 10)

2  Round the following numbers to the number of decimal places shown in brackets:
   a  6.78 (1 d.p.)  b  4.438 (2 d.p.)  c  7.975 (1 d.p.)
   d  63.084 (2 d.p.)  e  0.0567 (3 d.p.)  f  3.95 (2 d.p.)

3  Round the following numbers to the number of significant figures shown in brackets:
   a  42.6 (1 s.f.)  b  5.432 (2 s.f.)  c  0.0574 (1 s.f.)
   d  48572 (2 s.f.)  e  687453 (1 s.f.)  f  687453 (3 s.f.)

4  1 mile is 1760 yards. Estimate the number of yards in 19 miles.

5  Estimate the area of the figure below:

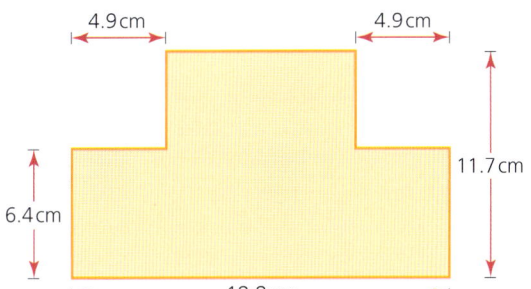

NB: The diagram is not drawn to scale.

6  Estimate the answers to the following. Do *not* work out an exact answer.
   a  $\dfrac{3.9 \times 26.4}{4.85}$  b  $\dfrac{(3.2)^3}{(5.4)^2}$  c  $\dfrac{2.8 \times (7.3)^2}{(3.2)^2 \times 6.2}$

7  A cuboid's dimensions are given as 3.973 m by 2.4 m by 3.16 m. Calculate its volume, giving your answer to an appropriate degree of accuracy.

8  A girl runs a race in 14.2 seconds. If she rounds her time down to 14 seconds, what is her error as a percentage of her actual time?

9  a  Use a calculator to find the exact answer to Q.5.
   b  Calculate your error as a percentage of the real area.

10  Show that the following numbers are rational:
   a  0.875  b  $3\sqrt{25}$  c  0.44

# STUDENT ASSESSMENTS 1

## Student assessment 5

1. Write the following numbers in standard form:
   a  6 million        b  0.0045        c  3 800 000 000
   d  0.000 000 361    e  460 million   f  3

2. Write the following numbers in order of magnitude, starting with the largest:
   $3.6 \times 10^2$       $2.1 \times 10^{-3}$      $9 \times 10^1$
   $4.05 \times 10^8$      $1.5 \times 10^{-2}$      $7.2 \times 10^{-3}$

3. Write the following numbers:
   a  in standard form
   b  in order of magnitude, starting with the smallest.
      15 million   430 000   0.000 435   4.8   0.0085

4. Write the answers to the following calculations in standard form:
   a  $50\,000 \times 2400$                  b  $(3.7 \times 10^6) \times (4.0 \times 10^4)$
   c  $(5.8 \times 10^7) + (9.3 \times 10^6)$  d  $(4.7 \times 10^6) - (8.2 \times 10^5)$

5. The speed of light is $3 \times 10^8$ m/s. Jupiter is 778 million km from the Sun. Calculate the number of minutes it takes for sunlight to reach Jupiter.

6. A star is 300 light years away from Earth. If the speed of light is $3 \times 10^5$ km/s, calculate the distance from the star to Earth. Give your answer in kilometres and written in standard form.

7. Blue paint costs $4.50 per litre. Black paint costs $6.20 per litre. Blue paint and black paint are mixed in the ratio 5:1.
   Work out the cost of 24 litres of this paint mixture.

8. Calculate the following:
   a  $\frac{4}{7} + \frac{2}{5}$    b  $2\frac{1}{3} - 1\frac{2}{5}$    c  $3\frac{1}{4} \times \frac{1}{2}$    d  $5\frac{2}{3} \div 1\frac{2}{3}$

**Note**

Core students will be allowed to use calculators for any calculations involving standard form.

## Student assessment 6

1. Using your calculator, convert the following times written as decimals into times using hours, minutes and seconds.
   a  2.4 h        b  0.8 h        c  3.56 h

2. A woman climbs to the top of a hill at an average vertical speed of 1.5 km/h. If she climbs for 4 hours and 40 minutes, with two half-hour breaks, calculate the height of the hill.

3. A cyclist completes a journey of 240 km in 8 hours.
   a  Calculate his average speed.
   b  If his average speed was 25% faster, how long would the journey have taken him?

4. Land is split into three regions in the ratio 2:5:9 for different crops. If the largest region has an area of 6.75 km², calculate the area of each of the other two regions.

5. At a school, the ratio of boys to girls is 6:5. There are 80 fewer girls than boys. Calculate the total number of students in the school.

## Student assessment 7

1. Describe the following sets in words.
   a  {2, 4, 6, 8}
   b  {2, 4, 6, 8, ...}
   c  {1, 4, 9, 16, 25, ...}
   d  {Arctic, Atlantic, Indian, Pacific}
2. Calculate the value of $n(A)$ for each of the sets shown below.
   a  $A$ = {days of the week}
   b  $A$ = {prime numbers between 50 and 60}
   c  $A = \{x : x$ is an integer and $5 \leqslant x \leqslant 10\}$
   d  $A$ = {days in a leap year}
3. Copy the Venn diagram below twice.

   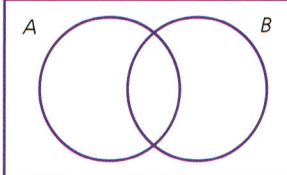

   a  On one copy, shade and label the region that represents $A \cap B$.
   b  On the other copy, shade and label the region that represents $A \cup B$.
4. If $A$ = {a, b}, list all the subsets of $A$.
5. If $\cup$ = {m, a, t, h, s} and $A$ = {a, s}, what set is represented by $A'$?

## Student assessment 8

1. If $A$ = {2, 4, 6, 8}, write all the subsets of $A$ with two or more elements.
2. $J$ = {London, Paris, Rome, Washington, Canberra, Ankara, Cairo}
   $K$ = {Cairo, Nairobi, Pretoria, Ankara}
   a  Draw a Venn diagram to represent the above information.
   b  Copy and complete the statement $J \cap K$ = {...}.
   c  Copy and complete the statement $J' \cap K$ = {...}.
3. $M = \{x : x$ is an integer and $2 \leqslant x \leqslant 20\}$
   $N$ = {prime numbers less than 30}
   a  Draw a Venn diagram to represent the above information.
   b  Copy and complete the statement $M \cap N$ = {...}.
   c  Copy and complete the statement $(M \cap N)'$ = {...}.
4. $\cup$ = {natural numbers}, $M$ = {even numbers} and $N$ = {multiples of 5}.
   a  Draw a Venn diagram and place the numbers 1, 2, 3, 4, 5, 6, 7, 8, 9, 10 in the appropriate places in it.
   b  If $X = M \cap N$, describe set $X$ in words.
5. In a region of mixed farming, farms keep goats, cattle or sheep. There are 77 farms altogether. 19 farms keep only goats, 8 keep only cattle and 13 keep only sheep. 13 keep both goats and cattle, 28 keep both cattle and sheep and 8 keep both goats and sheep.
   a  Draw a Venn diagram to show the above information.
   b  Calculate $n(G \cap C \cap S)$.

# TOPIC 2

# Algebra

## Contents

Chapter 13 Algebraic representation and manipulation (C2.1, C2.2, C2.5, E1.1, E2.1, E2.2, E2.5)
Chapter 14 Further algebraic representation and manipulation (C2.2, C2.3, E1.1, E2.2, E2.3, E2.5)
Chapter 15 Linear and simultaneous equations (C2.5, E2.5)
Chapter 16 Solving quadratic equations (C2.5, E2.5)
Chapter 17 Using a graphic display calculator to solve equations (C2.5, C3.2, E2.5, E3.2)
Chapter 18 Inequalities (C1.5, C2.5, C2.6, C3.2, E1.5, E2.5, E2.6, E3.2)
Chapter 19 Indices and algebra (C2.4, E2.4)
Chapter 20 Sequences (C2.7, E1.16, E2.7)
Chapter 21 Direct and inverse variation (E2.8)

### Learning objectives

**E1.1**
Identify and use:
- natural numbers (0, 1, 2, …)
- integers (positive, zero and negative)
- prime numbers
- square numbers
- cube numbers
- triangle numbers
- common factors
- common multiples
- rational and irrational numbers
- reciprocals

**C1.5      E1.5**
Order quantities by magnitude and demonstrate familiarity with the symbols $=, \neq, >, <, \geq$ and $\leq$

**E1.16**
Use exponential growth and decay

**C2.1      E2.1**
1   Know that letters can be used to represent generalised numbers
2   Substitute numbers into expressions and formulas

**C2.2**   **E2.2**
1 Simplify expressions **involving directed numbers**
2 Expand products of algebraic expressions
3 Factorise by extracting common factors
4 **Factorise expressions of the form:**
   $ax + bx + kay + kby$
   $a^2x^2 - b^2y^2$
   $a^2 + 2ab + b^2$
   $ax^2 + bx + c$
   $ax^3 + bx^2 + cx$

**C2.3**
Simplify algebraic fractions

**E2.3**
1 **Manipulate algebraic fractions**
2 **Factorise and simplify rational expressions**

**C2.4**   **E2.4**
1 Understand, use **and interpret** indices (positive, zero, negative **and fractional**)
2 Understand and use the rules of indices

**C2.5**   **E2.5**
1 Construct simple expressions, **expressions,** equations and formulas
2 Solve linear equations in one unknown
3 **Solve fractional equations with numerical and linear algebraic denominators**
4 Construct and solve simultaneous linear equations in two unknowns
5 **Construct and solve quadratic equations by factorisation, using a graphic display calculator and by use of the quadratic formula**
6 Use a graphic display calculator to solve equations, including those which may be unfamiliar
7 Change the subject of simple formulas **and formulas**

**C2.6**   **E2.6**
1 Represent and interpret inequalities, including on a number line
2 **Construct, solve and interpret linear inequalities**
3 **Solve inequalities using a graphic display calculator**
4 **Represent linear inequalities in two variables graphically**
5 **List inequalities that define a given region**

**C2.7**   **E2.7**
1 Continue a given number sequence or pattern
2 Recognise patterns in sequences, including the term-to-term rule, and relationships between different sequences
3 Find and use the $n$th term of the following sequences:
   (a) linear
   (b) simple quadratic
   (c) simple cubic
3 **Find and use the nth term of sequences**

**E2.8**
1 **Express direct and inverse proportion in algebraic terms and use this form of expression to find unknown quantities**
2 **Identify the best variation model for given data**

**C3.2**   **E3.2**
Use of a graphic display calculator (GDC) to:
(a) sketch the graph of a function
(b) produce a table of values
(c) plot points on GDC
(d) find zeros, local maxima or minima
(e) find the intersection of the graphs of functions
(f) find the vertex of a quadratic

Elements in purple refer to the Extended curriculum only.

# The Persians

Abu Ja'far Muhammad Ibn Musa al-Khwarizmi is called the 'father of algebra'. He was born in Baghdad in AD790. He wrote the book *Hisab al-jabr w'al-muqabala* in AD830 when Baghdad had the greatest university in the world and the greatest mathematicians studied there. He gave us the word 'algebra' and worked on quadratic equations. He also introduced the decimal system from India.

Muhammad al-Karaji was born in North Africa in what is now Morocco. He lived in the eleventh century and worked on the theory of indices. He also worked on an algebraic method of calculating square and cube roots. He may also have travelled to the University of Granada (then part of the Moorish Empire) where works of his can be found in the University library.

*Al-Khwarizmi (790–850)*

The poet Omar Khayyam is known for his long poem *The Rubaiyat*. He was also a fine mathematician working on the binomial theorem. He introduced the symbol 'shay', which became our '$x$'.

# 13 Algebraic representation and manipulation

Algebra is a mathematical language and is at the heart of mathematics. As well as numbers, letters are also used. The letters are used to represent unknown quantities or a variety of possible different values.

Using algebra may at first seem complicated, but as with any language, the more you use it and the more you understand its rules, the easier it becomes.

This topic deals with those rules.

## Expanding brackets

When removing brackets, every term inside the bracket must be multiplied by whatever is outside the bracket.

*As there are two x terms, they can be combined. This is known as simplifying the expression.*

### → Worked examples

a  $3(x + 4) - x$
  $= 3x + 12 - x$
  $= 2x + 12$

b  $5x(2y + 3) - 9xy$
  $= 10xy + 15x - 9xy$
  $= xy + 15x$

*As there are two xy terms, they can be combined. The expression is now simplified.*

c  $2a(3a + 2b - 3c)$
  $= 6a^2 + 4ab - 6ac$

d  $-4p(2p - q + r^2)$
  $= -8p^2 + 4pq - 4pr^2$

e  $-2x^2\left(x + 3y - \frac{1}{x}\right)$
  $= -2x^3 - 6x^2y + 2x$

f  $\frac{-2}{x}\left(-x + 4y + \frac{1}{x}\right)$
  $= 2 - \frac{8y}{x} - \frac{2}{x^2}$

 **Exercise 2.1**

Expand the following:

1  a  $4(x - 3)$    b  $5(2p - 4)$    c  $-6(7x - 4y)$
   d  $3(2a - 3b - 4c)$    e  $-7(2m - 3n)$    f  $-2(8x - 3y)$

2  a  $3x(x - 3y)$    b  $a(a + b + c)$    c  $4m(2m - n)$
   d  $-5a(3a - 4b)$    e  $-4x(-x + y)$    f  $-8p(-3p + q)$

3  a  $-(2x^2 - 3y^2)$    b  $-(-a + b)$    c  $-(-7p + 2q)$
   d  $\frac{1}{2}(6x - 8y + 4z)$    e  $\frac{3}{4}(4x - 2y)$    f  $\frac{1}{5}x(10x - 15y)$

4  a  $3r(4r^2 - 5s + 2t)$    b  $a^2(a + b + c)$    c  $3a^2(2a - 3b)$
   d  $pq(p + q - pq)$    e  $m^2(m - n + nm)$    f  $a^3(a^3 + a^2b)$

 **Exercise 2.2**

Expand and simplify the following:

1  a  $3a - 2(2a + 4)$    b  $8x - 4(x + 5)$
   c  $3(p - 4) - 4$    d  $7(3m - 2n) + 8n$
   e  $6x - 3(2x - 1)$    f  $5p - 3p(p + 2)$

# Simple factorising

2
a $7m(m + 4) + m^2 + 2$
b $3(x - 4) + 2(4 - x)$
c $6(p + 3) - 4(p - 1)$
d $5(m - 8) - 4(m - 7)$
e $3a(a + 2) - 2(a^2 - 1)$
f $7a(b - 2c) - c(2a - 3)$

3
a $\frac{1}{2}(6x + 4) + \frac{1}{3}(3x + 6)$
b $\frac{1}{4}(2x + 6y) + \frac{3}{4}(6x - 4y)$
c $\frac{1}{8}(6x - 12y) + \frac{1}{2}(3x - 2y)$
d $\frac{1}{5}(15x + 10y) + \frac{3}{10}(5x - 5y)$
e $\frac{2}{3}(6x - 9y) + \frac{1}{3}(9x + 6y)$
f $\frac{x}{7}(14x - 21y) - \frac{x}{2}(4x - 6y)$

## Simple factorising

When factorising fully, the **highest common factor** is removed from each of the terms and placed outside the brackets.

*When asked to factorise, the expectation is to factorise fully.*

e.g. The highest common factor of $6x$ and $2x^2$ is $2x$, so factorising $6x + 2x^2$ fully gives $2x(3 + x)$.

### → Worked examples

Factorise the following expressions:

a $10x + 15$
 $= 5(2x + 3)$

b $8p - 6q + 10r$
 $= 2(4p - 3q + 5r)$

c $-2q - 6p + 12$
 $= 2(-q - 3p + 6)$

d $2a^2 + 3ab - 5ac$
 $= a(2a + 3b - 5c)$

e $6ax - 12ay - 18a^2$
 $= 6a(x - 2y - 3a)$

f $3b + 9ba - 6bd$
 $= 3b(1 + 3a - 2d)$

### Exercise 2.3

Factorise the following:

1
a $4x - 6$
b $18 - 12p$
c $6y - 3$
d $4a + 6b$
e $3p - 3q$
f $8m + 12n + 16r$

2
a $3ab + 4ac - 5ad$
b $8pq + 6pr - 4ps$
c $a^2 - ab$
d $4x^2 - 6xy$
e $abc + abd + fab$
f $3m^2 + 9m$

3
a $3pqr - 9pqs$
b $5m^2 - 10mn$
c $8x^2y - 4xy^2$
d $2a^2b^2 - 3b^2c^2$
e $12p - 36$
f $42x - 54$

4
a $18 + 12y$
b $14a - 21b$
c $11x + 11xy$
d $4s - 16t + 20r$
e $5pq - 10qr + 15qs$
f $4xy + 8y^2$

5
a $m^2 + mn$
b $3p^2 - 6pq$
c $pqr + qrs$
d $ab + a^2b + ab^2$
e $3p^3 - 4p^4$
f $7b^3c + b^2c^2$

6
a $m^3 - m^2n + mn^2$
b $4r^3 - 6r^2 + 8r^2s$
c $56x^2y - 28xy^2$
d $72m^2n + 36mn^2 - 18m^2n^2$

# 13 ALGEBRAIC REPRESENTATION AND MANIPULATION

## Substitution

### → Worked examples

Find the value of the following expressions if $a = 3$, $b = 4$, $c = -5$:

a  $2a + 3b - c$
   $= 6 + 12 + 5$
   $= 23$

b  $3a - 4b + 2c$
   $= 9 - 16 - 10$
   $= -17$

c  $-2a + 2b - 3c$
   $= -6 + 8 + 15$
   $= 17$

d  $a^2 + b^2 + c^2$
   $= 9 + 16 + 25$
   $= 50$

e  $3a(2b - 3c)$
   $= 9(8 + 15)$
   $= 9 \times 23$
   $= 207$

f  $-2c(-a + 2b)$
   $= 10(-3 + 8)$
   $= 10 \times 5$
   $= 50$

Graphic display calculators have a large number of memory channels. These can be used to store numbers which can then be substituted into an expression.

Using $a = 3$, $b = 4$, $c = -5$ as above, use your graphic display calculator to work out $2a - 3b + c$.

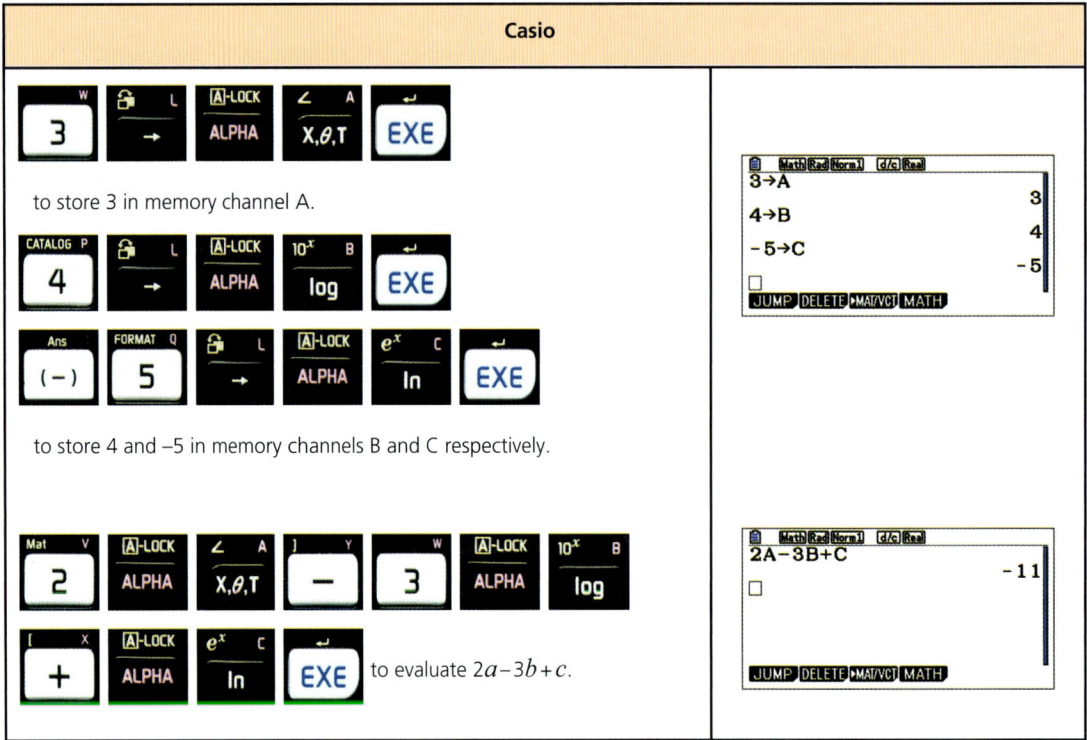

## Changing the subject of a formula

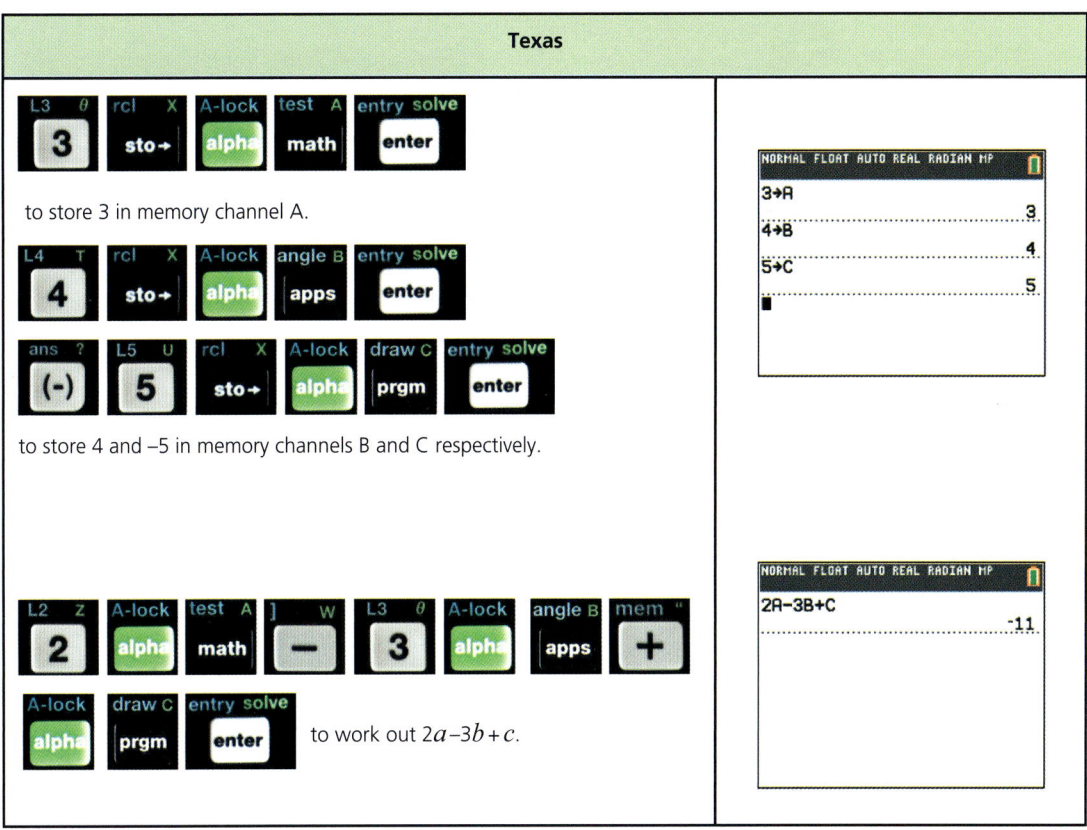

### Exercise 2.4

Find the value of the following expressions if $p = 4$, $q = -2$, $r = 3$ and $s = -5$:

1. a $2p + 4q$
   b $5r - 3s$
   c $3q - 4s$
   d $6p - 8q + 4s$
   e $3r - 3p + 5q$
   f $-p - q + r + s$

2. a $2p - 3q - 4r + s$
   b $3s - 4p + r + q$
   c $p^2 + q^2$
   d $r^2 - s^2$
   e $p(q - r + s)$
   f $r(2p - 3q)$

3. a $2s(3p - 2q)$
   b $pq + rs$
   c $2pr - 3rq$
   d $q^3 - r^2$
   e $s^3 - p^3$
   f $r^4 - q^5$

4. a $-2pqr$
   b $-2p(q + r)$
   c $-2rq + r$
   d $(p + q)(r - s)$
   e $(p + s)(r - q)$
   f $(r + q)(p - s)$

5. a $(2p + 3q)(p - q)$
   b $(q + r)(q - r)$
   c $q^2 - r^2$
   d $p^2 - r^2$
   e $(p + r)(p - r)$
   f $(-s + p)q^2$

## Changing the subject of a formula

In the formula $a = 2b + c$, $a$ is the subject. In order to make either $b$ or $c$ the subject, the formula has to be rearranged.

# 13 ALGEBRAIC REPRESENTATION AND MANIPULATION

> **Worked examples**

Rearrange the following formulas to make the red letter the subject:

**a** $a = 2b + c$
$a - 2b = c$

**b** $2r + p = q$
$p = q - 2r$

**c** $ab = cd$
$\dfrac{ab}{d} = c$

**d** $\dfrac{a}{b} = \dfrac{c}{d}$
$ad = cb$
$d = \dfrac{cb}{a}$

## Exercise 2.5

In the following questions, make the letter in red the subject of the formula:

**1 a** $m + n = r$  **b** $m + n = p$  **c** $2m + n = 3p$
 **d** $3x = 2p + q$  **e** $ab = cd$  **f** $ab = cd$

**2 a** $3xy = 4m$  **b** $7pq = 5r$  **c** $3x = c$
 **d** $3x + 7 = y$  **e** $5y - 9 = 3r$  **f** $5y - 9 = 3x$

**3 a** $6b = 2a - 5$  **b** $6b = 2a - 5$  **c** $3x - 7y = 4z$
 **d** $3x - 7y = 4z$  **e** $3x - 7y = 4z$  **f** $2pr - q = 8$

**4 a** $\dfrac{p}{4} = r$  **b** $\dfrac{4}{p} = 3r$  **c** $\dfrac{1}{5}n = 2p$
 **d** $\dfrac{1}{5}n = 2p$  **e** $p(q + r) = 2t$  **f** $p(q + r) = 2t$

**5 a** $3m - n = rt(p + q)$  **b** $3m - n = rt(p + q)$  **c** $3m - n = rt(p + q)$
 **d** $3m - n = rt(p + q)$  **e** $3m - n = rt(p + q)$  **f** $3m - n = rt(p + q)$

**6 a** $\dfrac{ab}{c} = de$  **b** $\dfrac{ab}{c} = de$  **c** $\dfrac{ab}{c} = de$
 **d** $\dfrac{a+b}{c} = d$  **e** $\dfrac{a}{c} + b = d$  **f** $\dfrac{a}{c} + b = d$

# 14 Further algebraic representation and manipulation

## Further expansion

When multiplying together expressions in brackets, it is necessary to multiply all the terms in one bracket by all the terms in the other bracket.

### → Worked examples

Expand the following:

**a** $(x + 3)(x + 5)$

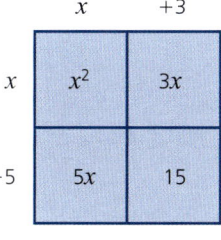

The result is a **trinomial** expression with one variable as it has three terms and only the variable $x$.

$= x^2 + 5x + 3x + 15$
$= x^2 + 8x + 15$

**b** $(x + y)(x + 2y)$

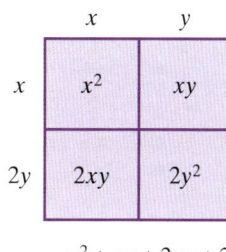

This is a **trinomial** expression with two variables as it has three terms and the two variables $x$ and $y$.

$= x^2 + xy + 2xy + 2y^2$
$= x^2 + 3xy + 2y^2$

**c** $(x + 4)^2$

This is known as a **perfect square** and is the same as the multiplication $(x + 4)(x + 4)$

A multiplication grid is not needed every time as long as you remember to multiply each term in one bracket by each term in the other bracket.

The result is known as a **perfect square trinomial**.

$(x + 4)(x + 4) = x^2 + 4x + 4x + 16$
$\phantom{(x + 4)(x + 4)} = x^2 + 8x + 16$

**d** $(x - 4)^2$

This is also known as a **perfect square** and is the same as the multiplication $(x - 4)(x - 4)$

$(x - 4)(x - 4) = x^2 - 4x - 4x + 16$
$\phantom{(x - 4)(x - 4)} = x^2 - 8x + 16$

 **Exercise 2.6**

Expand and simplify the following:

1  **a** $(x + 2)(x + 3)$ **b** $(x + 3)(x + 4)$
   **c** $(x + 5)(x + 2)$ **d** $(x + 6)(x + 1)$
   **e** $(x - 2)(x + 3)$ **f** $(x + 8)(x - 3)$

2  **a** $(x - 4)(x + 6)$ **b** $(x - 7)(x + 4)$
   **c** $(x + 5)(x - 7)$ **d** $(x + 3)(x - 5)$
   **e** $(x + 1)(x - 3)$ **f** $(x - 7)(x + 9)$

# 14 FURTHER ALGEBRAIC REPRESENTATION AND MANIPULATION

**Exercise 2.6 (cont)**

3  a  $(x-2)(x-3)$      b  $(x-5)(x-2)$
   c  $(x-4)(x-8)$      d  $(x+3)(x+3)$
   e  $(x-3)(x-3)$      f  $(x-7)(x-5)$

4  a  $(x+3)(x-3)$      b  $(x+7)(x-7)$
   c  $(x-8)(x+8)$      d  $(x+y)(x-y)$
   e  $(a+b)(a-b)$      f  $(p-q)(p+q)$

5  a  $(y+x)(2y+3)$      b  $(y-x)(2y+x)$
   c  $(2y+1)(x+y)$      d  $(2p+q)(2p-q)$
   e  $(3m+4n)(2m+5)$      f  $(6a+3b)(2a-b)$

6  a  $(2p-3)(p+8)$      b  $(4p-5)(p+7)$
   c  $(3p-4)(2p+3)$      d  $(4p-5)(3p+7)$
   e  $(6p+2)(3p-1)$      f  $(7p-3)(4p+8)$

7  a  $(2x-1)(2x-1)$      b  $(3x+1)^2$
   c  $(4x-2)^2$      d  $(5x-4)^2$
   e  $(2x+6)^2$      f  $(2x+3)(2x-3)$

8  a  $(3+2x)(3-2x)$      b  $(4x-3)(4x+3)$
   c  $(3+4x)(3-4x)$      d  $(7-5y)(7+5y)$
   e  $(3+2y)(4y-6)$      f  $(7-5y)^2$

9  a  $(x+3)(3x+1)(x+2)$      b  $(2x+4)(2x+1)(x-2)$
   c  $(-x+1)(3x-1)(4x+3)$      d  $(-2x-3)(-x+1)(-x+5)$
   e  $(2x^2-3x+1)(-x+4)$      f  $(4x-1)(-3x^2-3x-2)$

## Further factorisation

### Factorisation by grouping

> **Worked examples**

Factorise the following expressions:

a   $6x + 3 + 2xy + y$

Although there is no common factor to all four terms, pairs of terms can be factorised.

$= 3(2x + 1) + y(2x + 1) = (3 + y)(2x + 1)$

Note that $(2x + 1)$ was a common factor of both terms.

b   $ax + ay - bx - by$
$= a(x + y) - b(x + y)$
$= (a - b)(x + y)$

c   $2x^2 - 3x + 2xy - 3y$
$= x(2x - 3) + y(2x - 3)$
$= (x + y)(2x - 3)$

## Further factorisation

### Exercise 2.7

Factorise the following by grouping:
1. a  $ax + bx + ay + by$
   b  $ax + bx - ay - by$
   c  $3m + 3n + mx + nx$
   d  $4m + mx + 4n + nx$
   e  $3m + mx - 3n - nx$
   f  $6x + xy + 6z + zy$

2. a  $pr - ps + qr - qs$
   b  $pq - 4p + 3q - 12$
   c  $pq + 3q - 4p - 12$
   d  $rs + rt + 2ts + 2t^2$
   e  $rs - 2ts + rt - 2t^2$
   f  $ab - 4cb + ac - 4c^2$

3. a  $xy + 4y + x^2 + 4x$
   b  $x^2 - xy - 2x + 2y$
   c  $ab + 3a - 7b - 21$
   d  $ab - b - a + 1$
   e  $pq - 4p - 4q + 16$
   f  $mn - 5m - 5n + 25$

4. a  $mn - 2m - 3n + 6$
   b  $mn - 2mr - 3rn - 6r^2$
   c  $pr - 4p - 4qr + 16q$
   d  $ab - a - bc + c$
   e  $x^2 - 2xz - 2xy + 4yz$
   f  $2a^2 + 2ab + b^2 + ab$

## Difference of two squares

On expanding: $(x + y)(x - y)$
$= x^2 - xy + xy - y^2$
$= x^2 - y^2$

The reverse is that $x^2 - y^2$ factorises to $(x + y)(x - y)$. $x^2$ and $y^2$ are both square and therefore $x^2 - y^2$ is known as the **difference of two squares**.

### Worked examples

a  $p^2 - q^2$
$= (p + q)(p - q)$

b  $4a^2 - 9b^2$
$= (2a)^2 - (3b)^2$
$= (2a + 3b)(2a - 3b)$

c  $(mn)^2 - 25k^2$
$= (mn)^2 - (5k)^2$
$= (mn + 5k)(mn - 5k)$

d  $4x^2 - (9y)^2$
$= (2x)^2 - (9y)^2$
$= (2x + 9y)(2x - 9y)$

### Exercise 2.8

Factorise the following:
1. a  $a^2 - b^2$
   b  $m^2 - n^2$
   c  $x^2 - 25$
   d  $m^2 - 49$
   e  $81 - x^2$
   f  $100 - y^2$

2. a  $144 - y^2$
   b  $q^2 - 169$
   c  $m^2 - 1$
   d  $1 - t^2$
   e  $4x^2 - y^2$
   f  $25p^2 - 64q^2$

3. a  $9x^2 - 4y^2$
   b  $16p^2 - 36q^2$
   c  $64x^2 - y^2$
   d  $x^2 - 100y^2$
   e  $(pq)^2 - 4p^2$
   f  $(ab)^2 - (cd)^2$

4. a  $m^2n^2 - 9y^2$
   b  $\frac{1}{4}x^2 - \frac{1}{9}y^2$
   c  $p^4 - q^4$
   d  $4m^4 - 36y^4$
   e  $16x^4 - 81y^4$
   f  $(2x)^2 - (3y)^4$

## Evaluation

Once factorised, numerical expressions can be evaluated.

# 14 FURTHER ALGEBRAIC REPRESENTATION AND MANIPULATION

→ **Worked examples**

Find the value of the following expressions:

a $13^2 - 7^2$
$= (13 + 7)(13 - 7)$
$= 20 \times 6$
$= 120$

b $6.25^2 - 3.75^2$
$= (6.25 + 3.75)(6.25 - 3.75)$
$= 10 \times 2.5$
$= 25$

**Exercise 2.9**

By factorising, work out the following:

1. a $8^2 - 2^2$    b $16^2 - 4^2$    c $49^2 - 1$
   d $17^2 - 3^2$    e $88^2 - 12^2$    f $96^2 - 4^2$

2. a $45^2 - 25$    b $99^2 - 1$    c $27^2 - 23^2$
   d $66^2 - 34^2$    e $999^2 - 1$    f $225 - 8^2$

3. a $8.4^2 - 1.6^2$    b $9.3^2 - 0.7^2$    c $42.8^2 - 7.2^2$
   d $\left(8\tfrac{1}{2}\right)^2 - \left(1\tfrac{1}{2}\right)^2$    e $\left(7\tfrac{3}{4}\right)^2 - \left(2\tfrac{1}{4}\right)^2$    f $5.25^2 - 4.75^2$

4. a $8.62^2 - 1.38^2$    b $0.9^2 - 0.1^2$    c $3^4 - 2^4$
   d $2^4 - 1$    e $1111^2 - 111^2$    f $2^8 - 2^5$

## Factorising quadratic and cubic expressions

$x^2 + 5x + 6$ is known as a quadratic expression as the highest power of any of its terms is squared, in this case $x^2$.

It can be factorised by writing it as a product of two brackets.

→ **Worked examples**

a Factorise $x^2 + 5x + 6$.

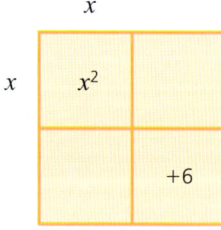 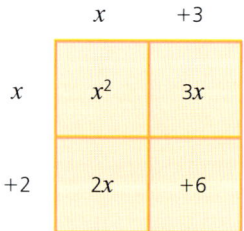

On setting up a $2 \times 2$ grid, some of the information can immediately be entered.

As there is only one term in $x^2$, this can be entered, as can the constant $+6$. The only two values which multiply to give $x^2$ are $x$ and $x$. These too can be entered.

You now need to find two values which multiply to give $+6$ and which add to give $+5x$.

The only two values which satisfy both these conditions are $+3$ and $+2$.
Therefore $x^2 + 5x + 6 = (x + 3)(x + 2)$.

**b** Factorise $x^2 + 2x - 24$.

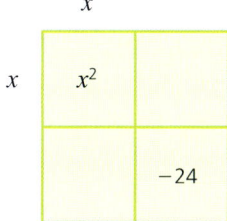

Therefore $x^2 + 2x - 24 = (x + 6)(x - 4)$.

**c** Factorise $2x^2 + 11x + 12$.

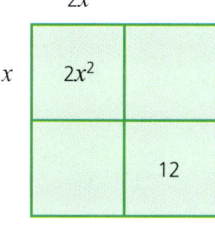 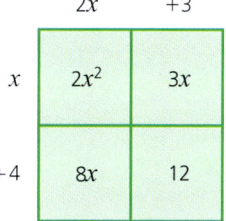

Therefore $2x^2 + 11x + 12 = (2x + 3)(x + 4)$.

**d** Factorise $3x^2 + 7x - 6$.

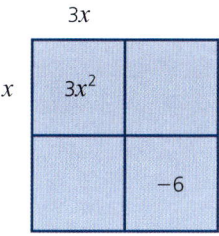 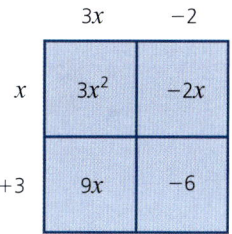

Therefore $3x^2 + 7x - 6 = (3x - 2)(x + 3)$.

**e** Factorise $x^3 + x^2 - 2x$
Note that this is a **cubic** expression and not a quadratic. However, as all three terms are powers of $x$, the first step is to factorise out $x$ from each of the terms as shown:
$x^3 + x^2 - 2x = x(x^2 + x - 2)$
The expression $x^2 + x - 2$ is a quadratic and it can be factorised as $(x + 2)(x - 1)$
Therefore $x^3 + x^2 - 2x = x(x + 2)(x - 1)$

**f** Factorise $2x^3 - 5x^2 - 3x$
This is a **cubic** expression. However, as all three terms are powers of $x$, the first step is to factorise out $x$ from each of the terms as shown:
$2x^3 - 5x^2 - 3x = x(2x^2 - 5x - 3)$
The expression $2x^2 - 5x - 3$ is a quadratic and it can be factorised as $(2x + 1)(x - 3)$
Therefore $2x^3 - 5x^2 - 3x = x(2x + 1)(x - 3)$.

# 14 FURTHER ALGEBRAIC REPRESENTATION AND MANIPULATION

 **Exercise 2.10**

Factorise the following expressions:

1.  a. $x^2 + 7x + 12$
    b. $x^2 + 8x + 12$
    c. $x^2 + 13x + 12$
    d. $x^2 - 7x + 12$
    e. $x^2 - 8x + 12$
    f. $x^2 - 13x + 12$

2.  a. $x^2 + 6x + 5$
    b. $x^2 + 6x + 8$
    c. $x^2 + 6x + 9$
    d. $x^2 + 10x + 25$
    e. $x^2 + 22x + 121$
    f. $x^2 - 13x + 42$

3.  a. $x^2 + 14x + 24$
    b. $x^2 + 11x + 24$
    c. $x^2 - 10x + 24$
    d. $x^2 + 15x + 36$
    e. $x^2 + 20x + 36$
    f. $x^2 - 12x + 36$

4.  a. $x^2 + 2x - 15$
    b. $x^2 - 2x - 15$
    c. $x^2 + x - 12$
    d. $x^2 - x - 12$
    e. $x^2 + 4x - 12$
    f. $x^2 - 15x + 36$

5.  a. $x^2 - 2x - 8$
    b. $x^2 - x - 20$
    c. $x^2 + x - 30$
    d. $x^2 - x - 42$
    e. $x^2 - 2x - 63$
    f. $x^2 + 3x - 54$

6.  a. $2x^2 + 4x + 2$
    b. $2x^2 + 7x + 6$
    c. $2x^2 + x - 6$
    d. $2x^2 - 7x + 6$
    e. $3x^2 + 8x + 4$
    f. $3x^2 + 11x - 4$
    g. $4x^2 + 12x + 9$
    h. $9x^2 - 6x + 1$
    i. $6x^2 - x - 1$

7.  a. $x^3 - x^2 - 2x$
    b. $x^3 - x$
    c. $x^3 + 3x^2 + 2x$
    d. $x^3 + 2x^2 - 3x$
    e. $2x^3 - 10x^2 + 8x$
    f. $x^3 + 10x^2 + 25x$
    g. $3x^3 - 2x^2 - x$
    h. $12x^3 + 10x^2 - 12x$

## Changing the subject of complex formulas

> **Worked examples**
>
> Make the letters in **red** the subject of each formula:
>
> a. $C = 2\pi r$
>
> $\dfrac{C}{2\pi} = r$
>
> b. $A = \pi r^2$
>
> $\dfrac{A}{\pi} = r^2$
>
> $\sqrt{\dfrac{A}{\pi}} = r$
>
> c. $Rx^2 = p$
>
> $x^2 = \dfrac{p}{R}$
>
> $x = \sqrt{\dfrac{p}{R}}$
>
> d. $x^2 + y^2 = h^2$
>
> $y^2 = h^2 - x^2$
>
> $y = \sqrt{h^2 - x^2}$
>
> Note: not $y = h - x$
>
> e. $\sqrt{x} = tv$
>
> $x = t^2v^2$
>
> or $x = (tv)^2$
>
> f. $f = \sqrt{\dfrac{x}{k}}$
>
> $f^2 = \dfrac{x}{k}$
>
> $f^2 k = x$

## Changing the subject of complex formulas

**g** $m = 3a\sqrt{\dfrac{p}{x}}$

Square both sides

$m^2 = \dfrac{9a^2 p}{x}$

$m^2 x = 9a^2 p$

$x = \dfrac{9a^2 p}{m^2}$

**h** $A = \dfrac{y + x}{p + q^2}$

$A(p + q^2) = y + x$

$p + q^2 = \dfrac{y + x}{A}$

$q^2 = \dfrac{y + x}{A} - p$

$q = \sqrt{\dfrac{y + x}{A} - p}$

**i** $y = \dfrac{x + 5}{x}$

$yx = x + 5$

$yx - x = 5$

$x(y - 1) = 5$

$x = \dfrac{5}{y - 1}$

*Note: in this example the subject $x$ appears twice so will need to be combined.*

**j** $2(x + y) = a(y - x)$

$2x + 2y = ay - ax$

$2x + ax = ay - 2y$

$x(2 + a) = y(a - 2)$

$x = \dfrac{y(a - 2)}{2 + a}$

### Exercise 2.11

In the formulas below, make $x$ the subject:

**1 a** $P = 2mx$    **b** $T = 3x^2$    **c** $mx^2 = y^2$
    **d** $x^2 + y^2 = p^2 - q^2$    **e** $m^2 + x^2 = y^2 - n^2$    **f** $p^2 - q^2 = 4x^2 - y^2$

**2 a** $\dfrac{P}{Q} = rx$    **b** $\dfrac{P}{Q} = rx^2$    **c** $\dfrac{P}{Q} = \dfrac{x^2}{r}$
    **d** $\dfrac{m}{n} = \dfrac{1}{x^2}$    **e** $\dfrac{r}{st} = \dfrac{w}{x^2}$    **f** $\dfrac{p + q}{r} = \dfrac{w}{x^2}$

**3 a** $\sqrt{x} = rp$    **b** $\dfrac{mn}{p} = \sqrt{x}$    **c** $g = \sqrt{\dfrac{k}{x}}$
    **d** $r = 2\pi\sqrt{\dfrac{x}{g}}$    **e** $p^2 = \dfrac{4m^2 r}{x}$    **f** $p = 2m\sqrt{\dfrac{r}{x}}$

**4 a** $\dfrac{3x - 2y}{6} = x + y$    **b** $4(x + y) = a(x + y)$    **c** $\dfrac{2 - 3x}{x} = y$

### Exercise 2.12

In the following questions, make the letter in **red** the subject of the formula:

**1 a** $v = u + a\textcolor{red}{t}$    **b** $v^2 = \textcolor{red}{u}^2 + 2as$    **c** $v^2 = u^2 + 2a\textcolor{red}{s}$
    **d** $s = u\textcolor{red}{t} + \tfrac{1}{2}at^2$    **e** $s = ut + \tfrac{1}{2}\textcolor{red}{a}t^2$    **f** $s = ut + \tfrac{1}{2}a\textcolor{red}{t}^2$

**2 a** $A = \pi r\sqrt{\textcolor{red}{s}^2 + t^2}$    **b** $A = \pi r\sqrt{\textcolor{red}{h}^2 + r^2}$    **c** $\dfrac{1}{f} = \dfrac{1}{u} + \dfrac{1}{\textcolor{red}{v}}$
    **d** $\dfrac{1}{f} = \dfrac{1}{\textcolor{red}{u}} + \dfrac{1}{v}$    **e** $t = 2\pi\sqrt{\dfrac{\textcolor{red}{l}}{g}}$    **f** $t = 2\pi\sqrt{\dfrac{l}{\textcolor{red}{g}}}$

# 14 FURTHER ALGEBRAIC REPRESENTATION AND MANIPULATION

> **Note**
>
> Core students need to be able to simplify algebraic fractions involving one step.
>
> For example, $\frac{5x}{10}$ simplifies to $\frac{x}{2}$ and $\frac{y}{y^2}$ simplifies to $\frac{1}{y}$

## Algebraic fractions

### Simplifying algebraic fractions

The rules for fractions involving algebraic terms are the same as those for numeric fractions. However, the actual calculations are often easier when using algebra.

### ➡ Worked examples

a  $\frac{3}{4} \times \frac{5}{7} = \frac{15}{28}$

b  $\frac{a}{c} \times \frac{b}{d} = \frac{ab}{cd}$

c  $\frac{\cancel{2}}{4} \times \frac{5}{\cancel{2}^2} = \frac{5}{8}$

d  $\frac{\cancel{a}}{c} \times \frac{b}{2\cancel{a}} = \frac{b}{2c}$

e  $\frac{m^2}{m} = \frac{m \times \cancel{m}}{\cancel{m}} = m$

f  $\frac{x^5}{x^3} = \frac{\cancel{x} \times \cancel{x} \times \cancel{x} \times x \times x}{\cancel{x} \times \cancel{x} \times \cancel{x}} = x^2$

g  $\frac{\cancel{a}b}{e\cancel{c}} \times \frac{\cancel{c}d}{f\cancel{a}} = \frac{bd}{ef}$

**Exercise 2.13**   Simplify the following algebraic fractions:

1  a  $\frac{x}{y} \times \frac{p}{q}$   b  $\frac{x}{y} \times \frac{q}{x}$   c  $\frac{p}{q} \times \frac{q}{r}$

   d  $\frac{m^3}{m}$   e  $\frac{r^7}{r^2}$   f  $\frac{x^9}{x^3}$

   g  $\frac{2}{b} \times \frac{a}{3}$   h  $\frac{4}{x} \times \frac{y}{2}$   i  $\frac{8}{x} \times \frac{x}{4}$

2  a  $\frac{ab}{c} \times \frac{d}{ab}$   b  $\frac{ab}{c} \times \frac{d}{ac}$   c  $\frac{p^2}{q^2} \times \frac{q^2}{p}$

   d  $\frac{x^2 y^4}{xy^2}$   e  $\frac{a^2 b^3 c^4}{ab^2 c}$   f  $\frac{pq^2 r^4}{p^2 q^3 r}$

3  a  $\frac{4ax}{2ay}$   b  $\frac{12pq^2}{3p}$   c  $\frac{15mn^2}{3mn}$

   d  $\frac{24x^5 y^3}{8x^2 y^2}$   e  $\frac{36p^2 qr}{12pqr}$   f  $\frac{16m^2 n}{24m^3 n^2}$

4  a  $\frac{9y}{2} \times \frac{2x}{3}$   b  $\frac{12x}{7} \times \frac{7}{4x}$   c  $\frac{4x^3}{3y} \times \frac{9y^2}{2x^2}$

5  a  $\frac{2ax}{3bx} \times \frac{4by}{a}$   b  $\frac{3p^2}{2q} \times \frac{5q}{3p}$   c  $\frac{p^2 q}{rs} \times \frac{pr}{q}$

   d  $\frac{a^2 b}{fc^2} \times \frac{cd}{bd} \times \frac{ef^2}{ca^2}$   e  $\frac{2pq^2}{3rs} \times \frac{5m}{4q} \times \frac{8rs}{15p^2}$   f  $\frac{x^4}{wy^2} \times \frac{yz^2}{x^2} \times \frac{wx}{z^3}$

## Addition and subtraction of algebraic fractions

In arithmetic it is easy to add or subtract fractions with the same denominator. It is the same process when dealing with algebraic fractions.

### Worked examples

a $\frac{4}{11} + \frac{3}{11} = \frac{7}{11}$  b $\frac{a}{11} + \frac{b}{11} = \frac{a+b}{11}$  c $\frac{4}{x} + \frac{3}{x} = \frac{7}{x}$

If the denominators are different, the fractions need to be changed to form fractions with the same denominator.

d $\frac{2}{9} + \frac{1}{3} = \frac{2}{9} + \frac{3}{9} = \frac{5}{9}$  e $\frac{a}{9} + \frac{b}{3} = \frac{a}{9} + \frac{3b}{9} = \frac{a+3b}{9}$

f $\frac{4}{5a} + \frac{7}{10a} = \frac{8}{10a} + \frac{7}{10a} = \frac{15}{10a} = \frac{3}{2a}$

Similarly, with subtraction, the denominators need to be the same.

g $\frac{7}{a} - \frac{1}{2a} = \frac{14}{2a} - \frac{1}{2a} = \frac{13}{2a}$  h $\frac{p}{3} - \frac{q}{15} = \frac{5p}{15} - \frac{q}{15} = \frac{5p-q}{15}$

i $\frac{5}{3b} - \frac{8}{9b} = \frac{15}{9b} - \frac{8}{9b} = \frac{7}{9b}$

**Exercise 2.14**   Simplify the following fractions:

1  a $\frac{1}{7} + \frac{3}{7}$   b $\frac{a}{7} + \frac{b}{7}$   c $\frac{5}{13} + \frac{6}{13}$

   d $\frac{c}{13} + \frac{d}{13}$   e $\frac{x}{3} + \frac{y}{3} + \frac{z}{3}$   f $\frac{p^2}{5} + \frac{q^2}{5}$

2  a $\frac{5}{11} - \frac{2}{11}$   b $\frac{c}{11} - \frac{d}{11}$   c $\frac{6}{a} - \frac{2}{a}$

   d $\frac{2a}{3} - \frac{5b}{3}$   e $\frac{2x}{7} - \frac{3y}{7}$   f $\frac{3}{4x} - \frac{5}{4x}$

3  a $\frac{5}{6} - \frac{1}{3}$   b $\frac{5}{2a} - \frac{1}{a}$   c $\frac{2}{3c} + \frac{1}{c}$

   d $\frac{2}{x} + \frac{3}{2x}$   e $\frac{5}{2p} - \frac{1}{p}$   f $\frac{1}{w} - \frac{3}{2w}$

4  a $\frac{p}{4} - \frac{q}{12}$   b $\frac{x}{4} - \frac{y}{2}$   c $\frac{m}{3} - \frac{n}{9}$

   d $\frac{x}{12} - \frac{y}{6}$   e $\frac{r}{2} + \frac{m}{10}$   f $\frac{s}{3} - \frac{t}{15}$

5  a $\frac{3x}{4} - \frac{2x}{12}$   b $\frac{3x}{5} - \frac{2y}{15}$   c $\frac{3m}{7} + \frac{m}{14}$

   d $\frac{4m}{3p} - \frac{3m}{9p}$   e $\frac{4x}{3y} - \frac{5x}{6y}$   f $\frac{3r}{7s} + \frac{2r}{14s}$

Often one denominator is not a multiple of the other. In these cases the **lowest common multiple** of both denominators has to be found.

# 14 FURTHER ALGEBRAIC REPRESENTATION AND MANIPULATION

> **Worked examples**
>
> a $\frac{1}{4}+\frac{1}{3}=\frac{3}{12}+\frac{4}{12}=\frac{7}{12}$
>
> b $\frac{1}{5}+\frac{2}{3}=\frac{3}{15}+\frac{10}{15}=\frac{13}{15}$
>
> c $\frac{a}{3}+\frac{b}{4}=\frac{4a}{12}+\frac{3b}{12}=\frac{4a+3b}{12}$
>
> d $\frac{2a}{3}+\frac{3b}{5}=\frac{10a}{15}+\frac{9b}{15}=\frac{10a+9b}{15}$

 **Exercise 2.15** Simplify the following fractions:

1. a $\frac{1}{2}+\frac{1}{3}$  b $\frac{1}{3}+\frac{1}{5}$  c $\frac{1}{4}+\frac{1}{7}$
   d $\frac{2}{5}+\frac{1}{3}$  e $\frac{1}{4}+\frac{5}{9}$  f $\frac{2}{7}+\frac{2}{5}$

2. a $\frac{a}{2}+\frac{b}{3}$  b $\frac{a}{3}+\frac{b}{5}$  c $\frac{p}{4}+\frac{q}{7}$
   d $\frac{2a}{5}+\frac{b}{3}$  e $\frac{x}{4}+\frac{5y}{9}$  f $\frac{2x}{7}+\frac{2y}{5}$

3. a $\frac{a}{2}-\frac{a}{3}$  b $\frac{a}{3}-\frac{a}{5}$  c $\frac{p}{4}+\frac{p}{7}$
   d $\frac{2a}{5}+\frac{a}{3}$  e $\frac{x}{4}+\frac{5x}{9}$  f $\frac{2x}{7}+\frac{2x}{5}$

4. a $\frac{3m}{5}-\frac{m}{2}$  b $\frac{3r}{5}-\frac{r}{2}$  c $\frac{5x}{4}-\frac{3x}{2}$
   d $\frac{2x}{7}+\frac{3x}{4}$  e $\frac{11x}{2}-\frac{5x}{3}$  f $\frac{2p}{3}-\frac{p}{2}$

5. a $p-\frac{p}{2}$  b $c-\frac{c}{3}$  c $x-\frac{x}{5}$
   d $m-\frac{2m}{3}$  e $q-\frac{4q}{5}$  f $w-\frac{3w}{4}$

6. a $2m-\frac{m}{2}$  b $3m-\frac{2m}{3}$  c $2m-\frac{5m}{2}$
   d $4m-\frac{3m}{2}$  e $2p-\frac{5p}{3}$  f $6q-\frac{6q}{7}$

7. a $p-\frac{p}{r}$  b $\frac{x}{y}+x$  c $m+\frac{m}{n}$
   d $\frac{a}{b}+a$  e $2x-\frac{x}{y}$  f $2p-\frac{3p}{q}$

With more complex algebraic fractions, the method of getting a common denominator is still required.

> **Worked examples**
>
> a $\frac{2}{x}+\frac{1}{x^2}$
>
> $=\frac{2x}{x^2}+\frac{1}{x^2}$
>
> $=\frac{2x+1}{x^2}$
>
> b $\frac{1}{x^3}-\frac{3}{x}$
>
> $=\frac{1}{x^3}-\frac{3x^2}{x^3}$
>
> $=\frac{1-3x^2}{x^3}$

*Algebraic fractions*

c $\dfrac{2}{x+1} + \dfrac{3}{x+2}$

$= \dfrac{2(x+2)}{(x+1)(x+2)} + \dfrac{3(x+1)}{(x+1)(x+2)}$

$= \dfrac{2(x+2) + 3(x+1)}{(x+1)(x+2)}$

$= \dfrac{5x+7}{(x+1)(x+2)}$

d $\dfrac{5}{p+3} - \dfrac{3}{p-5}$

$= \dfrac{5(p-5)}{(p+3)(p-5)} - \dfrac{3(p+3)}{(p+3)(p-5)}$

$= \dfrac{5(p-5) - 3(p+3)}{(p+3)(p-5)}$

$= \dfrac{5p - 25 - 3p - 9}{(p+3)(p-5)}$

$= \dfrac{2p - 34}{(p+3)(p-5)}$

e $\dfrac{x^2 - 2x}{x^2 + x - 6}$

$= \dfrac{x\cancel{(x-2)}}{(x+3)\cancel{(x-2)}}$

$= \dfrac{x}{x+3}$

f $\dfrac{x^2 - 3x}{x^2 + 2x - 15}$

$= \dfrac{x\cancel{(x-3)}}{\cancel{(x-3)}(x+5)}$

$= \dfrac{x}{x+5}$

**Exercise 2.16**  Simplify the following algebraic fractions:

1  a $\dfrac{3}{2x} + \dfrac{2}{x^2}$   b $\dfrac{5}{x^4} - \dfrac{1}{2x^2}$   c $\dfrac{4}{x} + \dfrac{3}{x+1}$

2  a $\dfrac{1}{x+1} + \dfrac{2}{x+2}$   b $\dfrac{3}{m+2} - \dfrac{2}{m-1}$   c $\dfrac{2}{p-3} + \dfrac{1}{p-2}$

   d $\dfrac{3}{w-1} - \dfrac{2}{w+3}$   e $\dfrac{4}{y+4} - \dfrac{1}{y+1}$   f $\dfrac{2}{m-2} - \dfrac{3}{m+3}$

3  a $\dfrac{x(x-4)}{(x-4)(x+2)}$   b $\dfrac{y(y-3)}{(y+4)(y-3)}$   c $\dfrac{(m+2)(m-2)}{(m-2)(m-3)}$

   d $\dfrac{p(p+5)}{(p-5)(p+5)}$   e $\dfrac{m(2m+3)}{(m+4)(2m+3)}$   f $\dfrac{(m+1)(m-1)}{(m+2)(m-1)}$

4  a $\dfrac{x^2 - 5x}{(x+3)(x-5)}$   b $\dfrac{x^2 - 3x}{(x+4)(x-3)}$   c $\dfrac{y^2 - 7y}{(y-7)(y-1)}$

   d $\dfrac{x(x-1)}{x^2 + 2x - 3}$   e $\dfrac{x(x+2)}{x^2 + 4x + 4}$   f $\dfrac{x(x+4)}{x^2 + 5x + 4}$

5  a $\dfrac{x^2 - x}{x^2 - 1}$   b $\dfrac{x^2 + 2x}{x^2 + 5x + 6}$   c $\dfrac{x^2 + 4x}{x^2 + x - 12}$

   d $\dfrac{x^2 - 5x}{x^2 - 3x - 10}$   e $\dfrac{x^2 + 3x}{x^2 - 9}$   f $\dfrac{x^2 - 7x}{x^2 - 49}$

# 15 Linear and simultaneous equations

An equation is formed when the value of an unknown quantity is needed.

## Simple linear equations

### Worked examples

Solve the following linear equations:

*When solving an equation, what is done to one side of the equation must also be done to the other side.*

a  $3x + 8 = 14$
   $3x = 6$
   $x = 2$

b  $12 = 20 + 2x$
   $-8 = 2x$
   $-4 = x$

c  $3(p + 4) = 21$
   $3p + 12 = 21$
   $3p = 9$
   $p = 3$

d  $4(x - 5) = 7(2x - 5)$
   $4x - 20 = 14x - 35$
   $4x + 15 = 14x$
   $15 = 10x$
   $1.5 = x$

### Exercise 2.17

Solve the following linear equations:

1.  a  $3x = 2x - 4$
    b  $5y = 3y + 10$
    c  $2y - 5 = 3y$
    d  $p - 8 = 3p$
    e  $3y - 8 = 2y$
    f  $7x + 11 = 5x$

2.  a  $3x - 9 = 4$
    b  $4 = 3x - 11$
    c  $6x - 15 = 3x + 3$
    d  $4y + 5 = 3y - 3$
    e  $8y - 31 = 13 - 3y$
    f  $4m + 2 = 5m - 8$

3.  a  $7m - 1 = 5m + 1$
    b  $5p - 3 = 3 + 3p$
    c  $12 - 2k = 16 + 2k$
    d  $6x + 9 = 3x - 54$
    e  $8 - 3x = 18 - 8x$
    f  $2 - y = y - 4$

4.  a  $\frac{x}{2} = 3$
    b  $\frac{1}{2}y = 7$
    c  $\frac{x}{4} = 1$
    d  $\frac{1}{4}m = 3$
    e  $7 = \frac{x}{5}$
    f  $4 = \frac{1}{2}p$

5.  a  $\frac{x}{3} - 1 = 4$
    b  $\frac{x}{5} + 2 = 1$
    c  $\frac{2}{3}x = 5$
    d  $\frac{3}{4}x = 6$
    e  $\frac{1}{5}x = \frac{1}{2}$
    f  $\frac{2x}{5} = 4$

6.  a  $\frac{x+1}{2} = 3$
    b  $4 = \frac{x-2}{3}$
    c  $\frac{x-10}{3} = 4$
    d  $8 = \frac{5x-1}{3}$
    e  $\frac{2(x-5)}{3} = 2$
    f  $\frac{3(x-2)}{4} = 4x - 8$

7.  a  $6 = \frac{2(y-1)}{3}$
    b  $2(x+1) = 3(x-5)$
    c  $5(x-4) = 3(x+2)$
    d  $\frac{3+y}{3} = \frac{y+1}{4}$
    e  $\frac{7+2x}{3} = \frac{9x-1}{7}$
    f  $\frac{2x+3}{4} = \frac{4x-2}{6}$

# Constructing simple equations

In many cases, when dealing with the practical applications of mathematics, equations need to be constructed first before they can be solved. Often the information is either given within the context of a problem or in a diagram.

## Worked examples

a   Find the size of each of the angles in the triangle by constructing an equation and solving it to find the value of $x$.

The sum of the angles of a triangle is 180°.

$(x + 30) + (x - 30) + 90 = 180$

$2x + 90 = 180$

$2x = 90$

$x = 45$

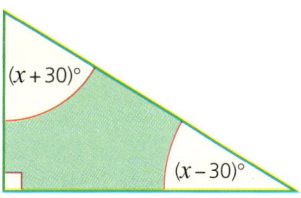

The three angles are therefore: $90°$, $x + 30 = 75°$, $x - 30 = 15°$.

Check: $90° + 75° + 15° = 180°$.

b   Find the size of each of the angles in the quadrilateral by constructing an equation and solving it to find the value of $x$.

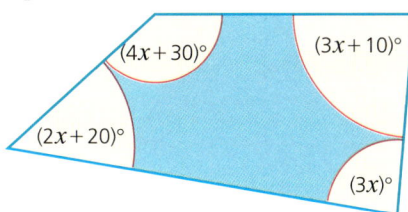

The sum of the angles of a quadrilateral is 360°.

$4x + 30 + 3x + 10 + 3x + 2x + 20 = 360$

$12x + 60 = 360$

$12x = 300$

$x = 25$

The angles are:

$4x + 30 = (4 \times 25) + 30 = 130°$

$3x + 10 = (3 \times 25) + 10 = 85°$

$3x = 3 \times 25 \quad\quad = 75°$

$2x + 20 = (2 \times 25) + 20 = 70°$

Total = 360°

# 15 LINEAR AND SIMULTANEOUS EQUATIONS

c  Construct an equation and solve it to find the value of $x$ in the diagram.

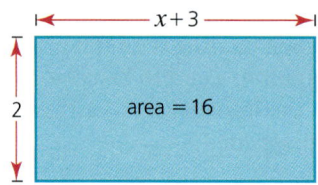

Area of rectangle = base × height

$2(x + 3) = 16$

$2x + 6 = 16$

$2x = 10$

$x = 5$

Check: $2(5 + 3) = 2 \times 8 = 16$

## Exercise 2.18

In Questions 1–3:
i construct an equation in terms of $x$
ii solve the equation
iii calculate the value of each of the angles
iv check your answers.

1
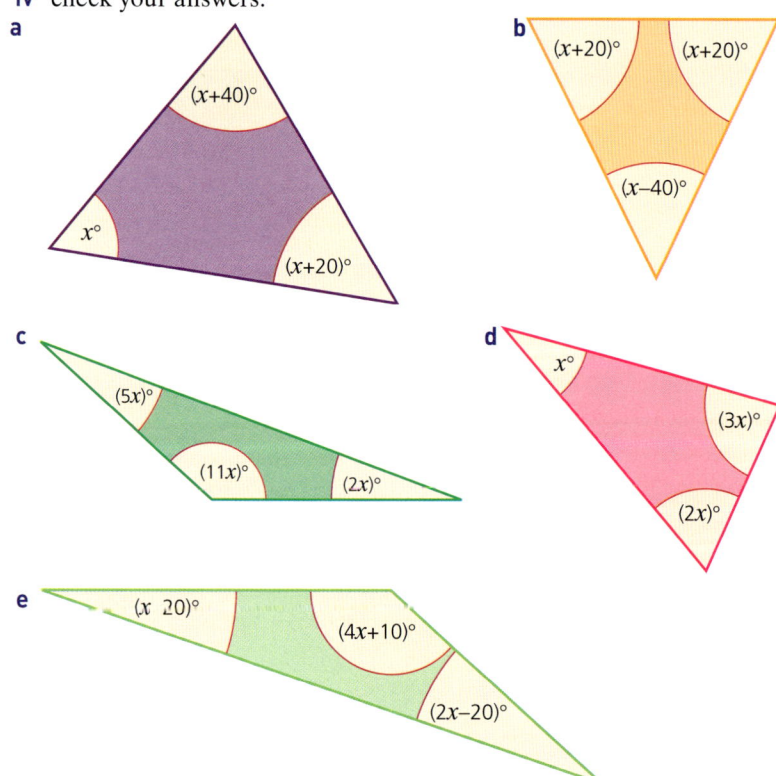

*Constructing simple equations*

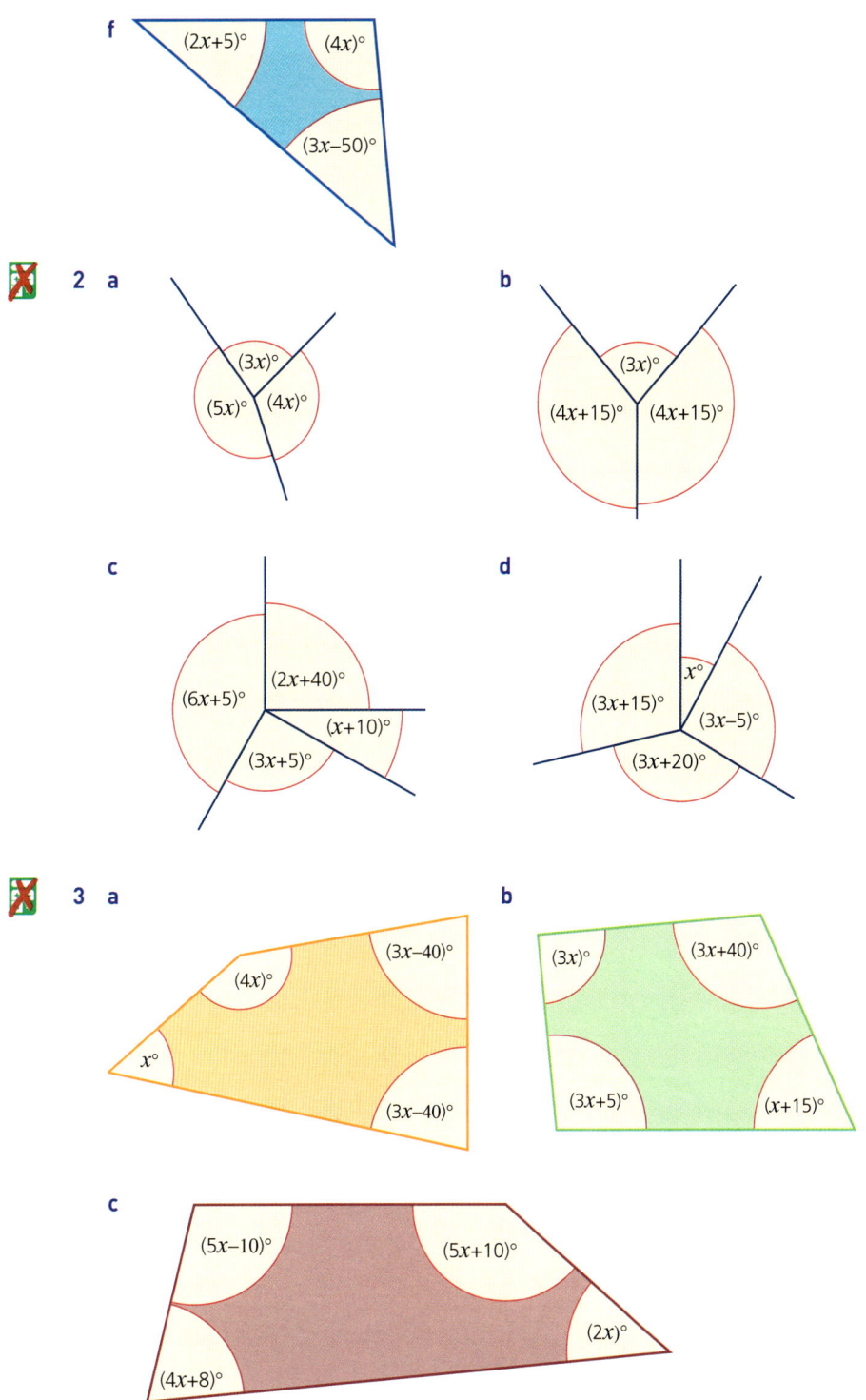

# 15 LINEAR AND SIMULTANEOUS EQUATIONS

**Exercise 2.18 (cont)**

d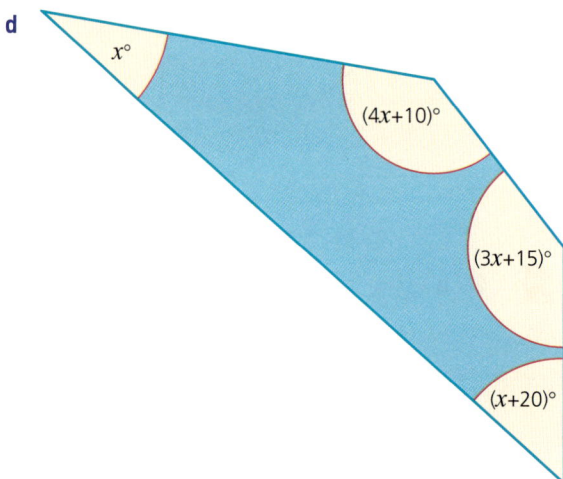

e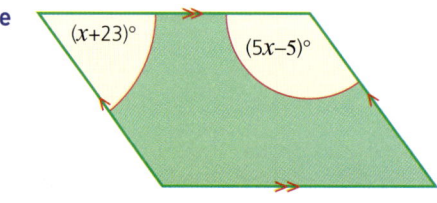

4 By constructing an equation and solving it, find the value of $x$ in each of these isosceles triangles:

a 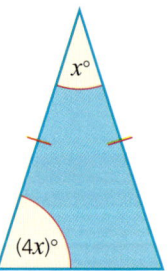   b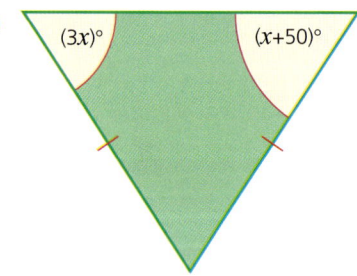

c 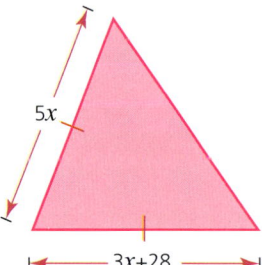   d

*Constructing simple equations*

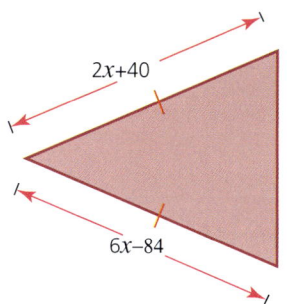
e

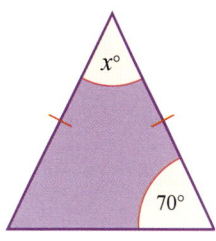
f

**5** Using angle properties, calculate the value of $x$ in each of these questions:

a   b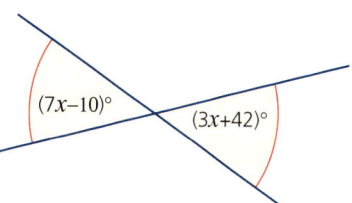

c 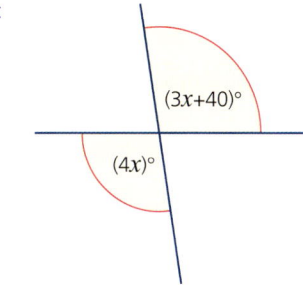  d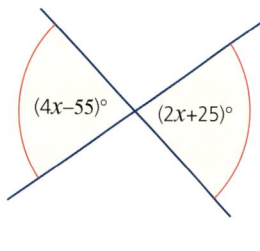

**6** Calculate the value of $x$:

a
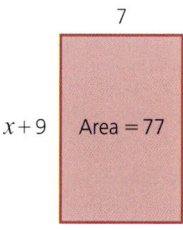

b
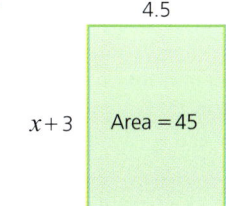

c
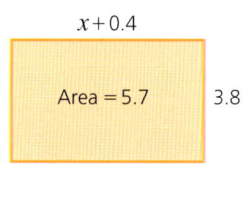

d

# 15 LINEAR AND SIMULTANEOUS EQUATIONS

**Exercise 2.18 (cont)**

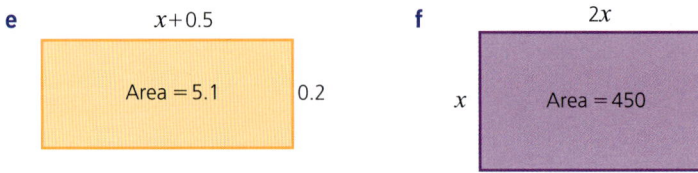

## Constructing formulas

$2x + 3 = 13$ is an equation. It is only true when $x = 5$.

$v = u + at$ is a formula because it describes a relationship between different variables which is true for all values of those variables. $v = u + at$ is a well-known formula for calculating the final velocity ($v$) of an object when its initial velocity ($u$), acceleration ($a$) and time taken ($t$) are known.

*Velocity is speed in a given direction.*

→ **Worked examples**

a  Using the formula $v = u + at$, calculate the final velocity ($v$) of a car in m/s if it started from rest and accelerated at a constant rate of $2\,\text{m/s}^2$ for 10 seconds.

Therefore $u = 0\,\text{m/s}$

$a = 2\,\text{m/s}^2$

$t = 10\,\text{s}$

$v = 0 + (2 \times 10)$

$v = 20\,\text{m/s}$

A formula can be rearranged to make different variables the subject of the formula.

b  Using the formula $v = u + at$ above, calculate the time it took for the car to reach a velocity of $30\,\text{m/s}$.

Rearrange the formula to make $t$ the subject: $t = \dfrac{v-u}{a}$.

Therefore $t = \dfrac{30-0}{2} = 15\,\text{s}$.

It is also important to be able to construct a formula from the information given.

c  Let $T$ be the temperature in °C at the base of a mountain. It is known that the temperature falls by 1 °C for each 200 m climbed.

i  Write a formula linking the temperature $t$ (°C) at any point on the mountain to the height climbed $h$ (m) and the base temperature $T$ (°C).

$t = T - \dfrac{h}{200}$

ii  Calculate the temperature on the mountain at a height of 4000 m if the base temperature is 25 °C.

$t = 25 - \dfrac{4000}{200} = 5$

The temperature at a height of 4000 m is 5 °C.

## Constructing formulas

**iii** At what height does the temperature become 0°C?
Rearrange the formula to make $h$ the subject:
$h = 200(T - t)$
$h = 200(25 - 0) = 5000$
The temperature is 0°C at a height of 5000m.

### Exercise 2.19

1. The area ($A$) of a circle is given by the formula $A = \pi r^2$, where $r$ is the radius.
   a. Calculate the area of a circle if its radius is 6.5 cm.
   b. Rearrange the formula to make $r$ the subject.
   c. Calculate the radius of a circle with an area of 500 cm².

2. The volume ($V$) of a cone is given by the formula $V = \frac{1}{3}\pi r^2 h$, where $r$ is the radius of its base and $h$ its perpendicular height.
   a. Calculate the volume of a cone if $r = 8$ cm and $h = 18$ cm.
   b. A cone has a volume of 600 cm³ and a height of 20 cm. Calculate the radius of its base.

3. To convert a temperature in °F ($F$) to °C ($C$), the following formula is used:
   $C = \frac{5}{9}(F - 32)$
   a. Convert 100°F to degrees Celsius.
   b. Rearrange the formula to make $F$ the subject.
   c. Convert 10°C to degrees Fahrenheit.

4. The distance, $s$ (m), travelled by a moving object can be calculated using the formula $s = ut + \frac{1}{2}at^2$, where $u$ represents the initial velocity (m/s), $t$ is the time taken (s) and $a$ is the acceleration (m/s²).
   a. Calculate the distance a car will travel in 10 seconds if its initial velocity is 5 m/s and its acceleration is 1.5 m/s².
   b. Assuming the car starts from rest, rearrange the formula to make $t$ the subject.
   c. Calculate the time taken for the car to travel 500 m if it starts from rest and $a = 1.5$ m/s².

5. To calculate the area ($A$) of the trapezium shown, the mean length of the parallel sides $a$ and $b$ is calculated and then multiplied by the distance between them, $h$.
   a. Write a formula for calculating the area $A$ of a trapezium.
   b. Calculate the area if $a = 9$ cm, $b = 7$ cm and $h = 2.5$ cm.
   c. If the area $A = 80$ cm², $a = 20$ cm and $b = 12$ cm, calculate the value of $h$.

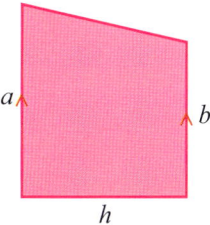

# 15 LINEAR AND SIMULTANEOUS EQUATIONS

**Exercise 2.19 (cont)**

6  The cost $C$ (€) of a taxi ride is €2.50/km plus a fixed charge of €5.
   a  Write a formula for calculating the cost of travelling $n$ kilometres in the taxi.
   b  Rearrange the formula to make $n$ the subject.
   c  A taxi journey cost €80. Calculate the length of the journey.

7  A bakery sells bread rolls for 20 cents each.
   a  Calculate the amount of change in dollars due if a customer buys three rolls and pays with $10.
   b  Write a formula for the amount of change given ($C$), when a $10 note is offered for $x$ bread rolls.
   c  Write a formula for the amount of change given ($C$), when a $10 note is offered for $x$ bread rolls costing $y$ cents each.

8  A coffee shop sells three types of coffee: espresso, latte and cappuccino. The cost of each is €$e$, €$l$ and €$c$ respectively.
   a  Write a formula for the total cost ($T$) of buying $x$ espresso coffees.
   b  Write a formula for the total cost ($T$) of buying $x$ espresso and $y$ latte coffees.
   c  A customer buys $x$ espresso, $y$ latte and $z$ cappuccino coffees. Write a formula to calculate the change due ($C$) if she pays with €20.

9  A dressmaker orders material online. The cost of the material is £15.50 per metre. The cost of delivery is £20 irrespective of the amount bought.
   a  Write a formula to calculate the total cost ($C$) of ordering $n$ metres of material.
   b  i  Rearrange the formula to make n the subject.
      ii  If the total cost of ordering material came to £384.25, calculate the length of material ordered.

10 Metal containers are made by cutting squares from the corners of rectangular pieces of metal and are then folded as shown below.

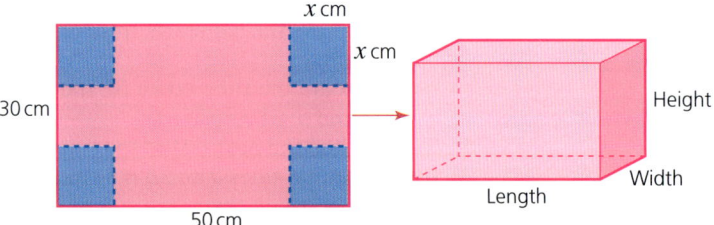

The metal sheet has dimensions 50 cm × 30 cm. Squares of side length $x$ cm are cut from each corner.
   a  Write a formula in terms of $x$ to calculate the length ($L$) of the container.
   b  Write a formula in terms of $x$ to calculate the width ($W$) of the container.
   c  Write a formula in terms of $x$ to calculate the height ($H$) of the container.
   d  Write a formula in terms of $x$ to calculate the volume ($V$) of the container.
   e  Calculate the volume of the container if a square of side length 12 cm is cut from each corner of the metal sheet.

# Simultaneous equations

When the values of two unknowns are needed, two equations need to be formed and solved. The process of solving two equations and finding a common solution is known as solving equations simultaneously.

The two most common ways of solving simultaneous equations algebraically are by **elimination** and by **substitution**.

## By elimination

The aim of this method is to eliminate one of the unknowns by either adding or subtracting the two equations.

### → Worked examples

Solve the following simultaneous linear equations by finding the values of $x$ and $y$ which satisfy both equations:

a  $3x + y = 9$    (1)
   $5x - y = 7$    (2)

By adding equations (1) + (2), you eliminate the variable $y$:
$8x = 16$
$x = 2$

To find the value of $y$, substitute $x = 2$ into either equation (1) or (2). Substituting $x = 2$ into equation (1):

$3x + y = 9$
$6 + y = 9$
$y = 3$

To check that the solution is correct, the values of $x$ and $y$ are substituted into equation (2). If it is correct then the left-hand side of the equation will equal the right-hand side.

$5x - y = 7$
$5 \times 2 - 3 = 7$
$10 - 3 = 7$
$7 = 7$

b  $4x + y = 23$    (1)
   $x + y = 8$     (2)

By subtracting the equations, i.e. (1) − (2), you eliminate the variable $y$:
$3x = 15$
$x = 5$

By substituting $x = 5$ into equation (2), $y$ can be calculated:
$x + y = 8$
$5 + y = 8$
$y = 3$

Check by substituting both values into equation (1):
$4x + y = 23$
$4 \times 5 + 3 = 23$
$20 + 3 = 23$
$23 = 23$

## 15 LINEAR AND SIMULTANEOUS EQUATIONS

### By substitution

The same equations can also be solved by the method known as **substitution**.

> **Worked examples**
>
> a  $3x + y = 9$     (1)
>    $5x - y = 7$     (2)
>
> Equation (2) can be rearranged to give: $y = 5x - 7$
>
> This can now be substituted into equation (1):
>
> $3x + (5x - 7) = 9$
> $3x + 5x - 7 = 9$
> $8x - 7 = 9$
> $8x = 16$
> $x = 2$
>
> To find the value of $y$, $x = 2$ is substituted into either equation (1) or (2) as before, giving $y = 3$
>
> b  $4x + y = 23$     (1)
>    $x + y = 8$     (2)
>
> Equation (2) can be rearranged to give $y = 8 - x$. This can be substituted into equation (1):
>
> $4x + (8 - x) = 23$
> $4x + 8 - x = 23$
> $3x + 8 = 23$
> $3x = 15$
> $x = 5$
>
> $y$ can be found as before, giving a result of $y = 3$

Graphic display calculators can solve simultaneous equations both graphically and algebraically. The following instructions are for the graphical method.

> **Note**
>
> Using a graphic display calculator to work through the algebraic method to solve simultaneous equations can be helpful, but it is important to note that calculators with symbolic algebraic logic are not permitted in the examinations.

## Simultaneous equations

Graphically, simultaneous equations are solved by plotting both lines on the same pair of axes and finding the coordinates of the point where the two lines intersect. Your graphic display calculator can solve simultaneous equations graphically.

For example, to solve the simultaneous equations $3x + y = 9$ and $5x - y = 7$ graphically, first rearrange each equation into the form $y = ax + b$.

$3x + y = 9$ can be rearranged to $y = -3x + 9$

$5x - y = 7$ can be rearranged to $y = 5x - 7$

Use the following steps on your graphic display calculator:

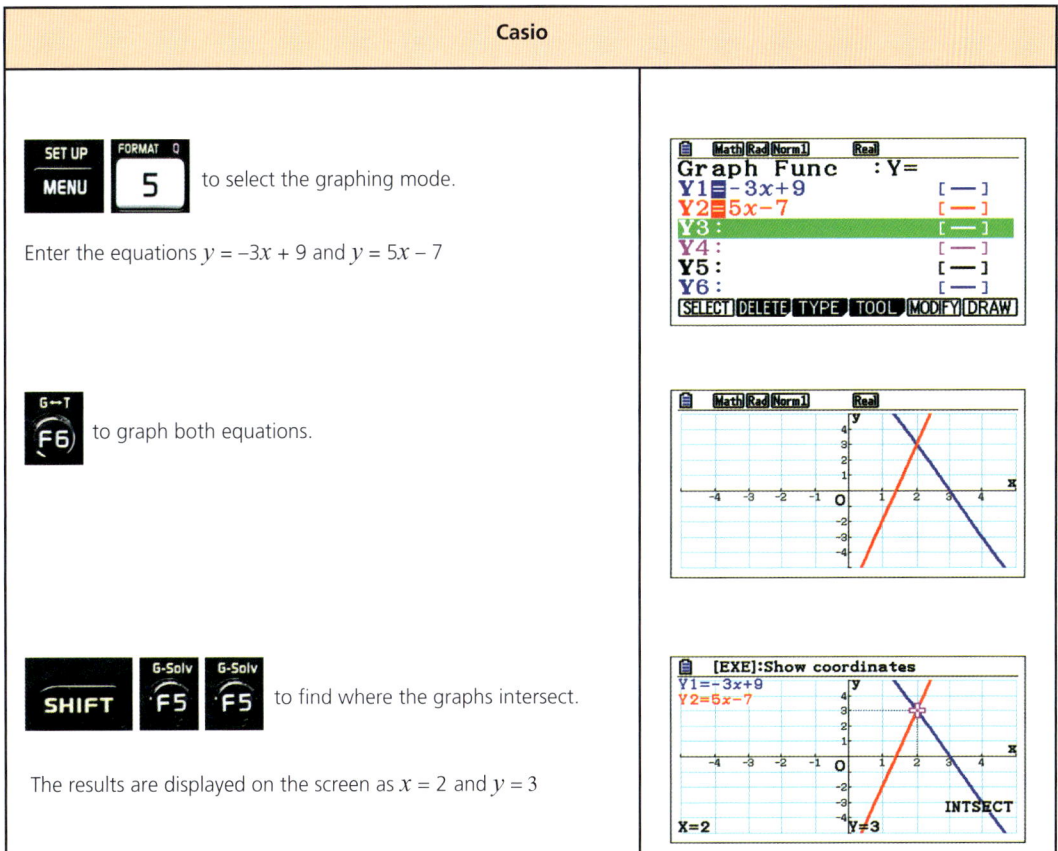

## 15 LINEAR AND SIMULTANEOUS EQUATIONS

| Texas |
|---|

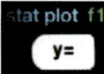 and enter the equations $y = -3x + 9$ and $y = 5x - 7$

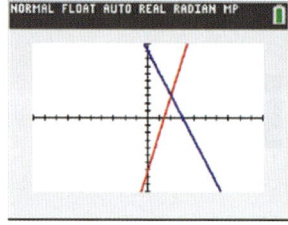

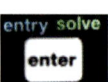 to graph both equations.

 followed by option '5' to select 'intersect'.

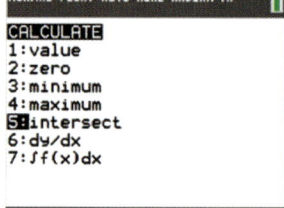

The screen cursor will be flashing on the first line.

enter to select it.

The screen cursor will now be flashing on the second line.

enter to select it.

The calculator will give the coordinates of the point of intersection as $x = 2$ and $y = 3$

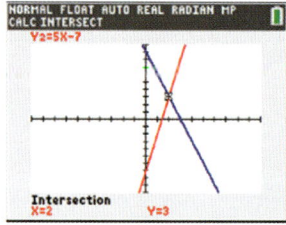

## Simultaneous equations

**Exercise 2.20** Solve the following simultaneous equations.

1. 
   a. $x + y = 6$
      $x - y = 2$
   b. $x + y = 11$
      $x - y - 1 = 0$
   c. $x + y = 5$
      $x - y = 7$
   d. $2x + y = 12$
      $2x - y = 8$
   e. $3x + y = 17$
      $3x - y = 13$
   f. $5x + y = 29$
      $5x - y = 11$

2. 
   a. $3x + 2y = 13$
      $4x = 2y + 8$
   b. $6x + 5y = 62$
      $4x - 5y = 8$
   c. $x + 2y = 3$
      $8x - 2y = 6$
   d. $9x + 3y = 24$
      $x - 3y = -14$
   e. $7x - y = -3$
      $4x + y = 14$
   f. $3x = 5y + 14$
      $6x + 5y = 58$

3. 
   a. $2x + y = 14$
      $x + y = 9$
   b. $5x + 3y = 29$
      $x + 3y = 13$
   c. $4x + 2y = 50$
      $x + 2y = 20$
   d. $x + y = 10$
      $3x = -y + 22$
   e. $2x + 5y = 28$
      $4x + 5y = 36$
   f. $x + 6y = -2$
      $3x + 6y = 18$

4. 
   a. $x - y = 1$
      $2x - y = 6$
   b. $3x - 2y = 8$
      $2x - 2y = 4$
   c. $7x - 3y = 26$
      $2x - 3y = 1$
   d. $x = y + 7$
      $3x - y = 17$
   e. $8x - 2y = -2$
      $3x - 2y = -7$
   f. $4x - y = -9$
      $7x - y = -18$

5. 
   a. $x + y = -7$
      $x - y = -3$
   b. $2x + 3y = -18$
      $2x = 3y + 6$
   c. $5x - 3y = 9$
      $2x + 3y = 19$
   d. $7x + 4y = 42$
      $9x - 4y = -10$
   e. $4x - 4y = 0$
      $8x + 4y = 12$
   f. $x - 3y = -25$
      $5x - 3y = -17$

6. 
   a. $2x + 3y = 13$
      $2x - 4y + 8 = 0$
   b. $2x + 4y = 50$
      $2x + y = 20$
   c. $x + y = 10$
      $3y = 22 - x$
   d. $5x + 2y = 28$
      $5x + 4y = 36$
   e. $2x - 8y = 2$
      $2x - 3y = 7$
   f. $x - 4y = 9$
      $x - 7y = 18$

7. 
   a. $-4x = 4y$
      $4x - 8y = 12$
   b. $3x = 19 + 2y$
      $-3x + 5y = 5$
   c. $3x + 2y = 12$
      $-3x + 9y = -12$
   d. $3x + 5y = 29$
      $3x + y = 13$
   e. $-5x + 3y = 14$
      $5x + 6y = 58$
   f. $-2x + 8y = 6$
      $2x = 3 - y$

If neither $x$ nor $y$ can be eliminated by simply adding or subtracting the two equations, then it is necessary to multiply one or both of the equations. The equations are multiplied by a number in order to make the coefficients of $x$ (or $y$) numerically equal.

## → Worked examples

a. $3x + 2y = 22$   (1)
   $x + y = 9$   (2)

Multiply equation (2) by 2 to be able to eliminate $y$:

$3x + 2y = 22$   (1)
$2x + 2y = 18$   (3)

Subtract (3) from (1) to eliminate the variable $y$:
$x = 4$

# 15 LINEAR AND SIMULTANEOUS EQUATIONS

Substitute $x = 4$ into equation (2):

$x + y = 9$
$4 + y = 9$
$\quad y = 5$

Check by substituting both values into equation (1):

$\quad 3x + 2y = 22$
$3 \times 4 + 2 \times 5 = 22$
$\quad 12 + 10 = 22$
$\quad\quad 22 = 22$

**b** $\quad 5x - 3y = 1 \quad (1)$
$\quad 3x + 4y = 18 \quad (2)$

Multiply equation (1) by 4 and equation (2) by 3 in order to eliminate the variable $y$:

$20x - 12y = 4 \quad (3)$
$9x + 12y = 54 \quad (4)$

Add equations (3) and (4) to eliminate the variable $y$:

$29x = 58$
$\quad x = 2$

Substitute $x = 2$ into equation (2):

$3x + 4y = 18$
$6 + 4y = 18$
$\quad 4y = 12$
$\quad\quad y = 3$

Check by substituting both values into equation (1):

$\quad 5x - 3y = 1$
$5 \times 2 - 3 \times 3 = 1$
$\quad 10 - 9 = 1$
$\quad\quad 1 = 1$

---

### Exercise 2.21

Solve the following simultaneous equations:

**1**
**a** $2x + y = 7$
$\quad 3x + 2y = 12$
**b** $5x + 4y = 21$
$\quad x + 2y = 9$
**c** $x + y = 7$
$\quad 3x + 4y = 23$
**d** $2x - 3y = -3$
$\quad 3x + 2y = 15$
**e** $4x = 4y + 8$
$\quad x + 3y = 10$
**f** $x + 5y = 11$
$\quad 2x - 2y = 10$

**2**
**a** $x + y = 5$
$\quad 3x - 2y + 5 = 0$
**b** $2x - 2y = 6$
$\quad x - 5y = -5$
**c** $2x + 3y = 15$
$\quad 2y = 15 - 3x$
**d** $x - 6y = 0$
$\quad 3x - 3y = 15$
**e** $2x - 5y = -11$
$\quad 3x + 4y = 18$
**f** $x + y = 5$
$\quad 2x - 2y = -2$

**3**
**a** $3y = 9 + 2x$
$\quad 3x + 2y = 6$
**b** $x + 4y = 13$
$\quad 3x - 3y = 9$
**c** $2x = 3y - 19$
$\quad 3x + 2y = 17$
**d** $2x - 5y = -8$
$\quad -3x - 2y = -26$
**e** $5x - 2y = 0$
$\quad 2x + 5y = 29$
**f** $8y = 3 - x$
$\quad 3x - 2y = 9$

**4**
**a** $4x + 2y = 5$
$\quad 3x + 6y = 6$
**b** $4x + y = 14$
$\quad 6x - 3y = 3$
**c** $10x - y = -2$
$\quad -15x + 3y = 9$
**d** $-2y = 0.5 - 2x$
$\quad 6x + 3y = 6$
**e** $x + 3y = 6$
$\quad 2x - 9y = 7$
**f** $5x - 3y = -0.5$
$\quad 3x + 2y = 3.5$

# Constructing more complex equations

Earlier in this section you looked at some simple examples of constructing and solving equations when you were given geometrical diagrams. Here this work is extended with more complicated formulas and equations.

## → Worked examples

Construct and solve the equations below:

a  Using the shape below, construct an equation for the perimeter in terms of $x$. Find the value of $x$ by solving the equation.

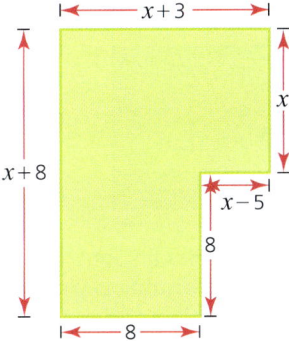

perimeter = 54

$$x + 3 + x + x - 5 + 8 + 8 + x + 8 = 54$$
$$4x + 22 = 54$$
$$4x = 32$$
$$x = 8$$

b  A number is doubled, 5 is subtracted from the result, and the total is 17. Find the number.
Let $x$ be the unknown number.
$$2x - 5 = 17$$
$$2x = 22$$
$$x = 11$$

c  3 is added to a number. The result is multiplied by 8. If the answer is 64, calculate the value of the original number.
Let $x$ be the unknown number.
$$8(x + 3) = 64$$
$$8x + 24 = 64$$
$$8x = 40$$
$$x = 5$$
or $8(x + 3) = 64$
$$x + 3 = 8$$
$$x = 5$$

# 15 LINEAR AND SIMULTANEOUS EQUATIONS

 **Exercise 2.22**

1. The sum of two numbers is 17 and their difference is 3. Find the two numbers by forming two equations and solving them simultaneously.
2. The difference between two numbers is 7. If their sum is 25, find the two numbers by forming two equations and solving them simultaneously.
3. Find the values of $x$ and $y$:

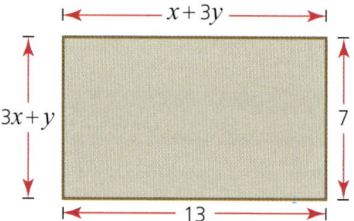

4. Find the values of $x$ and $y$:

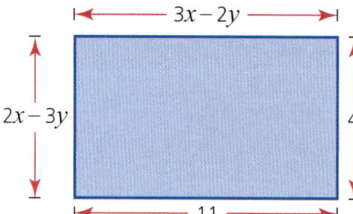

5. A man's age is three times his son's age. Ten years ago he was five times his son's age. By forming two equations and solving them simultaneously, find both of their ages.
6. A grandfather is ten times older than his granddaughter. He is also 54 years older than her. How old is each of them?

**Exercise 2.23**

1. Calculate the value of $x$:

   a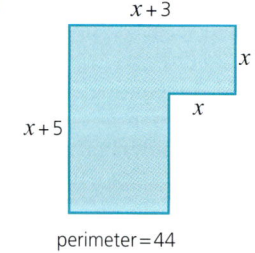
   
   perimeter = 44

   b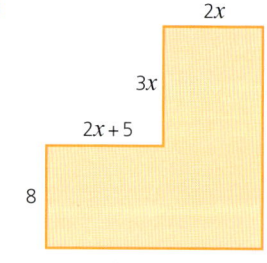
   
   perimeter = 68

   c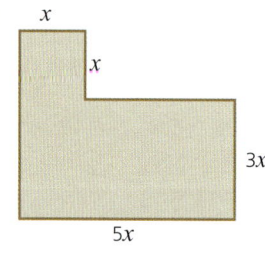

   perimeter = 108

   d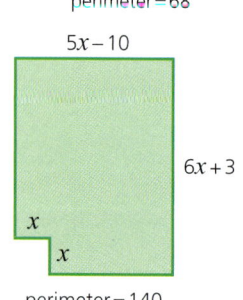
   
   perimeter = 140

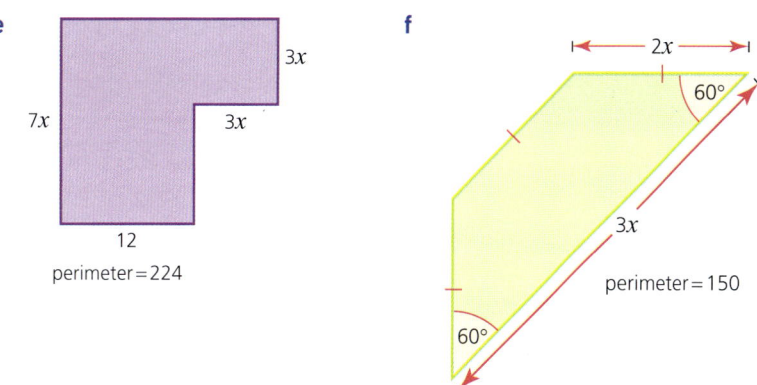

2  a  A number is trebled and then 7 is added to it. If the total is 28, find the number.
   b  Multiply a number by 4 and then add 5 to it. If the total is 29, find the number.
   c  If 31 is the result of adding 1 to 5 times a number, find the number.
   d  Double a number and then subtract 9. If the answer is 11, what is the number?
   e  If 9 is the result of subtracting 12 from 7 times a number, find the number.

3  a  Add 3 to a number and then double the result. If the total is 22, find the number.
   b  27 is the answer when you add 4 to a number and then treble it. What is the number?
   c  Subtract 1 from a number and multiply the result by 5. If the answer is 35, what is the number?
   d  Add 3 to a number. If the result of multiplying this total by 7 is 63, find the number.
   e  Add 3 to a number. Quadruple the result. If the answer is 36, what is the number?

4  a  Gabriella is $x$ years old. Her brother is 8 years older and her sister is 12 years younger than she is. If their total age is 50 years, how old is each of them?
   b  A series of Mathematics textbooks consists of four volumes. The first volume has $x$ pages, the second 54 pages more. The third and fourth volume each have 32 pages more than the second. If the total number of pages in all four volumes is 866, calculate the number of pages in each of the volumes.
   c  The five interior angles (in °) of a pentagon are $x$, $x + 30$, $2x$, $2x + 40$ and $3x + 20$. The sum of the interior angles of a pentagon is 540°. Calculate the size of each of the angles.
   d  A hexagon consists of three interior angles of equal size and a further three which are double this size. The sum of all six angles is 720°. Calculate the size of each of the angles.
   e  Four of the exterior angles of an octagon are the same size. The other four are twice as big. If the sum of the exterior angles is 360°, calculate the size of the interior angles.

# 16 Solving quadratic equations

An equation of the form $y = ax^2 + bx + c$, in which the highest power of the variable $x$ is $x^2$, is known as a **quadratic equation**. The following are all types of quadratic equations:

$$y = x^2 + 2x - 4 \qquad y = -3x^2 + x + 2 \qquad y = x^2 \qquad y = \tfrac{1}{2}x^2 + 2$$

There are a number of ways to solve quadratic equations and the most efficient method to use is largely dependent on the type of quadratic equation given. The main methods are explained later in this section; however, you can also use your graphic display calculator to solve quadratic equations and therefore check your answers.
Note: You should always show some working as to how you solve a quadratic equation; therefore you should use your calculator only as a tool for checking your answer.

## Solving quadratic equations by factorising

$x^2 - 3x - 10 = 0$ is a quadratic equation which when factorised can be written as $(x - 5)(x + 2) = 0$.
Therefore either $x - 5 = 0$ or $x + 2 = 0$ since, if two things multiply to make zero, then one of them must be zero.

$\quad x - 5 = 0 \quad$ or $\quad x + 2 = 0$
$\quad x = 5 \qquad\qquad\quad x = -2$

### ➡ Worked examples

Solve the following equations to give two solutions for $x$:

a $\quad x^2 - x - 12 = 0$
$\quad (x - 4)(x + 3) = 0$
$\quad$ so either $x - 4 = 0 \quad$ or $\quad x + 3 = 0$
$\qquad\qquad\qquad x = 4 \qquad\qquad\qquad x = -3$

b $\quad x^2 + 2x = 24$
$\quad$ This becomes $x^2 + 2x - 24 = 0$
$\quad (x + 6)(x - 4) = 0$
$\quad$ so either $x + 6 = 0 \quad$ or $\quad x - 4 = 0$
$\qquad\qquad\qquad x = -6 \qquad\qquad\quad x = 4$

c $\quad x^2 - 6x = 0$
$\quad x(x - 6) = 0$
$\quad$ so either $x = 0 \quad$ or $\quad x - 6 = 0$
$\qquad\qquad\qquad\qquad\qquad\quad x = 6$

## Solving quadratic equations by factorising

    **d** $x^2 - 4 = 0$
    $(x - 2)(x + 2) = 0$
    so either $x - 2 = 0$   or   $x + 2 = 0$
              $x = 2$            $x = -2$

###  Exercise 2.24

Solve the following quadratic equations by factorising. Check your solutions using a calculator.

**1**  **a** $x^2 + 7x + 12 = 0$   **b** $x^2 + 8x + 12 = 0$   **c** $x^2 + 13x + 12 = 0$
    **d** $x^2 - 7x + 10 = 0$   **e** $x^2 - 5x + 6 = 0$   **f** $x^2 - 6x + 8 = 0$

**2**  **a** $x^2 + 3x - 10 = 0$   **b** $x^2 - 3x - 10 = 0$   **c** $x^2 + 5x - 14 = 0$
    **d** $x^2 - 5x - 14 = 0$   **e** $x^2 + 2x - 15 = 0$   **f** $x^2 - 2x - 15 = 0$

**3**  **a** $x^2 + 5x = -6$   **b** $x^2 + 6x = -9$   **c** $x^2 + 11x = -24$
    **d** $x^2 - 10x = -24$   **e** $x^2 + x = 12$   **f** $x^2 - 4x = 12$

**4**  **a** $x^2 - 2x = 8$   **b** $x^2 - x = 20$   **c** $x^2 + x = 30$
    **d** $x^2 - x = 42$   **e** $x^2 - 2x = 63$   **f** $x^2 + 3x = 54$

### Exercise 2.25

Solve the following quadratic equations. Check your solutions using a calculator.

**1**  **a** $x^2 - 9 = 0$   **b** $x^2 - 16 = 0$   **c** $x^2 = 25$
    **d** $x^2 = 121$   **e** $x^2 - 144 = 0$   **f** $x^2 - 220 = 5$

**2**  **a** $4x^2 - 25 = 0$   **b** $9x^2 - 36 = 0$   **c** $25x^2 = 64$
    **d** $x^2 = \frac{1}{4}$   **e** $x^2 - \frac{1}{9} = 0$   **f** $16x^2 - \frac{1}{25} = 0$

**3**  **a** $x^2 + 5x + 4 = 0$   **b** $x^2 + 7x + 10 = 0$   **c** $x^2 + 6x + 8 = 0$
    **d** $x^2 - 6x + 8 = 0$   **e** $x^2 - 7x + 10 = 0$   **f** $x^2 + 2x - 8 = 0$

**4**  **a** $x^2 - 3x - 10 = 0$   **b** $x^2 + 3x - 10 = 0$   **c** $x^2 - 3x - 18 = 0$
    **d** $x^2 + 3x - 18 = 0$   **e** $x^2 - 2x - 24 = 0$   **f** $x^2 - 2x - 48 = 0$

**5**  **a** $x^2 + x = 12$   **b** $x^2 + 8x = -12$   **c** $x^2 + 5x = 36$
    **d** $x^2 + 2x = -1$   **e** $x^2 + 4x = -4$   **f** $x^2 + 17x = -72$

**6**  **a** $x^2 - 8x = 0$   **b** $x^2 - 7x = 0$   **c** $x^2 + 3x = 0$
    **d** $x^2 + 4x = 0$   **e** $x^2 - 9x = 0$   **f** $4x^2 - 16x = 0$

**7**  **a** $2x^2 + 5x + 3 = 0$   **b** $2x^2 - 3x - 5 = 0$   **c** $3x^2 + 2x - 1 = 0$
    **d** $2x^2 + 11x + 5 = 0$   **e** $2x^2 - 13x + 15 = 0$   **f** $12x^2 + 10x - 8 = 0$

**8**  **a** $x^2 + 12x = 0$   **b** $x^2 + 12x + 27 = 0$   **c** $x^2 + 4x = 32$
    **d** $x^2 + 5x = 14$   **e** $2x^2 = 72$   **f** $3x^2 - 12 = 288$

### Exercise 2.26

In the following questions, construct equations from the information given and then solve them to find the unknown.

**1** When a number $x$ is added to its square, the total is 12. Find two possible values for $x$.

**2** A number $x$ is equal to its own square minus 42. Find two possible values for $x$.

**3** If the area of the rectangle on the right is 10 cm², calculate the only possible value for $x$.

# 16 SOLVING QUADRATIC EQUATIONS

**Exercise 2.26 (cont)**

4  If the area of the rectangle is 52 cm², calculate the only possible value for $x$.

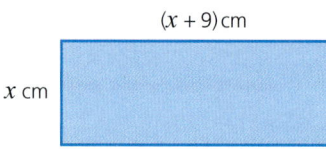

5  A triangle has a base length of $2x$ cm and a height of $(x - 3)$ cm. If its area is 18 cm², calculate its height and base length.
6  A triangle has a base length of $(x - 8)$ cm and a height of $2x$ cm. If its area is 20 cm², calculate its height and base length.
7  A right-angled triangle has a base length of $x$ cm and a height of $(x - 1)$ cm. If its area is 15 cm², calculate the base length and height.
8  A rectangular garden has a square flowerbed of side length $x$ m in one of its corners. The remainder of the garden consists of lawn and has dimensions as shown. If the total area of the lawn is 50 m²:

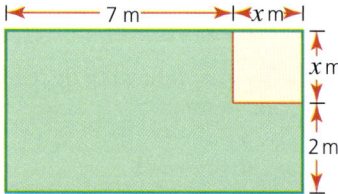

  a  form an equation in terms of $x$
  b  solve the equation
  c  calculate the length and width of the whole garden.

## The quadratic formula

In general a quadratic equation takes the form $ax^2 + bx + c = 0$ where $a$, $b$ and $c$ are integers. Quadratic equations can be solved by the use of the quadratic formula, which states that:

$$x = \frac{-b \pm \sqrt{b^2 - 4ac}}{2a}$$

*The quadratic formula will be given to you in the examination formula book.*

### → Worked examples

a  Solve the quadratic equation $x^2 + 7x + 3 = 0$.
   $a = 1$, $b = 7$ and $c = 3$
   Substituting these values into the quadratic formula gives:
   $x = \dfrac{-7 \pm \sqrt{7^2 - 4 \times 1 \times 3}}{2 \times 1}$
   $x = \dfrac{-7 \pm \sqrt{49 - 12}}{2}$
   $x = \dfrac{-7 \pm \sqrt{37}}{2}$
   Therefore $x = \dfrac{-7 + 6.08}{2}$  or  $x = \dfrac{-7 - 6.08}{2}$
   $x = -0.46$ (2 d.p.)  or  $x = -6.54$ (2 d.p.)

*The answers −0.46 and −6.54 are only approximations correct to 2 d.p.*

*The answer $\frac{-7 \pm \sqrt{37}}{2}$ are the exact solutions given in surd form.*

**b** Solve the quadratic equation $x^2 - 4x - 2 = 0$.
$a = 1, b = -4$ and $c = -2$

Substituting these values into the quadratic formula gives:

$$x = \frac{-(-4) \pm \sqrt{(-4)^2 - (4 \times 1 \times (-2))}}{2 \times 1}$$

$$x = \frac{4 \pm \sqrt{16 + 8}}{2}$$

$$x = \frac{4 \pm \sqrt{24}}{2}$$

Therefore $x = \frac{4 + 4.90}{2}$ or $x = \frac{4 - 4.90}{2}$

$x = 4.45$ (2 d.p.) or $x = -0.45$ (2 d.p.)

*If the exact answer is required, then $\frac{4 \pm \sqrt{24}}{2}$ needs to be given in its simplest form.*

$$\frac{4 \pm \sqrt{24}}{2} = \frac{4 \pm 2\sqrt{6}}{2}$$

$$= 2 \pm \sqrt{6}$$

## Completing the square

The method of completing the square is another form for writing a quadratic expression. This method often simplifies problems involving quadratics and their graphs.

Quadratics can also be solved by writing them in terms of a perfect square. Look once again at the quadratic $x^2 - 4x - 2 = 0$.

The perfect square $(x - 2)^2$ can be expanded to give $x^2 - 4x + 4$. Notice that the $x^2$ and $x$ terms are the same as those in the original quadratic.

Therefore $(x - 2)^2 - 6 = x^2 - 4x - 2$ and can be used to solve the quadratic.
$(x - 2)^2 - 6 = 0$
$\qquad (x - 2)^2 = 6$
$\qquad\quad x - 2 = \pm\sqrt{6}$
$\qquad\qquad\; x = 2 \pm \sqrt{6}$
$x = 4.45 \quad$ or $\quad x = -0.45$

The quadratic formula stated earlier can be derived using the method of completing the square as shown:
Solve $ax^2 + bx + c = 0$.

# 16 SOLVING QUADRATIC EQUATIONS

Divide all through by $a$: $x^2 + \frac{b}{a}x + \frac{c}{a} = 0$

Complete the square: $\left(x + \frac{b}{2a}\right)^2 - \frac{b^2}{4a^2} + \frac{c}{a} = 0$

Rearrange: $\left(x + \frac{b}{2a}\right)^2 = \frac{b^2}{4a^2} - \frac{c}{a}$

Arrange both fractions on the right-hand side with a common denominator of $4a^2$:

$$\left(x + \frac{b}{2a}\right)^2 = \frac{b^2}{4a^2} - \frac{4ac}{4a^2}$$

Simplify: $\left(x + \frac{b}{2a}\right)^2 = \frac{b^2 - 4ac}{4a^2}$

Take the square root of both sides: $x + \frac{b}{2a} = \pm\sqrt{\frac{b^2 - 4ac}{4a^2}}$

Simplify: $x + \frac{b}{2a} = \pm\frac{\sqrt{b^2 - 4ac}}{2a}$

Rearrange: $x = -\frac{b}{2a} \pm \frac{\sqrt{b^2 - 4ac}}{2a}$

Simplify to give the quadratic formula: $x = \frac{-b \pm \sqrt{b^2 - 4ac}}{2a}$

### Exercise 2.27

Solve the following quadratic equations using either the quadratic formula or by completing the square. Give your answers to 2 d.p.

1. 
   a. $x^2 - x - 13 = 0$
   b. $x^2 + 4x - 11 = 0$
   c. $x^2 + 5x - 7 = 0$
   d. $x^2 + 6x + 6 = 0$
   e. $x^2 + 5x - 13 = 0$
   f. $x^2 - 9x + 19 = 0$

2. 
   a. $x^2 + 7x + 9 = 0$
   b. $x^2 - 35 = 0$
   c. $x^2 + 3x - 3 = 0$
   d. $x^2 - 5x - 7 = 0$
   e. $x^2 + x - 18 = 0$
   f. $x^2 - 8 = 0$

3. 
   a. $x^2 - 2x - 2 = 0$
   b. $x^2 - 4x - 11 = 0$
   c. $x^2 - x - 5 = 0$
   d. $x^2 + 2x - 7 = 0$
   e. $x^2 - 3x + 1 = 0$
   f. $x^2 - 8x + 3 = 0$

4. 
   a. $2x^2 - 3x - 4 = 0$
   b. $4x^2 + 2x - 5 = 0$
   c. $5x^2 - 8x + 1 = 0$
   d. $-2x^2 - 5x - 2 = 0$
   e. $3x^2 - 4x - 2 = 0$
   f. $-7x^2 - x + 15 = 0$

# 17 Using a graphic display calculator to solve equations

As seen earlier, a linear equation, when plotted, produces a straight line.

The following are all examples of linear equations:

$$y = x + 1 \qquad y = 2x - 1 \qquad y = 3x \qquad y = -x - 2 \qquad y = 4$$

They all have a similar form, i.e. $y = mx + c$.

In the equation

$y = x + 1$, $m = 1$ and $c = 1$
$y = 2x - 1$, $m = 2$ and $c = -1$
$y = 3x$, $m = 3$ and $c = 0$
$y = -x - 2$, $m = -1$ and $c = -2$
$y = 4$, $m = 0$ and $c = 4$

Their graphs are shown below:

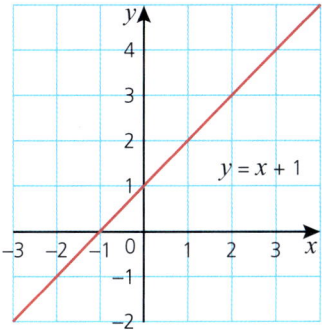
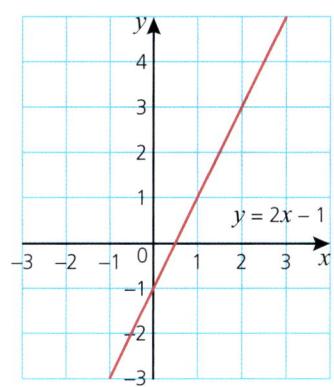
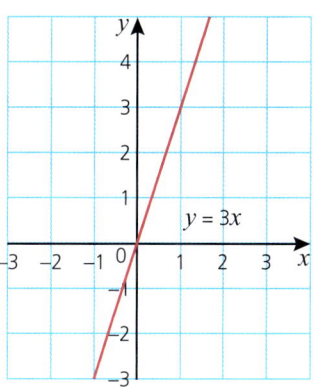
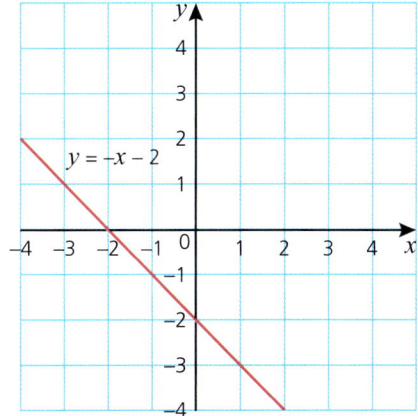
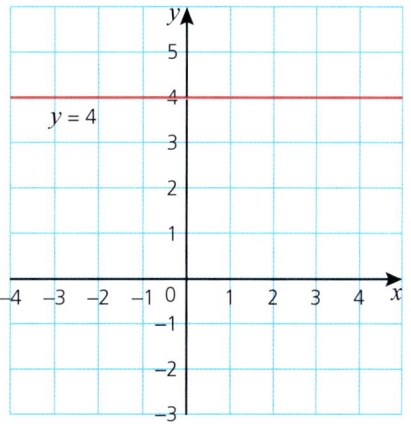

# 17 USING A GRAPHIC DISPLAY CALCULATOR TO SOLVE EQUATIONS

## Using a graphic display calculator to plot a linear equation

In Topic 0 you learned how to plot a single linear equation using your graphic display calculator. For example, to graph the linear equation $y = 2x + 3$:

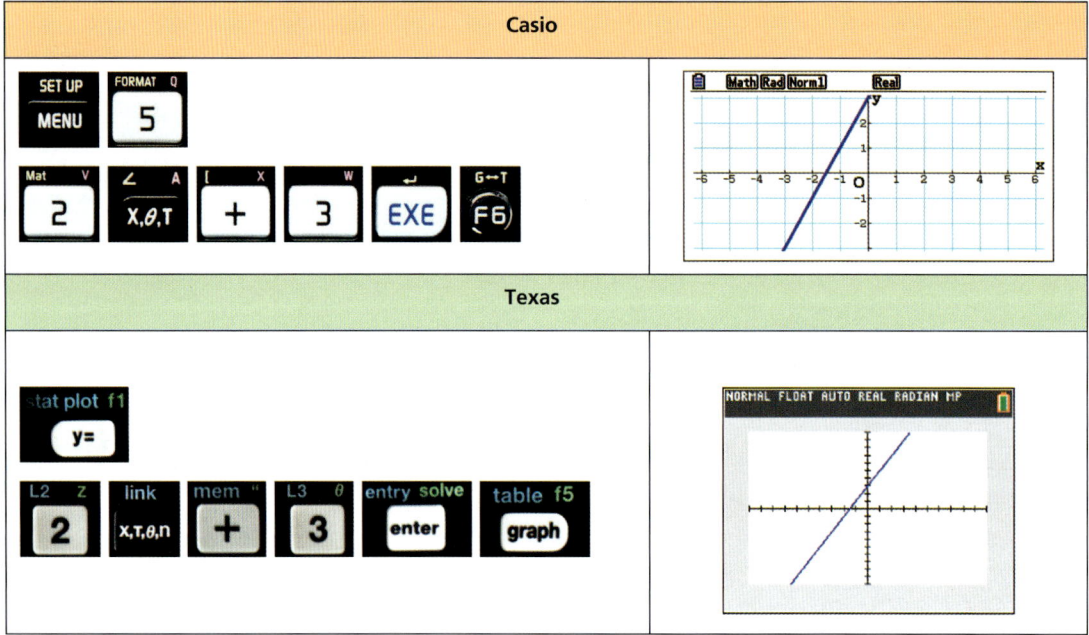

Unless they are parallel to each other, when two linear graphs are plotted on the same axes, they will intersect at one point. Solving the equations simultaneously will give the coordinates of the point of intersection. Your graphic display calculator will be able to work out the coordinates of the point of intersection.

### ➡ Worked example

Find the point of intersection of these linear equations:
$$y = 2x - 1 \text{ and } y = \tfrac{1}{2}x + 2$$

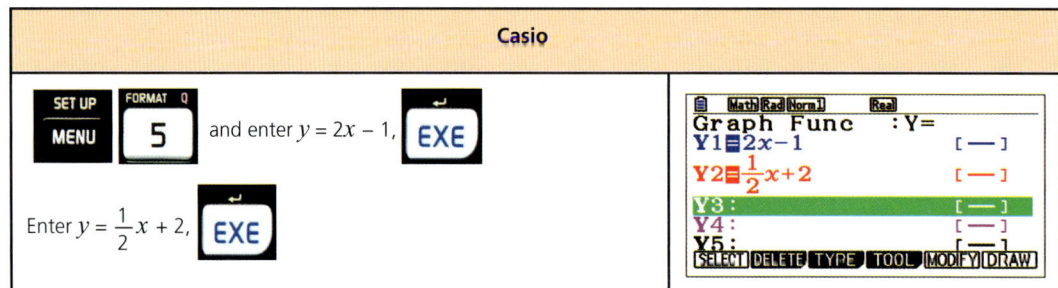

*Using a graphic display calculator to plot a linear equation*

 to graph the equations.

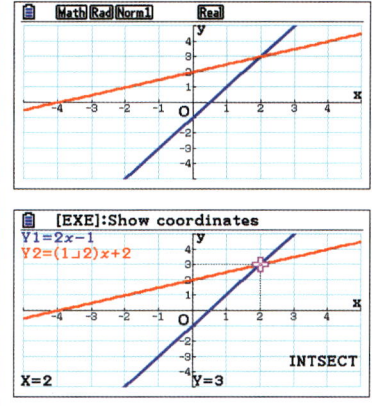

  followed by  to select 'intersect'

in the 'graph solve' menu.

The results $x = 2$ and $y = 3$ are displayed at the bottom of the screen.

Note: Equations of lines have to be entered in the form $y = \ldots$

An equation like $2x - 3y = 9$, for example, would need to be rearranged to make $y$ the subject, $y = \dfrac{2x - 9}{3}$ or $y = \dfrac{2}{3}x - 3$.

| Texas |
|---|

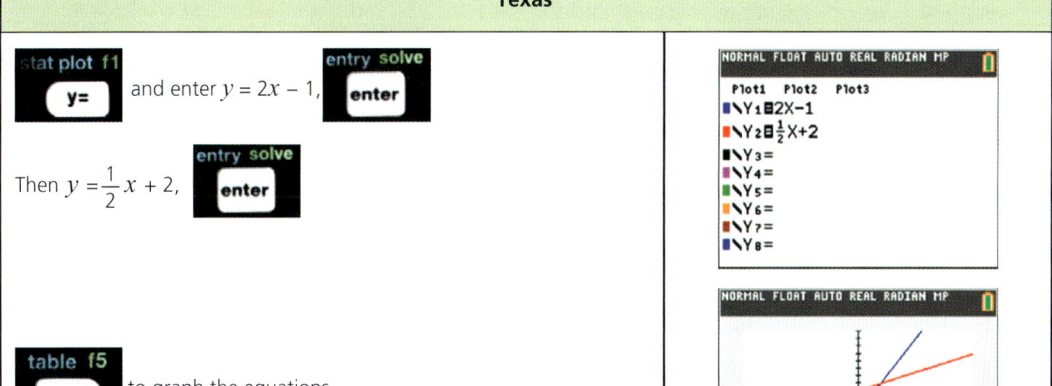

# 17 USING A GRAPHIC DISPLAY CALCULATOR TO SOLVE EQUATIONS

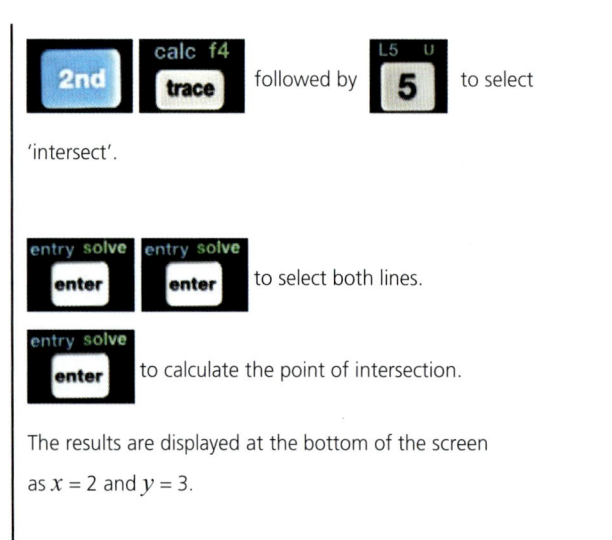

followed by 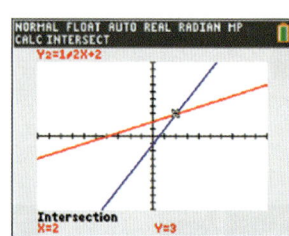 to select 'intersect'.

to select both lines.

to calculate the point of intersection.

The results are displayed at the bottom of the screen as $x = 2$ and $y = 3$.

Note: See the note for the Casio above.

**Exercise 2.28**  Use a graphic display calculator to find the coordinates of the points of intersection of the following pairs of linear graphs:

1   a  $y = 5 - x$ and $y = x - 1$  
    b  $y = 7 - x$ and $y = x - 3$  
    c  $y = -2x + 5$ and $y = x - 1$  
    d  $y = -x + 3$ and $y = 2x - 3$  
    e  $x + 3y = -1$ and $y = \frac{1}{2}x + 3$  
    f  $x - y = 6$ and $x + y = 2$

2   a  $3x - 2y = 13$ and $2x + y = 4$  
    b  $4x - 5y = 1$ and $2x + y = -3$  
    c  $x + 5 = y$ and $2x + 3y - 5 = 0$  
    d  $x = y$ and $x + y + 6 = 0$  
    e  $2x + y = 4$ and $4x + 2y = 8$  
    f  $y - 3x = 1$ and $y = 3x - 3$

3   By referring to the lines, explain your answers to Q.2 e and f above.

## Quadratic equations

As you will recall from Chapter 16, an equation of the form $y = ax^2 + bx + c$, in which the highest power of the variable $x$ is $x^2$, is known as a **quadratic equation**.

When plotted, a quadratic graph has a specific shape known as a parabola. This will look like

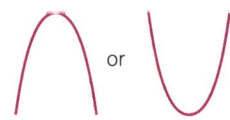 or

# Quadratic equations

Depending on the values of $a$, $b$ and $c$, the position and shape of the graph will vary, for example:

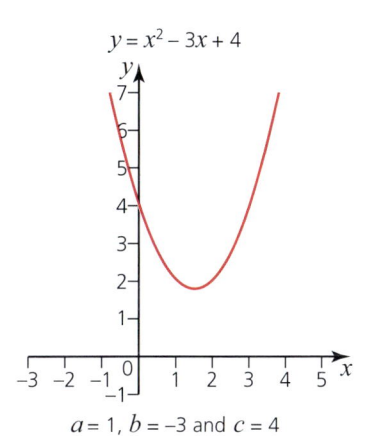

$y = x^2 - 3x + 4$

$a = 1$, $b = -3$ and $c = 4$

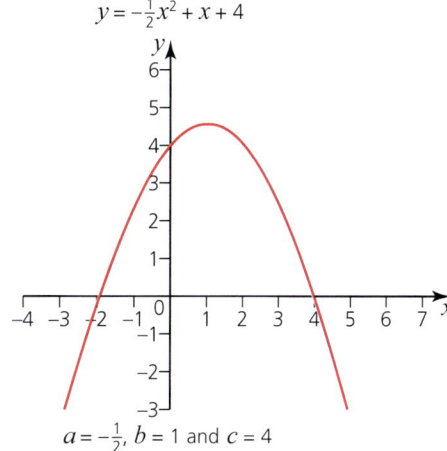

$y = -\frac{1}{2}x^2 + x + 4$

$a = -\frac{1}{2}$, $b = 1$ and $c = 4$

Solving a quadratic equation of the form $ax^2 + bx + c = 0$ implies finding where the graph crosses the $x$-axis, because $y = 0$ on the $x$-axis.

In the case of $-\frac{1}{2}x^2 + x + 4 = 0$ above, it can be seen that the graph crosses the $x$-axis at $x = -2$ and $x = 4$. These are therefore the solutions to, or **zeros** (or roots) of, the equation.

In the case of $x^2 - 3x + 4 = 0$ above, the graph does not cross the $x$-axis. Therefore the equation has no real solutions. (Note: There are imaginary solutions, but these are not dealt with in this textbook.)

A graphic display calculator can be used to find the solution to quadratic equations.

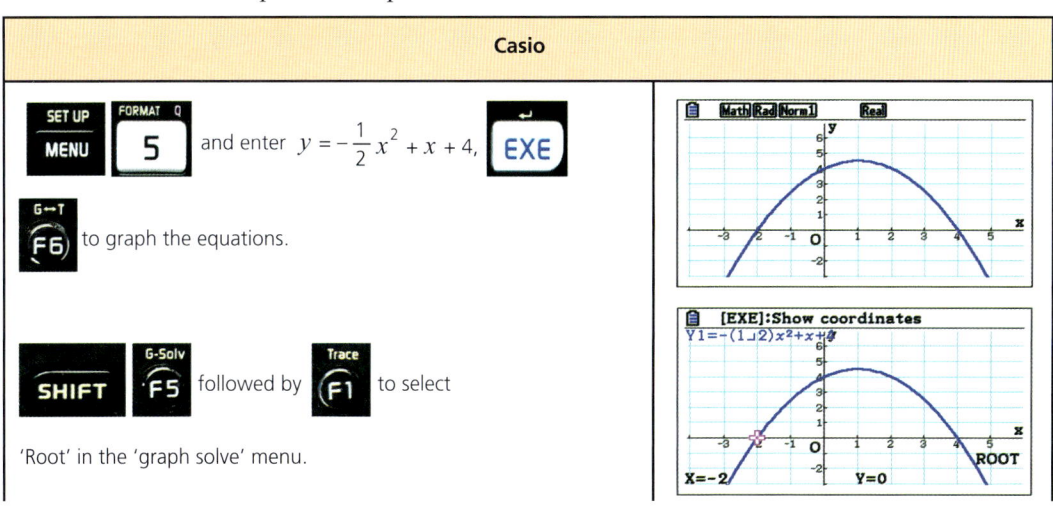

# 17 USING A GRAPHIC DISPLAY CALCULATOR TO SOLVE EQUATIONS

Use  to find the second root.

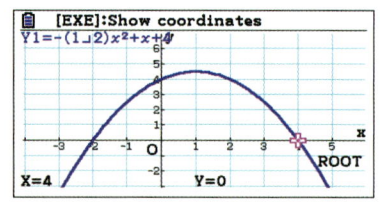

| Texas |
|---|

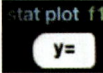 and enter $y = -\frac{1}{2}x^2 + x + 4$,

 to graph the equation.

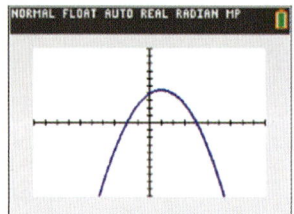

 followed by  to select 'zero' in the 'graph calc' menu.

Use  and follow the on-screen prompts to identify a point to the left and a point to the right of the root, in order for the calculator to give the point where $y = 0$.

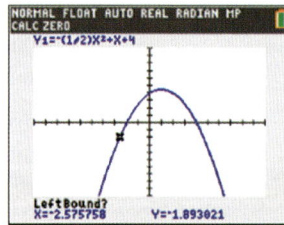

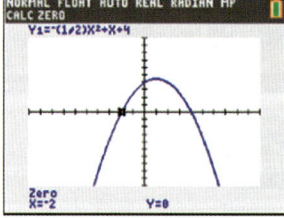

  followed by  again and the same steps in order to find the second point where $y = 0$.

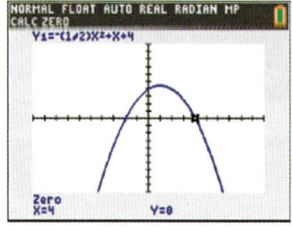

## Exercise 2.29

Using a graphic display calculator:
i graph the following quadratic equations
ii find the coordinates of any zeros.

1. 
   a $y = x^2 - 3x + 2$
   b $y = x^2 + 4x - 12$
   c $y = -x^2 + 8x - 15$
   d $y = x^2 + 2x + 6$
   e $y = -x^2 + x - 4$
   f $y = x^2 - 6x + 9$

2. 
   a $y = \frac{1}{2}x^2 - \frac{1}{2}x - 3$
   b $y = -2x^2 + 20x - 48$
   c $4y = -x^2 + 6x + 16$
   d $-2y = x^2 + 10x + 25$

# 18 Inequalities

< means 'is less than'

> means 'is greater than'

⩽ means 'is less than or equal to'

⩾ means 'is greater than or equal to'

= means 'is equal to'

≠ means 'is not equal to'

The statement

6 is less than 8 can be written as:

$6 < 8$

This inequality can be manipulated in the following ways:

| | | |
|---|---|---|
| adding 2 to each side: | $8 < 10$ | this inequality is still true |
| subtracting 2 from each side: | $4 < 6$ | this inequality is still true |
| multiplying both sides by 2: | $12 < 16$ | this inequality is still true |
| dividing both sides by 2: | $3 < 4$ | this inequality is still true |
| multiplying both sides by −2: | $-12 < -16$ | this inequality is not true |
| dividing both sides by −2: | $-3 < -4$ | this inequality is not true |

As can be seen, when both sides of an inequality are either multiplied or divided by a negative number, the inequality is no longer true. For it to be true, the inequality sign needs to be changed around:

i.e.   $-12 > -16$   and   $-3 > -4$

The method used to solve linear inequalities is very similar to that used to solve linear equations.

## → Worked examples

Remember:

○——→ implies that the number is not included in the solution. It is associated with > and <.

●——→ implies that the number is included in the solution. It is associated with ⩾ and ⩽.

Solve the following inequalities and represent the solution on a number line:

a  $15 + 3x < 6$
   $3x < -9$
   $x < -3$

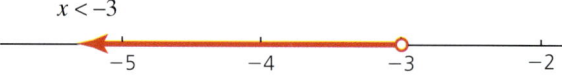

b  $17 \leqslant 7x + 3$
   $14 \leqslant 7x$
   $2 \leqslant x$ that is $x \geqslant 2$

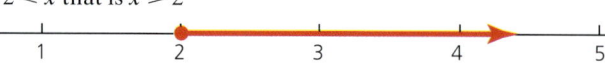

**Note**

Students following the core syllabus need to be able represent inequalities on a number line. For example, $-2 \leqslant x < 1$ is

# Inequalities

**c** $9 - 4x \geqslant 17$
$-4x \geqslant 8$
$x \leqslant -2$

*Note the inequality sign has changed direction.*

 **Exercise 2.30** Solve the following inequalities and show your solution on a number line:
1. **a** $x + 3 < 7$ **b** $5 + x > 6$ **c** $4 + 2x \leqslant 10$
   **d** $8 \leqslant x + 1$ **e** $5 > 3 + x$ **f** $7 < 3 + 2x$
2. **a** $x - 3 < 4$ **b** $x - 6 \geqslant -8$ **c** $8 + 3x > -1$
   **d** $5 \geqslant -x - 7$ **e** $12 > -x - 12$ **f** $4 \leqslant 2x + 10$
3. **a** $\frac{x}{2} < 1$ **b** $4 \geqslant \frac{x}{3}$ **c** $1 \leqslant \frac{x}{2}$
   **d** $9x \geqslant -18$ **e** $-4x + 1 < 3$ **f** $1 \geqslant -3x + 7$

 **Worked example**

Find the range of values for which $7 < 3x + 1 \leqslant 13$ and show the solutions on a number line.

This is in fact two inequalities which can therefore be solved separately.

$7 < 3x + 1$ and $3x + 1 \leqslant 13$

$6 < 3x$ $3x \leqslant 12$

$2 < x$ that is $x > 2$ $x \leqslant 4$

 **Exercise 2.31** Find the range of values for which the following inequalities are satisfied. Show each solution on a number line:
1. **a** $4 < 2x \leqslant 8$ **b** $3 \leqslant 3x < 15$
   **c** $7 \leqslant 2x < 10$ **d** $10 \leqslant 5x < 21$
2. **a** $5 < 3x + 2 \leqslant 17$ **b** $3 \leqslant 2x + 5 < 7$
   **c** $12 < 8x - 4 < 20$ **d** $15 \leqslant 3(x - 2) < 9$

# 18 INEQUALITIES

The solution to an inequality can also be shown on a graph.

## ➜ Worked examples

**a** On a pair of axes, leave unshaded the region which satisfies the inequality $x \leqslant 3$.

First draw the line $x = 3$.

Shade the region that represents the inequality $x \geqslant 3$, i.e. the region to the right of $x = 3$.

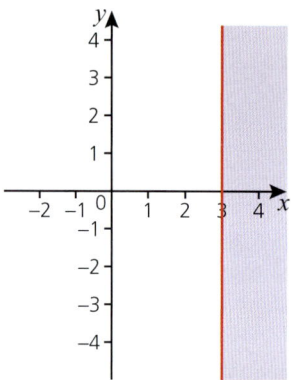

**b** On a pair of axes, leave unshaded the region which satisfies the inequality $y > 2$.

First draw the line $y = 2$ (in this case it is drawn as a broken line).
Note: A broken (dashed) line shows $<$ or $>$ and a solid line shows $\leqslant$ or $\geqslant$.

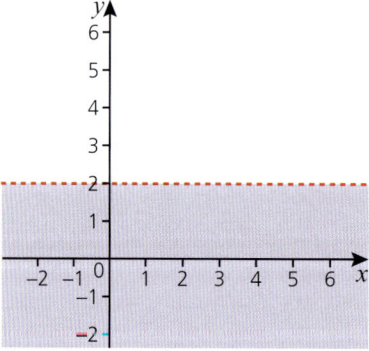

Shade the region that represents the inequality $y < 2$, i.e. the region below the line $y = 2$.

**c** On a pair of axes, leave unshaded the region which satisfies the inequality $y \geqslant x + 2$.

First draw the line $y = x + 2$ (since it is included, this line is solid).

To find the region that satisfies the inequality, and hence to know which side of the line to shade:

1. Choose any point which does not lie on the line, e.g. (1, 4).
2. Substitute those values of $x$ and $y$ into the inequality $y \geqslant x + 2$, i.e. $4 \geqslant 1 + 2$.

# Inequalities

3 If the inequality holds true, then the region in which the point lies satisfies the inequality and is therefore left unshaded.

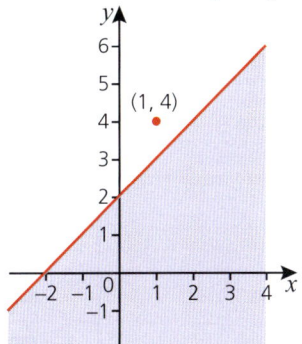

Graphic display calculators can also be used to plot and shade the graphs of inequalities. However, care must be taken as calculators shade the region which satisfies the inequality.

For example, identify, on a pair of axes, the region which satisfies the inequality $y \leq x + 2$

| Casio |
|---|

**MENU** **5** to select the graphing menu.

**F3** **F6** **F4** to change the graph type from

Y = to Y ≤.

Enter the inequality $y \leq x + 2$

**F6** to plot the inequality and shade the correct region.

Note: This calculator does distinguish graphically between '$y \leq$' and '$y <$' by using a solid line for '$y \leq$' and a dashed line for '$y <$'.
The inequality of $y < x + 2$ is shown opposite.
If the region that satisfies the inequality $y \leq x + 2$ is to be left **unshaded**, when using the calculator, the inequality $y \geq x + 2$ must be entered instead.

165

# 18 INEQUALITIES

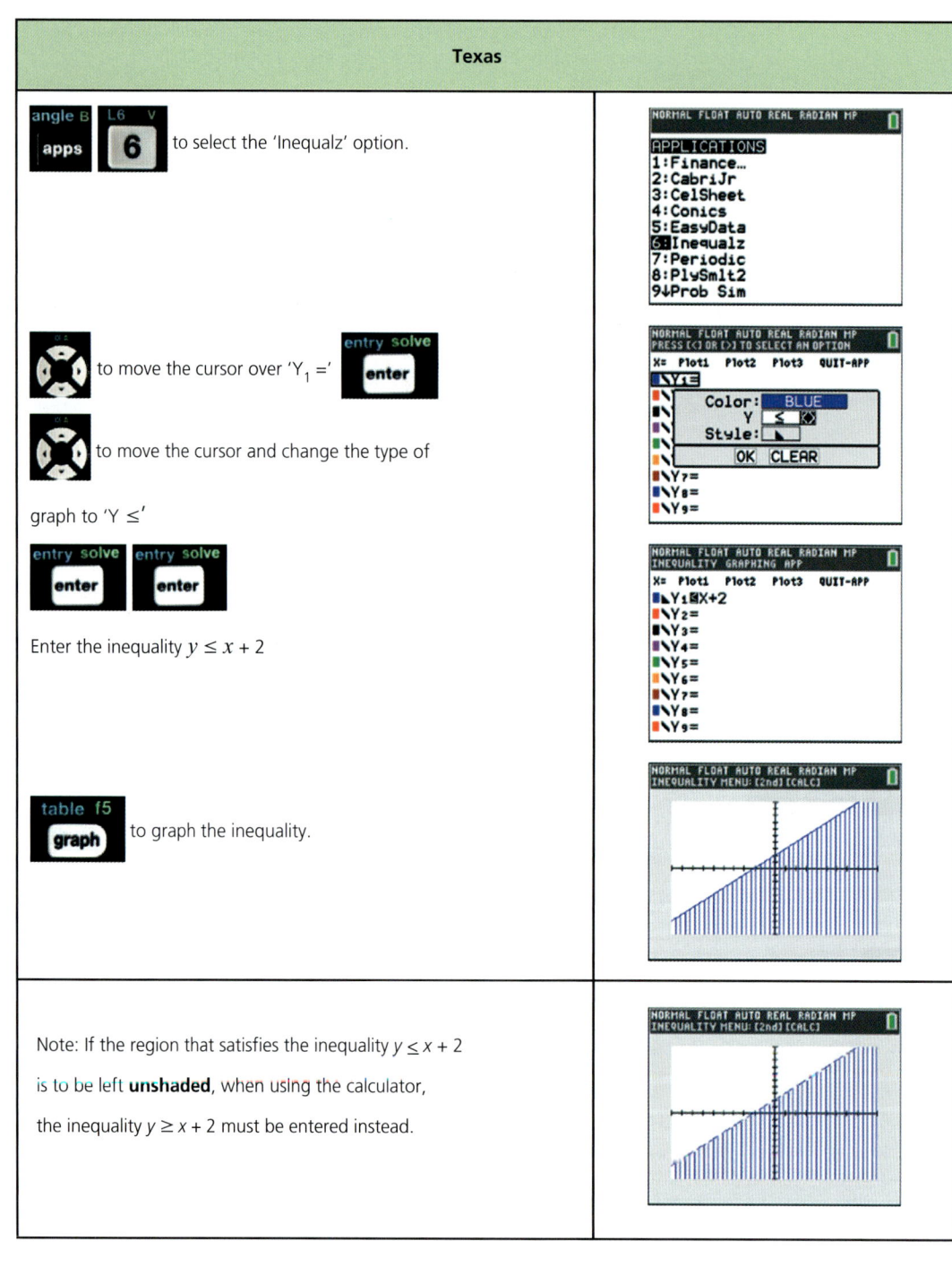

Exercise 2.32

1 By drawing axes, leave unshaded the region which satisfies each of the following inequalities:

a  $y < 2$
b  $x > 3$
c  $y \geq 4$
d  $x \leq -1$
e  $y < 2x + 1$
f  $y \geq x - 3$

Inequalities

**2** By drawing axes, leave unshaded the region which satisfies each of the following inequalities:

a  $y \geq -x$
b  $y \leq 2 - x$
c  $x \geq y - 3$
d  $x + y \geq 4$
e  $2x - y \geq 3$
f  $2y - x \leq 4$

Several inequalities can be graphed on the same set of axes. If the regions which satisfy each inequality are left unshaded, then a solution can be found which satisfies all the inequalities, i.e. the region left unshaded by all the inequalities.

## Worked example

On the same pair of axes leave unshaded the regions which satisfy the following inequalities simultaneously:

$x \leq 2$    $y > -1$    $y \leq 3$    $y \leq x + 2$

Hence find the region which satisfies all four inequalities.

If the four inequalities are graphed on separate axes, the solutions are as shown:

a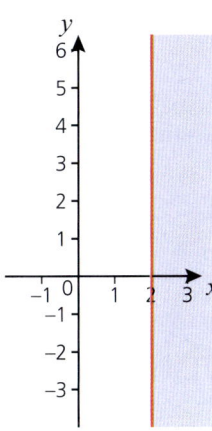
$x \leq 2$

b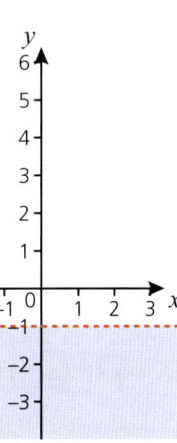
$y > -1$

c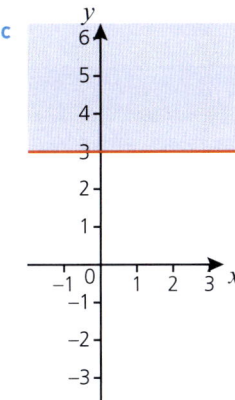
$y \leq 3$

d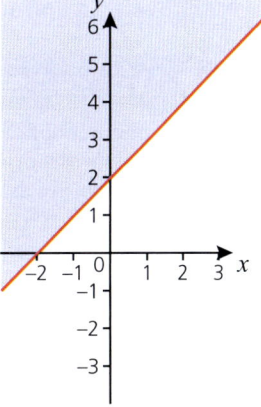
$y \leq x + 2$

167

# 18 INEQUALITIES

Combining all four on one pair of axes gives this diagram:

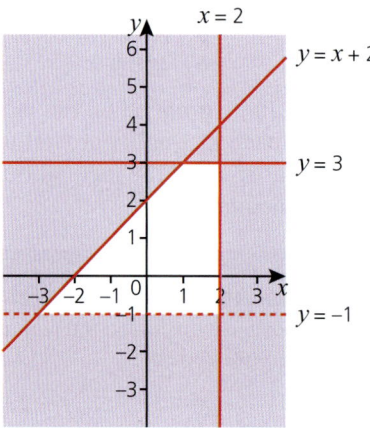

The unshaded region therefore gives a solution which satisfies all four inequalities.

### Exercise 2.33

For Q.1–4, plot, on the same pair of axes, all the inequalities given. Leave unshaded the region which satisfies all of them simultaneously:

1. $y \leqslant x$      $y > 1$      $x \leqslant 5$
2. $x + y \leqslant 6$      $y < x$      $y \geqslant 1$
3. $y \geqslant 3x$      $y \leqslant 5$      $x + y > 4$
4. $2y \geqslant x + 4$      $y \leqslant 2x + 2$      $y < 4$

## Practical problems and inequalities

Inequalities are sometimes used to define problems. Solving the inequalities simultaneously can provide a number of possible solutions to the problem. More importantly, their solution can sometimes provide an **optimum** solution to the problem.

*Optimum means most favourable.*

### → Worked example

The number of fields a farmer plants with wheat is $w$ and the number of fields he plants with corn is $c$. There are limits on how many fields he can plant of each. These are as follows:
- There must be at least two fields of corn.
- There must be at least two fields of wheat.
- Not more than 10 fields can be sown in total.

a   Construct three inequalities from the information given above.

    $c \geqslant 2$      $w \geqslant 2$      $c + w \leqslant 10$

**b** On one pair of axes, graph the three inequalities and leave **unshaded** the region which satisfies all three simultaneously.

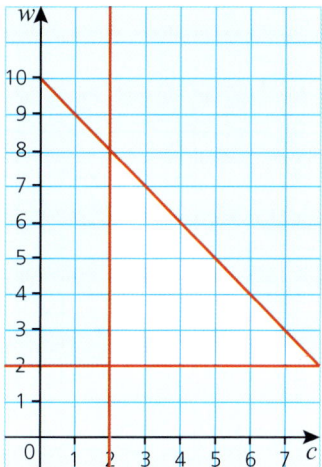

**c** Give one possible arrangement as to how the farmer should plant his fields.

Four fields of corn and four fields of wheat.

The practical application of constructing and solving linear inequalities is sometimes called linear programming.

# Solving quadratic inequalities

The inequalities covered so far have all been linear inequalities i.e. they involve straight lines. However, inequalities can be used with other functions too. This section will look at how to solve quadratic inequalities.

## ➡ Worked example

Solve the inequality $x^2 - 4x < 5$

Both sides of the inequality can be plotted on the same axes as the two separate graphs

$y = x^2 - 4x$ and $y = 5$.

# 18 INEQUALITIES

The points of intersection of the two graphs are the critical points for solving the inequality.

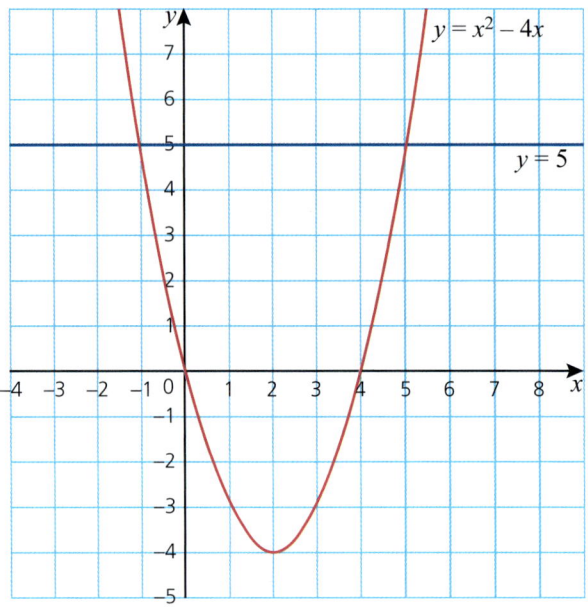

To solve the quadratic equation $x^2 - 4x = 5$

first rearrange to give $\qquad x^2 - 4x - 5 = 0$

then factorise to give $\qquad (x+1)(x-5) = 0$

Therefore $x = -1$ and $x = 5$ as can be seen on the graph are the $x$-coordinates of the points of intersection of the two lines.

Looking at the graph and knowing the points of intersection enables the quadratic $x^2 - 4x < 5$ to be solved.

The curve $y = x^2 - 4x$ is below the line $y = 5$ when $-1 < x < 5$.

*This inequality could also be represented on a number line.*

> **Note**
>
> If the inequality to be solved was $x^2 - 4x > 5$, the process would be exactly the same as shown above. By looking at the graph and points of intersection you would deduce that the curve $y = x^2 - 4x$ is above the line $y = 5$ when $x < -1$ and when $x > 5$.

A graphic display calculator cannot solve a quadratic inequality directly, but by graphing both sides of the inequality and finding their points of intersection, the solution can be deduced as shown.

# Solving quadratic inequalities

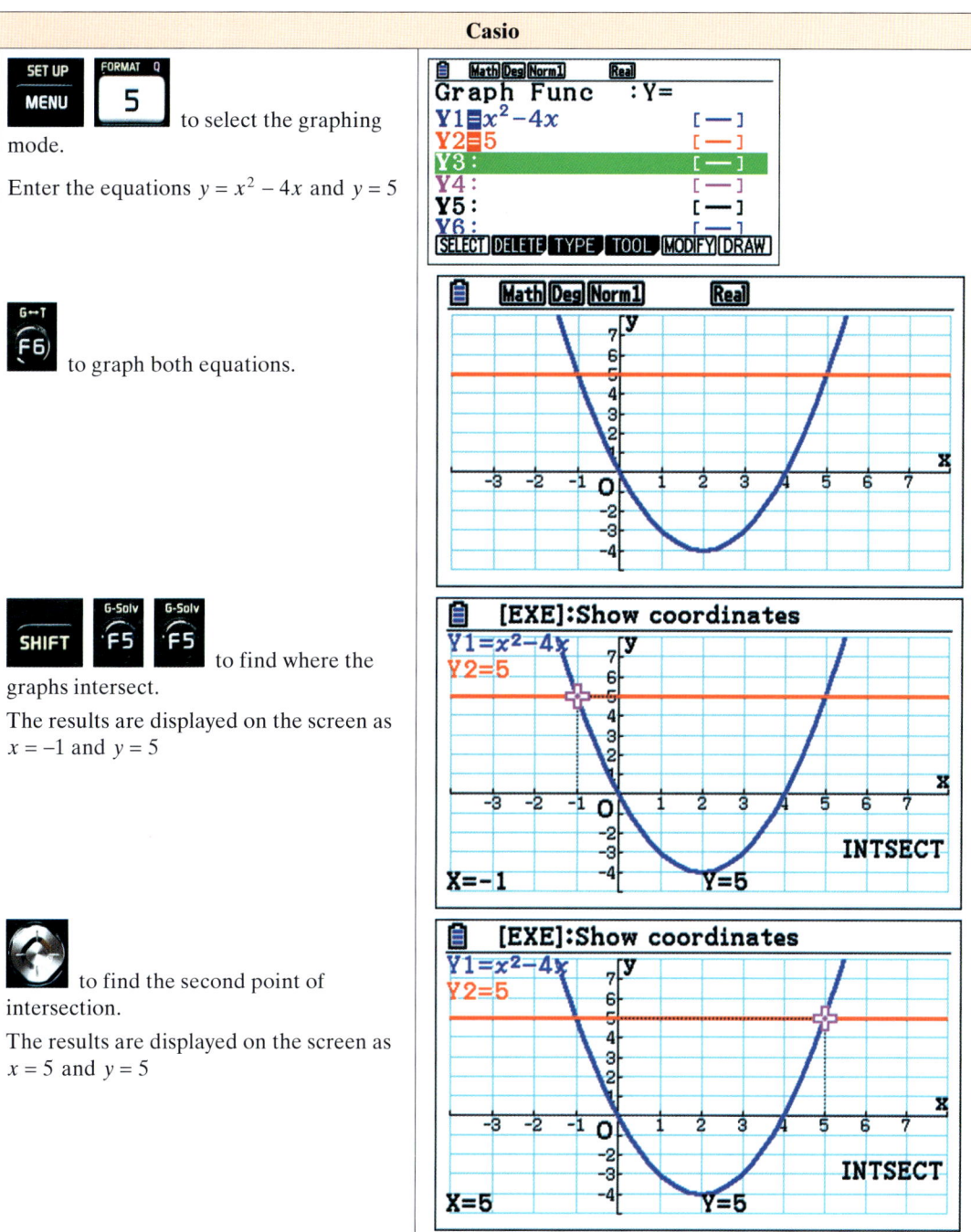

**SET UP** **MENU** **5** to select the graphing mode.

Enter the equations $y = x^2 - 4x$ and $y = 5$

**F6** to graph both equations.

**SHIFT** **F5** **F5** to find where the graphs intersect.

The results are displayed on the screen as $x = -1$ and $y = 5$

to find the second point of intersection.

The results are displayed on the screen as $x = 5$ and $y = 5$

The solution to the inequality $x^2 - 4x < 5$ can be deduced from the graphs as $-1 < x < 5$.

# 18 INEQUALITIES

| Texas |
|---|

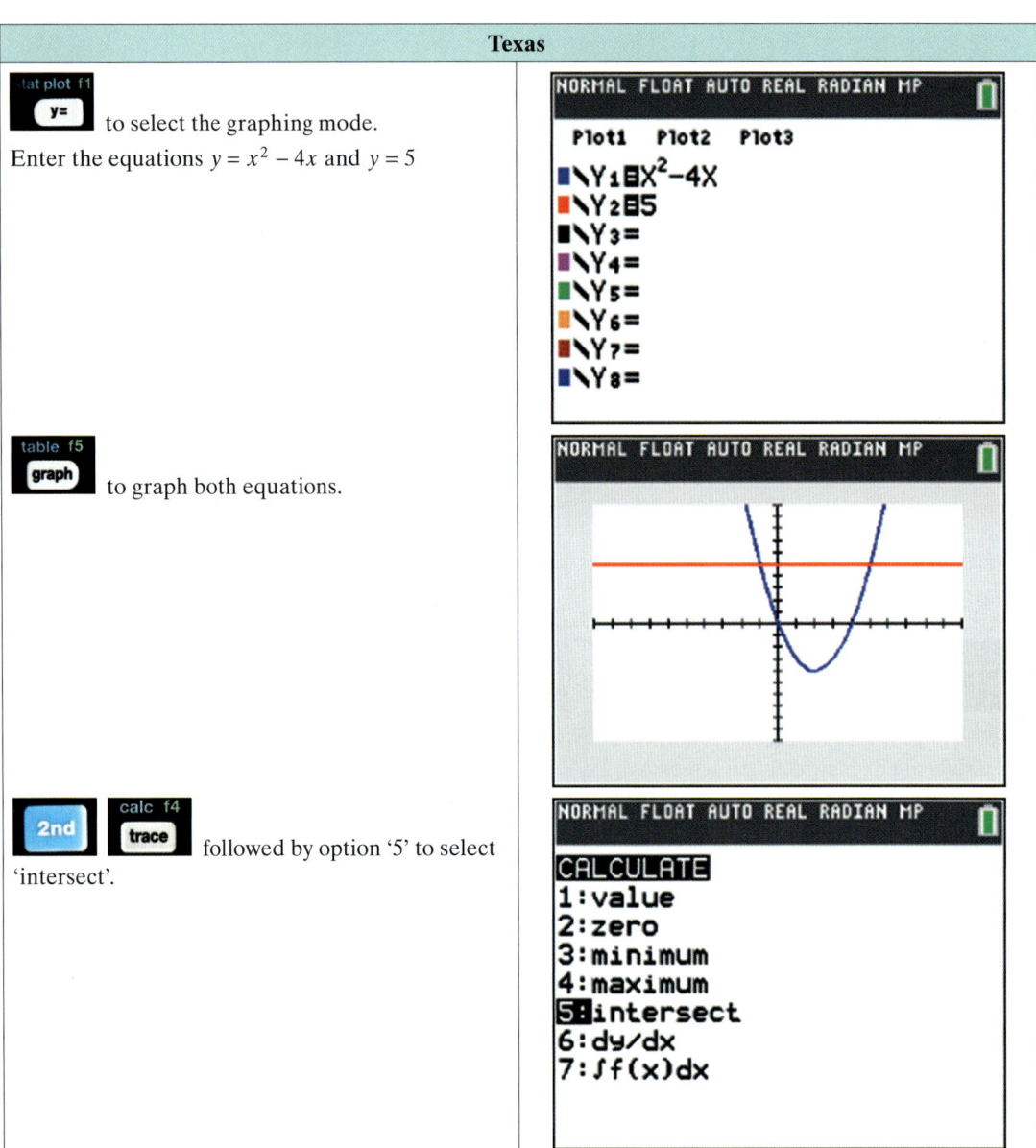

[y=] to select the graphing mode.
Enter the equations $y = x^2 - 4x$ and $y = 5$

[graph] to graph both equations.

[2nd] [trace] followed by option '5' to select 'intersect'.

Solving quadratic inequalities

The screen cursor will be flashing on the first line plotted.

 to select it.

The screen cursor will now be flashing on the second line plotted.

 to select it

 to move the cursor near the desired point of intersection, then

The calculator will give the coordinates of the point of intersection as $x = -1$ and $y = 5$

Repeat the above steps for the second point of intersection.

The results are displayed on the screen as $x = 5$ and $y = 5$

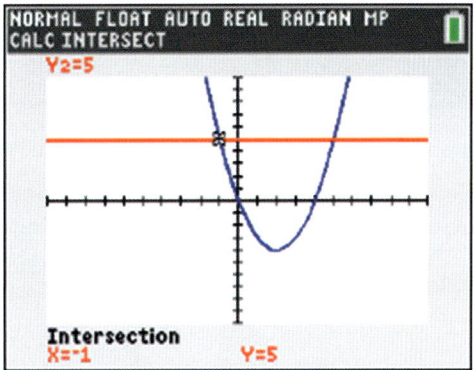

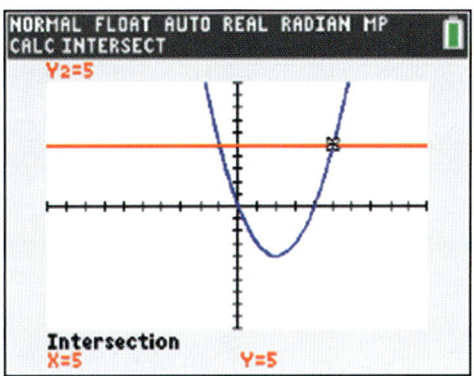

The solution to the inequality $x^2 - 4x < 5$ can be deduced from the graphs as $-1 < x < 5$.

**Exercise 2.34**

1 Using your graphic display calculator to help, find the solution to the following inequalities.
   **a** $x^2 + 4x < 21$              **b** $x^2 > 64$
   **c** $x^2 + 6x \geq -8$           **d** $2x^2 \leq 9x + 18$

# 19 Indices and algebra

You applied indices to numbers earlier in this book in Chapter 7 in Topic 1.

The rules involving numerical indices also apply to indices involving algebra, as shown below.

**Rules of indices**

| | Numeric | Algebraic |
|---|---|---|
| 1 | $2^3 \times 2^4 = 2^{3+4} = 2^7$ | $a^m \times a^n = a^{m+n}$ |
| 2 | $3^6 \div 3^2 = 3^{6-2} = 3^4$ | $a^m \div a^n = a^{m-n}$ |
| 3 | $(5^2)^3 = 5^{2\times3} = 5^6$ | $(a^m)^n = a^{mn}$ |
| 4 | $5^0 = 1$ | $a^0 = 1$ |
| 5 | $4^{-2} = \dfrac{1}{4^2}$ | $a^{-m} = \dfrac{1}{a^m}$ |
| 6 | $5^{\frac{1}{3}} = \sqrt[3]{5}$ | $a^{\frac{1}{n}} = \sqrt[n]{a}$ |
| 7 | $8^{\frac{2}{3}} = (\sqrt[3]{8})^2$ or $\sqrt[3]{8^2}$ | $a^{\frac{m}{n}} = (\sqrt[n]{a})^m$ or $\sqrt[n]{a^m}$ |
| 8 | $(2 \times 4)^3 = 2^3 \times 4^3 = 8^3$ | $(ab)^n = a^n b^n$ |
| 9 | $\left(\dfrac{3}{5}\right)^4 = \dfrac{3^4}{5^4}$ | $\left(\dfrac{a}{b}\right)^n = \dfrac{a^n}{b^n}$ |

### → Worked examples

Simplify the following:

a $p^3 \times p^5$
$= p^{3+5}$
$= p^8$

b $\dfrac{k^2 \times k^6}{k^4}$
$= \dfrac{k^{2+6}}{k^4}$
$= k^{8-4}$
$= k^4$

c $\dfrac{3a^3b^2}{6ab}$
$= \dfrac{a^2b}{2}$

d $(5m^2n)^3$
$= 5m^2n \times 5m^2n \times 5m^2n$
$= 125m^6n^3$

### Exercise 2.35

1 Simplify the following:

a $c^5 \times c^3$
b $m^4 \div m^2$
c $(b^3)^5 \div b^6$
d $\dfrac{m^4n^9}{mn^3}$
e $\dfrac{6a^6b^4}{3a^2b^3}$
f $\dfrac{12x^5y^7}{4x^2y^5}$
g $\dfrac{4u^3v^6}{8u^2v^3}$
h $\dfrac{3x^6y^5z^3}{9x^4y^2z}$

**2** Simplify the following:
  a  $4a^2 \times 3a^3$
  b  $2a^2b \times 4a^3b^2$
  c  $(2p^2)^3$
  d  $(4m^2n^3)^2$
  e  $(5p^2)^2 \times (2p^3)^3$
  f  $(4m^2n^2) \times (2mn^3)^3$
  g  $\dfrac{(6x^2y^4)^2 \times (2xy)^3}{12x^6y^8}$
  h  $(ab)^d \times (ab)^e$

**3** Simplify the following:
  a  $a^{\frac{2}{3}} \times a^{\frac{1}{3}}$
  b  $(x^{-4})^{-1}$
  c  $(\sqrt[5]{m})^2$
  d  $\sqrt{m^4n^2}$

# Exponential equations

Equations that involve indices as unknowns are known as **exponential equations**.

> ## Worked examples
>
> a  Find the value of $x$ if $2^x = 32$
>    Write 32 as a power of 2:
>    $32 = 2^5$
>    Therefore $2^x = 2^5$
>    $x = 5$
>
> b  Find the value of $m$ if $3^m = 81$
>    Write 81 as a power of 3:
>    $81 = 3^4$
>    Therefore $3^m = 3^4$
>    $m = 4$
>
> c  Solve $32^x = 2$
>    32 is $2^5$
>    so $(2^5)^x = 2$
>    or $2^{5x} = 2^1$
>    Therefore $5x = 1$
>    $x = \dfrac{1}{5}$
>
> d  Solve $125^x = 5$
>    125 is $5^3$
>    so $(5^3)^x = 5$
>    or $5^{3x} = 5^1$
>    Therefore $3x = 1$
>    $x = \dfrac{1}{3}$

**Exercise 2.36**

**1** Find the value of $x$ in each of the following:
  a  $2^x = 4$
  b  $2^x = 16$
  c  $4^x = 64$
  d  $10^x = 1000$
  e  $5^x = 625$
  f  $3^x = 1$

**2** Find the value of $z$ in each of the following:
  a  $2^{(z-1)} = 8$
  b  $3^{(z+2)} = 27$
  c  $4^{2z} = 64$
  d  $10^{(z+1)} = 1$
  e  $3^z = 9^{(z-1)}$
  f  $5^z = 125$

**3** Find the value of $n$ in each of the following:
  a  $\left(\dfrac{1}{2}\right)^n = 8$
  b  $\left(\dfrac{1}{3}\right)^n = 81$
  c  $\left(\dfrac{1}{2}\right)^n = 32$
  d  $\left(\dfrac{1}{2}\right)^n = 4^{(n+1)}$
  e  $\left(\dfrac{1}{2}\right)^{(n+1)} = 2$
  f  $\left(\dfrac{1}{16}\right)^n = 4$

**4** Find the value of $x$ in each of the following:
  a  $3^{-x} = 27$
  b  $2^{-x} = 128$
  c  $2^{(-x+3)} = 64$
  d  $4^{-x} = \dfrac{1}{16}$
  e  $2^{-x} = \dfrac{1}{256}$
  f  $3^{(-x+1)} = \dfrac{1}{81}$

# 19 INDICES AND ALGEBRA

**Exercise 2.36 (cont)**

For Q.5–6, solve each equation without the use of a calculator:

5  a  $16^x = 4$        b  $8^x = 2$          c  $9^x = 3$
   d  $27^x = 3$        e  $100^x = 10$      f  $64^x = 2$
6  a  $1000^x = 10$     b  $49^x = 7$         c  $81^x = 3$
   d  $343^x = 7$       e  $1\,000\,000^x = 10$   f  $216^x = 6$

**Exercise 2.37**

1  A tap is dripping at a constant rate into a container. The level ($l$ cm) of the water in the container is given by the equation $l = 2^t - 1$ where $t$ hours is the time taken.
   a  Calculate the level of the water after 3 hours.
   b  Calculate the level of the water in the container at the start.
   c  Calculate the time taken for the level of the water to reach 31 cm.
   d  Plot a graph showing the level of the water over the first 6 hours.
   e  From your graph, estimate the time taken for the water to reach a level of 45 cm.

2  Draw a graph of $y = 4^x$ for values of $x$ between $-1$ and 3. Use your graph to find approximate solutions to these equations:
   a  $4^x = 30$        b  $4^x = \frac{1}{2}$

3  Draw a graph of $y = 2^x$ for values of $x$ between $-2$ and 5. Use your graph to find approximate solutions to the following equations:
   a  $2^x = 20$        b  $2^{(x+2)} = 40$        c  $2^{-x} = 0.2$

4  Draw a graph of $y = 3^x$ for values of $x$ between $-1$ and 3. Use your graph to find approximate solutions to these equations:
   a  $3^{(x+2)} = 12$        b  $3^{(x-3)} = 0.5$

# 20 Sequences

A **sequence** is a collection of terms arranged in a specific order, where each term is obtained according to a rule. Examples of some simple sequences are given below:

$$2, 4, 6, 8, 10 \qquad 1, 4, 9, 16, 25 \qquad 1, 2, 4, 8, 16$$
$$1, 1, 2, 3, 5, 8 \qquad 1, 8, 27, 64, 125 \qquad 10, 5, \frac{5}{2}, \frac{5}{4}, \frac{5}{8}$$

You could discuss with another student the rules involved in producing the sequences above.

The terms of a sequence can be written as $T_1, T_2, T_3, \ldots, T_n$ where:

$T_1$ is the first term

$T_2$ is the second term

$T_n$ is the $n$th term

Therefore in the sequence 2, 4, 6, 8, 10, $T_1 = 2$, $T_2 = 4$, etc.

## Linear sequences

In a **linear sequence** there is a common difference ($d$) between successive terms. Examples of some linear sequences are given below:

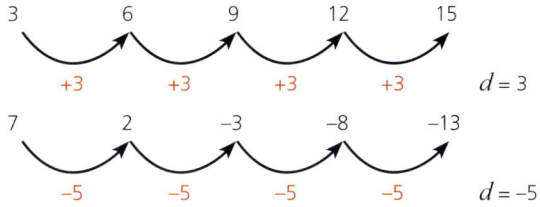

### Formulas for the terms of a linear sequence

There are two main ways of describing a sequence.

**1** A **term-to-term** rule

In the following sequence,

the term-to-term rule is +5, i.e. $T_2 = T_1 + 5$, $T_3 = T_2 + 5$, etc. The general form is therefore written as $T_{n+1} = T_n + 5$, where $T_1 = 7$, $T_n$ is the $n$th term and $T_{n+1}$ the term after the $n$th term.

Note: It is important to give one of the terms, e.g. $T_1$, so that the exact sequence can be generated.

*Subscript notation, $T_n$, is relevant to the Extended syllabus only.*

# 20 SEQUENCES

**2  A formula for the nth term of a sequence**
This type of rule links each term to its position in the sequence, e.g.

| Position | 1 | 2 | 3 | 4 | 5 | n |
|---|---|---|---|---|---|---|
| Term | 7 | 12 | 17 | 22 | 27 | |

You can deduce from the figures above that each term can be calculated by multiplying its position number by 5 and adding 2. Algebraically this can be written as the formula for the $n$th term:

$T_n = 5n + 2$

This textbook focuses on the generation and use of the rule for the $n$th term.

With a linear sequence, the rule for the $n$th term can be deduced by looking at the common difference, e.g.

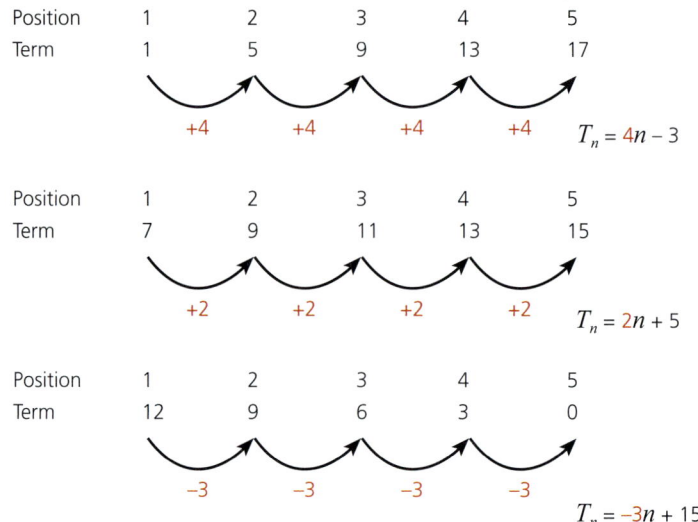

The common difference is the coefficient of $n$ (i.e. the number by which $n$ is multiplied). The constant is then worked out by calculating the number needed to make the term.

## ➡ Worked example

Find the rule for the $n$th term of the sequence 12, 7, 2, −3, −8,

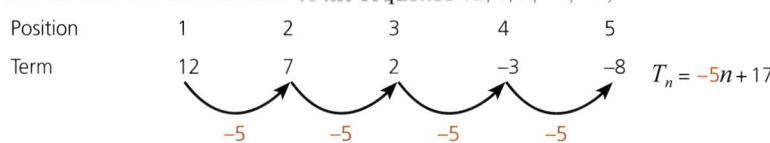

*Linear sequences*

**Exercise 2.38**

1. For each of the following sequences:
   i  find the formula for the $n$th term
   ii calculate the 10th term.
   a  5, 8, 11, 14, 17
   b  0, 4, 8, 12, 16
   c  $\frac{1}{2}, 1\frac{1}{2}, 2\frac{1}{2}, 3\frac{1}{2}, 4\frac{1}{2}$
   d  6, 3, 0, −3, −6
   e  −7, −4, −1, 2, 5
   f  −9, −13, −17, −21, −25

2. Copy and complete each of the following tables of linear sequences:

a

| Position | 1 | 2 | 5 |    | 50 | $n$ |
|---|---|---|---|---|---|---|
| Term |   |   | 45 |   |   | $4n - 3$ |

b

| Position | 1 | 2 | 5 |   |   | $n$ |
|---|---|---|---|---|---|---|
| Term |   |   | 59 | 449 |   | $6n - 1$ |

c

| Position | 1 |   |   |   | 100 | $n$ |
|---|---|---|---|---|---|---|
| Term |   | 0 | −5 | −47 |   | $-n + 3$ |

d

| Position | 1 | 2 | 3 |   |   | $n$ |
|---|---|---|---|---|---|---|
| Term | 3 | 0 | −3 | −24 | −294 |   |

e

| Position |   | 5 | 7 |   |   | $n$ |
|---|---|---|---|---|---|---|
| Term | 1 | 10 | 16 | 25 | 145 |   |

f

| Position | 1 | 2 | 5 |   | 50 | $n$ |
|---|---|---|---|---|---|---|
| Term | −5.5 | −7 |   | −34 |   |   |

3. For each of the following linear sequences:
   i   find the common difference $d$
   ii  give the formula for the $n$th term
   iii calculate the 50th term.
   a  5, 9, 13, 17, 21
   b  0, ..., 2, ..., 4
   c  −10, ..., ..., ..., 2
   d  $T_1 = 6, T_9 = 10$
   e  $T_3 = -50, T_{20} = 18$
   f  $T_5 = 60, T_{12} = 39$

A graphic display calculator can be used to check your answers by using it to generate a sequence based on the rule for the $n$th term. In the previous worked example, the $n$th term for the sequence 12, 7, 2, −3, −8, ... was calculated as being $T_n = -5n + 17$. The instructions below show how this can be checked.

179

# 20 SEQUENCES

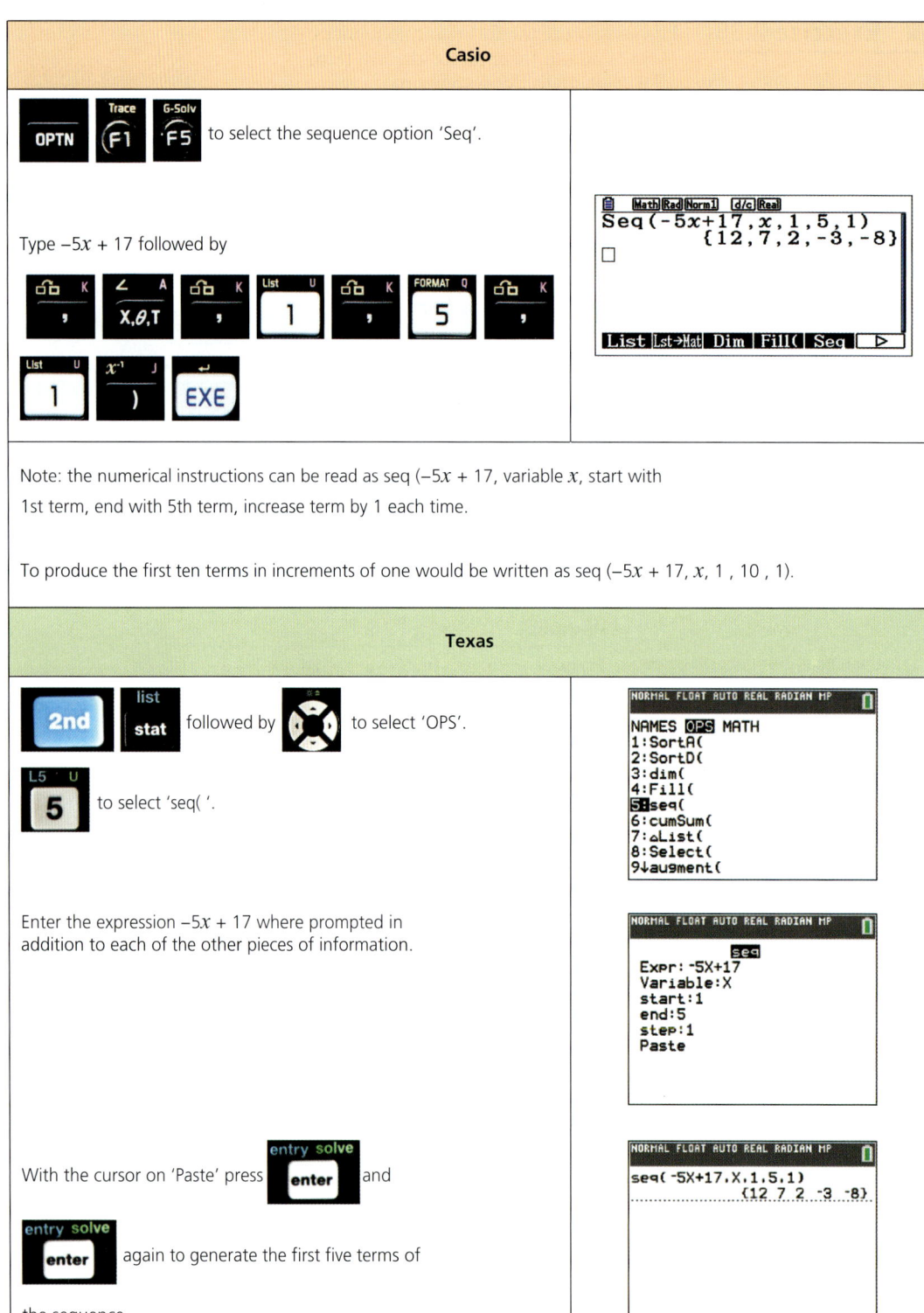

## Casio

OPTN (F1) (F5) to select the sequence option 'Seq'.

Type $-5x + 17$ followed by

[,] [X,θ,T] [,] [1] [,] [5] [,] [1] [)] [EXE]

Note: the numerical instructions can be read as seq ($-5x + 17$, variable $x$, start with 1st term, end with 5th term, increase term by 1 each time.

To produce the first ten terms in increments of one would be written as seq ($-5x + 17$, $x$, 1, 10, 1).

## Texas

2nd stat followed by ◄► to select 'OPS'.

5 to select 'seq( '.

Enter the expression $-5x + 17$ where prompted in addition to each of the other pieces of information.

With the cursor on 'Paste' press enter and

enter again to generate the first five terms of

the sequence.

# Sequences with quadratic and cubic rules

So far all the sequences you have looked at have been arithmetic, i.e. the rule for the $n$th term is linear and takes the form $T_n = an + b$. The rule for the $n$th term can be found algebraically using the method of differences and this method is particularly useful for more complex sequences.

## Worked examples

a Find the rule for the $n$th term for the sequence 4, 7, 10, 13, 16, … .
Firstly, produce a table of the terms and their positions in the sequence:

| Position | 1 | 2 | 3 | 4 | 5 |
|---|---|---|---|---|---|
| Term | 4 | 7 | 10 | 13 | 16 |

Extend the table to look at the differences:

| Position | 1 | 2 | 3 | 4 | 5 |
|---|---|---|---|---|---|
| Term | 4 | 7 | 10 | 13 | 16 |
| 1st Difference | | 3 | 3 | 3 | 3 |

As the row of 1st differences is constant, the rule for the $n$th term is linear and takes the form $T_n = an + b$.
By substituting the values of $n$ into the rule, each term can be written in terms of $a$ and $b$:

| Position | 1 | 2 | 3 | 4 | 5 |
|---|---|---|---|---|---|
| Term | $a+b$ | $2a+b$ | $3a+b$ | $4a+b$ | $5a+b$ |
| 1st Difference | | $a$ | $a$ | $a$ | $a$ |

Compare the two tables in order to find the values of $a$ and $b$:

$a = 3$
$a + b = 4$; therefore $b = 1$

The rule for the $n$th term $T_n = an + b$ can be written as $T_n = 3n + 1$.
For a linear rule, this method is perhaps overcomplicated. However, it is very efficient for quadratic and cubic rules.

b Find the rule for the $n$th term for the sequence 0, 7, 18, 33, 52, … .
Entering the sequence in a table gives:

| Position | 1 | 2 | 3 | 4 | 5 |
|---|---|---|---|---|---|
| Term | 0 | 7 | 18 | 33 | 52 |

> **Note**
> Students following the core syllabus will be only expected to use the difference method with a simple quadratic sequence. For example, the $n$th term of 3, 6, 11, 18, … is $n^2 + 2$

## 20 SEQUENCES

Extending the table to look at the differences gives:

| Position | 1 | 2 | 3 | 4 | 5 |
|---|---|---|---|---|---|
| Term | 0 | 7 | 18 | 33 | 52 |
| 1st Difference | | 7 | 11 | 15 | 19 |

The row of 1st differences is not constant, and so the rule for the $n$th term is not linear. Extend the table again to look at the row of 2nd differences:

| Position | 1 | 2 | 3 | 4 | 5 |
|---|---|---|---|---|---|
| Term | 0 | 7 | 18 | 33 | 52 |
| 1st Difference | | 7 | 11 | 15 | 19 |
| 2nd Difference | | | 4 | 4 | 4 |

The row of 2nd differences is constant, and so the rule for the $n$th term is therefore a quadratic which takes the form $T_n = an^2 + bn + c$.
By substituting the values of $n$ into the rule, each term can be written in terms of $a$, $b$ and $c$ as shown:

| Position | 1 | 2 | 3 | 4 | 5 |
|---|---|---|---|---|---|
| Term | $a+b+c$ | $4a+2b+c$ | $9a+3b+c$ | $16a+4b+c$ | $25a+5b+c$ |
| 1st Difference | | $3a+b$ | $5a+b$ | $7a+b$ | $9a+b$ |
| 2nd Difference | | | $2a$ | $2a$ | $2a$ |

Comparing the two tables, the values of $a$, $b$ and $c$ can be deduced:

$2a = 4$ and therefore $a = 2$

$3a + b = 7$ and therefore $6 + b = 7$, giving $b = 1$

$a + b + c = 0$ and therefore $2 + 1 + c = 0$, giving $c = -3$

The rule for the $n$th term $T_n = an^2 + bn + c$ can be written as $T_n = 2n^2 + n - 3$.

c Find the rule for the $n$th term for the sequence $-6, -8, -6, 6, 34, \ldots$ .
Entering the sequence in a table gives:

| Position | 1 | 2 | 3 | 4 | 5 |
|---|---|---|---|---|---|
| Term | −6 | −8 | −6 | 6 | 34 |

Extending the table to look at the differences:

| Position | 1 | 2 | 3 | 4 | 5 |
|---|---|---|---|---|---|
| Term | −6 | −8 | −6 | 6 | 34 |
| 1st Difference | | −2 | 2 | 12 | 28 |

## Sequences with quadratic and cubic rules

The row of 1st differences is not constant, and so the rule for the $n$th term is not linear. Extend the table again to look at the row of 2nd differences:

| Position | 1 | 2 | 3 | 4 | 5 |
|---|---|---|---|---|---|
| Term | −6 | −8 | −6 | 6 | 34 |
| 1st Difference | | −2 | 2 | 12 | 28 |
| 2nd Difference | | | 4 | 10 | 16 |

The row of 2nd differences is not constant either, and so the rule for the $n$th term is not quadratic. Extend the table by a further row to look at the row of 3rd differences:

| Position | 1 | 2 | 3 | 4 | 5 |
|---|---|---|---|---|---|
| Term | −6 | −8 | −6 | 6 | 34 |
| 1st Difference | | −2 | 2 | 12 | 28 |
| 2nd Difference | | | 4 | 10 | 16 |
| 3rd Difference | | | | 6 | 6 |

The row of 3rd differences is constant, and so the rule for the $n$th term is therefore a cubic which takes the form $T_n = an^3 + bn^2 + cn + d$.
By substituting the values of $n$ into the rule, each term can be written in terms of $a$, $b$, $c$ and $d$ as shown:

| Position | 1 | 2 | 3 | 4 | 5 |
|---|---|---|---|---|---|
| Term | $a+b+c+d$ | $8a+4b+2c+d$ | $27a+9b+3c+d$ | $64a+16b+4c+d$ | $125a+25b+5c+d$ |
| 1st Difference | | $7a+3b+c$ | $19a+5b+c$ | $37a+7b+c$ | $61a+9b+c$ |
| 2nd Difference | | | $12a+2b$ | $18a+2b$ | $24a+2b$ |
| 3rd Difference | | | | $6a$ | $6a$ |

By comparing the two tables, equations can be formed and the values of $a$, $b$, $c$ and $d$ can be found:

$6a = 6$  Therefore  $a = 1$

$12a + 2b = 4$  Therefore  $12 + 2b = 4$,  giving  $b = -4$

$7a + 3b + c = -2$  Therefore  $7 - 12 + c = -2$,  giving  $c = 3$

$a + b + c + d = -6$  Therefore  $1 - 4 + 3 + d = -6$,  giving  $d = -6$

Therefore the equation for the $n$th term is $T_n = n^3 - 4n^2 + 3n - 6$.

# 20 SEQUENCES

**Exercise 2.39** By using a table if necessary, find the formula for the *n*th term of each of the following sequences. Check your formula for the *n*th term using your graphic display calculator.

1  2, 5, 10, 17, 26
2  0, 3, 8, 15, 24
3  6, 9, 14, 21, 30
4  9, 12, 17, 24, 33
5  −2, 1, 6, 13, 22
6  4, 10, 20, 34, 52
7  0, 6, 16, 30, 48
8  5, 14, 29, 50, 77
9  0, 12, 32, 60, 96
10  1, 16, 41, 76, 121

**Exercise 2.40** Use a table to find the formula for the *n*th term of the following sequences:

1  11, 18, 37, 74, 135
2  0, 6, 24, 60, 120
3  −4, 3, 22, 59, 120
4  2, 12, 36, 80, 150
5  7, 22, 51, 100, 175
6  7, 28, 67, 130, 223
7  1, 10, 33, 76, 145
8  13, 25, 49, 91, 157

## Exponential sequences

So far you have looked at sequences where there is a common difference between successive terms. There are, however, other types of sequences, e.g. 2, 4, 8, 16, 32. There is clearly a pattern to the way the numbers are generated as each term is double the previous term, but there is no common difference.

A sequence where there is a **common ratio** (*r*) between successive terms is known as an **exponential sequence** (or also a **geometric sequence**). For example:

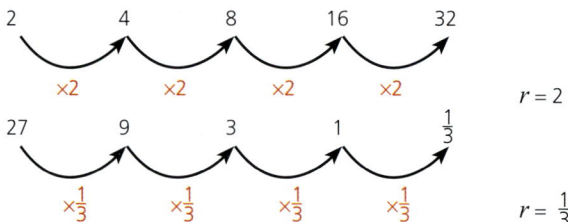

As with a linear sequence, there are two main ways of describing an exponential sequence.

1  The term-to-term rule

For example, for the following sequence,

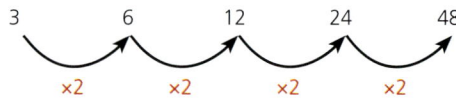

the general rule is $T_{n+1} = 2T_n$; $T_1 = 3$.

## Applications of exponential sequences

**2  The formula for the $n$th term of an exponential sequence**

As with a linear sequence, this rule links each term to its position in the sequence,

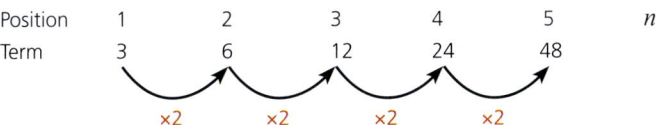

to reach the second term, the calculation is $3 \times 2$ or $3 \times 2^1$
to reach the third term, the calculation is $3 \times 2 \times 2$ or $3 \times 2^2$
to reach the fourth term, the calculation is $3 \times 2 \times 2 \times 2$ or $3 \times 2^3$
In general, therefore

$$T_n = ar^{n-1}$$

where $a$ is the first term and $r$ is the common ratio.

# Applications of exponential sequences

In Topic 1, simple and compound interest were shown as different ways that interest could be earned on money left in a bank account for a period of time. You will now look at compound interest as an example of an exponential sequence.

## Compound interest

For example, $100 is deposited in a bank account and left untouched. After 1 year the amount has increased to $110 as a result of interest payments. To work out the interest rate, calculate the multiplier from $100 → $110:

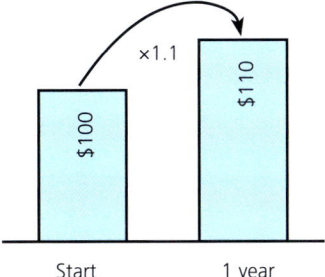

$\dfrac{110}{100} = 1.1$

The multiplier is 1.1. This corresponds to a 10% increase. Therefore the simple interest rate is 10% in the first year.

Assume the money is left in the account and that the interest rate remains unchanged. Calculate the amount in the account after five years.

# 20 SEQUENCES

This is an example of an exponential sequence.

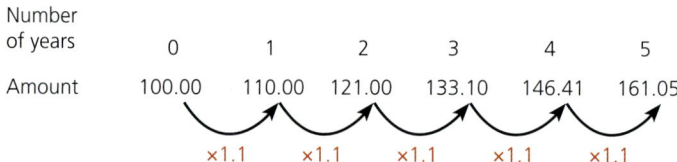

Alternatively the amount after 5 years can be calculated using a variation of $T_n = ar^{n-1}$, i.e. $T_5 = 100 \times 1.1^5 = 161.05$. Note: As the number of years starts at 0, ×1.1 is applied five times to get to the fifth year.

This is an example of compound interest, as the previous year's interest is added to the total and included in the following year's calculation.

## ➜ Worked examples

**a** Alex deposits $1500 in his savings account. The interest rate offered by the savings account is 6% each year for a 10-year period. Assuming Alex leaves the money in the account, calculate how much interest he has gained after the 10 years.

An interest rate of 6% implies a common ratio of 1.06
Therefore $T_{10} = 1500 \times 1.06^{10} = 2686.27$
The amount of interest gained is $2686.27 - 1500 = \$1186.27$

**b** Adrienne deposits $2000 in her savings account. The interest rate offered by the bank for this account is 8% compound interest per year. Calculate the number of years Adrienne needs to leave the money in her account for it to double in value.

An interest rate of 8% implies a common ratio of 1.08

The amount each year can be found using the term-to-term rule
$T_{n+1} = 1.08 \times T_n$
$T_1 = 2000 \times 1.08 = 2160$
$T_2 = 2160 \times 1.08 = 2332.80$
$T_3 = 2332.80 \times 1.08 = 2519.42$
...
$T_9 = 3998.01$
$T_{10} = 4317.85$

Adrienne needs to leave the money in the account for 10 years in order for it to double in value.

# Combinations of sequences

**Exercise 2.41**

1. Identify which of the following are exponential sequences and which are not.
   - **a** 2, 6, 18, 54
   - **b** 25, 5, 1, $\frac{1}{5}$
   - **c** 1, 4, 9, 16
   - **d** −3, 9, −27, 81
   - **e** $\frac{1}{2}, \frac{2}{3}, \frac{3}{4}, \frac{4}{5}$
   - **f** $\frac{1}{2}, \frac{2}{4}, \frac{4}{8}, \frac{8}{16}$

2. For the sequences in Q.1 that are exponential, calculate:
   - **i** the common ratio $r$
   - **ii** the next two terms
   - **iii** a formula for the $n$th term.

3. The $n$th term of an exponential sequence is given by the formula $T_n = -6 \times 2^{n-1}$.
   - **a** Calculate $T_1$, $T_2$ and $T_3$.
   - **b** What is the value of $n$, if $T_n = -768$?

4. Part of an exponential sequence is given below:

   ..., −1, ..., ..., 64, ...   where $T_2 = -1$ and $T_5 = 64$.

   Calculate:
   - **a** the common ratio $r$
   - **b** the value of $T_1$
   - **c** the value of $T_{10}$.

5. Ayse takes out a loan with a mortgage company for €200 000. The interest rate is 6% per year. If she is unable to repay any of the loan during the first 3 years, calculate the extra amount she will have to pay by the end of the third year because of interest.

6. A car is bought for $10 000. It loses value at a rate of 20% each year.
   - **a** Explain why the car is not worthless after 5 years.
   - **b** Calculate its value after 5 years.
   - **c** Explain why a depreciation of 20% per year means, in theory, that the car will never be worthless.

# Combinations of sequences

So far, you have looked at arithmetic and exponential sequences and explored different ways of working out the term-to-term rule and the rule for the $n$th term. However, sometimes sequences are just variations of well-known ones. Being aware of these can often save a lot of time and effort.

Consider the sequence 2, 5, 10, 17, 26. By looking at the differences it could be established that the 2nd differences are constant and, therefore, the formula for the nth term will involve an $n^2$ term, but this takes quite a lot of time.

On closer inspection, when compared with the sequence 1, 4, 9, 16, 25, you can see that each of the terms in the sequence 2, 5, 10, 17, 26 has had 1 added to it.

## 20 SEQUENCES

1, 4, 9, 16, 25 is the well-known sequence of square numbers. The formula for the $n$th term for the sequence of square numbers is $T_n = n^2$; therefore the formula for the $n$th term for the sequence 2, 5, 10, 17, 26 is $T_n = n^2 + 1$.

> **Note**
>
> It is good practice to be aware of key sequences and to always check whether a sequence you are looking at is a variation of one of those.
>
> The key sequences include:
>
> Square numbers    1, 4, 9, 16, 25, … where $T_n = n^2$
>
> Cube numbers      1, 8, 27, 64, 125, … where $T_n = n^3$
>
> Powers of two     2, 4, 8, 16, 32, … where $T_n = 2^n$
>
> Triangle numbers   1, 3, 6, 10, 15, … where $T_n = \frac{1}{2}n(n+1)$

### ➡ Worked examples

a  Consider the sequence below:
   2, 8, 18, 32
   i  By inspection, write the rule for the $n$th term of the sequence.
      The terms of the sequence are double those of the sequence of square numbers, therefore $T_n = 2n^2$.
   ii  Write down the next two terms.
      50, 72

b  For the sequence below, by inspection, write down the rule for the $n$th term and the next two terms
   1, 3, 7, 15, 31
   The terms of the sequence are one less than the terms of the sequence of powers of two. Therefore $T_n = 2^n - 1$
   The next two terms are 63, 127.

 **Exercise 2.42**  In each of the questions below:
i  write down the rule for the $n$th term of the sequence by inspection
ii  write down the next two terms of the sequence.

1  2, 5, 10, 17          2  3, 10, 29, 66       3  $\frac{1}{2}, 2, 4\frac{1}{2}, 8$
4  1, 2, 4, 8             5  2, 6, 12, 20         6  2, 12, 36, 80
7  0, 4, 18, 48          8  6, 12, 24, 48

# 21 Direct and inverse variation

## Direct variation

Consider the tables below:

| x | 0 | 1 | 2 | 3 | 5  | 10 |
|---|---|---|---|---|----|----|
| y | 0 | 2 | 4 | 6 | 10 | 20 |

$y = 2x$

| x | 0 | 1 | 2 | 3 | 5  | 10 |
|---|---|---|---|---|----|----|
| y | 0 | 3 | 6 | 9 | 15 | 30 |

$y = 3x$

| x | 0 | 1   | 2 | 3   | 5    | 10 |
|---|---|-----|---|-----|------|----|
| y | 0 | 2.5 | 5 | 7.5 | 12.5 | 25 |

$y = 2.5x$

In each case $y$ is directly proportional to $x$. This is written $y \propto x$. If any of these three tables is shown on a graph, the graph will be a straight line passing through the origin $(0, 0)$.

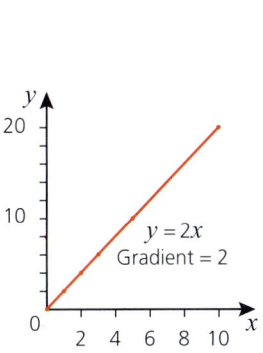
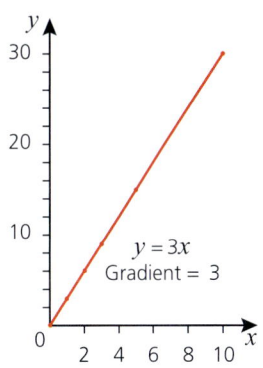
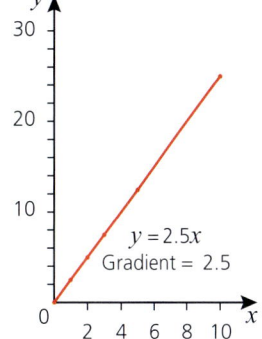

For any statement where $y \propto x$,

$y = kx$

where $k$ is a constant equal to the gradient of the graph and is called the **constant of proportionality** or **constant of variation**.

Consider the tables below:

| x | 1 | 2 | 3  | 4  | 5  |
|---|---|---|----|----|----|
| y | 2 | 8 | 18 | 32 | 50 |

$y = 2x^2$

# 21 DIRECT AND INVERSE VARIATION

| $x$ | 1 | 2 | 3 | 4 | 5 | $y = \frac{1}{2}x^3$ |
|---|---|---|---|---|---|---|
| $y$ | $\frac{1}{2}$ | 4 | $13\frac{1}{2}$ | 32 | $62\frac{1}{2}$ | |

| $x$ | 1 | 2 | 3 | 4 | 5 | $y = \sqrt{x} = x^{\frac{1}{2}}$ |
|---|---|---|---|---|---|---|
| $y$ | 1 | $\sqrt{2}$ | $\sqrt{3}$ | 2 | $\sqrt{5}$ | |

In the cases above, $y$ is directly proportional to $x^n$, where $n > 0$. This can be written as $y \propto x^n$.

The graphs of each of the three equations are shown below:

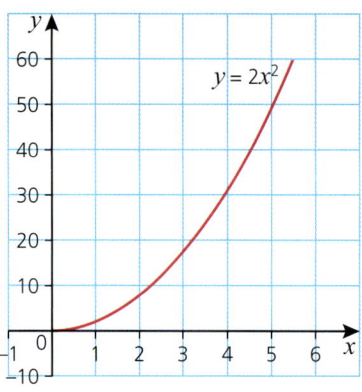
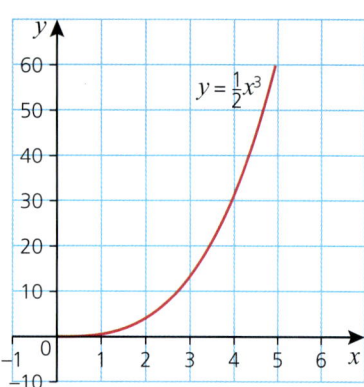

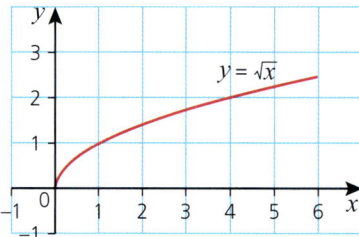

The graphs above, with $(x, y)$ plotted, are not linear. However, if the graph of $y = 2x^2$ is plotted as $(x^2, y)$, then the graph is linear and passes through the origin, demonstrating that $y \propto x^2$ as shown.

| $x$ | 1 | 2 | 3 | 4 | 5 |
|---|---|---|---|---|---|
| $x^2$ | 1 | 4 | 9 | 16 | 25 |
| $y$ | 2 | 8 | 18 | 32 | 50 |

*Inverse variation*

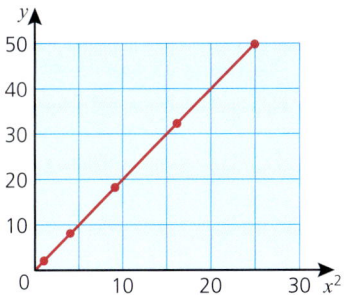

Similarly, the graph of $y = \frac{1}{2}x^3$ is curved when plotted as $(x, y)$, but is linear and passes through the origin if it is plotted as $(x^3, y)$ as shown:

| $x$   | 1             | 2 | 3               | 4  | 5               |
|-------|---------------|---|-----------------|----|-----------------|
| $x^3$ | 1             | 8 | 27              | 64 | 125             |
| $y$   | $\frac{1}{2}$ | 4 | $13\frac{1}{2}$ | 32 | $62\frac{1}{2}$ |

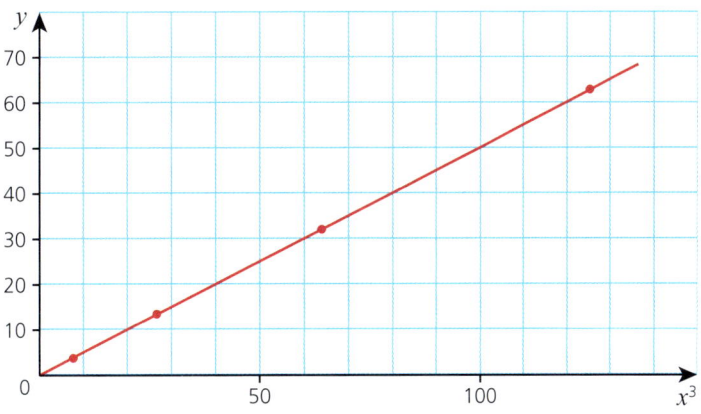

The graph of $y = \sqrt{x}$ is also linear if plotted as $(\sqrt{x}, y)$.

# Inverse variation

If $y$ is **inversely proportional** to $x$, then $y \propto \frac{1}{x}$ and $y = \frac{k}{x}$.
If a graph of $y$ against $\frac{1}{x}$ is plotted, this too will be a straight line passing through the origin.

## 21 DIRECT AND INVERSE VARIATION

### → Worked examples

a  $y \propto x$. If $y = 7$ when $x = 2$, find $y$ when $x = 5$.

$y = kx$
$7 = k \times 2$
$k = 3.5$
When $x = 5$,
$y = 3.5 \times 5$
$= 17.5$

b  $y \propto \frac{1}{x}$. If $y = 5$ when $x = 3$, find $y$ when $x = 30$.

$y = \frac{k}{x}$
$5 = \frac{k}{3}$
$k = 15$
When $x = 30$,
$y = \frac{15}{30}$
$= 0.5$

### Exercise 2.43

1  $y$ is directly proportional to $x$. If $y = 6$ when $x = 2$, find:
   a  the constant of proportionality
   b  the value of $y$ when $x = 7$
   c  the value of $y$ when $x = 9$
   d  the value of $x$ when $y = 9$
   e  the value of $x$ when $y = 30$.

2  $y$ is directly proportional to $x^2$. If $y = 18$ when $x = 6$, find:
   a  the constant of proportionality
   b  the value of $y$ when $x = 4$
   c  the value of $y$ when $x = 7$
   d  the value of $x$ when $y = 32$
   e  the value of $x$ when $y = 128$.

3  $y$ is inversely proportional to $x^3$. If $y = 3$ when $x = 2$, find:
   a  the constant of proportionality
   b  the value of $y$ when $x = 4$
   c  the value of $y$ when $x = 6$
   d  the value of $x$ when $y = 24$.

4  $y$ is inversely proportional to $x^2$. If $y = 1$ when $x = 0.5$, find:
   a  the constant of proportionality
   b  the value of $y$ when $x = 0.1$
   c  the value of $y$ when $x = 0.25$
   d  the value of $x$ when $y = 64$.

### Exercise 2.44

1  Write each of the following in the form:
   i   $y \propto x$
   ii  $y = kx$.
      a  $y$ is directly proportional to $x^3$
      b  $y$ is inversely proportional to $x^3$
      c  $t$ is directly proportional to $P$

## Inverse variation

    **d**  $s$ is inversely proportional to $t$
    **e**  $A$ is directly proportional to $r^2$
    **f**  $T$ is inversely proportional to the square root of $g$

**2**  If $y \propto x$ and $y = 6$ when $x = 2$, find $y$ when $x = 3.5$.

**3**  If $y \propto \frac{1}{x}$ and $y = 4$ when $x = 2.5$, find:
    **a**  $y$ when $x = 20$
    **b**  $x$ when $y = 5$.

**4**  If $p \propto r^2$ and $p = 2$ when $r = 2$, find $p$ when $r = 8$.

**5**  If $m \propto \frac{1}{r^3}$ and $m = 1$ when $r = 2$, find:
    **a**  $m$ when $r = 4$
    **b**  $r$ when $m = 125$.

**6**  If $y \propto x^2$ and $y = 12$ when $x = 2$, find $y$ when $x = 5$.

### Exercise 2.45

**1**  If a stone is dropped off the edge of a cliff, the height ($h$ metres) of the cliff is proportional to the square of the time ($t$ seconds) taken for the stone to reach the ground.
A stone takes 5 seconds to reach the ground when dropped off a cliff 125 m high.
    **a**  Write down a relationship between $h$ and $t$, using $k$ as the constant of variation.
    **b**  Calculate the constant of variation.
    **c**  Find the height of a cliff if a stone takes 3 seconds to reach the ground.
    **d**  Find the time taken for a stone to fall from a cliff 180 m high.

**2**  The velocity ($v$ metres per second) of a body is known to be proportional to the square root of its kinetic energy ($e$ joules). When the velocity of a body is 120 m/s, its kinetic energy is 1600 J.
    **a**  Write down a relationship between $v$ and $e$, using $k$ as the constant of variation.
    **b**  Calculate the value of $k$.
    **c**  If $v = 21$, calculate the kinetic energy of the body in joules.

**3**  The length ($l$ cm) of an edge of a cube is proportional to the cube root of its mass ($m$ grams). It is known that if $l = 15$, then $m = 125$. Let $k$ be the constant of variation.
    **a**  Write down the relationship between $l$, $m$ and $k$.
    **b**  Calculate the value of $k$.
    **c**  Calculate the value of $l$ when $m = 8$.

**4**  The power ($P$) generated in an electrical circuit is proportional to the square of the current ($I$ amps). When the power is 108 watts, the current is 6 amps.
    **a**  Write down a relationship between $P$, $I$ and the constant of variation, $k$.
    **b**  Calculate the value of $I$ when $P = 75$ watts.

# 21 DIRECT AND INVERSE VARIATION

**Exercise 2.45 (cont)**

5 The following data shows the relationship between the two variables $x$ and $y$.

| $x$ | 0 | 1 | 2 | 3 | 4 |
|---|---|---|---|---|---|
| $y$ | 0 | −1.5 | −6 | −13.5 | −24 |

   **a** Plot a graph of $y$ against $x$.
   **b** **i** Describe a possible variation model between $y$ and $x$.
      **ii** Write the relationship between $y$ and $x$ algebraically using $k$ as the constant of variation.
   **c** Find the value of $k$.
   **d** Check that your value of $k$ works for the data above.

6 The following data shows the relationship between the two variables $x$ and $y$.

| $x$ | 0 | 1 | 4 | 9 | 16 |
|---|---|---|---|---|---|
| $y$ | 0 | 2 | 4 | 6 | 8 |

   **a** Plot a graph of $y$ against $x$.
   **b** **i** Describe a possible variation model between $y$ and $x$.
      **ii** Write the relationship between $y$ and $x$ algebraically using $k$ as the constant of variation.
   **c** Find the value of $k$.
   **d** Check that your value of $k$ works for the data above.

# Investigations, modelling and ICT 2

## Investigation: House of cards

The drawing shows a house of cards three layers high. 15 cards are needed to construct it.

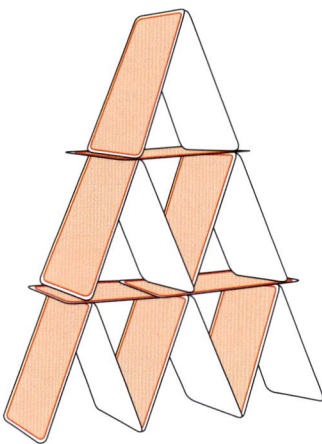

1. How many cards are needed to construct a house 10 layers high?
2. The world record is for a house 61 layers high. How many cards are needed to construct this house of cards?
3. Show that the general formula for a house $n$ layers high requiring $c$ cards is:

$$c = \tfrac{1}{2} n(3n + 1)$$

## Investigation: Chequered board

A chessboard is an $8 \times 8$ square grid consisting of alternating black and white squares as shown on the left.

There are 64 unit squares of which 32 are black and 32 are white.

Consider boards of different sizes. The examples below show rectangular boards, each consisting of alternating black and white unit squares.

Total number of unit squares is 30
Number of black squares is 15
Number of white squares is 15

# INVESTIGATIONS, MODELLING AND ICT 2

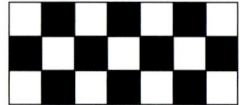

Total number of unit squares is 21
Number of black squares is 10
Number of white squares is 11

1. Investigate the number of black and white unit squares on different rectangular boards. For consistency, you may find it helpful to always keep the bottom right-hand square the same colour.
2. What is the number of black and white squares on a board $m \times n$ units?

## Modelling: Freefall

A skydiver jumps out of a plane and immediately begins to freefall. Her distance fallen ($D$ metres) at a time ($t$ seconds) after jumping out of the plane is recorded and presented in the table below.

| Time (s)     | 0 | 1   | 2    | 3    | 4    |
|--------------|---|-----|------|------|------|
| Distance (m) | 0 | 4.6 | 19.4 | 43.7 | 77.8 |

1. Plot a graph of time against distance fallen.
2. It is suspected that the relationship is of the form $D = \frac{1}{2}kt^2$ where $k$ is a constant.
   a. Plot a graph of time² against distance fallen.
   b. Explain how the graph of time² against distance fallen enables the value of $k$ to be calculated.
   c. Show that the value of $k \approx 9.8$.
3. Use the model to predict how far she will have fallen after:
   a. 5 seconds
   b. 10 seconds

When a person jumps out of a plane, their velocity does not increase indefinitely. Due to air resistance, there is a limit to the velocity at which a person will fall. This is known as the terminal velocity.

The terminal velocity is known to be approximately 195 km/h.

4. The formula for the skydiver's velocity ($v$ m/s) is known to be $v^2 = 2kD$.
   a. Convert the terminal velocity to metres/second.
   b. Calculate the distance she drops before she reaches the terminal velocity.
   c. How long does it take for her to reach the terminal velocity?
5. Refine your answer to 3b above as a result of your findings in Q.4.

## ICT Activity 1

For each question, use a graphing package to plot the inequalities on the same pair of axes. Leave unshaded the region which satisfies all of them simultaneously.

1. $y \leqslant x$     $y > 0$     $x \leqslant 3$
2. $x + y > 3$     $y \leqslant 4$     $y - x > 2$
3. $2y + x \leqslant 5$     $y - 3x - 6 < 0$     $2y - x > 3$

## ICT Activity 2

You have seen that it is possible to solve some exponential equations by applying the laws of indices.

Use a graphic display calculator and appropriate graphs to solve the following exponential equations:

1. $4^x = 40$
2. $3^x = 17$
3. $5^{x-1} = 6$
4. $3^{-x} = 0.5$

# Student assessments 2

## Student assessment 1

1. Expand the following and simplify where possible:
   a. $3(2x - 3y + 5z)$
   b. $4p(2m - 7)$
   c. $-4m(2mn - n^2)$
   d. $4p^2(5pq - 2q^2 - 2p)$
   e. $4x - 2(3x + 1)$
   f. $4x(3x - 2) + 2(5x^2 - 3x)$
   g. $\frac{1}{5}(15x - 10) - \frac{1}{3}(9x - 12)$
   h. $\frac{x}{2}(4x - 6) + \frac{x}{4}(2x + 8)$

2. Factorise the following:
   a. $16p - 8q$
   b. $p^2 - 6pq$
   c. $5p^2q - 10pq^2$
   d. $9pq - 6p^2q + 12q^2p$

3. If $a = 4$, $b = 3$ and $c = -2$, work out the following:
   a. $3a - 2b + 3c$
   b. $5a - 3b^2$
   c. $a^2 + b^2 + c^2$
   d. $(a + b)(a - b)$
   e. $a^2 - b^2$
   f. $b^3 - c^3$

4. Rearrange the following formulas to make the **red** letter the subject:
   a. $p = 4m + n$
   b. $4x - 3\mathbf{y} = 5z$
   c. $2x = \frac{3\mathbf{y}}{5p}$
   d. $m(x + \mathbf{y}) = 3w$
   e. $\frac{pq}{4r} = \frac{mn}{t}$
   f. $\frac{p + \mathbf{q}}{r} = m - n$

## Student assessment 2

1. Factorise the following fully:
   a. $pq - 3rq + pr - 3r^2$
   b. $1 - t^4$
   c. $875^2 - 125^2$
   d. $7.5^2 - 2.5^2$

2. Expand the following and simplify where possible:
   a. $(x - 4)(x + 2)$
   b. $(x - 8)^2$
   c. $(x + y)^2$
   d. $(x - 11)(x + 11)$
   e. $(3x - 2)(2x - 3)$
   f. $(5 - 3x)^2$

3. Factorise the following:
   a. $x^2 - 4x - 77$
   b. $x^2 - 6x + 9$
   c. $x^2 - 144$
   d. $3x^2 + 3x - 18$
   e. $2x^2 + 5x - 12$
   f. $4x^2 - 20x + 25$

4. Make the letter in **red** the subject of the formula:
   a. $m\mathbf{f}^2 = p$
   b. $m = 5\mathbf{t}^2$
   c. $A = \pi r\sqrt{\mathbf{p} + q}$
   d. $\frac{1}{\mathbf{x}} + \frac{1}{y} = \frac{1}{t}$

5. Simplify the following algebraic fractions:
   a. $\frac{x^7}{x^3}$
   b. $\frac{mn}{p} \times \frac{pq}{m}$
   c. $\frac{(y^3)^3}{(y^2)^3}$
   d. $\frac{28pq^2}{7pq^3}$

**6** Simplify the following algebraic fractions:
  a $\dfrac{m}{11} + \dfrac{3m}{11} - \dfrac{2m}{11}$
  b $\dfrac{3p}{8} - \dfrac{9p}{16}$
  c $\dfrac{4x}{3y} - \dfrac{7x}{12y}$
  d $\dfrac{3m}{15p} + \dfrac{4n}{5p} - \dfrac{11n}{30p}$

**7** Simplify the following:
  a $\dfrac{p}{5} + \dfrac{p}{4}$
  b $\dfrac{3m}{5} - \dfrac{2m}{4}$
  c $\dfrac{2p}{3} - \dfrac{3p}{4}$

**8** Simplify the following:
  a $\dfrac{4}{(x-5)} + \dfrac{3}{(x-2)}$
  b $\dfrac{a^2 - b^2}{(a+b)^2}$
  c $\dfrac{x-2}{x^2 + x - 6}$

## Student assessment 3

**1** The volume of a cylinder is given by the formula $V = \pi r^2 h$, where $h$ is the height of the cylinder and $r$ is the radius.
  a Find the volume of a cylindrical post of length 7.5 m and with a diameter of 30 cm.
  b Make $r$ the subject of the formula.
  c A cylinder of height 75 cm has a volume of 6000 cm³. Find its radius correct to three significant figures.

**2** The formula $C = \dfrac{5}{9}(F - 32)$ can be used to convert temperatures in degrees Fahrenheit (°F) into degrees Celsius (°C).
  a What temperature in °C is equivalent to 150 °F?
  b What temperature in °C is equivalent to 12 °F?
  c Make $F$ the subject of the formula.
  d Use your rearranged formula to find what temperature in °F is equivalent to 160 °C.

**3** The height of Mount Kilimanjaro is given as 5900 m. The formula for the time taken, $T$ hours, to climb to a height $H$ metres is:

$$T = \dfrac{H}{1200} + k$$

where $k$ is a constant.
  a Calculate the time taken, to the nearest hour, to climb to the top of the mountain if $k = 9.8$.
  b Make $H$ the subject of the formula.
  c How far up the mountain, to the nearest 100 m, could you expect to be after 14 hours?

**4** The formula for the volume $V$ of a sphere is given as $V = \dfrac{4}{3}\pi r^3$.
  a Find $V$ when $r = 5$ cm.
  b Make $r$ the subject of the formula.
  c Find the radius of a sphere of volume 2500 m³.

**5** The cost $x$ of printing $n$ newspapers is given by the formula $x = 1.50 + 0.05n$.
  a Calculate the cost of printing 5000 newspapers.
  b Make $n$ the subject of the formula.
  c How many newspapers can be printed for $25?

# STUDENT ASSESSMENTS 2

## Student assessment 4

For Q.1–4, solve the equations.

1. 
   a $y + 9 = 3$
   b $3x - 5 = 13$
   c $12 - 5p = -8$
   d $2.5y + 1.5 = 7.5$

2. 
   a $5 - p = 4 + p$
   b $8m - 9 = 5m + 3$
   c $11p - 4 = 9p + 15$
   d $27 - 5r = r - 3$

3. 
   a $\frac{p}{-2} = -3$
   b $6 = \frac{2}{5}x$
   c $\frac{m-7}{5} = 3$
   d $\frac{4t-3}{3} = 7$

4. 
   a $\frac{2}{5}(t - 1) = 3$
   b $5(3 - m) = 4(m - 6)$
   c $5 = \frac{2}{3}(x - 1)$
   d $\frac{4}{5}(t - 2) = \frac{1}{4}(2t + 8)$

5. Solve the following simultaneous equations:
   a $x + y = 11$
      $x - y = 3$
   b $5p - 3q = -1$
      $-2p - 3q = -8$
   c $3x + 5y = 26$
      $x - y = 6$
   d $2m - 3n = -9$
      $3m + 2n = 19$

## Student assessment 5

1. The angles of a quadrilateral are $x$, $3x$, $(2x - 40)$ and $(3x - 50)$ degrees.
   a Construct an equation in terms of $x$.
   b Solve the equation.
   c Calculate the size of the four angles.

2. Three is subtracted from seven times a number. The result is multiplied by 5. If the answer is 55, calculate the value of the number by constructing an equation and solving it.

3. The interior angles of a pentagon are $9x$, $5x + 10$, $6x + 5$, $8x - 25$ and $10x - 20$ degrees. If the sum of the interior angles of a pentagon is 540°, find the size of each of the angles.

4. Solve $x^2 - x = 20$ by factorisation.

5. Solve $2x^2 - 7 = 3x$ by using the quadratic formula.

6. Solve the inequality $6 < 2x \leqslant 10$ and show your answer on a number line.

7. For what values of $m$ is $\frac{1}{m^2} > 0$ true?

## Student assessment 6

1. The angles of a triangle are $x°$, $y°$ and $40°$. The difference between the two unknown angles is $30°$.
   a. Write down two equations from the information given above.
   b. What are the sizes of the two unknown angles?
2. The interior angles of a pentagon increase by $10°$ as you progress clockwise.
   a. Draw a diagram to show this information.
   b. Write an expression for the sum of the interior angles.
   c. The sum of the interior angles of a pentagon is $540°$. Use this to calculate the largest **exterior** angle of the pentagon.
   d. Show, on your diagram, the size of each of the five exterior angles.
   e. Show that the sum of the exterior angles is $360°$.
3. A flat sheet of card measures 12 cm by 10 cm. It is made into an open box by cutting a square of side $x$ cm from each corner and then folding up the sides.
   a. Show the flat card and its dimensions on a simple two-dimensional sketch, and the box and its dimensions on a simple three-dimensional sketch.
   b. Write an expression for the surface area of the outside of the box.
   c. If the surface area is 56 cm², form and solve a quadratic equation to find the value of $x$.
4. a. Show that $x - 2 = \frac{4}{x-3}$ can be written as $x^2 - 5x + 2 = 0$.
   b. Use the quadratic formula to solve $x - 2 = \frac{4}{x-3}$.
5. A right-angled triangle ABC has side lengths as follows: AB = $x$ cm, AC is 2 cm shorter than AB, and BC is 2 cm shorter than AC.
   a. Draw a diagram to show this information.
   b. Using this information, show that $x^2 - 12x + 20 = 0$.
   c. Solve the above quadratic equation and hence find the length of each of the three sides of the triangle.
6. Solve the following inequalities:
   a. $5 + 6x \leq 47$
   b. $4 \geq \frac{y+3}{3}$
7. Find the range of values for which:
   a. $3 \leq 3p < 12$
   b. $24 < 8(x - 1) \leq 48$

## Student assessment 7

1. Using indices, simplify the following:
   a. $3 \times 2 \times 2 \times 3 \times 27$
   b. $2 \times 2 \times 4 \times 4 \times 4 \times 2 \times 32$

2. Write the following out in full:
   a. $6^5$
   b. $2^{-5}$

3. Work out the value of the following without using a calculator:
   a. $3^3 \times 10^3$
   b. $1^{-4} \times 5^3$

4. Simplify the following using indices:
   a. $2^4 \times 2^3$
   b. $7^5 \times 7^2 \times 3^4 \times 3^8$
   c. $\dfrac{4^8}{2^{10}}$
   d. $\dfrac{(3^3)^4}{27^3}$
   e. $\dfrac{7^6 \times 4^2}{4^3 \times 7^6}$
   f. $\dfrac{8^{-2} \times 2^6}{2^{-2}}$

5. Without using a calculator, work out the following:
   a. $5^2 \times 5^{-1}$
   b. $\dfrac{4^5}{4^3}$
   c. $\dfrac{7^{-5}}{7^{-7}}$
   d. $\dfrac{3^{-5} \times 4^2}{3^{-6}}$

6. Find the value of $x$ in each of the following:
   a. $2^{(2x+2)} = 128$
   b. $\dfrac{1}{4^{-x}} = \dfrac{1}{2}$
   c. $3^{(-x+4)} = 81$
   d. $8^{-3x} = \dfrac{1}{4}$

## Student assessment 8

1. Find the value of the following:
   a. $64^{\frac{1}{6}}$
   b. $27^{\frac{4}{3}}$
   c. $9^{-\frac{1}{2}}$
   d. $512^{\frac{2}{3}}$
   e. $\sqrt[3]{27}$
   f. $\sqrt[4]{16}$
   g. $\dfrac{1}{36^{-\frac{1}{2}}}$
   h. $\dfrac{2}{64^{-\frac{2}{3}}}$

2. Find the value of the following:
   a. $\dfrac{25^{\frac{1}{2}}}{9^{-\frac{1}{2}}}$
   b. $\dfrac{4^{\frac{5}{2}}}{2^3}$
   c. $\dfrac{27^{\frac{4}{3}}}{3^3}$
   d. $25^{\frac{3}{2}} \times 5^2$
   e. $4^{\frac{6}{4}} \times 4^{-\frac{1}{2}}$
   f. $\dfrac{27^{\frac{2}{3}} \times 3^{-3}}{9^{-\frac{1}{2}}}$
   g. $\dfrac{(4^2)^{-\frac{1}{4}} \times 9^{\frac{3}{2}}}{\left(\frac{1}{4}\right)^{\frac{1}{2}}}$
   h. $\dfrac{(5^{\frac{1}{3}})^{\frac{1}{2}} \times 5^{\frac{5}{6}}}{4^{-\frac{1}{2}}}$

3. Draw a pair of axes with $x$ from $-4$ to $4$ and $y$ from $0$ to $18$.
   a. Plot a graph of $y = 4^{-\frac{x}{2}}$.
   b. Use your graph to estimate when $4^{-\frac{x}{2}} = 6$.

4. Using your graphic display calculator, solve the inequality $3x^2 \leq 6 - 17x$. Give your answers in exact form.

## Student assessment 9

1. For each of the following linear sequences:
   i write down a formula for the $n$th term
   ii calculate the 10th term.
   a 1, 5, 9, 13, ...
   b 1, −2, −5, −8, ...

2. For both of the following, calculate $T_5$ and $T_{100}$:
   a $T_n = 6n - 3$
   b $T_n = -\frac{1}{2}n + 4$

3. Copy and complete both of the following tables of linear sequences:
   a
   | Position | 1 | 2 | 3 | 10 | | $n$ |
   |---|---|---|---|---|---|---|
   | Term | 17 | 14 | | | −55 | |

   b
   | Position | 2 | 6 | 10 | | $n$ |
   |---|---|---|---|---|---|
   | Term | −4 | −2 | | 35 | |

4. Anya deposits $300 in a bank account. The bank offers 7% interest per year.
   Assuming she does not take any money out of the account, calculate:
   a the amount of money in the account after 8 years
   b the minimum number of years the money must be left in the account for the amount to be greater than $350.

5. A computer loses 35% of its value each year. If the computer cost €600 new, calculate:
   a its value after 2 years
   b its value after 10 years.

6. Part of an exponential sequence is given below:
   ...., ...., 27, ...., ...., −1
   where $T_3 = 27$ and $T_6 = -1$.
   Calculate:
   a the common ratio $r$
   b the value $T_1$
   c the value of $n$ if $T_n = -\frac{1}{81}$

7. Using a table of differences if necessary, calculate the rule for the $n$th term of the sequence 8, 24, 58, 116, 204, ... .

8. Using a table of differences, calculate the rule for the $n$th term of the sequence 10, 23, 50, 97, 170, ... .

# STUDENT ASSESSMENTS 2

## Student assessment 10

1. $y = kx$. When $y = 12$, $x = 8$.
   a. Calculate the value of $k$.
   b. Calculate $y$ when $x = 10$.
   c. Calculate $y$ when $x = 2$.
   d. Calculate $x$ when $y = 18$.

2. $y = \frac{k}{x}$. When $y = 2$, $x = 5$.
   a. Calculate the value of $k$.
   b. Calculate $y$ when $x = 4$.
   c. Calculate $x$ when $y = 10$.
   d. Calculate $x$ when $y = 0.5$.

3. $p = kq^3$. When $p = 9$, $q = 3$.
   a. Calculate the value of $k$.
   b. Calculate $p$ when $q = 6$.
   c. Calculate $p$ when $q = 1$.
   d. Calculate $q$ when $p = 576$.

4. $m = \frac{k}{\sqrt{n}}$. When $m = 1$, $n = 25$.
   a. Calculate the value of $k$.
   b. Calculate $m$ when $n = 16$.
   c. Calculate $m$ when $n = 100$.
   d. Calculate $n$ when $m = 5$.

5. $y = \frac{k}{x^2}$. When $y = 3$, $x = \frac{1}{3}$.
   a. Calculate the value of $k$.
   b. Calculate $y$ when $x = 0.5$.
   c. Calculate both values of $x$ when $y = \frac{1}{12}$.
   d. Calculate both values of $x$ when $y = \frac{1}{3}$.

## Student assessment 11

1. $y$ is inversely proportional to $x$.
   a. Copy and complete the table below:

   | $x$ | 1 | 2 | 4 | 8 | 16 | 32 |
   |---|---|---|---|---|---|---|
   | $y$ | | | 4 | | | |

   b. What is the value of $x$ when $y = 20$?

2. Copy and complete the tables below:
   a. $y \propto x$

   | $x$ | 1 | 2 | 4 | 5 | 10 |
   |---|---|---|---|---|---|
   | $y$ | | 10 | | | |

   b. $y \propto \frac{1}{x}$

   | $x$ | 1 | 2 | 4 | 5 | 10 |
   |---|---|---|---|---|---|
   | $y$ | 20 | | | | |

**c** $y \propto \sqrt{x}$

| $x$ | 4 | 16 | 25 | 36 | 64 |
|---|---|---|---|---|---|
| $y$ | 4 | | | | |

**3** The pressure ($P$) of a given mass of gas is inversely proportional to its volume ($V$) at a constant temperature. If $P = 4$ when $V = 6$, calculate:
  **a** $P$ when $V = 30$
  **b** $V$ when $P = 30$.

**4** The gravitational force ($F$) between two masses is inversely proportional to the square of the distance ($d$) between them. If $F = 4$ when $d = 5$, calculate:
  **a** $F$ when $d = 8$
  **b** $d$ when $F = 25$.

# TOPIC 3

# Functions

## Contents

Chapter 22  Function notation (C3.3, E3.3)
Chapter 23  Recognising graphs of common functions (C3.1, C3.2, E3.1, E3.2, E3.5)
Chapter 24  Transformation of graphs (E3.5, E3.6)
Chapter 25  Using a graphic display calculator to sketch and analyse functions (C2.5, C3.2, E2.5, E3.2)
Chapter 26  Finding a quadratic function from key values (E3.1, E3.4)
Chapter 27  Finding the equation of other common functions (E3.1)
Chapter 28  Composite functions (E3.3)
Chapter 29  Inverse functions (E3.3)
Chapter 30  Logarithmic functions (E1.16, E3.7)

### Learning objectives

**E1.16**
**Use exponential growth and decay**

**C2.5    E2.5**
1   Construct simple expressions, **expressions**, equations and formulas
2   Solve linear equations in one unknown
3   **Solve fractional equations with numerical and linear algebraic denominators**
4   Construct and solve simultaneous linear equations in two unknowns
5   **Construct and solve quadratic equations by factorisation, using a graphic display calculator and by use of the quadratic formula**
6   Use a graphic display calculator to solve equations, including those which may be unfamiliar
7   Change the subject of simple formulas **and formulas**

**C3.1    E3.1**
1   Recognise the following function types from the shape of their graphs:
   (a)  linear           $f(x) = ax + b$
   (b)  quadratic        $f(x) = ax^2 + bx + c$
   (c)  **cubic**        $f(x) = ax^3 + bx^2 + cx + d$
   (d)  reciprocal       $f(x) = a/x$
   (e)  exponential      $f(x) = ax$ with $0 < a < 1$ or $a > 1$
   (f)  trigonometric    $f(x) = a\sin(bx); a\cos(bx); \tan x$
2   Determine one or two of $a$, $b$, $c$ or $d$ for the graphs above
3   **Determine values in a function from its graph**

**C3.2      E3.2**
Use of a graphic display calculator (GDC) to:
(a)   sketch the graph of a function
(b)   produce a table of values
(c)   plot points on GDC
(d)   find zeros, local maxima or minima
(e)   find the intersection of the graphs of functions
(f)   find the vertex of a quadratic

**C3.3      E3.3**
1   Understand **functions, domain and range**, and use function notation
2   **Understand and find inverse functions $f^{-1}(x)$**
3   **Form composite functions as defined by $gf(x) = g(f(x))$**

**E3.4**
Find the quadratic function given:
(a)   vertex and another point
(b)   $x$-intercepts and a point
(c)   vertex or $x$-intercepts in the case where $a = 1$

**E3.5**
Understand the concept of asymptotes, and graphical identification of simple examples parallel to the axes

**E3.6**
Describe and identify transformations to a graph of $y = f(x)$
when $y = f(x) + k$,
$y = f(x + k)$

**E3.7**
Understand and use the logarithmic function as the inverse of the exponential function
$y = a^x$ equivalent to $x = \log_a y$
Solution to
$a^x = b$ as $x = \log b / \log a$

Elements in purple refer to the Extended curriculum only.

# The Chinese

Chinese mathematicians were the first to discover various algebraic and geometric principles. The textbook *Nine Chapters on the Mathematical Art* has special importance. *Nine Chapters* (known in Chinese as *Jiu Zhang Suan Shu* or *Chiu Chang Suan Shu*) was probably written during the early Han Dynasty (about 165 BC) by Chang Tshang.

Chang's book gives methods of arithmetic (including cube roots) and algebra (including a solution of simultaneous equations), uses the decimal system with zero and negative numbers, proves Pythagoras' theorem and includes a clever geometric proof that the perimeter of a right-angled triangle multiplied by the radius of its inscribing circle equals the area of its circumscribing rectangle.

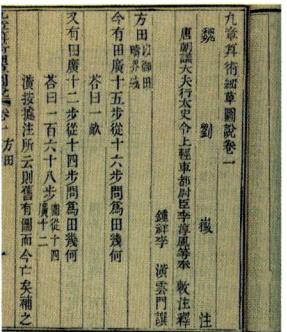

A page from *Nine Chapters*

Chang was concerned with the ordinary lives of the people. He wrote, 'For a civilization to endure and prosper, it must give its citizens order and fairness', so three chapters were concerned with ratio and proportion, so that 'rice and other cereals can be planted in the correct proportion to our needs, and the ratio of taxes could be paid fairly'.

*Nine Chapters* was probably based on earlier books but, even so, this book had great historical importance. It was the main Chinese mathematical text for centuries, and had great influence throughout the Far East. Some of the teachings made their way to India and from there to the Islamic world and Europe. The Hindus may have borrowed the decimal system itself from books like *Nine Chapters*.

In AD 600, Wang Xiaotong wrote *The Continuation of Ancient Mathematics*, which included work on squares, cubes and their roots.

# 22 Function notation

## Functions as a mapping

Consider the equation $y = 2x + 3$. It describes the relationship between two variables $x$ and $y$. In this case, 3 is added to twice the value of $x$ to produce $y$.

A function is a particular type of relationship between two variables. It has certain characteristics.

Consider the equation $y = 2x + 3$ for values of $x$ within $-1 \leqslant x \leqslant 3$.

A table of results can be constructed and a mapping drawn.

| $x$ | $y$ |
|---|---|
| −1 | 1 |
| 0 | 3 |
| 1 | 5 |
| 2 | 7 |
| 3 | 9 |

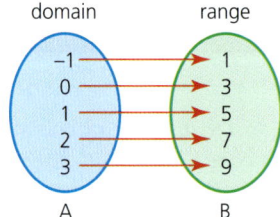

The relationship can be written as a function:

$f(x) = 2x + 3; -1 \leqslant x \leqslant 3$

*It is also usual to include the domain after the function, as a different domain will produce a different range. If a domain is not given, it is assumed to be for all values.*

With a function, each value in set B (the **range**) is produced from one value in set A (the **domain**).

The mapping from A to B can be a one-to-one mapping or a many-to-one mapping.

The function above, $f(x) = 2x + 3; -1 \leqslant x \leqslant 3$, is a **one-to-one function** as one value in the domain maps onto one value in the range. However, the function $f(x) = x^2; -3 \leqslant x \leqslant 3$ is a **many-to-one function**, as a value in the range can be generated by more than one value in the domain, as shown.

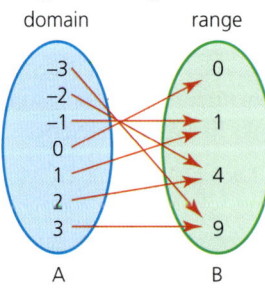

It is important to understand that one value in the domain (set A) maps to only one value in the range (set B). Therefore the mapping shown is the function $f(x) = x^2; -3 \leqslant x \leqslant 3$.

Some mappings will not represent functions; for example, consider the relationship $y = \pm\sqrt{x}$.

*Calculating the range from the domain*

The following table and mapping diagram can be produced:

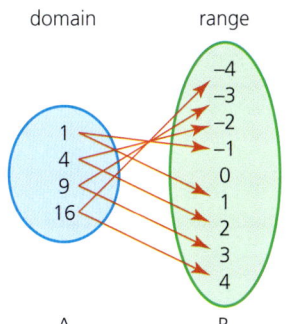

| x | y |
|---|---|
| 1 | ±1 |
| 4 | ±2 |
| 9 | ±3 |
| 16 | ±4 |

This relationship is not a function as a value in the domain produces more than one value in the range.

# Calculating the range from the domain

The domain is the set of input values and the range is the set of output values for a function. (Note that the range is not the difference between the greatest and least values as in statistics.) The range is therefore not only dependent on the function itself, but also on the domain.

## ➡ Worked example

Calculate the range for the following functions:

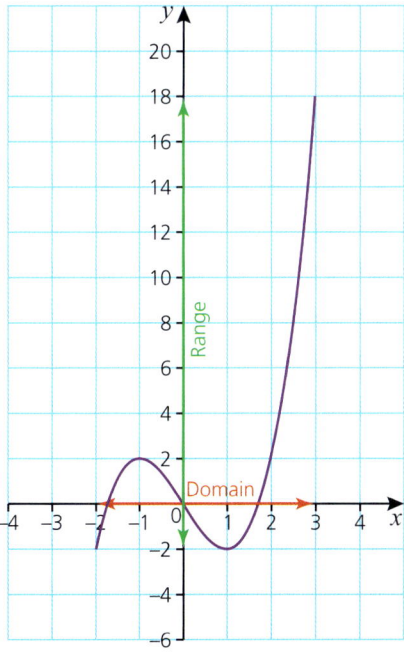

# 22 FUNCTION NOTATION

**a** $f(x) = x^3 - 3x; -2 \leq x \leq 3$

The graph of the function is shown above. As the domain is restricted to $-2 \leq x \leq 3$, the range is limited from $-2$ to $18$.

This is written as: Range $-2 \leq f(x) \leq 18$.

**b** $f(x) = x^3 - 3x$

The graph will be similar to the one above except that the domain is not restricted. As the domain is for all values of $x$, this implies that any number can be an input value. As a result, the range will also be all values.

This is written as: Range $f(x) \in$ {all values}.

 **Exercise 3.1**

1 Which of the following mappings show a function?

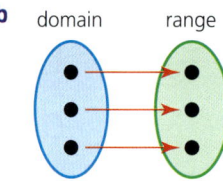

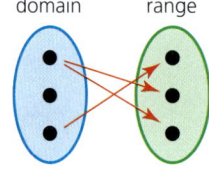

Give the domain and range of each of the functions in Q.2–8.
2 $f(x) = 2x - 1; -1 \leq x \leq 3$
3 $f(x) = 3x + 2; -4 \leq x \leq 0$
4 $f(x) = -x + 4; -4 \leq x \leq 4$
5 $f(x) = x^2 + 2; -3 \leq x \leq 3$
6 $f(x) = x^2 + 2; x \in$ {all real values}
7 $f(x) = -x^2 + 2; 0 \leq x \leq 4$
8 $f(x) = x^3 - 2; -3 \leq x \leq 1$

# 23 Recognising graphs of common functions

Graphs of functions take many different forms. It is important to be able to identify common functions and their graphs.

## The linear function

A linear function takes the form $f(x) = ax + b$ and when graphed produces a straight line.

Three different linear functions are shown below:

$f(x) = x - 3$

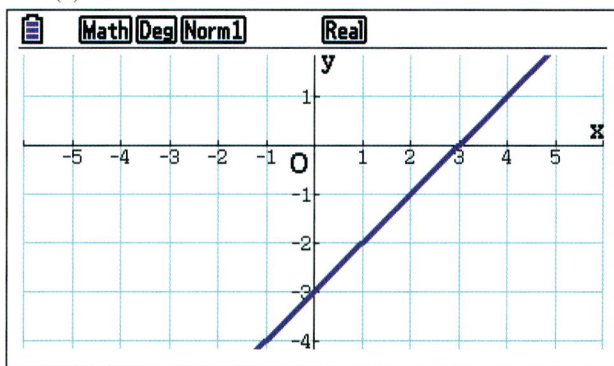

$f(x) = 2x - 4$

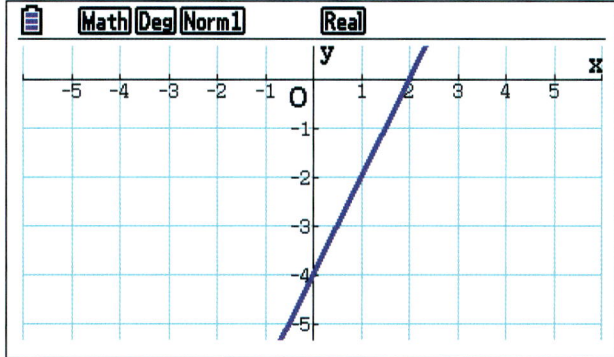

# 23 RECOGNISING GRAPHS OF COMMON FUNCTIONS

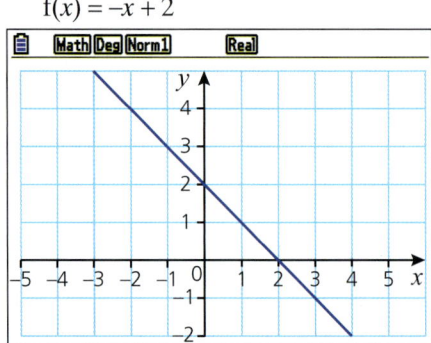

$f(x) = -x + 2$

The values of $a$ and $b$ affect the orientation and position of the line.

For the function $f(x) = x - 3$, $a = 1$ and $b = -3$.

For the function $f(x) = 2x - 4$, $a = 2$ and $b = -4$.

For the function $f(x) = -x + 2$, $a = -1$ and $b = 2$.

You can use your graphic display calculator to investigate linear functions and determine the effects that different values of $a$ and $b$ have on the graph.

The instructions below will remind you how to graph the function $f(x) = 2x - 4$ using your graphic display calculator:

### Casio

**MENU** **5** to select the graphing menu.

**2** **X,θ,T** **−** **4** **EXE** to enter the function.

**F6** to graph the function.

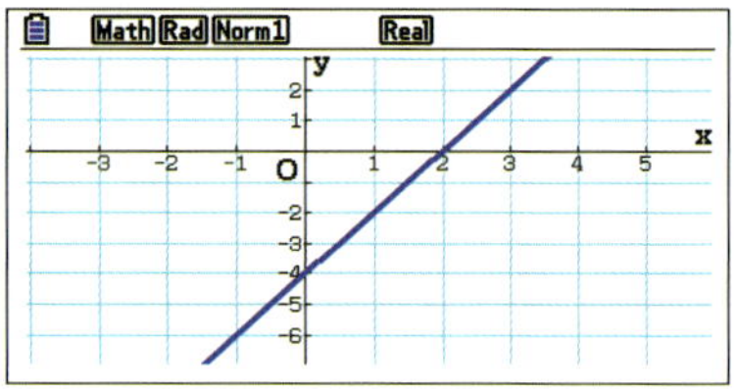

### Texas

**y=** to select the function.

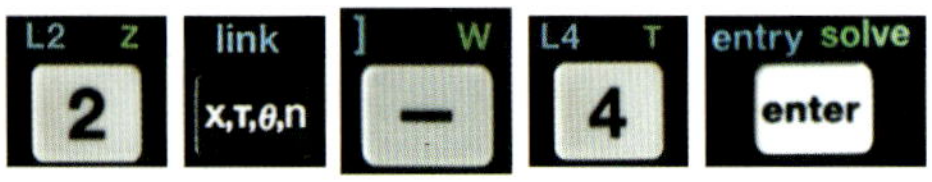

**2** **X,T,θ,n** **−** **4** **enter** to enter the function.

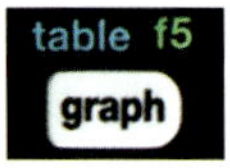

**graph** to graph the function.

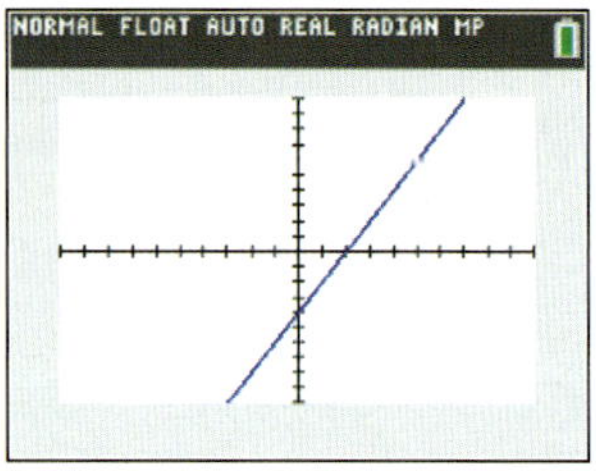

*The linear function*

**Exercise 3.2**

1. Use your graphic display calculator to investigate the effect of *b* on the orientation or position of functions of the type f(x) = ax + b.
   a. By keeping the value of *a* constant and changing the value of *b*, write down five different linear functions.
   b. Using your graphic display calculator, graph each of the five functions.
   c. Sketch your functions, labelling each clearly.
   d. Write a short conclusion about the effect of *b* on the graph.

2. Use your graphic display calculator to investigate the effect of *a* on the orientation or position of functions of the type f(x) = ax + b.
   a. By keeping the value of *b* constant and changing the value of *a*, write down five different linear functions.
   b. Using your graphic display calculator, graph each of the five functions.
   c. Sketch your functions, labelling each clearly.
   d. Write a short conclusion about the effect of *a* on the graph.

3. Use your graphic display calculator to produce a **similar** screen to those shown below. The equation of one of the functions is given each time.

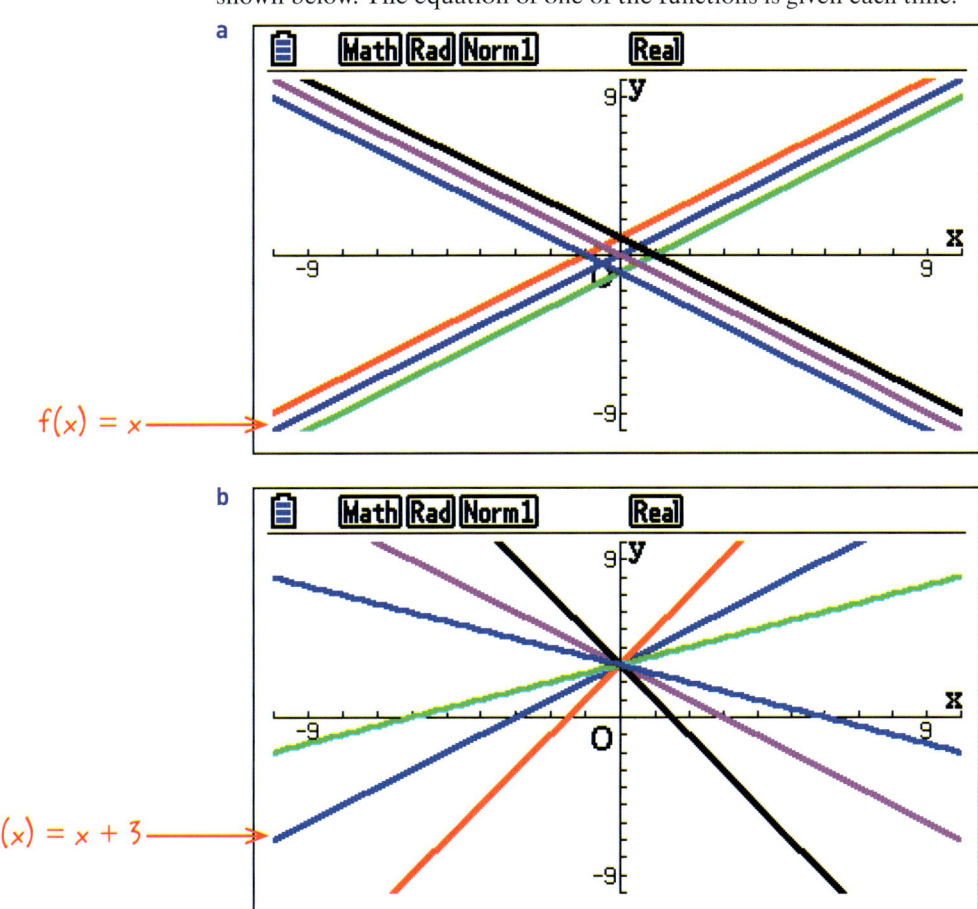

# 23 RECOGNISING GRAPHS OF COMMON FUNCTIONS

## The quadratic function

A quadratic function takes the form $f(x) = ax^2 + bx + c$ where $a \neq 0$. The graph of a quadratic function also has a characteristic shape that you can use to identify that a function is quadratic.

Two quadratic functions are shown below:

$f(x) = x^2 + 2x - 4$

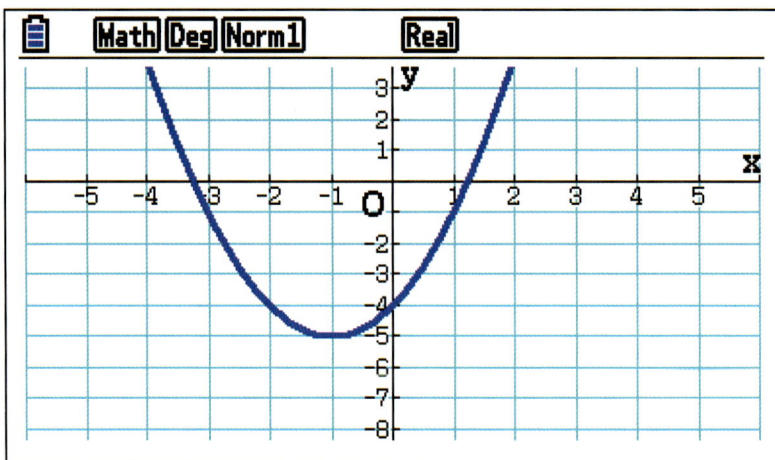

$f(x) = -x^2 + 2x + 4$

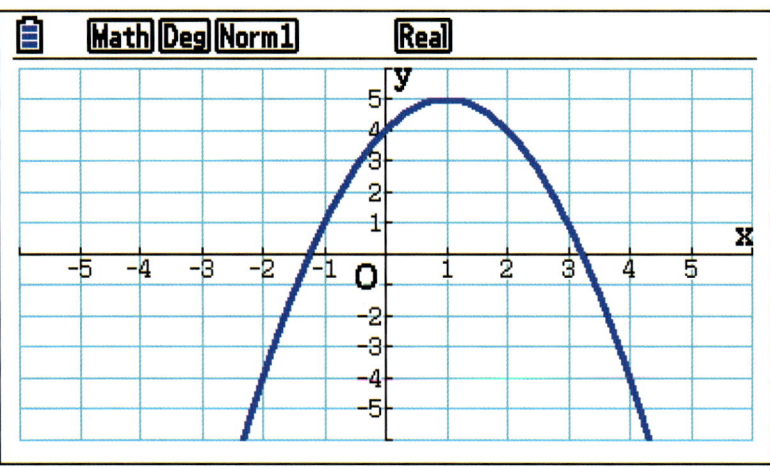

The graphs of quadratic functions are always either U-shaped or upside down U-shaped. The values of $a$, $b$ and $c$ affect the shape and position of the graph. The graph also has a line of symmetry.

*The quadratic function*

**Exercise 3.3**

1 Use your graphic display calculator to investigate the effect of $c$ on the shape or position of functions of the type $f(x) = ax^2 + bx + c$.
  **a** By keeping the values of $a$ and $b$ constant and changing the value of $c$, write down five different quadratic functions.
  **b** Using your graphic display calculator, graph each of the five functions.
  **c** Sketch your functions, labelling each clearly.
  **d** Write a short conclusion about the effect of $c$ on the graph.

2 Use your graphic display calculator to investigate the effect of $a$ on the shape or position of functions of the type $f(x) = ax^2 + bx + c$.
  **a** By keeping the values of $b$ and $c$ constant and changing the value of $a$, write down five different quadratic functions.
  **b** Using your graphic display calculator, graph each of the five functions.
  **c** Sketch your functions, labelling each clearly.
  **d** Write a short conclusion about the effect of $a$ on the graph.

3 Use your graphic display calculator to investigate the effect of $b$ on the shape or position of functions of the type $f(x) = ax^2 + bx + c$.
  **a** By keeping the values of $a$ and $c$ constant and changing the value of $b$, write down five different quadratic functions.
  **b** Using your graphic display calculator, graph each of the five functions.
  **c** Sketch your functions, labelling each clearly.
  **d** Write a short conclusion about the effect of $b$ on the graph.

4 Use your graphic display calculator to produce a **similar** screen to those shown below. The equation of one of the functions is given each time.
  **a**

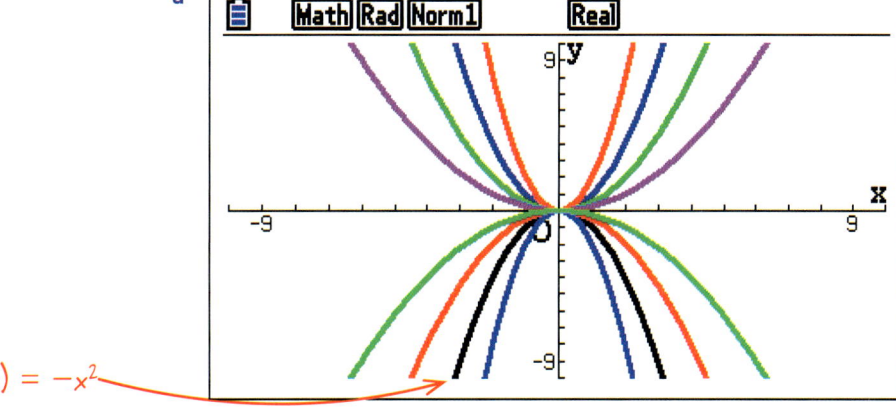

$f(x) = -x^2$

*Note that these quadratic curves have one line of symmetry.*

  **b**

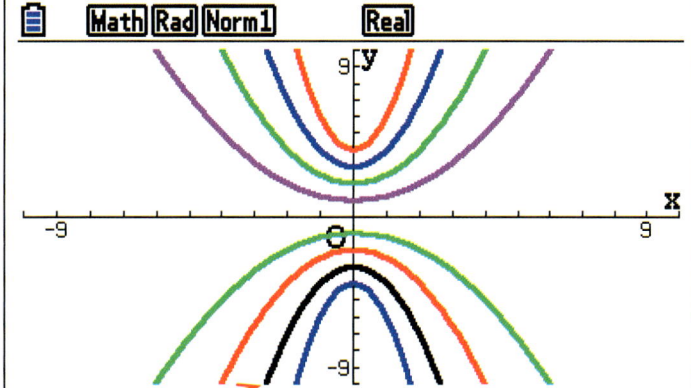

$f(x) = -x^2 - 3$

## 23 RECOGNISING GRAPHS OF COMMON FUNCTIONS

## The cubic function

A cubic function takes the form $f(x) = ax^3 + bx^2 + cx + d$ where $a \neq 0$.

They also have an identifiable shape.

Two examples are shown below:

$f(x) = x^3 - 3x^2 - 10x$

where $a = 1$, $b = -3$, $c = -10$ and $d = 0$

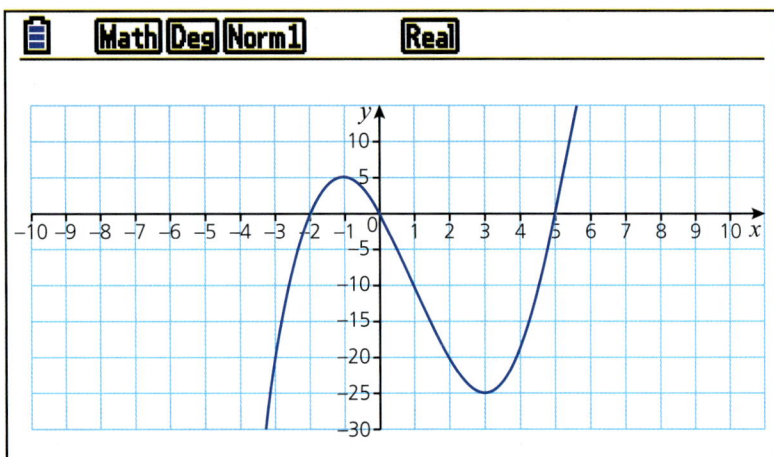

$f(x) = -x^3 - 2$

where $a = -1$, $b = 0$, $c = 0$ and $d = -2$

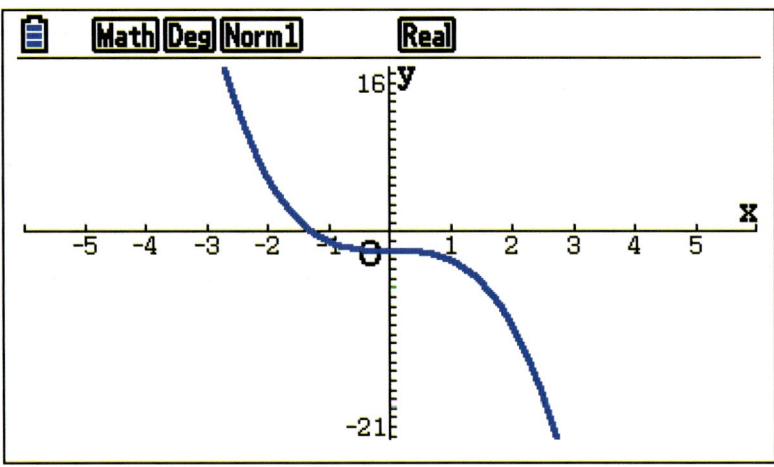

The shape of a cubic function has a characteristic 'S' shape. It can be tighter as shown in the first example, or more stretched as shown in the second example. The graph also has rotational symmetry.

# The reciprocal function

**Exercise 3.4**

1. **a** Use your graphic display calculator to investigate the effect of *a* on the shape or position of functions of the type f(x) = $ax^3 + bx^2 + cx + d$.
   **b** Write a short conclusion about the effect of *a* on the graph.
2. **a** Use your graphic display calculator to investigate the effect of *d* on the shape or position of functions of the type f(x) = $ax^3 + bx^2 + cx + d$.
   **b** Write a short conclusion about the effect of *d* on the graph.
3. Use your graphic display calculator to produce a **similar** screen to those shown below. The equation of one of the functions is given each time.

   **a**

   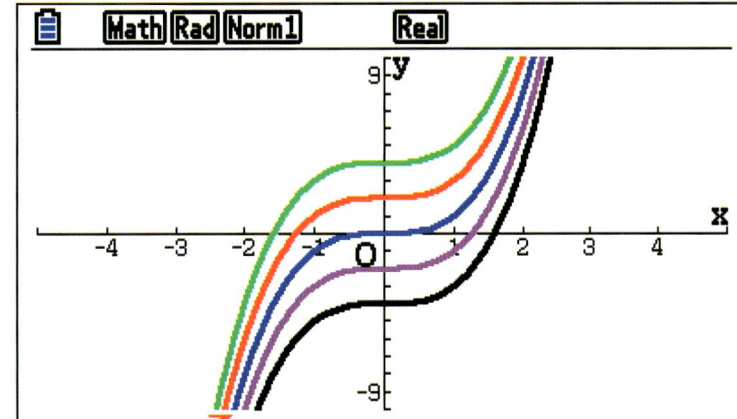

   f(x) = $x^3$

   **b**

   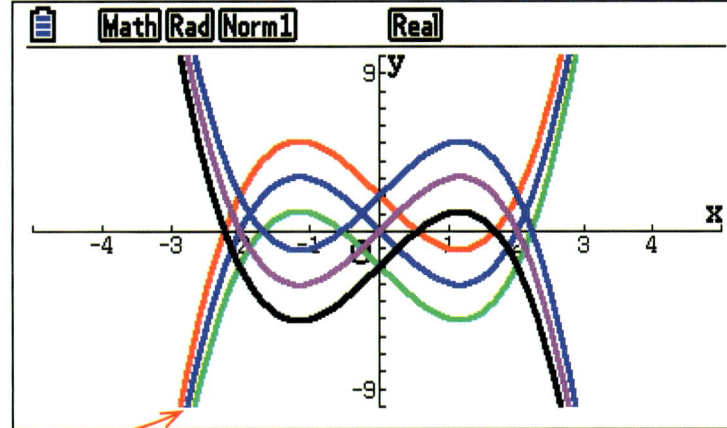

   f(x) = $x^3 - 4x$

## The reciprocal function

You will have encountered the term reciprocal before; for example, the reciprocal of 5 is $\frac{1}{5}$ and the reciprocal of $\frac{2}{3}$ is $\frac{1}{\frac{2}{3}}$, which simplifies to $\frac{3}{2}$. The reciprocal of $x$ is therefore $\frac{1}{x}$.

Functions where $x$ appears in the denominator are reciprocal functions and take the form f($x$) = $\frac{a}{x}$, where $a \neq 0$.

The graphs of reciprocal functions have particular characteristics as shown in the two examples below:

# 23 RECOGNISING GRAPHS OF COMMON FUNCTIONS

$f(x) = \frac{1}{x}$

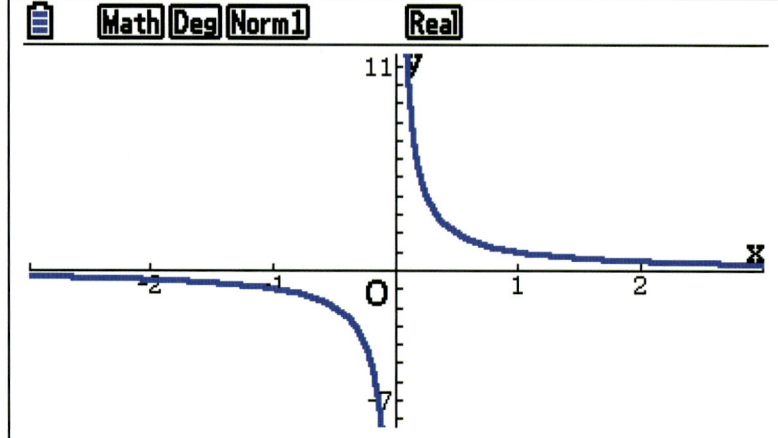

*Note that these reciprocal graphs have two lines of symmetry.*

$f(x) = -\frac{2}{x}$

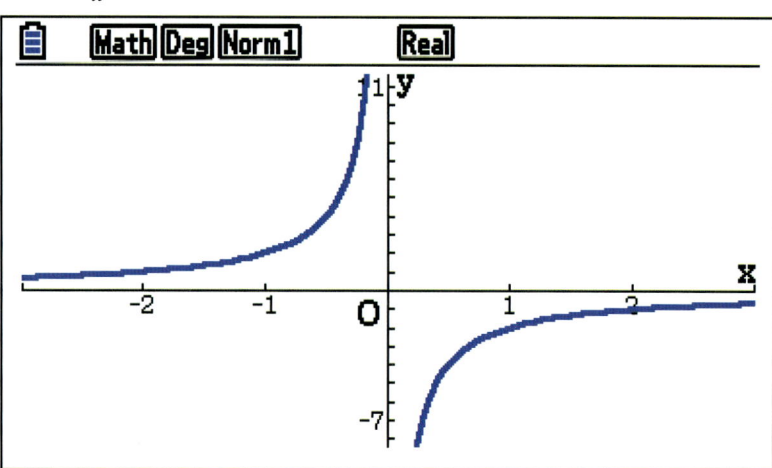

In each case, the graphs get closer to the axes but do not actually meet or cross them. This is because for the function $f(x) = \frac{a}{x}$, as $x \to \pm \infty$ then $\frac{a}{x} \to 0$ and as $x \to 0$ then $\frac{a}{x} \to \pm \infty$.

The axes are known as **asymptotes**. An asymptote is a line to which a curve gets closer and closer but never actually meets. The graph also has rotational symmetry.

## The exponential function

Until now all the functions you have encountered have a variable raised to a power; for example, $f(x) = x$, $f(x) = x^2$, $f(x) = x^3$ and $f(x) = \frac{1}{x} = x^{-1}$. With an exponential function, the variable is the power.

*The exponential function*

An exponential function will typically take the form f(x) = $a^x$ where $0 < a < 1$ or $a > 1$. The graphs of two exponential functions are shown below:

f(x) = $2^x$

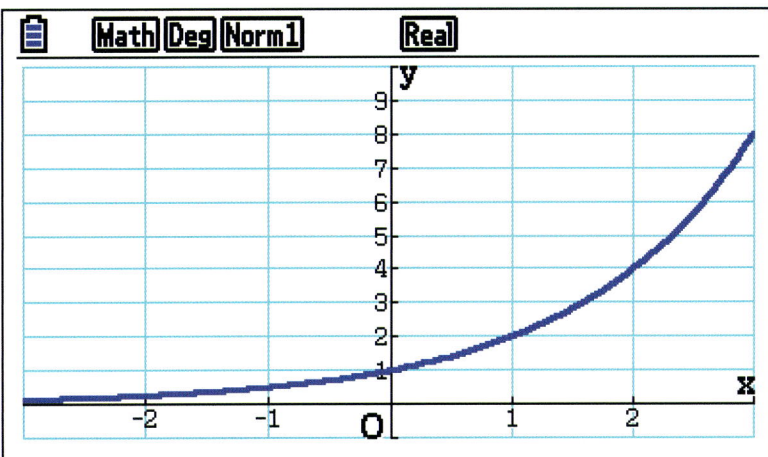

f(x) = $\left(\frac{1}{2}\right)^x$

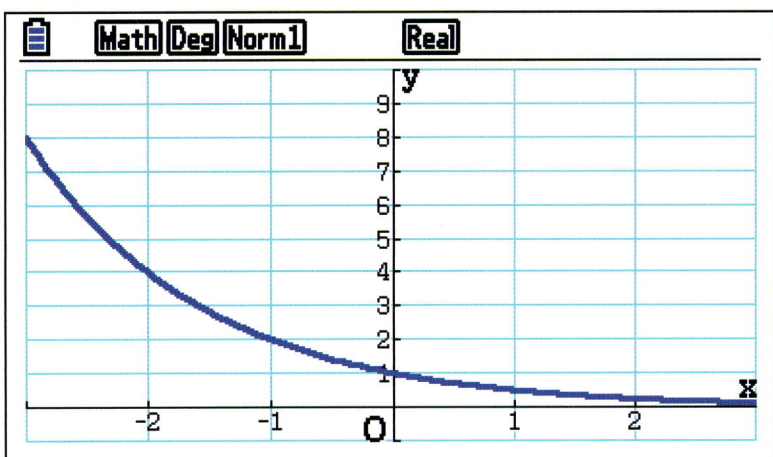

The graphs show that the x-axis is an asymptote to each of the curves. This is because for the function f(x) = $2^x$, as $x \to -\infty$ then $2^x \to 0$. This can be seen by applying the laws of indices where $2^{-x} = \frac{1}{2^x}$.

Similarly, for the function f(x) = $\left(\frac{1}{2}\right)^x$, as $x \to \infty$ then f(x) = $\left(\frac{1}{2}\right)^x \to 0$.

# 23 RECOGNISING GRAPHS OF COMMON FUNCTIONS

**Exercise 3.5**

1. **a** Use your graphic display calculator to investigate the effect of $a$ on the shape or position of functions of the type $f(x) = \frac{a}{x}$. Remember to include negative and positive values of $a$.
   **b** Write a short conclusion about the effect of $a$ on the graph.

2. **a** Use your graphic display calculator to graph the following functions on the same screen:
   $f(x) = 2^x$     $f(x) = 3^x$     $f(x) = 4^x$
   **b** Describe two characteristics that the graphs of all three functions share.

3. **a** Use your graphic display calculator to graph each of the following pairs of functions simultaneously:
   
   **i** $f(x) = 3^x$ and $f(x) = \left(\frac{1}{3}\right)^x$
   
   **ii** $f(x) = 4^x$ and $f(x) = \left(\frac{1}{4}\right)^x$
   
   **iii** $f(x) = \left(\frac{3}{2}\right)^x$ and $f(x) = \left(\frac{2}{3}\right)^x$
   
   **b** Comment on the relationship between each pair of graphs.

4. Use your graphic display calculator to produce a **similar** screen to those shown below. The equation of one of the functions is given each time.

   **a**
   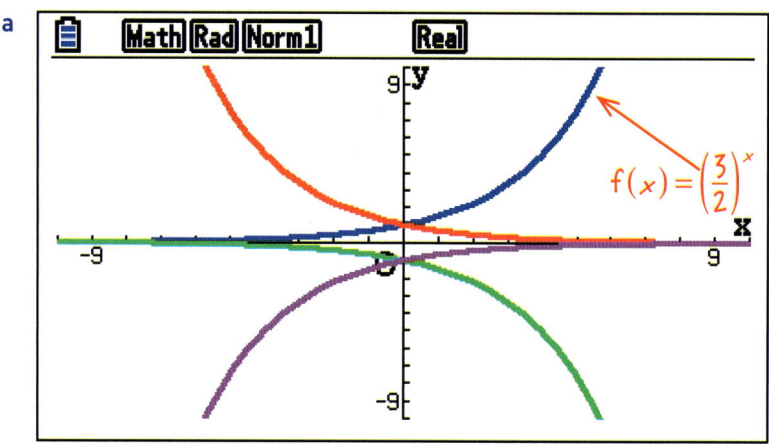

   **b**
   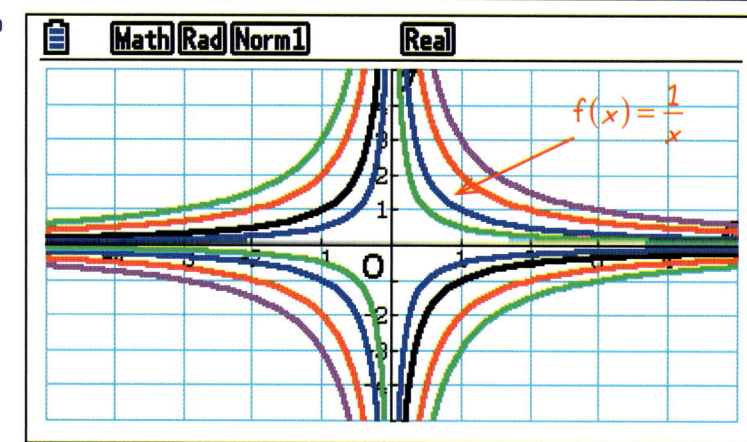

Trigonometric functions and their properties are dealt with in Topic 7.

# 24 Transformation of graphs

When a function undergoes a transformation, the shape or position of its graph changes. This change in shape or position depends on the type of transformation. This chapter focuses on the two types of translation that can occur.

Let $f(x) = x^2$. The graph $y = f(x)$ is therefore the graph of $y = x^2$ as shown below:

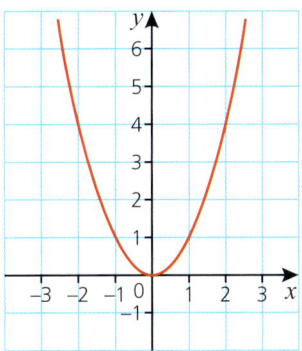

The graph of the function $y = f(x) + 1$ is therefore the graph of $y = x^2 + 1$.

The graph of the function $y = f(x) + 2$ is the graph of $y = x^2 + 2$ and the graph of the function $y = f(x) + 3$ is the graph of $y = x^2 + 3$.

These four functions are plotted on the same axes as shown:

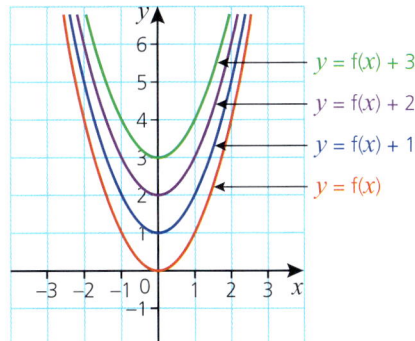

These graphs show that $y = f(x)$ is translated $\begin{pmatrix} 0 \\ 1 \end{pmatrix}$ to map onto $y = f(x) + 1$.

Similarly, $y = f(x)$ is translated $\begin{pmatrix} 0 \\ 2 \end{pmatrix}$ to map onto $y = f(x) + 2$ and translated $\begin{pmatrix} 0 \\ 3 \end{pmatrix}$ to map onto $y = f(x) + 3$.

Therefore $y = f(x) + k$ is a translation of $\begin{pmatrix} 0 \\ k \end{pmatrix}$ of the function $y = f(x)$.

# 24 TRANSFORMATION OF GRAPHS

In the examples above, the constant $k$ acted externally to the original function $y = f(x)$. If the constant is incorporated within the original function, a different translation occurs.

Let $f(x) = x^2$ be the original function. This is represented by the equation $y = x^2$. If $x$ is substituted by $(x - 2)$, the function becomes $f(x - 2) = (x - 2)^2$. This is represented by the equation $y = (x - 2)^2$. Graphing both on the same axes produces the graphs below:

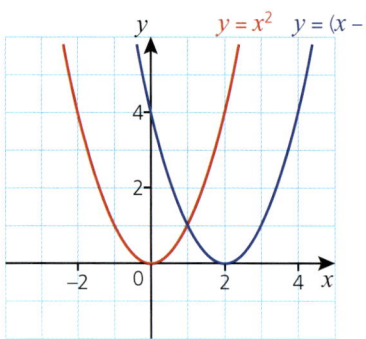

$y = x^2$ is mapped onto $y = (x - 2)^2$ by the translation $\begin{pmatrix} 2 \\ 0 \end{pmatrix}$.

The graphs of the functions $f(x) = x^2$, $f(x - 2) = (x - 2)^2$, $f(x - 4) = (x - 4)^2$, $f(x + 1) = (x + 1)^2$ and $f(x + 3) = (x + 3)^2$ are shown on the same axes below:

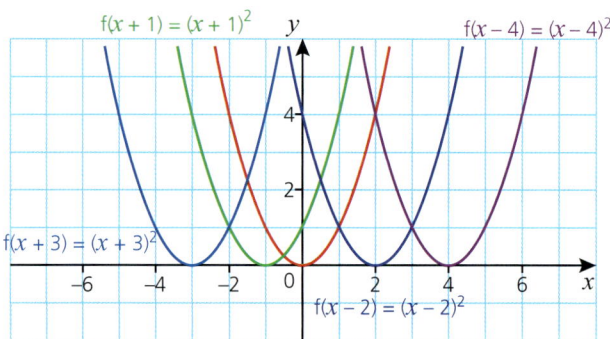

The transformations are each horizontal translations.

In general, therefore, mapping $y = f(x)$ onto $y = f(x + k)$ is a translation of $\begin{pmatrix} -k \\ 0 \end{pmatrix}$.

## Transformation of graphs

**Exercise 3.6**

1. Sketch the graph of $y = x^2$. Use transformations to sketch, on the same axes, both of the following. Label each graph clearly.
   a  $y = (x - 1)^2$  
   b  $y = x^2 - 3$

2. Sketch the graph of $y = x^3$. Use transformations to sketch $y = (x + 3)^3$

3. a  Sketch the graph of $y = 2^x$. Use transformations to sketch:
      i  $y = 2^x + 2$   ii  $y = 2^{(x-2)}$
   b  Give the equation of any asymptote in Q.3ai.
   c  Give the equation of any asymptote in Q.3aii.

4. a  Sketch the graph of $y = \frac{1}{x}$. Use transformations to sketch $y = \frac{1}{x+2}$
   b  Give the equation of any asymptote of $y = \frac{1}{x+2}$.

5. Describe mathematically the transformation that maps the graph of $y = f(x)$ onto:
   a  $y = f(x) - 2$  
   b  $y = f(x - 4)$

### → Worked example

The sketch shows the graph of $y = f(x)$. Points A, B and C have coordinates as shown:

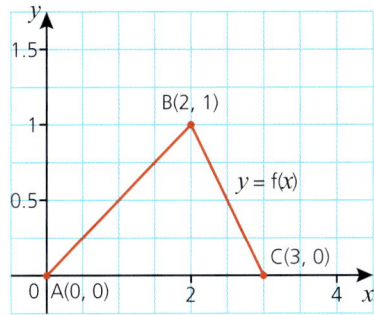

i  Sketch the graph of $y = f(x) + 2$. Mark the images of A, B and C under the transformation and state their coordinates.

The transformation $y = f(x) + 2$ is a translation of $\begin{pmatrix} 0 \\ 2 \end{pmatrix}$. Each point on $y = f(x)$ is therefore translated 2 units vertically upwards. The graph of $y = f(x) + 2$ and the images of A, B and C, labelled A', B' and C', are therefore as shown:

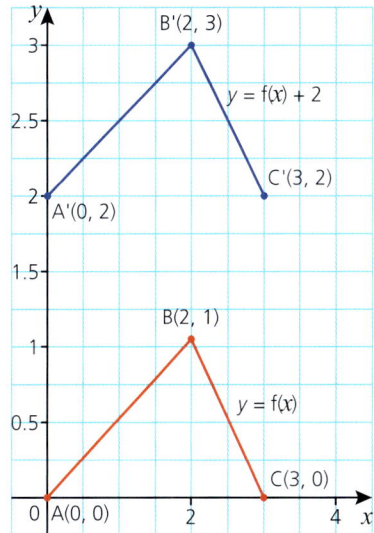

# 24 TRANSFORMATION OF GRAPHS

**ii** Sketch the graph of $y = f(x + 2)$. Mark the images of A, B and C under the transformation and state their coordinates.

The transformation $y = f(x + 2)$ is a translation of $\begin{pmatrix} -2 \\ 0 \end{pmatrix}$. Each point on $y = f(x)$ is therefore translated 2 units horizontally to the left. The graph of $y = f(x + 2)$ and images of A, B and C, labelled A', B' and C', are therefore as shown:

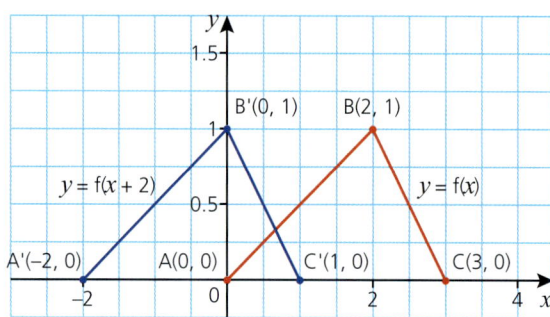

## Exercise 3.7

**1** Sketch the graph of $y = f(x)$ shown below:

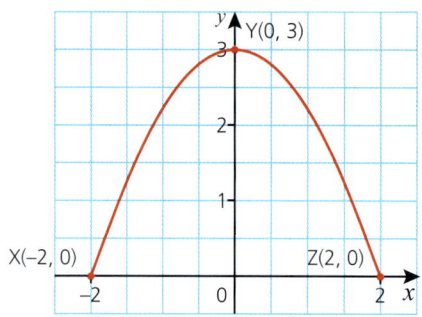

**a** On the same axes, sketch the graph of $y = f(x - 3)$, stating clearly the coordinates of the images of the points X, Y and Z.

**b** Describe the transformation that maps $y = f(x)$ onto $y = f(x - 3)$.

**2** Sketch the graph of $y = f(x)$ shown below:

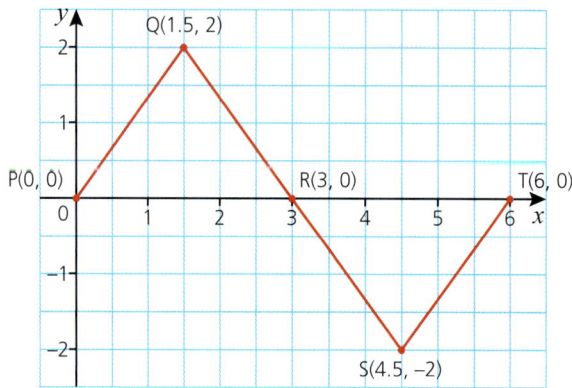

**a** On the same axes, sketch the graph of $y = f(x) - 2$, stating clearly the coordinates of the images of the points P, Q, R, S and T.

**b** Describe the transformation that maps $y = f(x)$ onto $y = f(x) - 2$.

**3 a** Given that $f(x) = x^2$, sketch the graphs of each of the following functions on a separate pair of axes:

    **i** $y = f(x) + 4$     **ii** $y = f(x + 2)$

**b** Write the equation of $y$ in terms of $x$ for each of the functions in Q.3a.

**4 a** Given that $f(x) = \frac{1}{x}$, sketch the graphs of each of the following functions on a separate pair of axes.

    **i** $y = f(x - 4)$     **ii** $y = f(x) - 2$

**b** Write the equation of $y$ in terms of $x$ for each of the functions in Q.4a.

**c** Write the equation of any asymptotes in the graphs of the functions.

# 25 Using a graphic display calculator to sketch and analyse functions

The graphic display calculator, introduced in the introductory topic, is a powerful tool to help understand graphs of functions and their properties. This chapter recaps some of the features that are particularly useful for checking your answers to some questions in the latter part of this topic.

## ➔ Worked example

Consider the function $f(x) = -x^3 + 9x^2 - 24x + 16$.

i Plot the function on a graphic display calculator.

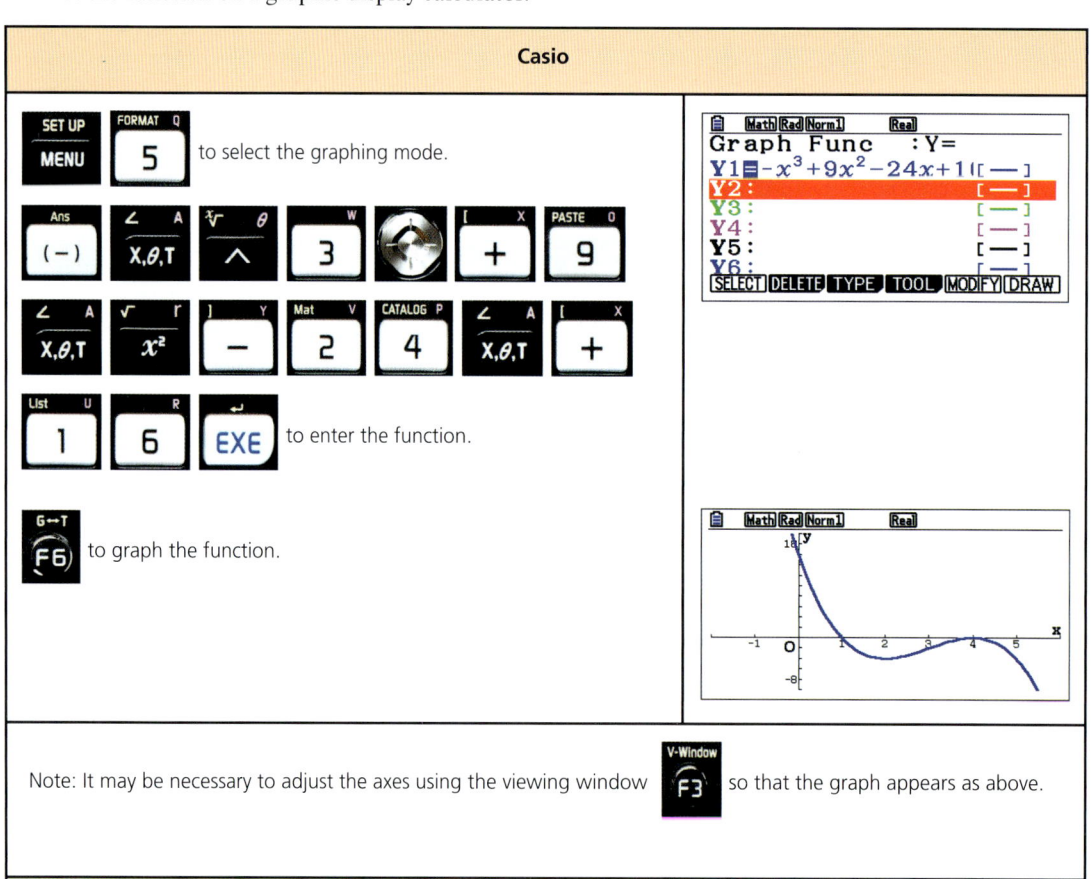

*Using a graphic display calculator to sketch and analyse functions*

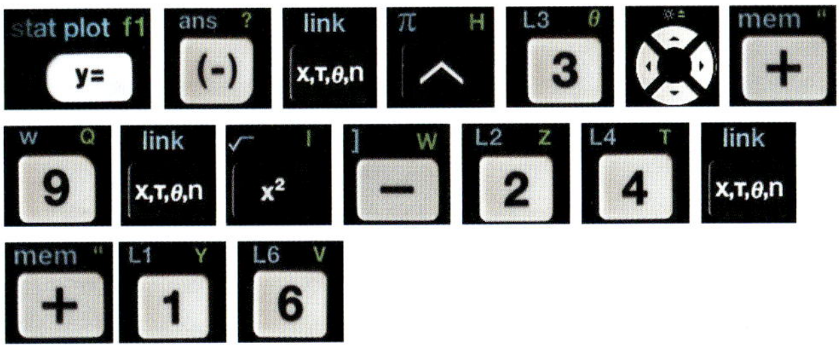

| Texas | |
|---|---|
| [y=] [(-)] [x,T,θ,n] [^] [3] [9] [x,T,θ,n] [x²] [−] [2] [4] [x,T,θ,n] [+] [1] [6] | Plot1 Plot2 Plot3<br>■Y1=-X³+9X²-24X+16<br>■Y2=<br>■Y3=<br>■Y4=<br>■Y5=<br>■Y6=<br>■Y7=<br>■Y8= |
| [enter] to enter the function. | |
| [graph] to graph the function. | (graph display) |
| Note: It may be necessary to adjust the axes using the  [window] button so that the graph appears as above. | |

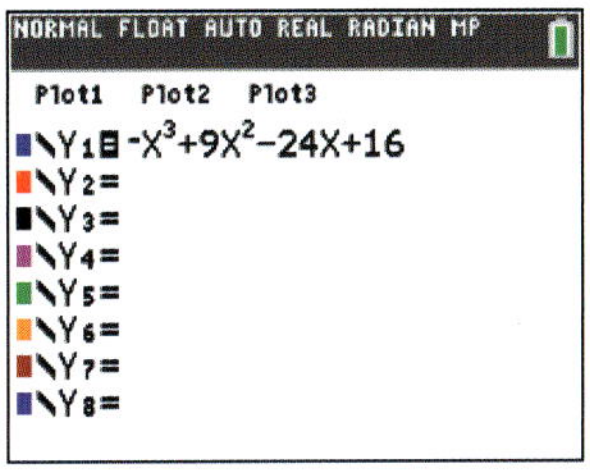

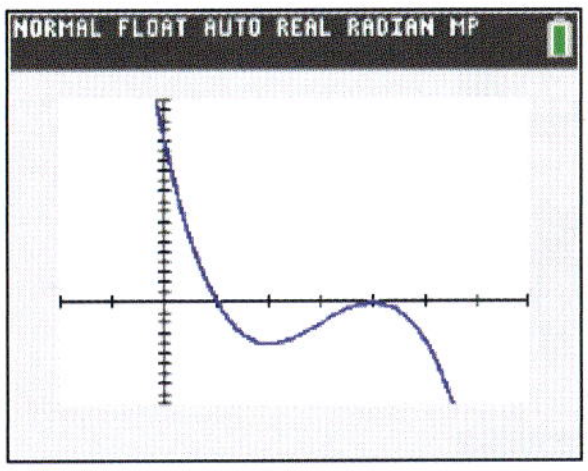

**ii** Find the zeros (roots) of the function, i.e. where its graph intersects the *x*-axis.

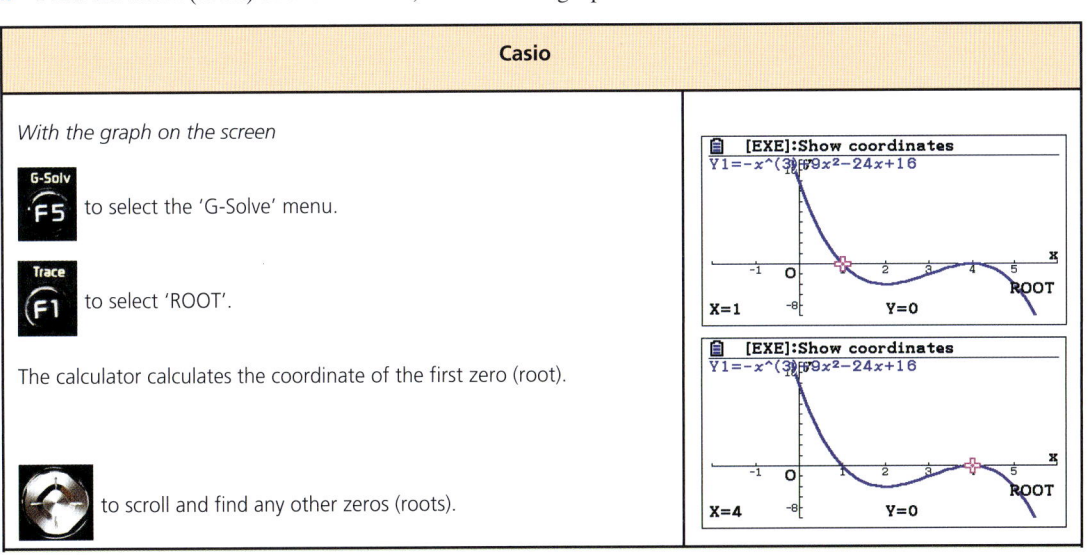

| Casio | |
|---|---|
| *With the graph on the screen*<br><br>[F5] to select the 'G-Solv' menu.<br><br>[F1] to select 'ROOT'.<br><br>The calculator calculates the coordinate of the first zero (root).<br><br>to scroll and find any other zeros (roots). | |

## 25 USING A GRAPHIC DISPLAY CALCULATOR TO SKETCH AND ANALYSE FUNCTIONS

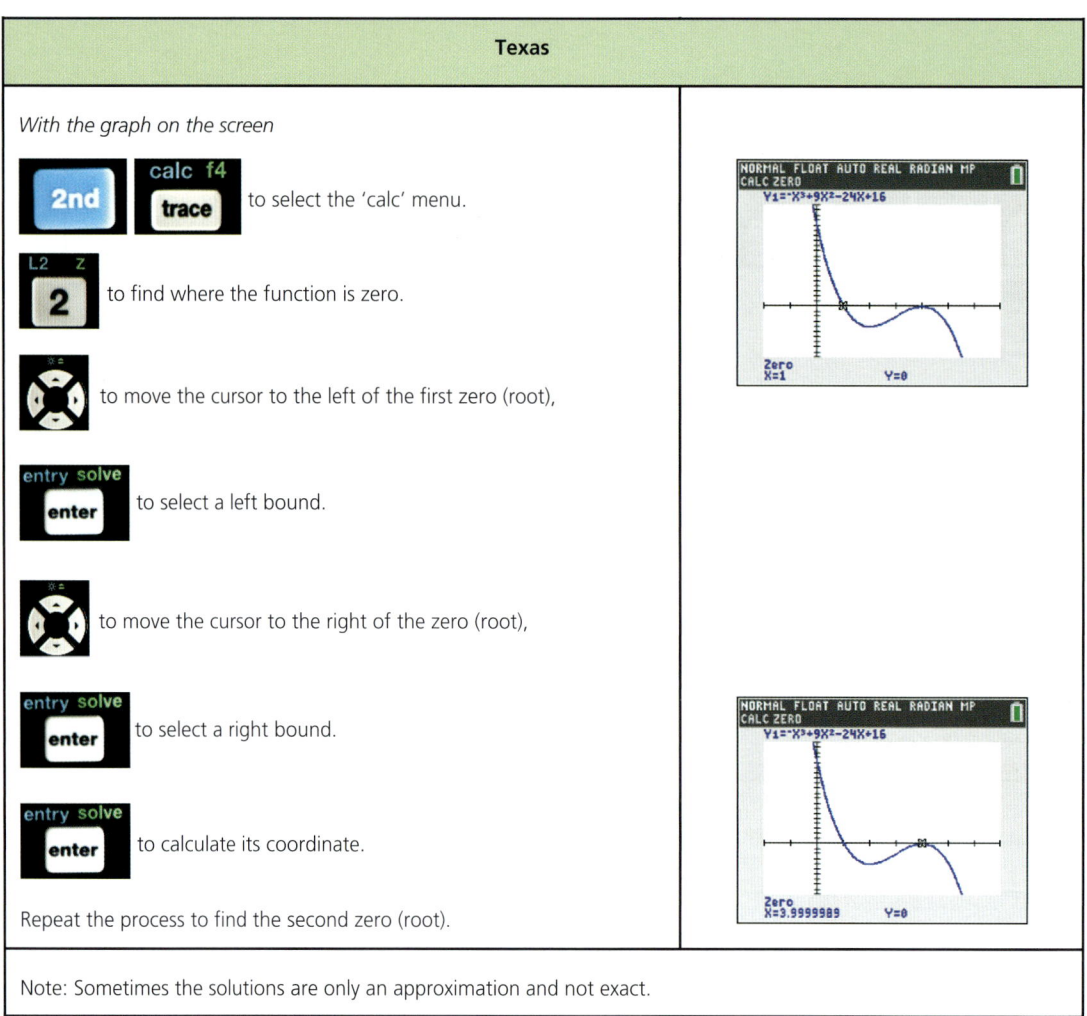

iii Find where the graph intersects the $y$-axis.

This can be done without a graphic display calculator, as the intersection with the $y$-axis occurs when $x = 0$. Substituting $x = 0$ into the equation gives the solution $y = 16$.

The graphic display calculator can be used as follows:

*Using a graphic display calculator to sketch and analyse functions*

| Casio |
|---|

*With the graph on the screen*

**F5** to select the 'G-Solv' menu.

**F4** to select 'Y-ICPT'.

The calculator gives the coordinate of the $y$-intercept.

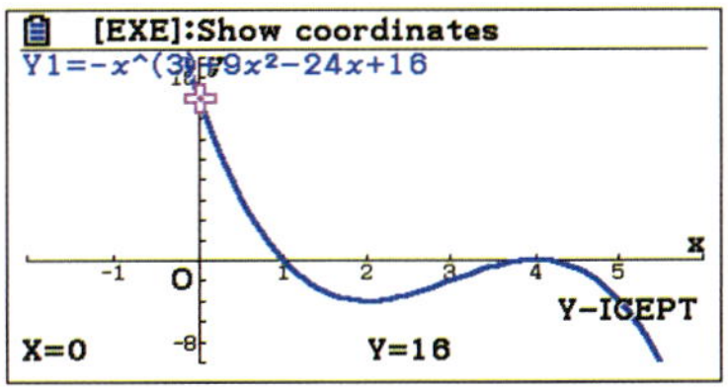

| Texas |
|---|

*With the graph on the screen*

**2nd** **trace** (calc f4) to select the 'calc' menu.

**1** to find where the $x$-value is zero.

Type 0 and press **enter**.

The calculator gives the coordinate of the $y$-intercept.

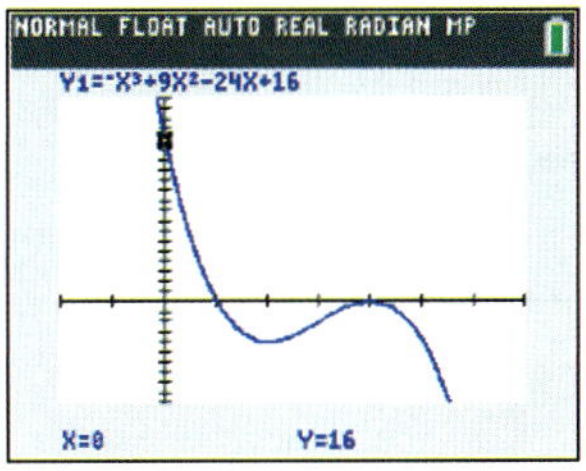

**iv** Find the coordinates of the points where the graph has local maxima or minima. Local maxima and minima refer to the peaks and dips of the graph respectively, as shown in the diagram below.

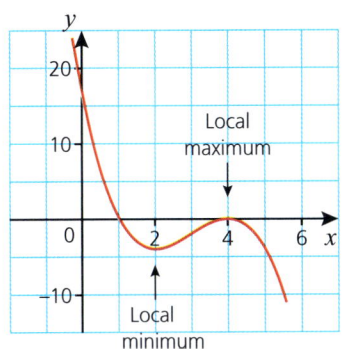

*The coordinates of the turning point (or **vertex**) of a quadratic can also be found in the same way.*

229

# 25 USING A GRAPHIC DISPLAY CALCULATOR TO SKETCH AND ANALYSE FUNCTIONS

| Casio |
|---|
| *With the graph on the screen*<br><br>**F5** (G-Solv) to select the 'G-Solv' menu.<br><br>**F3** (V-Window) to find any minimum points. The results are displayed in the graph.<br><br>**F5** (G-Solv) **F2** (Zoom) to find any maximum points. |

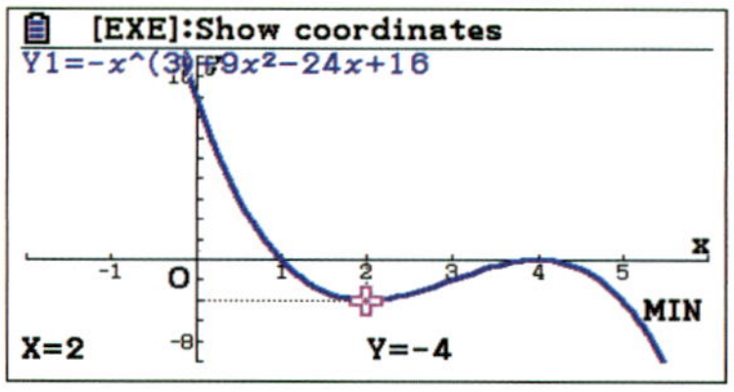

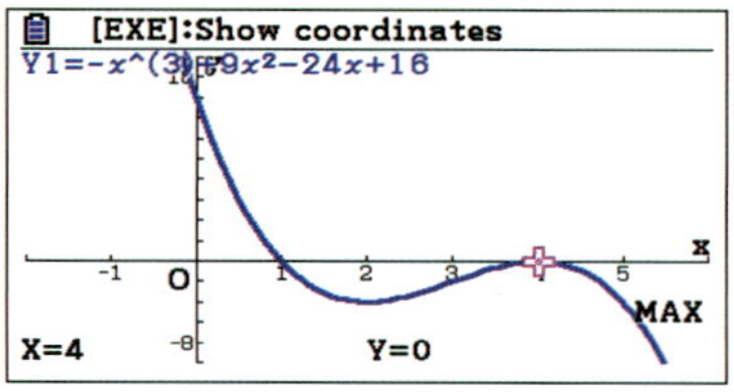

| Texas |
|---|
| *With the graph on the screen*<br><br>**2nd** **trace** (calc f4) to select the 'calc' menu.<br><br>**3** to find the local minimum<br><br>◄ to move the cursor to the left of the minimum point,<br><br>**enter** (entry solve) to select the left bound.<br><br>► to move the cursor to the right of the minimum point,<br><br>**enter** (entry solve) to select the right bound.<br><br>**enter** (entry solve) to find the coordinates of the minimum point. |

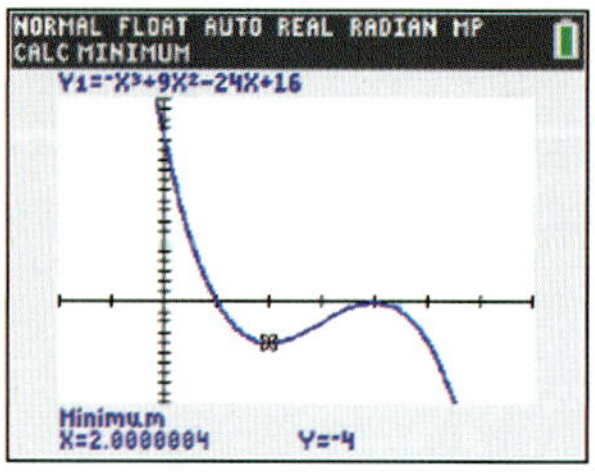

*Using a graphic display calculator to sketch and analyse functions*

 to search for the maximum point

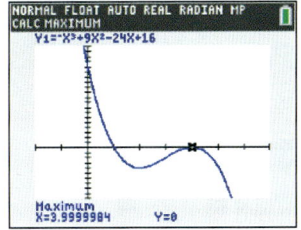

followed by the same procedure described above for the minimum point.

Note: Sometimes the solutions are only an approximation and not exact.

v   Complete the following table of values for the coordinates of some points on the graph.

| $x$ | −1 | 0 | 1 | 2 | 3 | 4 | 5 |
|---|---|---|---|---|---|---|---|
| $f(x)$ | | | | | | | |

The graphic display calculator can produce a table of values for a given function, within a given range of values of $x$. This is shown below:

| Casio |
|---|

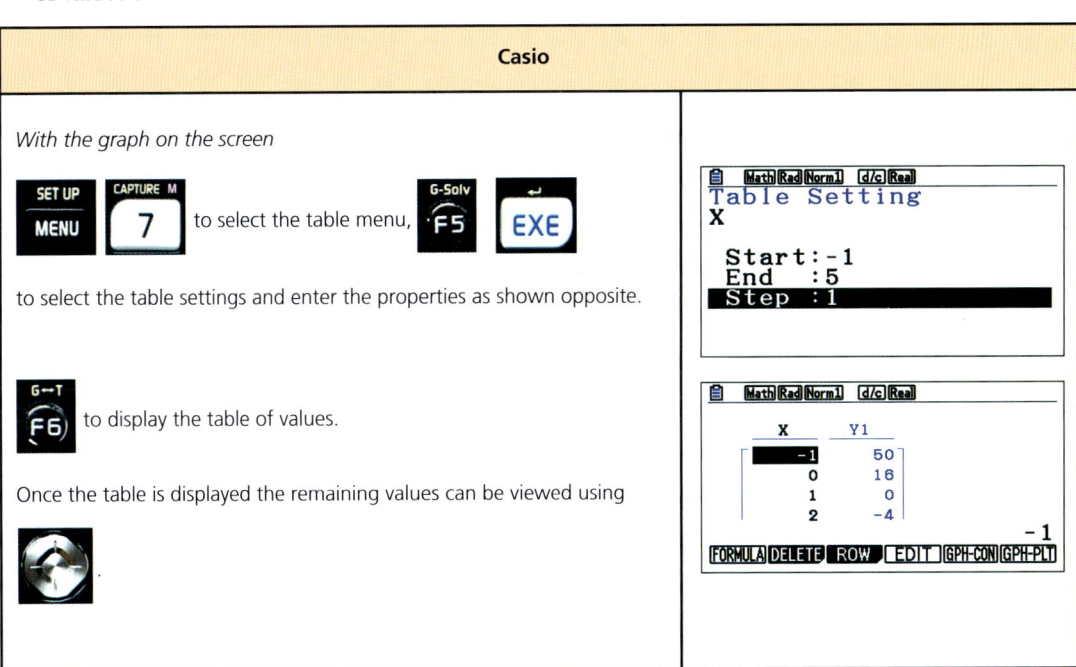

# 25 USING A GRAPHIC DISPLAY CALCULATOR TO SKETCH AND ANALYSE FUNCTIONS

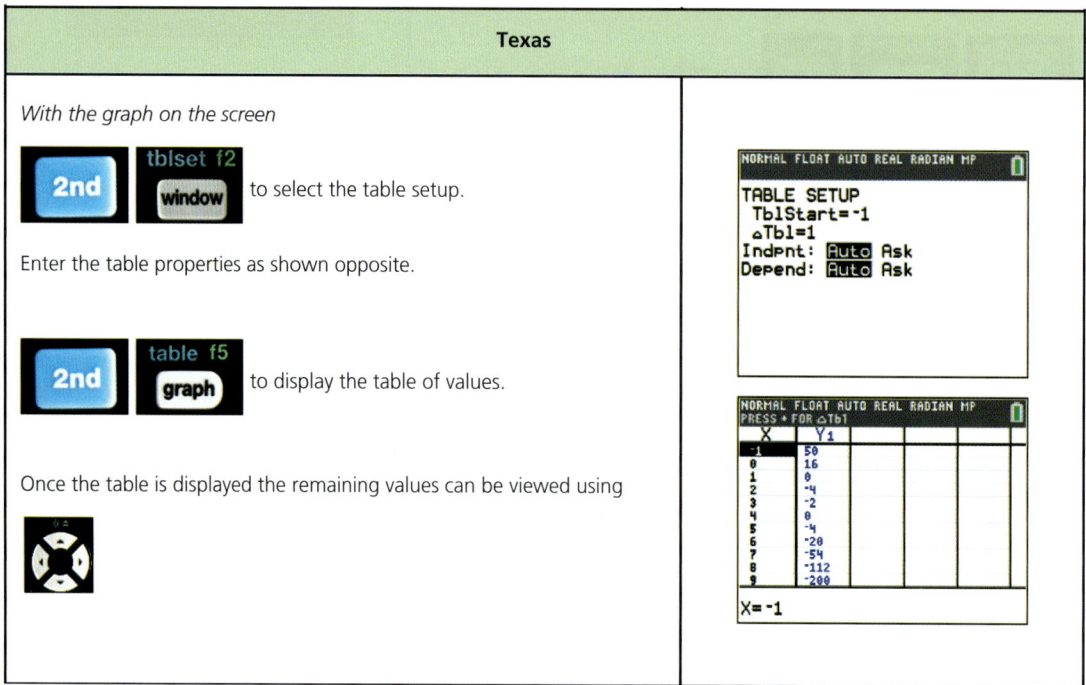

The instructions shown in the examples above will be useful when checking your solutions to various exercises throughout this textbook.

### Exercise 3.8

Use a graphic display calculator to help you to answer the following questions.

1. **a** Sketch a graph of the function $y = x^3 - 7x + 6$
   **b** Find the zeros (roots) of the function and label them clearly on your sketch.
   **c** Find where the graph intersects the $y$-axis. Label this clearly on your sketch.
   **d** Find the coordinates of any local maxima or minima.

2. **a** Sketch a graph of the function $y = 2x^3 - 8x^2 - 22x - 12$
   **b** Find the zeros (roots) of the function and label them clearly on your sketch.
   **c** Find where the graph intersects the $y$-axis. Label this clearly on your sketch.
   **d** Find the coordinates of any local maxima or minima.
   **e** Copy and complete the table of values below, for the coordinates of some points on the graph.

| $x$ | −3 | −2 | −1 | 0 | 1 | 2 | 3 | 4 | 5 | 6 | 7 |
|---|---|---|---|---|---|---|---|---|---|---|---|
| $y$ | | | | | | | | | | | |

*Using a graphic display calculator to sketch and analyse functions*

**3 a** Sketch a graph of the function $y = (x + 4)^2(x - 2)^2$
  **b** Find the zeros (roots) of the function and label them clearly on your sketch.
  **c** Find where the graph intersects the y-axis. Label this clearly on your sketch.
  **d** Find the coordinates of any local maxima or minima.
  **e** Copy and complete the table of values below, for the coordinates of some points on the graph.

| x | −5 | −4 | −3 | −2 | −1 | 0 | 1 | 2 | 3 |
|---|---|---|---|---|---|---|---|---|---|
| y |   |   |   |   |   |   |   |   |   |

**4** Repeat Q.3 above for the function $y = (x + 4)^2(x - 2)^2 - 14$

**5 a** Sketch a graph of the function $y = -\left(\frac{1}{2}x - 1\right)^2(x + 1)^2$
  **b** Find the zeros (roots) of the function and label them clearly on your sketch.
  **c** Find where the graph intersects the y-axis. Label this clearly on your sketch.
  **d** Find the coordinates of any local maxima or minima.
  **e** Copy and complete the table of values below, for the coordinates of some points on the graph.

| x | −4 | −3 | −2 | −1 | 0 | 1 | 2 | 3 |
|---|---|---|---|---|---|---|---|---|
| y |   |   |   |   |   |   |   |   |

Use your graphic display calculator to solve the following equations:

**6** $2^x - 1 = \frac{1}{x^3}$      **7** $3^{\frac{1}{2}x} = \frac{1}{4}x^4$      **8** $\frac{1}{3^x} = -x^2 + \sqrt{x} + \frac{1}{2}$

# 26 Finding a quadratic function from key values

## Using factorised form

It is not necessary to see the graph of a quadratic function or to know the coordinates of a large number of points on the curve in order to determine its equation. All that is needed are the coordinates of certain key points.

In Topic 2, you learned how to factorise a quadratic expression and therefore also a quadratic function. Factorised forms of a quadratic function give key information about its properties.

For example, $f(x) = x^2 - 5x + 6$ can be factorised to give $f(x) = (x - 3)(x - 2)$.

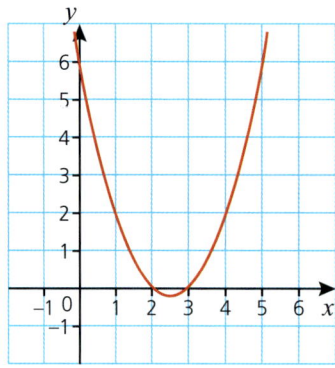

The graph of the function is shown above.

There are features of the graph that relate directly to the equation of the function. Exercise 3.9 looks at these relationships.

### Exercise 3.9

1 For each of the following quadratics:
   i write the function in factorised form
   ii with the aid of a graphic display calculator if necessary, sketch the function and identify clearly where it crosses both axes.
   
   a  $f(x) = x^2 + 3x + 2$
   b  $f(x) = x^2 + 3x - 4$
   c  $f(x) = x^2 + 3x - 10$
   d  $f(x) = x^2 + 13x + 42$
   e  $f(x) = x^2 - 9$
   f  $f(x) = x^2 - 64$

2 Using your solutions to Q.1, describe any relationship that you can see between a function written in factorised form and its graph.

Exercise 3.9 shows that there is a direct link between a function written in factorised form and where its graph crosses the $x$-axis. The reason for this is explained using the earlier example of $y = (x - 3)(x - 2)$.

## Using factorised form

You know that where a graph crosses the $x$-axis, the $y$-coordinates are zero.

Hence $(x-3)(x-2) = 0$.

To solve the equation, either $(x-2) = 0$ or $(x-3) = 0$, and therefore $x = 2$ or $x = 3$ respectively. These are the $x$-coordinates of the points where the graph crosses the $x$-axis.

### → Worked examples

a  The graph of a quadratic function of the form $f(x) = x^2 + bx + c$ crosses the $x$-axis at $x = 4$ and $x = 5$. Determine the equation of the quadratic and state the values of $b$ and $c$.

As the graph intercepts the $x$-axis at 4 and 5, the equation of the quadratic can be written in factorised form as $f(x) = (x-4)(x-5)$.

When expanded, the equation is written as $f(x) = x^2 - 9x + 20$.

Therefore $b = -9$ and $c = 20$

b  The graph of a quadratic function of the form $f(x) = ax^2 + bx + c$ crosses the $x$-axis at $x = -2$ and $x = 5$. It also passes through the point $(0, -20)$.

Determine the equation of the quadratic and state the values of $a$, $b$ and $c$.

This example is slightly more difficult than the first one. Although the graph crosses the $x$-axis at $-2$ and 5, this does not necessarily imply that the quadratic function is $f(x) = (x+2)(x-5)$. This is because more than one quadratic can pass through the points $(-2, 0)$ and $(5, 0)$, as shown:

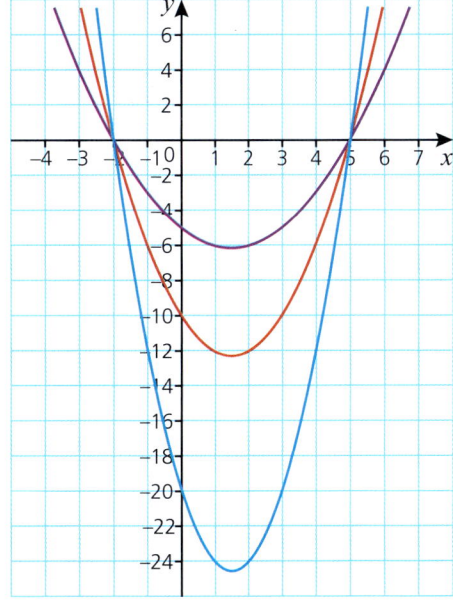 or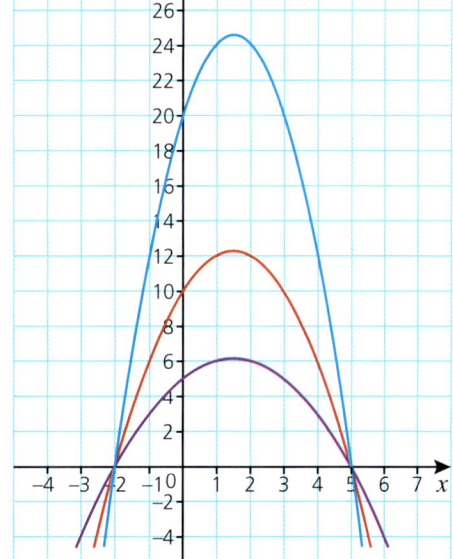

# 26 FINDING A QUADRATIC FUNCTION FROM KEY VALUES

However, it is also known that the graph passes through the point $(0, -20)$. By substituting this into the equation $y = (x + 2)(x - 5)$ it can be seen that the equation is incorrect:

$$-20 \neq (0 + 2)(0 - 5)$$

$$-20 \neq -10$$

This implies that the coefficient of $x^2$, $a$, is not equal to 1 in the function $f(x) = ax^2 + bx + c$.

However, as can be seen in the graphs above, because all possible quadratics that intersect the $x$-axis at $x = -2$ and $x = 5$ are stretches of $y = (x + 2)(x - 5)$ parallel to the $y$-axis, the quadratic takes the form $y = a(x + 2)(x - 5)$.

The value of $a$ can be calculated by substituting the coordinate $(0, -20)$ into this equation.

Therefore $\quad -20 = a(0 + 2)(0 - 5)$

$$-20 = -10a$$

$$a = 2$$

The quadratic function can therefore be written as $f(x) = 2(x + 2)(x - 5)$ which, when expanded, becomes $f(x) = 2x^2 - 6x - 20$.

Therefore $a = 2$, $b = -6$, $c = -20$

## Exercise 3.10

1 Find the equation of the quadratic graphs that intersect the $x$-axis at each of the following points. Give your answers in the form $f(x) = x^2 + bx + c$ clearly stating the values of $b$ and $c$.
   a $\quad x = 0$ and $x = 2$ 
   b $\quad x = -1$ and $x = -6$
   c $\quad x = -3$ and $4$
   d $\quad x = -\frac{1}{2}$ and $3$

2 The graphs of six quadratic functions are shown below. In each case the function takes the form $f(x) = ax^2 + bx + c$ where $a = \pm 1$.
   From the graphs, find the equations and state clearly the values of $a$, $b$ and $c$.

a
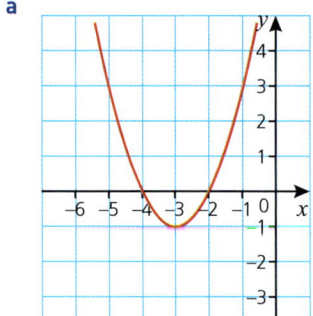

b
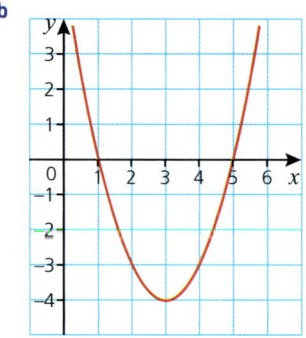

c
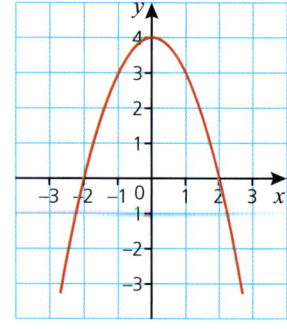

## Using factorised form

d 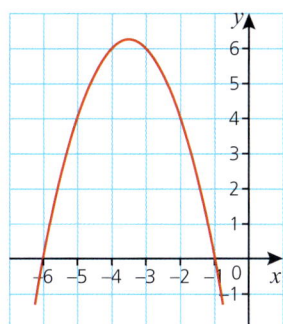  e 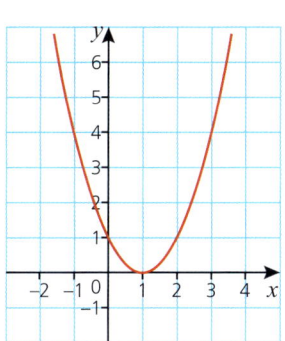  f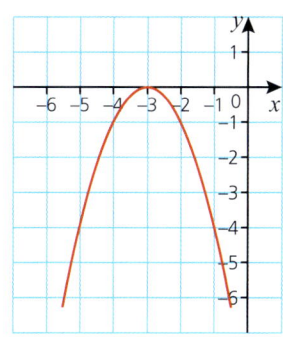

**3** Find the equations of the quadratic functions graphed below, giving your answers in the form $f(x) = ax^2 + bx + c$. In each case the graph shows where the function intersects the $x$-axis and the coordinates of one other point on the graph.

a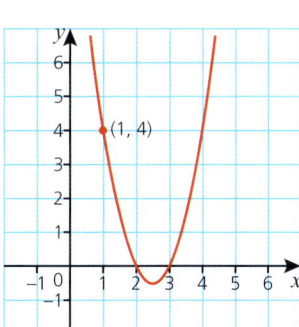

b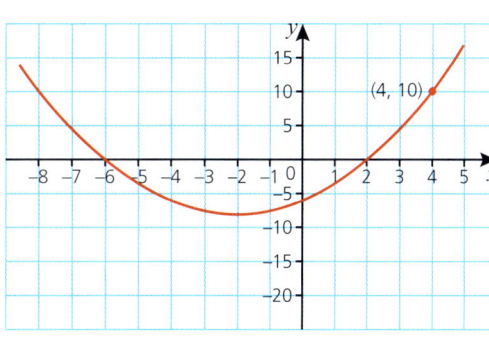

c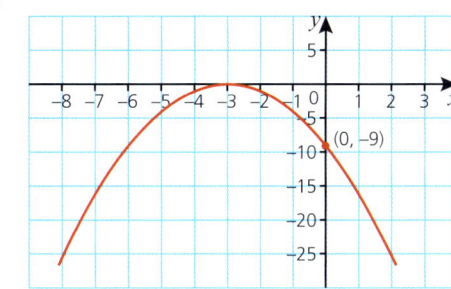

d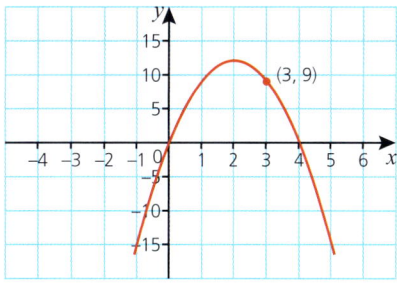

e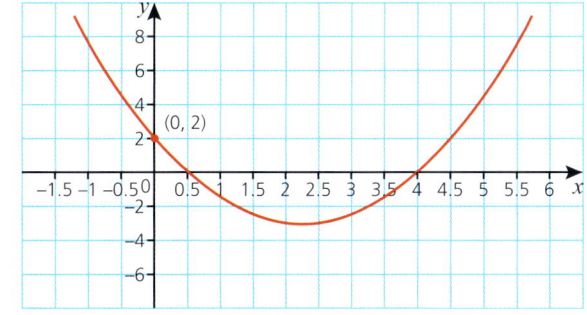

f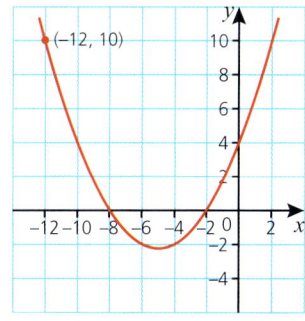

## 26 FINDING A QUADRATIC FUNCTION FROM KEY VALUES

# Using the vertex to find the quadratic function

The vertex of a quadratic function refers to the point at which the graph of the function is either a maximum or a minimum, as shown below:

a 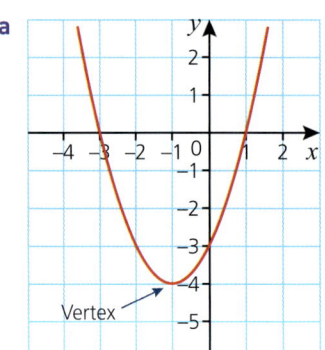  b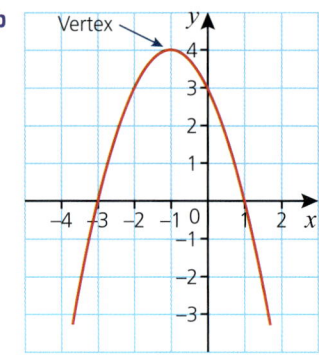

The coordinates of the vertex are useful for finding the quadratic function. You saw earlier in this topic (Chapter 24) how to transform functions. You have also studied how to factorise a quadratic by the method of completing the square. Both techniques are useful when deriving the quadratic function from the coordinates of its vertex.

### ➜ Worked example

A quadratic function is given as $f(x) = (x + 1)^2 - 9$.

i  Sketch the graph of the function by finding where it intersects each of the axes.

The intercept with the $y$-axis occurs when $x = 0$

Therefore $f(0) = (0 + 1)^2 - 9$
$\Rightarrow f(0) = 1^2 - 9$
$\Rightarrow f(0) = -8$

i.e. the $y$-intercept occurs at the coordinate $(0, -8)$

Intercepts with the $x$-axes occur when $y = 0$

Therefore $(x + 1)^2 - 9 = 0$
$\Rightarrow (x + 1)^2 = 9$
$\Rightarrow x + 1 = \pm \sqrt{9}$
$\Rightarrow x + 1 = \pm 3$
$\Rightarrow x = 2 \text{ or } -4$

## Using the vertex to find the quadratic function

The function can therefore be sketched:

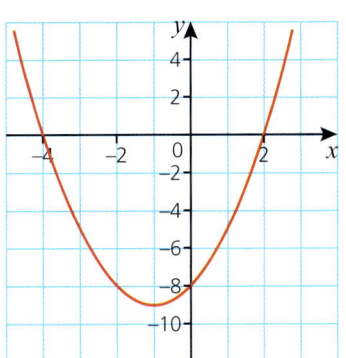

**ii** Find the coordinates of the graph's vertex.

There are two methods for approaching this.

**Method 1**

Because of symmetry, the $x$-coordinate of the vertex must be midway between the points where the graph intersects the $x$-axis.

Therefore the $x$-coordinate of the vertex is $-1$.

To find the $y$-coordinate of the vertex, substitute $x = -1$ into the function $f(x) = (x + 1)^2 - 9$:

$f(-1) = (-1 + 1)^2 - 9 \quad \Rightarrow \quad f(-1) = -9$

Therefore the coordinates of the vertex are $(-1, -9)$.

This can be checked using the graphic display calculator to find the coordinates of the minimum point.

**Method 2**

Look at the function written in completed square form as a series of transformations of $f(x) = x^2$.

The transformation that maps $f(x) = x^2$ to $f(x) = (x + 1)^2 - 9$ is the translation $\begin{pmatrix} -1 \\ -9 \end{pmatrix}$. As $f(x) = x^2$ has a vertex at $(0, 0)$, $f(x) = (x + 1)^2 - 9$ has a vertex at $(-1, -9)$.

# 26 FINDING A QUADRATIC FUNCTION FROM KEY VALUES

**Exercise 3.11**

1. In each of the following, the quadratic functions are of the form $f(x) = (x - h)^2 + k$.
   i   Find where the graph of the function intersects each axis.
   ii  Sketch the function.
   iii Find the coordinates of the vertex.
   iv  Check your answers to Q.1i–iii using a graphic display calculator.
   
   a  $f(x) = (x - 2)^2 - 9$  
   b  $f(x) = (x + 5)^2 - 1$  
   c  $f(x) = (x - 3)^2 - 4$  
   d  $f(x) = (x - \frac{1}{2})^2 - 16$  
   e  $f(x) = (x + 4)^2 - 10$  
   f  $f(x) = (x - 5)^2$

2. In each of the following, the quadratic function is of the form $f(x) = ax^2 + bx + c$, where $a = \pm 1$ and $b$ and $c$ are rational.
   From the graph of each function:
   i   find the equation of the quadratic in the form $f(x) = ax^2 + bx + c$
   ii  find the coordinates of the vertex
   iii write the equation in the form $f(x) = a(x - h)^2 + k$
   iv  expand your answer to Q.2iii and check it is the same as your answer to Q.2i.

a
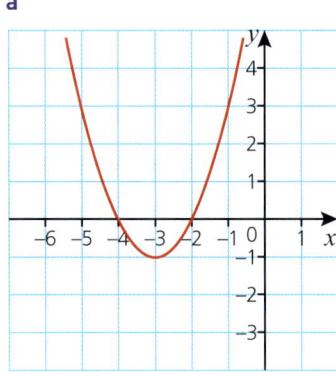

b
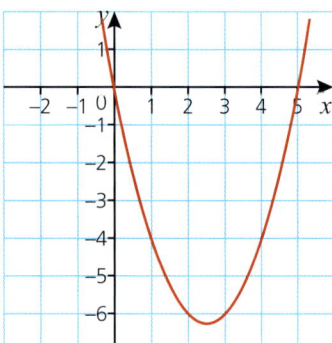

c
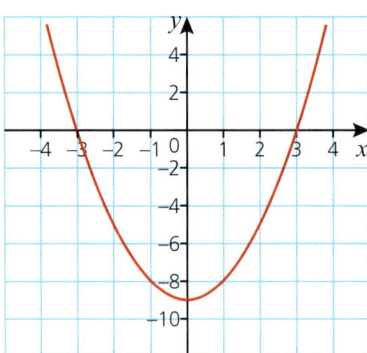

d
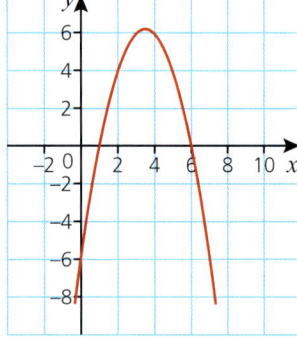

## Using the vertex to find the quadratic function

e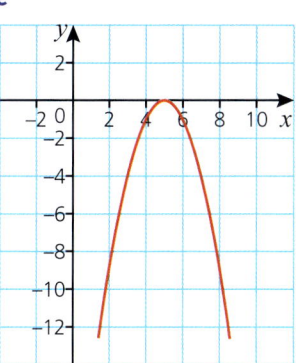

f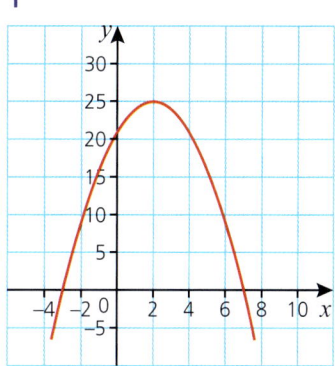

**3 a** Copy the table below and enter the relevant answers to Q.1 and Q.2 in the rows.

|  | In the form $f(x) = a(x - h)^2 + k$ | Coordinates of vertex |
|---|---|---|
| Example | $f(x) = (x + 1)^2 - 9$ | $(-1, -9)$ |
| 1 a |  |  |
| 1 b |  |  |
| 1 c |  |  |
| 1 d |  |  |
| 1 e |  |  |
| 1 f |  |  |
| 2 a |  |  |
| 2 b |  |  |
| 2 c |  |  |
| 2 d |  |  |
| 2 e |  |  |
| 2 f |  |  |

**b** Describe in your own words the relationship between the equation of a quadratic written in the form $f(x) = a(x - h)^2 + k$ and the coordinates of its vertex.

**4** The following quadratics are of the form $f(x) = a(x - h)^2 + k$ where $a = \pm 1$ and $h$ and $k$ are rational. Find the coordinates of the vertex of each function.

  **a** $f(x) = (x - 2)^2 + 4$     **b** $f(x) = (x + 5)^2 - 3$
  **c** $f(x) = (x - 6)^2 + 4$     **d** $f(x) = x^2 - 3$
  **e** $f(x) = -(x + 5)^2 + 3$     **f** $f(x) = -(x - 4)^2$

**5** Sketch each of the quadratics in Q.4. Clearly label the coordinates of the vertex and where the graph intersects the $y$-axis.

# 26 FINDING A QUADRATIC FUNCTION FROM KEY VALUES

**Exercise 3.11 (cont)**

6  The vertices of quadratic functions are given below. In each case the function is of the form $f(x) = ax^2 + bx + c$ where $a = 1$.
   i  Work out, from the vertex, the equation in the form $f(x) = a(x - h)^2 + k$.
   ii Expand your equation from Q.6i to write the equation in the form $f(x) = ax^2 + bx + c$.
   iii Check your answers to parts i and ii using a graphic display calculator.
   
   **a** $(-3, 6)$     **b** $(2, -4)$     **c** $(-1, 6)$     **d** $(-4, 0)$

7  Repeat Q.6 where each of the quadratics is of the form $f(x) = ax^2 + bx + c$ and $a = -1$.

You will have seen that, in general, if a quadratic is written in the form $f(x) = a(x - h)^2 + k$, the coordinates of its vertex are $(h, k)$.

## Finding the equation of a quadratic function given a vertex and another point

It was shown earlier that if a quadratic is of the form $f(x) = ax^2 + bx + c$ and $a$ can be any real number, then additional information is needed in order to deduce its exact equation. This is also the case regarding the coordinates of its vertex. If $a$ can be any real number, then simply knowing the coordinates of the vertex is insufficient to deduce its equation, as more than one quadratic can be drawn with the same vertex, as shown:

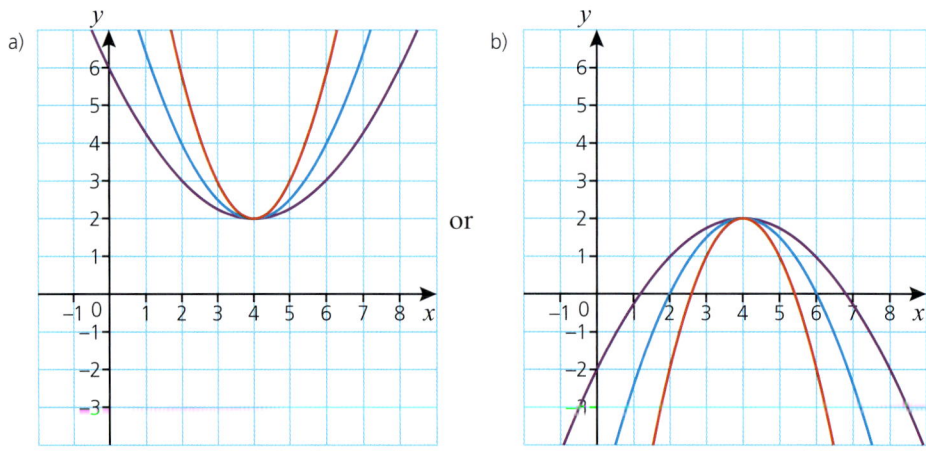

The coordinates of an additional point are also needed.

*Finding the equation of a quadratic function given a vertex and another point*

### ➜ Worked example

The graph of a quadratic function passes through the point (−1, −25). Its vertex has coordinates (2, 2).

**i** Find the equation of the quadratic in the form $f(x) = a(x - h)^2 + k$.

If the quadratic was of the form $f(x) = (x - h)^2 + k$, the coordinates of its vertex suggest that its equation would be $f(x) = (x - 2)^2 + 2$. However, the point (−1, −25) does not fit this equation. Therefore the quadratic function is of the form $f(x) = a(x - h)^2 + k$ where $a \neq 1$.

The function must be written as $f(x) = a(x - 2)^2 + 2$ and $a$ can be calculated by substituting the values of the point (−1, −25) into the function for $x$ and $y$.

Therefore $-25 = a(-1 - 2)^2 + 2$

$\Rightarrow \quad -25 = 9a + 2$

$\Rightarrow \quad -27 = 9a$

$\Rightarrow \quad a = -3$

Therefore the quadratic function is $f(x) = -3(x - 2)^2 + 2$.

**ii** Find the equation of the quadratic in the form $f(x) = ax^2 + bx + c$, stating clearly the values of $a$, $b$ and $c$.

The function $f(x) = -3(x - 2)^2 + 2$ can be expanded to give:

$f(x) = -3(x^2 - 4x + 4) + 2$

$\Rightarrow \quad f(x) = -3x^2 + 12x - 12 + 2$

$\Rightarrow \quad f(x) = -3x^2 + 12x - 10$

Therefore $a = -3$, $b = 12$ and $c = -10$.

**iii** Work out where the graph intercepts the $y$-axis.

The graph intersects the $y$-axis when $x = 0$. Substituting $x = 0$ into the function $f(x) = -3x^2 + 12x - 10$ gives:

$f(0) = -3(0)^2 + 12(0) - 10$

$f(0) = -10$

Therefore the graph intersects the $y$-axis at the point (0, −10).

**iv** Find the points of intersection with the $x$-axis, giving your answer in surd form.

The intercept with the $x$-axis occurs when $y = 0$. Substituting $y = 0$ into the equation gives:

$0 = -3x^2 + 12x - 10$

Using the quadratic formula: $x = \dfrac{-b \pm \sqrt{b^2 - 4ac}}{2a}$

where $a = -3$, $b = 12$ and $c = -10$

Therefore $x = \dfrac{-12 \pm \sqrt{12^2 - 4(-3)(-10)}}{2(-3)}$

# 26 FINDING A QUADRATIC FUNCTION FROM KEY VALUES

$$x = \frac{-12 \pm \sqrt{144-120}}{-6} = \frac{-12 \pm \sqrt{24}}{-6}$$

$$x = \frac{-12}{-6} \pm \frac{\sqrt{24}}{-6} = 2 \pm \sqrt{\frac{24}{36}}$$

Therefore $x = 2 \pm \sqrt{\frac{2}{3}}$

**v** Sketch the graph of the function.

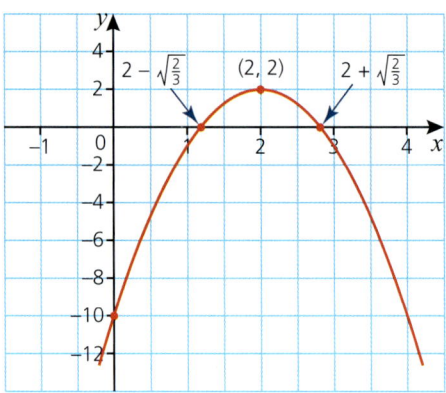

### Exercise 3.12

1. Find the coordinates of the vertex of each of the following quadratic functions:
   **a** $f(x) = 2(x-1)^2 + 6$      **b** $f(x) = 3(x+4)^2 - 6$
   **c** $f(x) = 3x^2 - 4$      **d** $f(x) = -2(x+2)^2 - 6$
   **e** $f(x) = 2[(x-6)^2 + 1]$      **f** $f(x) = -\frac{1}{2}(x+1)^2$

2. Four quadratic functions are given below. Each is written in the form $f(x) = a(x-h)^2 + k$ and its expanded form. Find the matching pairs.
   **a** $f(x) = (x-3)^2 - 2$      **b** $f(x) = 2(x+1)^2 - 3$
   **c** $f(x) = 2x^2 - 4x + 3$      **d** $f(x) = x^2 - 6x + 7$
   **e** $f(x) = 2x^2 + 4x$      **f** $f(x) = 2(x-1)^2 + 1$
   **g** $f(x) = 2x^2 + 4x - 1$      **h** $f(x) = 2[(x+1)^2 - 1]$

3. The graphs of three of the four quadratic functions below pass through the point (1, 1). Which is the odd one out?
   **a** $f(x) = (x-2)^2$      **b** $f(x) = \frac{1}{2}(x+3)^2 - 7$
   **c** $f(x) = 3(x-3)^2 - 11$      **d** $f(x) = (x-1)^2 - 1$

## Finding the equation of a quadratic function given a vertex and another point

**4** The graph below shows four quadratic functions each with a vertex at (3, 6). The coordinates of one other point on each graph is also given.

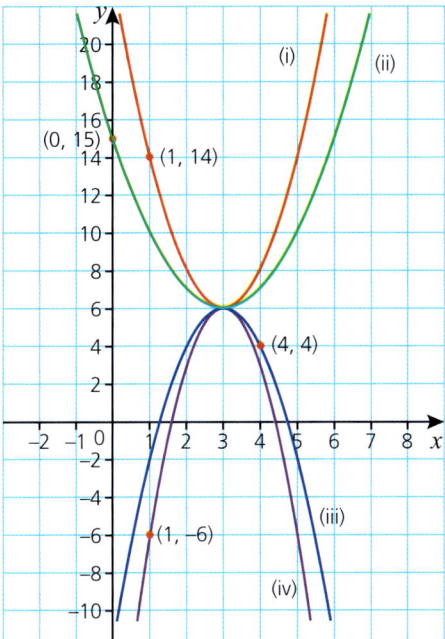

Match each of the equations below with the correct graph.
- **a** $f(x) = (x - 3)^2 + 6$
- **b** $f(x) = -2(x - 3)^2 + 6$
- **c** $f(x) = 2[(x - 3)^2 + 3]$
- **d** $f(x) = -3(x - 3)^2 + 6$

**5** The coordinates of the vertex of a quadratic function and one other point are given in parts a–f below. Work out the quadratic function:
- **i** in the form $f(x) = a(x - h)^2 + k$
- **ii** in the form $f(x) = ax^2 + bx + c$, stating clearly the values of $a$, $b$ and $c$.

|   | Vertex | Other point |
|---|---|---|
| **a** | (−1, −5) | (1, 3) |
| **b** | (1, 4) | (2, 2) |
| **c** | (−4, −4) | (−3, −1) |
| **d** | (−5, −2) | (−3, 0) |
| **e** | (3, 0) | (0, −9) |
| **f** | (−5, −12) | (−7, −6) |

**6** For each of the quadratic functions in Q.5:
- **i** find the y-intercept
- **ii** work out where/if the graph intersects the x-axis, giving your answer in surd form
- **iii** check your answers using your graphic display calculator.

# 26 FINDING A QUADRATIC FUNCTION FROM KEY VALUES

**Exercise 3.12 (cont)**

7  The quadratic function shown below has a vertex at $(p, 4)$ where $p > 0$, and an intercept with the $y$-axis at $(0, 13)$.

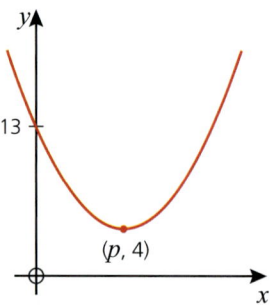

If the function is of the form $f(x) = x^2 + bx + c$, find the value of $p$.

8  The quadratic function shown below is of the form $f(x) = ax^2 + bx + c$ where $a = -1$. The graph of the function has a vertex at $(-2, t)$ and also passes through the point $(-3, 1)$.

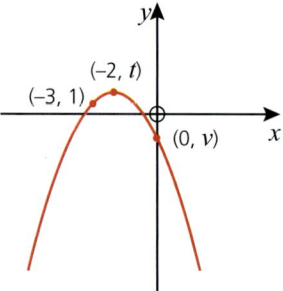

a  Find the value of $t$.
b  Work out the value of the $y$-intercept, $v$.

# 27 Finding the equation of other common functions

Chapter 26 showed that it is possible to find the equation of a quadratic function from some of its properties, such as the intercepts with the axes and the coordinates of the vertex.

Chapter 23 showed that it is possible to recognise the type of function by the shape of its graph.

e.g.

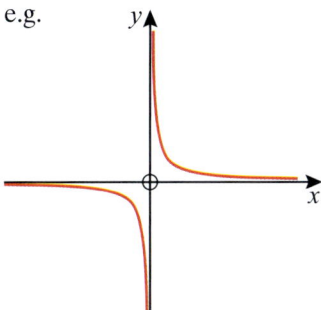

is a reciprocal function of the form $f(x) = \dfrac{a}{x}$ where $a \neq 0$

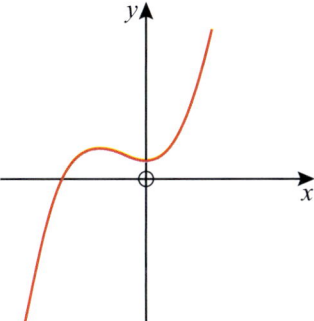

is a cubic function of the form $f(x) = ax^3 + bx^2 + cx + d$, where $a \neq 0$

It is therefore possible to find the equation of a graph from its shape and the coordinates of some of the points that lie on the graph.

# 27 FINDING THE EQUATION OF OTHER COMMON FUNCTIONS

## → Worked examples

**a** The following function passes through the points (0, 4) and (2, 0). Work out its equation.

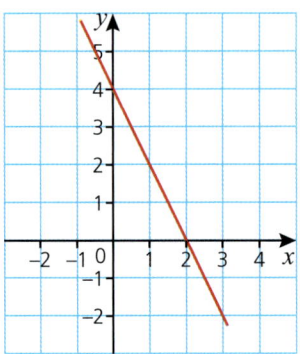

From the graph above, you can recognise that the function is linear; therefore it is of the form $f(x) = ax + b$, where $a$ represents the gradient and $b$ represents the $y$-intercept.

Gradient $= \dfrac{4-0}{0-2} = \dfrac{4}{-2} = -2$

$y$-intercept $= 4$

Therefore $f(x) = -2x + 4$.

**b** The graph below has the equation $f(x) = x^3 + bx^2 - 11x + d$. It intercepts the $y$-axis at (0, 12) and the $x$-axis in three places, one of which is the point (4, 0).

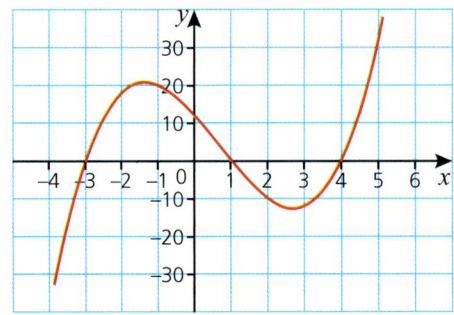

Find the values of $b$ and $d$ in the function $f(x) = x^3 + bx^2 - 11x + d$.

At the intercept with the $y$-axis, $x = 0$. This can be substituted into the equation to work out $d$:

$f(0) = 0^3 + b(0)^2 - 11(0) + d$

$\Rightarrow \quad 12 = 0 + 0 - 0 + d$

$\Rightarrow \quad 12 = d$

## Finding the equation of other common functions

When $x = 4$, $y = 0$. This can be substituted into the equation to find the value of $b$:

$$f(4) = 4^3 + b(4)^2 - 11(4) + 12$$
$$\Rightarrow \quad 0 = 64 + 16b - 44 + 12$$
$$\Rightarrow \quad 0 = 32 + 16b$$
$$\Rightarrow \quad -32 = 16b$$
$$\Rightarrow \quad b = -2$$

Therefore $f(x) = x^3 - 2x^2 - 11x + 12$.

c   The function below passes through the point $(-5, -1)$.

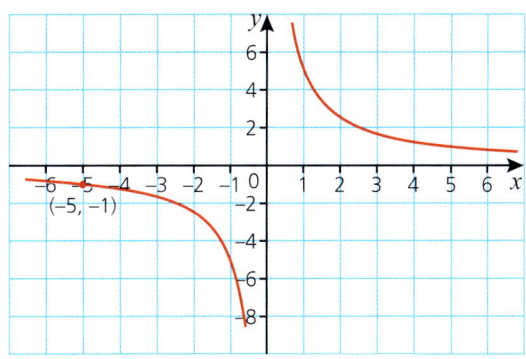

Given that the function is of the form $f(x) = \dfrac{a}{x}$, work out the value of $a$.

As its equation is of the form $f(x) = \dfrac{a}{x}$, the value of $a$ can be calculated by substituting in the value of $x$ and $y$ of a point on the graph:

$$\Rightarrow -1 = \dfrac{a}{-5},$$
$$\Rightarrow \quad a = 5$$

Therefore $f(x) = \dfrac{5}{x}$.

 **Exercise 3.13**   In Q.1–8, linear functions take the form $f(x) = ax + b$, cubic functions take the form $f(x) = ax^3 + bx^2 + cx + d$ and reciprocal functions take the form $f(x) = \dfrac{a}{x}$.

For each question:
a   identify the type of function from the shape of the graph
b   find the values of the unknown coefficients $a$, $b$, $c$ or $d$.

1
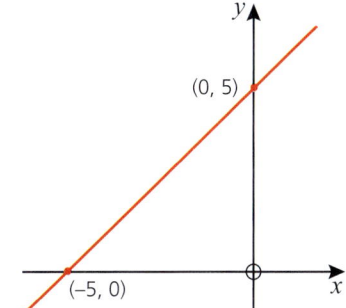

2
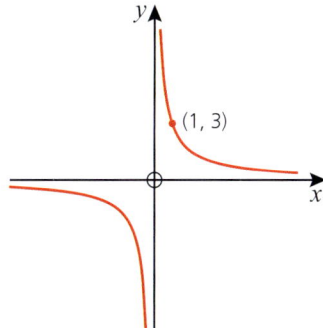

# 27 FINDING THE EQUATION OF OTHER COMMON FUNCTIONS

**Exercise 3.13 (cont)**

3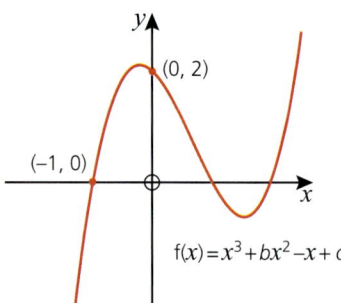

4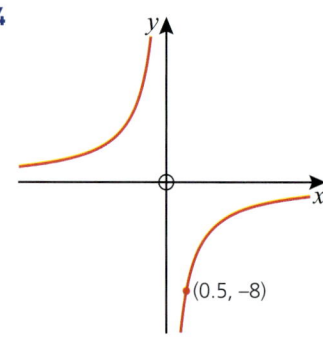

5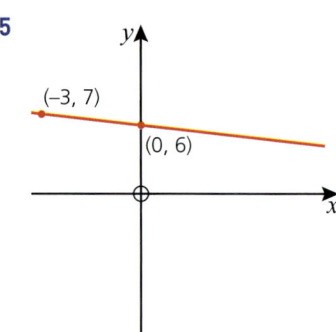

6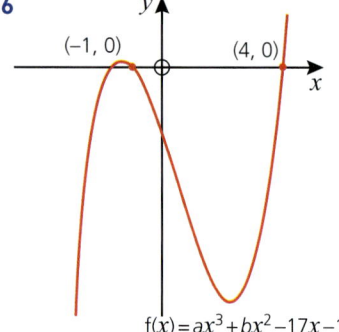

7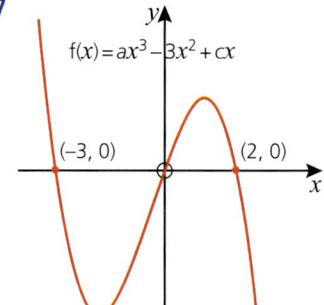

8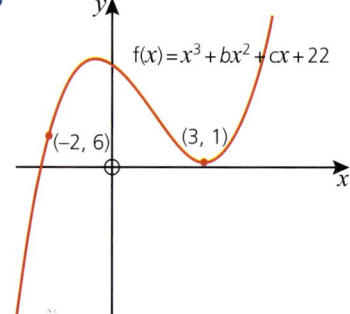

# Finding the equation of other common functions

A similar method can be used to work out the equations of exponential functions. Chapter 23 showed that exponential functions take the form $f(x) = a^x$ where $0 < a < 1$ or $a > 1$ and in general have a graph as shown:

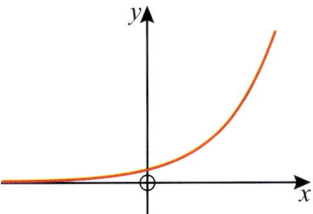

## → Worked example

The graph below passes through the point $(-1, 2)$. Identify the type of function and work out its equation.

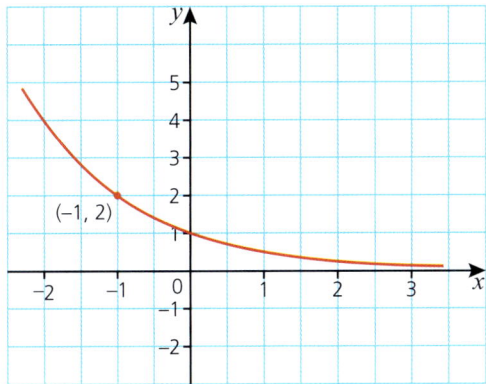

From the shape of the graph, you can identify that the function is of the form $f(x) = a^x$ where $0 < a < 1$. To find the value of $a$, substitute the values of $x$ and $y$ of a point on the curve.

Substituting $(-1, 2)$ into $y = a^x$

$$\Rightarrow \quad 2 = a^{-1}$$
$$\Rightarrow \quad 2 = \frac{1}{a}$$
$$\Rightarrow \quad a = \frac{1}{2}$$

Therefore $f(x) = \left(\frac{1}{2}\right)^x$.

# 27 FINDING THE EQUATION OF OTHER COMMON FUNCTIONS

**Exercise 3.14**

1  Find the equation of the following functions:

a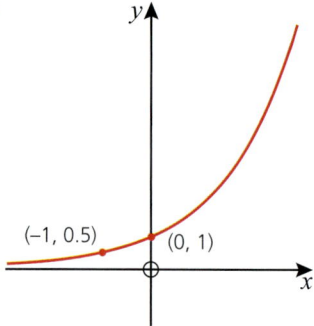

b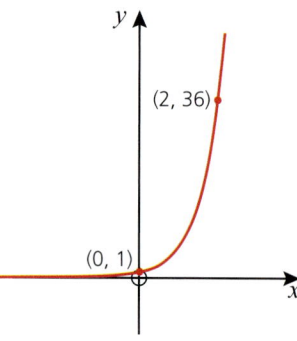

c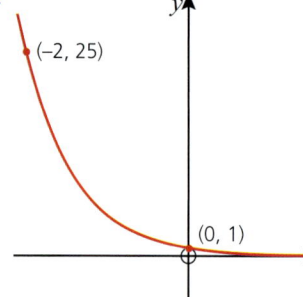

2  A graph of the form $f(x) = a^x$ is shown below:

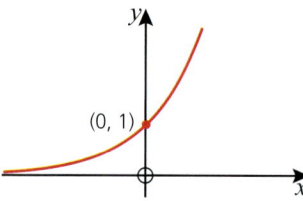

Explain clearly why the information given is insufficient to find the equation of the graph.

# 28 Composite functions

So far you have dealt with functions individually. However, if two functions are a function of $x$, it is possible to combine them to create a further function of $x$. These are known as **composite** functions.

### → Worked examples

**a** $f(x) = 3x$ and $g(x) = 2x - 1$

**i** Write an expression for $f(g(x))$.
$f(g(x))$ means that $g(x)$ is substituted for $x$ in the function $f(x)$,
i.e. $f(g(x)) = f(2x - 1)$
$= 3(2x - 1)$
$= 6x - 3$

> **Note**
> $f(g(x))$ may also be written as $fg(x)$

**ii** Write an expression for $g(f(x))$.
$g(f(x))$ means that $f(x)$ is substituted for $x$ in the function $g(x)$,
i.e. $g(f(x)) = g(3x)$
$= 2(3x) - 1$
$= 6x - 1$

It can be seen from the answers to parts i and ii that $f(g(x))$ and $g(f(x))$ are not equal.

**b** Given that $f(x) = \dfrac{2}{x - 3}$ and $g(x) = (2x - 1)^2$

**i** find $f(g(x))$. Give your answer as a fraction in its simplest form.

$f(g(x)) = \dfrac{2}{(2x - 1)^2 - 3}$

$= \dfrac{2}{(4x^2 - 4x + 1) - 3}$

$= \dfrac{2}{4x^2 - 4x - 2}$

$= \dfrac{1}{2x^2 - 2x - 1}$

**ii** Find the value of $f(g(-1))$

Substituting $x = -1$ into $f(g(x)) = \dfrac{1}{2x^2 - 2x - 1}$ gives

$f(g(-1)) = \dfrac{1}{2(-1)^2 - 2(-1) - 1}$

$= \dfrac{1}{2 + 2 - 1}$

$= \dfrac{1}{3}$

# 28 COMPOSITE FUNCTIONS

   **c** $f(x) = x - 4$ and $g(x) = 2x + 1$
     **i** Write an expression equivalent to $f(g(x))$.
       Substitute $g(x)$ for $x$ in the function $f(x)$:
$$f(g(x)) = f(2x + 1)$$
$$= (2x + 1) - 4$$
$$= 2x - 3$$
     **ii** Find the value of $f(g(3))$.
       From part i, $f(g(x)) = 2x - 3$. Substitute $x = 3$ into the function $f(g(x))$.
       Therefore $f(g(3)) = 2 \times 3 - 3$
$$= 3$$
     **iii** Solve $f(g(x)) = 0$,
       i.e. find the value of $x$ which produces $f(g(x)) = 0$.
       Therefore solve $2x - 3 = 0$
$$\Rightarrow x = \frac{3}{2}$$

 **Exercise 3.15**

**1** For the function $f(x) = 3x + 4$, work out
   **a** $f(0)$         **b** $f(2)$         **c** $f(-1)$

**2** If $g(x) = \frac{1}{2}x + 2$, work out
   **a** $f(4)$         **b** $f(-4)$         **c** $f\left(\frac{1}{2}\right)$

**3** $f(x)$ and $g(x)$ are given for each part below. In each case, find a simplified expression for $f(g(x))$.
   **a** $f(x) = 2x$        **b** $f(x) = 4x$        **c** $f(x) = 3x - 1$
      $g(x) = 3x$           $g(x) = x - 3$        $g(x) = 3x$
   **d** $f(x) = x - 2$      **e** $f(x) = 2x - 2$     **f** $f(x) = 2 - 2x$
      $g(x) = 3x + 1$         $g(x) = 3x + 1$      $g(x) = x + 4$

**4** $f(x) = 5x - 3$ and $g(x) = 2x + 3$
   **a** Solve $f(x) = 0$.
   **b** Solve $g(x) = 0$.
   **c** Find the value of $x$ where $f(x) = g(x)$.

**5** $f(x) = 4x - 2$ and $g(x) = 2x - 4$
   **a** Find a single expression for $f(g(x))$.
   **b** Find a single expression for $g(f(x))$.
   **c** **i** Find the value of $f(g(2))$.
      **ii** Find the value of $g(f(1))$.

**6** $f(x) = 2x$ and $g(x) = 2x - 3$
   **a** Write a single expression for:
     **i** $f(g(x))$         **ii** $g(f(x))$
   **b** Find the value of:
     **i** $f(g(0))$         **ii** $g(f(0))$
   **c** Explain whether, in this case, $f(g(x))$ can ever be equal to $g(f(x))$.

**7** Given that $f(x) = \dfrac{5}{x-1}$ and $h(x) = (x+1)^2$
   **a** show that $f(h(x)) = \dfrac{5}{x(x+2)}$
   **b** solve the equation $f(h(x)) = 1$. Give your solution(s) in exact form.

## Composite functions

> **Worked example**
>
> $f(x) = x^2 - 2$ and $g(x) = x + 1$
>
> **a** Find a simplified expression for $f(g(x))$.
>
> $g(x)$ is substituted for $x$ in the function $f(x)$:
>
> $f(x + 1) = (x + 1)^2 - 2$
>
> $\qquad\qquad = x^2 + 2x + 1 - 2$
>
> $\qquad\qquad = x^2 + 2x - 1$
>
> **b** Find the value of $f(g(2))$.
>
> $\qquad f(g(x)) = x^2 + 2x - 1$
>
> So $\quad f(g(2)) = (2)^2 + 2(2) - 1$
>
> $\qquad\qquad\quad = 4 + 4 - 1$
>
> $\qquad\qquad\quad = 7$
>
> **c** Solve $f(g(x)) = 0$ giving your answer in exact form.
>
> $\qquad x^2 + 2x - 1 = 0$
>
> As the quadratic does not factorise, it can be solved either by completing the square or using the quadratic formula. Using the quadratic formula gives:
>
> $x = \dfrac{-2 \pm \sqrt{4+4}}{2} = \dfrac{-2 \pm \sqrt{8}}{2} = \dfrac{-2 \pm 2\sqrt{2}}{2}$
>
> $x = -1 \pm \sqrt{2}$

**Exercise 3.16**

**1** $f(x)$ and $g(x)$ are given for each part below. In each case, find a simplified expression for $f(g(x))$.

   **a** $f(x) = x^2$                 $g(x) = x + 1$
   **b** $f(x) = 2x^2$               $g(x) = x - 1$
   **c** $f(x) = x^2 + 1$            $g(x) = 2x$
   **d** $f(x) = x^2 - 1$            $g(x) = 2x + 1$
   **e** $f(x) = 2x^2$               $g(x) = 3x - 2$
   **f** $f(x) = 4x^2$               $g(x) = \frac{1}{2}x + 1$
   **g** $f(x) = x^2$                 $g(x) = 3 - x$
   **h** $f(x) = \frac{1}{2}x^2$            $g(x) = 2 - 3x$

**2** $f(x)$ and $g(x)$ are given for each part below.
   **i**  Find the simplified form for $f(g(x))$.
   **ii**  Find the value of $f(g(0))$.
   **iii** Find the value of $f(g(1))$.
   **iv** Find the value of where possible, $f(g(-1))$.

      **a** $f(x) = \dfrac{3}{x}$            $g(x) = x + 1$
      **b** $f(x) = x^3$             $g(x) = 2x - 1$
      **c** $f(x) = 2^x$             $g(x) = 3x$

# 28 COMPOSITE FUNCTIONS

**Exercise 3.16 (cont)**

3  $f(x) = x^2 - 3$ and $g(x) = x + 1$
   a  Find $f(g(x))$.
   b  Find the value of $f(g(-1))$.
   c  Solve $f(g(x)) = 0$, leaving your answer in surd form.

4  $f(x) = 2x^2$ and $g(x) = x - 1$
   a  Find $f(g(x))$.
   b  Find the value of $f(g(2))$.
   c  Solve $f(g(x)) = 18$.

# 29 Inverse functions

The **inverse** of a function is its reverse, i.e. it 'undoes' the function's effects. The inverse of the function f(x) is written as $f^{-1}(x)$.

> **Worked examples**

*The inverse function takes a value from the original range and finds the value from which it came from on the domain. Therefore for the inverse function the range and domain of the original function are effectively swapped.*

a Find the inverse of the following functions:

  i  f(x) = x + 2
     This can be done by following these steps.
     Write the equation in terms of y:   y = x + 2
     Swap x and y:                        x = y + 2
     Rearrange to make y the subject:     y = x − 2
     Therefore $f^{-1}(x) = x − 2$.

  ii g(x) = 2x − 3
     Write the equation in terms of y:   y = 2x − 3
     Swap x and y:                        x = 2y − 3
     Rearrange to make y the subject: $x + 3 = 2y \Rightarrow y = \frac{x+3}{2}$
     Therefore $g^{-1}(x) = \frac{x+3}{2}$.

b If $f(x) = \frac{x-3}{3}$, calculate:

  i  $f^{-1}(2)$
     $f^{-1}(x) = 3x + 3$
     $f^{-1}(2) = 9$

  ii $f^{-1}(-3)$
     $f^{-1}(x) = 3x + 3$
     $f^{-1}(-3) = -6$

**Exercise 3.17**

Find the inverse of each of the following functions:

1 a f(x) = x + 3           b f(x) = x + 6
  c f(x) = x − 5           d g(x) = x
  e h(x) = 2x              f $p(x) = \frac{x}{3}$

2 a f(x) = 4x              b f(x) = 2x + 5
  c f(x) = 3x − 6          d $f(x) = \frac{x+4}{2}$
  e $g(x) = \frac{3x-2}{4}$   f $g(x) = \frac{8x+7}{5}$

3 a $f(x) = \frac{1}{2}x + 3$    b $g(x) = \frac{1}{4}x - 2$
  c h(x) = 4(3x − 6)       d p(x) = 6(x + 3)
  e q(x) = −2(−3x + 2)     f $f(x) = \frac{2}{3}(4x - 5)$
  g $p(x) = \frac{2}{x} + 3$     h $q(x) = \frac{1}{5x} - 2$

# 29 INVERSE FUNCTIONS

 **Exercise 3.18**

1. If $f(x) = x - 4$, work out:
   a  $f^{-1}(2)$
   b  $f^{-1}(0)$
   c  $f^{-1}(-5)$

2. If $f(x) = 2x + 1$, work out:
   a  $f^{-1}(5)$
   b  $f^{-1}(0)$
   c  $f^{-1}(-11)$

3. If $g(x) = 6(x - 1)$, work out:
   a  $g^{-1}(12)$
   b  $g^{-1}(3)$
   c  $g^{-1}(6)$

4. If $g(x) = \frac{2x + 4}{3}$, work out:
   a  $g^{-1}(4)$
   b  $g^{-1}(0)$
   c  $g^{-1}(-6)$

5. If $h(x) = \frac{1}{3}x - 2$, work out:
   a  $h^{-1}(-\frac{1}{2})$
   b  $h^{-1}(0)$
   c  $h^{-1}(-2)$

6. If $f(x) = \frac{4x - 2}{5}$, work out:
   a  $f^{-1}(6)$
   b  $f^{-1}(-2)$
   c  $f^{-1}(0)$

# 30 Logarithmic functions

## Log buttons

Earlier in this topic, exponential functions were investigated, i.e. functions of the form $y = a^x$ where the variable $x$ is the exponent (power). Solving exponential equations was done primarily using knowledge of indices. For example:

Solve $2^x = 32$

$\Rightarrow x = 5$ because $2^5 = 32$

However, had $x$ not been an integer value, the solution would have been more difficult to find. This chapter looks at the inverse of the exponential function.

Your calculator has a logarithm button.

| Casio | Texas |
|---|---|
| $10^x$  B <br> log | $10^x$  N <br> log |

This will be used throughout this chapter.

### Exercise 3.19

*The logarithm of a number is the inverse of the exponential of that number.*

1. Use the log button on your calculator to find the value of
   a. log 1
   b. log 10
   c. log 100
   d. log 1000
2. Explain clearly, referring to your answers to Q.1, what you think the log button does.
3. a. The log of what number will give an answer of −1?
   b. Explain your answer to part a.

You will have concluded from the exercise above that the log button is related to powers of 10. That is:

$\log_{10} 100 = 2 \Leftrightarrow 10^2 = 100$
       ↑
The base

Logarithms can have a base other than 10, but the relationship still holds. For example:

$\log_5 125 = 3 \Leftrightarrow 5^3 = 125$

In the example above, the relationship can be explained as the logarithm of a positive number (125) is the power (3) to which the base (5) must be raised to give that number.

In general, therefore, $\log_a y = x \Leftrightarrow a^x = y$.

# 30 LOGARITHMIC FUNCTIONS

*In this syllabus, log refers to $\log_{10}$.*

This generates three important results:
1. $\log_a a = 1$      as $a^1 = 1$
2. $\log_a 1 = 0$      as $a^0 = 1$
3. $\log_a \left(\frac{1}{a}\right) = -1$      as $a^{-1} = \frac{1}{a}$

Note: The log button on your calculator means log to the base 10, i.e. $\log_{10} x$.

### ➜ Worked examples

**a** Write $216 = 6^3$ in logarithmic notation.

$\log_6 216 = 3$

**b** Write $\log_2 128 = 7$ in index notation.

$2^7 = 128$

**c** Solve the equation $\log_4 1024 = x$.

$4^x = 1024$
$\Rightarrow x = 5$

**d** Find the value of $x$ in the equation $\log_{16} 2 = x$.

$16^x = 2$
$\Rightarrow (2^4)^x = 2 \Rightarrow 2^{4x} = 2^1$
$\Rightarrow 4x = 1 \Rightarrow x = \frac{1}{4}$

 **Exercise 3.20**

1. Write the following using logarithmic notation:
   a $2^3 = 8$    b $7^3 = 343$    c $10^5 = 100\,000$
   d $3^0 = 1$    e $2^{-2} = \frac{1}{4}$    f $p^4 = q$

2. Write the following using index notation:
   a $\log_2 64 = 6$    b $\log_3 81 = 4$    c $\log_5 625 = 4$
   d $\log_3 \frac{1}{9} = -2$    e $\log_x 1 = 0$    f $\log_5 \frac{1}{125} = -3$

3. Solve the following equations:
   a $\log_{10} 10 = x$    b $\log_{10} 10^6 = y$    c $\log_5 3125 = a$
   d $\log_7 1 = b$    e $\log_2 x = 5$    f $\log_3 x = -3$
   g $\log_5 m = -1$    h $\log_9 n = \frac{1}{2}$

Because a logarithm is an index (power), it follows the same rules as indices. The three basic rules of logarithms, which work for all bases, are as follows:

*The rules of logarithms are not covered in this syllabus.*

1. $\log_a x + \log_a y = \log_a (xy)$,    e.g. $\log_{10} 5 + \log_{10} 3 = \log_{10} (5 \times 3) = \log_{10} 15$

2. $\log_a x - \log_a y = \log_a \left(\frac{x}{y}\right)$,    e.g. $\log_{10} 12 - \log_{10} 4 = \log_{10} \left(\frac{12}{4}\right) = \log_{10} 3$

3. $\log_a x^y = y \log_a x$,    e.g. $\log_{10} 3^2 = 2\log_{10} 3$

Below is a mathematical proof to show why the first rule is true:

To prove that $\log_a x + \log_a y = \log_a (xy)$, let $p = \log_a x$ and let $q = \log_a y$.

Therefore, $\quad x = a^p$ and $y = a^q$

so $\quad\quad\quad xy = a^p \times a^q = a^{p+q}$.

Substituting $a^{p+q}$ for $xy$ in $\log_a (xy)$ gives: $\log_a a^{p+q}$

This simplifies to: $p + q$

Therefore $\log_a (xy) = p + q$

But $p = \log_a x$ and $q = \log_a y$.

Therefore $\log_a x + \log_a y = \log_a (xy)$

A similar proof can be used for the second rule. The third rule can be proved as follows:

To prove that $\log_a x^y = y \log_a x$

Let $m = \log_a x$

Write using powers $\quad\quad\quad\quad x = a^m$

Raise both sides to power of $y \quad\quad x^y = (a^m)^y$ or $x^y = a^{ym}$

Convert back to logarithms $\quad\quad \log_a x^y = ym$

Substitute back for $m = \log_a x \quad \log_a x^y = y \log_a x$

##  Worked example

Simplify $\dfrac{\log_{10} 32}{\log_{10} 128}$.

Both 32 and 128 are powers of 2. Written in terms of powers of 2, the fraction can be written as $\dfrac{\log_{10} 2^5}{\log_{10} 2^7} = \dfrac{5 \log_{10} 2}{7 \log_{10} 2} = \dfrac{5}{7}$

 **Exercise 3.21**

1  Simplify each of the following:

a  $\log_{10} 100$

b  $2 \log_4 64$

c  $\frac{1}{2} \log_{10} 4$

d  $\frac{1}{4} \log_{10} 625$

e  $-\frac{1}{3} \log_{10} 27$

f  $\dfrac{\log_{10} 64}{\log_{10} 16}$

g  $\dfrac{3 \log_{10} 3}{\log_{10} 27}$

h  $\dfrac{\log_{10} \sqrt{a}}{2 \log_{10} a}$

## 30 LOGARITHMIC FUNCTIONS

# Solving exponential equations using logarithms

Logarithms can be used to solve more complex exponential equations than the ones dealt with so far.

For example, to solve the exponential equation $3^x = 12$ is problematic because the solution is not an integer value. The solution must lie between $x = 2$ and $x = 3$ because $3^2 = 9$ and $3^3 = 27$. By trial and error, the solution could be found to one decimal place; however, this is a laborious process and not practical, especially if the solution is required to two or more decimal places.

However, the equation can be solved using logs.

$$3^x = 12$$

Take logs of both sides of the equation: $\quad \log_{10} 3^x = \log_{10} 12$

Using the rule $\log_a x^y = y \log_a x$: $\quad x\log_{10} 3 = \log_{10} 12$

Divide both sides by $\log_{10} 3$: $\quad x = \dfrac{\log_{10} 12}{\log_{10} 3}$

The solution can either be left in the exact form $x = \dfrac{\log_{10} 12}{\log_{10} 3}$ or given to the required number of decimal places or significant figures, e.g. $x = 2.26$ (3 s.f.).

### ➡ Worked examples

**a** Solve the equation $5^x = 100$, giving your answer to 3 s.f.

$$5^x = 100$$
$$\Rightarrow \log_{10} 5^x = \log_{10} 100$$
$$\Rightarrow x\log_{10} 5 = 2$$
$$\Rightarrow x = \dfrac{2}{\log_{10} 5} = 2.86$$

**b** Solve the equation $\log_3 x = 4$.

$$\log_3 x = 4$$
$$\Rightarrow 3^4 = x$$
$$\Rightarrow x = 81$$

**c** Find the smallest integer value of $x$ such that $5^x > 100\,000$.

First solve the equation $5^x = 100\,000$ to find the critical value of $x$:

$$5^x = 100\,000$$
$$\Rightarrow \log_{10} 5^x = \log_{10} 100\,000$$
$$\Rightarrow x\log_{10} 5 = 5$$
$$\Rightarrow x = \dfrac{5}{\log_{10} 5} = 7.15 \text{ (3 s.f.)}$$

Therefore the smallest integer value of $x$ which satisfies the inequality $5^x > 100\,000$ is 8.

# Applications of exponential equations

**Exercise 3.22**

1  Solve these equations. Give decimal answers correct to 3 s.f.
   a  $4^x = 50$
   b  $9^a = 34$
   c  $3^y = \frac{1}{5}$
   d  $\left(\frac{1}{10}\right)^p = 4$
   e  $5^{x+1} = 100$
   f  $8^{y-6} = 500$
   g  $7^{2x+1} = 6000$
   h  $5^{-m} = 20$
   i  $\log_5 x = 6$
   j  $\log_2 b = 5$
   k  $3\log_{10} x = \log_{10} 2$
   l  $\frac{1}{2}\log_{10} x = \log_{10} 6 - 1$

2  Solve each of the following inequalities:
   a  $6^x > 1000$
   b  $5^y > 10$
   c  $2^{x+1} > 50$
   d  $0.3^{2x+1} < 8$

3  Find the smallest integer $x$ such that $3^{\frac{1}{2}x-2} > 120$.

4  Find the largest integer $m$ such that $0.4^m > \frac{1}{1000}$.

# Applications of exponential equations

In Chapters 20 and 23 you have studied exponential functions, their properties and their graphs. One real-life application of an exponential function occurs with compound interest calculations.

This section covers a few of the occurrences of exponential equations in real life.

## ➔ Worked example

$1200 is deposited in a savings account with a compound interest rate of 4% per year.

a  Write an equation to show the total amount in the account ($S$) after $t$ years, assuming no money is withdrawn.

A 4% increase is equivalent to a multiplier of 1.04

Therefore $S = 1200 \times 1.04^t$

b  Construct a table of values showing the total amount in the account after each of the first five years.

| $t$ | 0 | 1 | 2 | 3 | 4 | 5 |
|---|---|---|---|---|---|---|
| $S$ | 1200.00 | 1248.00 | 1297.92 | 1349.84 | 1403.83 | 1459.98 |

× 1.04   × 1.04

*Money calculations should be rounded to 2 d.p. However, when multiplying by 1.04, the full value on the calculator should be used to avoid rounding errors.*

## 30 LOGARITHMIC FUNCTIONS

**c** Use your equation in part a to calculate the time taken for the amount in the account to double.

$S = 2400$, therefore the equation to be solved is

$$2400 = 1200 \times 1.04^t$$
$$1.04^t = 2$$
$$\log_{10} 1.04^t = \log_{10} 2$$
$$t \log_{10} 1.04 = \log_{10} 2$$
$$t = \frac{\log_{10} 2}{\log_{10} 1.04}$$
$$t = 17.7$$

Therefore, the number of years is 18

*As the interest is calculated at the end of each year, the answer is 18 years not 17.7.*

### Exercise 3.23

1 The population ($P$) of flamingos arriving at a lake over a 10 day period can be approximately modelled using the formula $P = 3 \times 1.8^t$, where $t$ is the number of days after the first flamingos arrived.
   **a** How many flamingos were at the lake at the start of the count?
   **b** How many flamingos were at the lake after 5 days?
   **c** Plot a graph to show the population of flamingos at the lake over the first five days.
   **d** At one point 40 flamingos were counted at the lake.
      **i** Using your graph estimate how long after the start this occurred.
      **ii** Find the value of $t$ by solving the equation $40 = 3 \times 1.8^t$.
   **e** Give a reason why the formula may not be an accurate model for the population of flamingos after ten days.

2 Two friends Pedro and Marta deposit some money in two separate long-term savings accounts.
   Pedro deposits $2500 in an account which pays interest compounded at 5% per year.
   Marta deposits $4000 in an account which pays interest compounded at 2% per year.
   Neither of them withdraws any money from the account.
   **a** Write a formula showing the total amount ($T$) in Pedro's savings account $n$ years after the initial deposit.
   **b** After how many years does the amount in Pedro's account exceed the amount in Marta's account?

## Applications of exponential equations

**3** A pan of hot water is removed from the heat source and allowed to cool naturally in the kitchen. The temperature ($T°C$), $t$ minutes after being taken off the heat, is given by the formula
$T = 76 \times 1.4^{-t} + 21$

  **a** Copy and complete the table below for the temperature of the water in the first 6 minutes.

| $t$ (min) | 0 | 1 | 2 | 3 | 4 | 5 | 6 |
|---|---|---|---|---|---|---|---|
| ($T°C$) | | | | | | | |

  **b** Plot a graph of your table of results and draw a smooth curve through the points.

  **c** After $n$ minutes the temperature of the water is 55°C.
   **i** Use your graph to estimate the value of $n$.
   **ii** Use the formula to calculate the value of $n$.

  **d** Find the room temperature of the kitchen. Justify your answer.

# Investigations, modelling and ICT 3

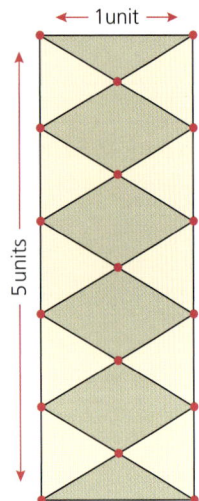

## Investigation: Paving blocks

*Isometric dot paper is needed for this investigation.*

A company offers to block pave driveways for homeowners. The design is always rectangular. An example is shown on the left:

The design shown is 1 unit by 5 units.

It consists of three different shapes of block paving:

    10 pieces in the form of an equilateral triangle

    2 pieces in the form of an isosceles triangle

    4 pieces in the form of a rhombus

Different-sized rectangular designs will have different numbers of each of the different-shaped blocks.

1. Draw a design of width 2 units and height 4 units. Count the number of each type of block.
2. Draw at least six different-sized rectangular designs, each time counting the number of different-shaped blocks.
3. Enter your results in a table similar to the one shown:

| Dimensions | | Number of blocks | | |
|---|---|---|---|---|
| Width | Height | Equilateral | Isosceles | Rhombus |
| 1 | 5 | 10 | 2 | 4 |
|   |   |   |   |   |

4. Investigate, by drawing more designs if necessary, the relationship between the width and height of a design and the number of different blocks needed.
5. a Describe in words the relationship between the width ($w$) and the number of isosceles triangles ($i$).
   b Write your rule from Q.5a using algebra.
6. a Describe in words the relationship between the height ($h$) and the number of equilateral triangles ($e$).
   b Write your rule from Q.6a using algebra.
7. a Describe in words the relationship between the width, the height and the number of rhombuses ($r$).
   b Write your rule from Q.7a using algebra.

# Investigation: Regions and intersections

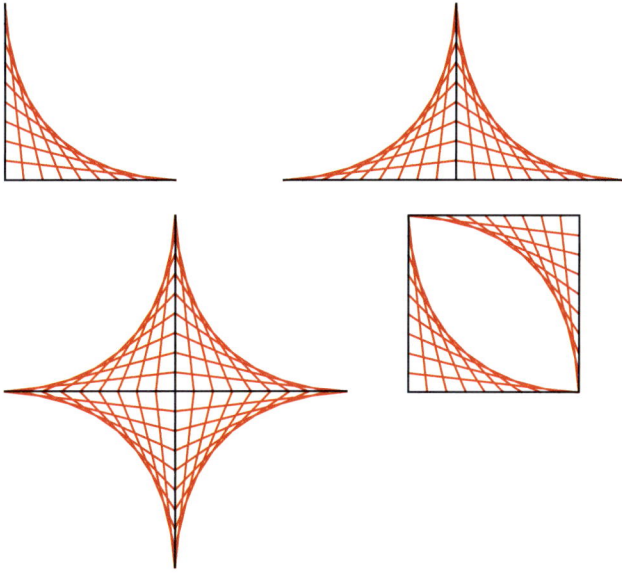

The patterns above are examples of 'curve stitching'. Although the patterns produce a curved effect, all the lines used are in fact straight.

Below is how to construct a simple one:

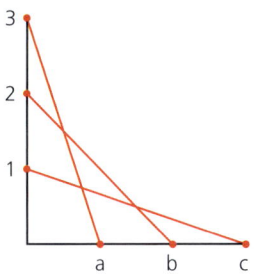

Lines are drawn from points

$a \to 3$

$b \to 2$

$c \to 1$

This 3 × 3 curve stitch has produced three points of intersection and six enclosed regions.

Different-sized curve stitches will produce a different number of points of intersection and a different number of enclosed regions.

## INVESTIGATIONS, MODELLING AND ICT 3

1. Investigate the number regions and points of intersection for different-sized curved stitching patterns. Record your results in a table similar to the one shown below:

| Dimension of curve stitch | Number of points of intersection | Number of enclosed regions |
|---|---|---|
| 1 × 1 | | |
| 2 × 2 | | |
| 3 × 3 | 3 | 6 |
| 4 × 4 | | |
| etc | | |

2. For an $m \times m$ pattern, write an algebraic rule for the number of points of intersection $p$.
3. For an $m \times m$ pattern, write an algebraic rule for the number of enclosed regions $r$.

## Modelling: Parking error

A driver parks his car on a road at the top of a hill with a constant gradient. Unfortunately, he does not put the handbrake on properly and as a result the car starts to roll down the hill.

The incident was captured on CCTV, so the distance travelled (m) and time (s) were both recorded. These results are presented in the table below:

| Time (s) | 0 | 2 | 4 | 6 | 8 | 10 | 12 |
|---|---|---|---|---|---|---|---|
| Distance rolled (m) | 0 | 0.2 | 0.8 | 1.8 | 3.2 | 5 | 7.2 |

1. Plot a graph of the data, with time on the $x$-axis and distance on the $y$-axis and draw a curve through the points.
2. Describe the relationship between the time and distance rolled.
3. Find the equation of the curve.
4. Use your model to predict how far the car will have rolled after 15 seconds.
5. If the road is 120 m long, use your equation to estimate how long it will take the car to reach the bottom of the hill.
6. What factors may affect the validity of your answer to Q.5?

# ICT Activity

A type of curve you have not encountered on this course is a rectangular hyperbola.

The equation of a rectangular hyperbola is given as $x^2 - y^2 = a^2$ where $a$ is a constant.

1. Investigate, using the internet if necessary, what a hyperbola is and how it relates to cones.
2. a  Rearrange the equation of a rectangular hyperbola to make $y$ the subject.
   b  By letting $a = 1$, plot a graph of a rectangular hyperbola using a graphic display calculator.
   c  Sketch the graph.
   d  By changing the value of $a$, determine its effect on the shape of the graph.
3. Using your graphs of a rectangular hyperbola as reference, determine the equation of any asymptotes.

# Student assessments 3

## Student assessment 1

1. Calculate the range of these functions:
   a. $f(x) = 3x - x^3$; $-3 \leqslant x \leqslant 3$
   b. $f(x) = 2x^2 - 4$; $x \in$ {all real numbers}
2. Linear functions take the form $f(x) = ax + b$. Explain the effect $a$ and $b$ have on the shape and/or position of the graph of the function.
3. The diagram below shows four linear functions. These are
   $y = x + 3$, $y = x - 2$, $y = -\frac{1}{2}x - 4$ and $y = 2x + 1$.
   State, giving reasons, which line corresponds to which function.

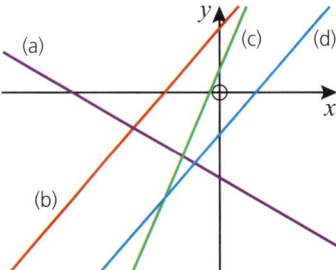

4. The diagram below shows four quadratic functions. The function $f(x) = x^2$ is highlighted.

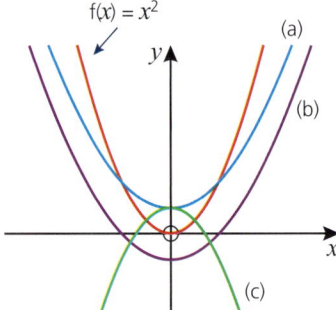

Give possible equations for each of the quadratic curves a, b and c.

5  The graph of the function $f(x) = \left(\frac{1}{2}\right)^x$ is shown below.

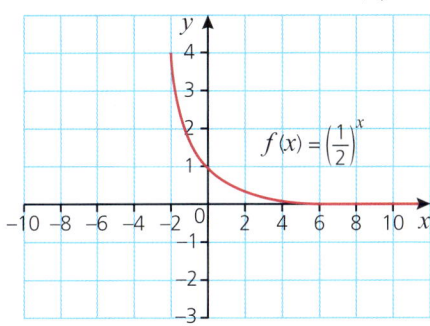

a  Write down the equation of any asymptote.
b  Copy the graph. Sketch and clearly label the following exponential functions on the same axes.

  i  $f(x) = \left(\frac{1}{2}\right)^x - 3$      ii  $f(x) = \left(\frac{1}{2}\right)^{x+4}$

c  Write the equation of any asymptotes in each of the functions in part b.

## Student assessment 2

1  The graph below shows the function $y = f(x)$.

Three points are labelled: $A(-2, 1)$, $B(0, -\frac{1}{2})$, $C(4, -3)$.

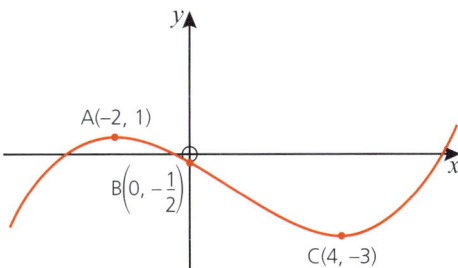

Give the coordinates of the points' images when mapped to the following:
a  $y = f(x) - 3$      b  $y = f(x - 3)$

2  Sketch the graph of $y = x^2$. On the same axes, sketch, using transformations of graphs, both of the following. Label each graph clearly.
a  $y = (x - 4)^2$      b  $y = x^2 - 4$

3 Sketch the graph of $y = \frac{1}{x}$. On the same axes, sketch, using transformations of graphs, both of the following. Label each graph clearly.

a $y = \frac{1}{x} + 2$ 

b $y = \frac{1}{x-4}$

  i Give the equation of any asymptotes in Q.3a.
  ii Give the equation of any asymptotes in Q.3b.

4 A function is given as $f(x) = x^3 + 6x^2 - 15x + 10$. Use your graphic display calculator to find:
  a the coordinates of the points where the function has a local maximum and minimum
  b the zero(s) of the function
  c the value of the y-intercept.

## Student assessment 3

1 The graph of a quadratic function of the form $f(x) = x^2 + bx + c$ intercepts the x-axis at $x = -3$ and $x = 2$. Determine the equation of the quadratic and state the values of b and c.

2 Find the equation of the quadratic function shown below with zeros (roots) at $x = -4$ and $x = 2$ which passes through the point $(-3, -15)$. Give your answer in the form $f(x) = ax^2 + bx + c$.

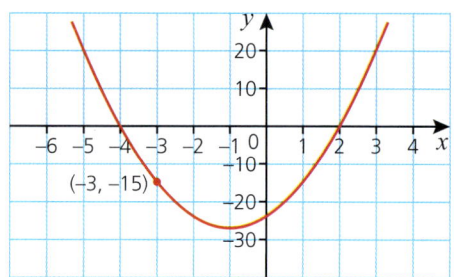

3 Find the coordinates of the vertices of these quadratic functions written in the form $y = a(x - h)^2 + k$.
  a $y = (x + 1)^2 - 3$
  b $y = 3(x - 5)^2 + 1$

4 The cubic function below is of the form $f(x) = ax^3 + bx^2 + cx + d$. Its zeros (roots) are at $x = 6$ and $x = 2$.

If the value of $a = -2$ and the value of $d = 48$, determine the values of b and c.

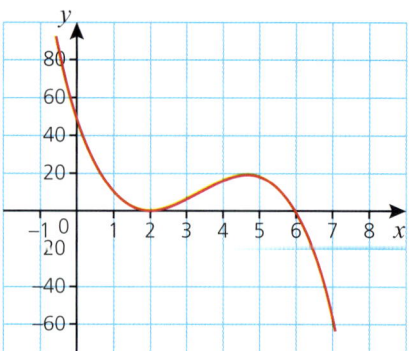

5. The graph below shows the function g(x) which is a translation of the function f(x) = $2^x$. It passes through the point (5, 1) as shown.

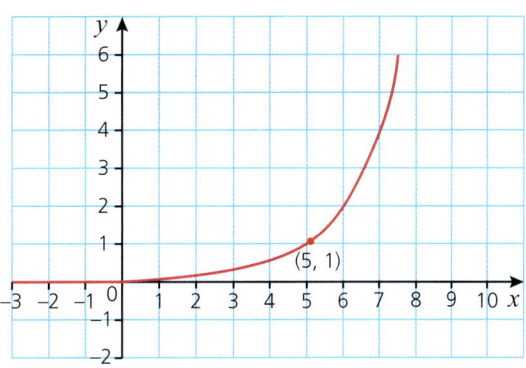

a Find the equation of the function g(x).
b Calculate where it intersects the y-axis.

## Student assessment 4

1. f(x) = 4x and g(x) = 3x + 2. Write an expression for:
   a f(g(x))  b g(f(x))
2. f(x) = 2x – 5 and g(x) = x + 1
   a Find the value of f(g(0)).  b Find the value of g(f(–2)).
   c Explain clearly why f(g(x)) ≠ g(f(x)).
3. f(x) = $x^2$ – 4 and g(x) = 2x + 1
   a Write an expression for f(g(x)).
   b Solve the equation f(g(x)) = 0.
4. Find the inverse of these functions:
   a $f(x) = \frac{5x+2}{4}$  b $h(x) = \frac{1}{2}x + 6$
5. Solve the following equations. Give your answers correct to 1 d.p.
   a $2^{x-4} = 20$  b $\frac{1}{2}\log_{10} x = \log_{10} 300 - 2$

# TOPIC 4

## Coordinate geometry

**Contents**

Chapter 31  Coordinates (C4.1, E4.1)
Chapter 32  Line segments (C4.1, C4.2, C4.3, C4.5, E4.1, E4.2, E4.3, E4.5, E4.6)
Chapter 33  Equation of a straight line (C4.4, E4.4)

**Learning objectives**

**C4.1     E4.1**
Use and interpret Cartesian coordinates in two dimensions

**C4.2     E4.2**
1   Find the gradient of a straight line
2   *Calculate the gradient of a straight line from the coordinates of two points on the line*

**C4.3**
1   Calculate the length of a line segment from the coordinates of its end points
2   Find the coordinates of the mid-point of a line segment from the coordinates of its end points

**E4.3**
1   *Calculate the length of a line segment*
2   *Find the coordinates of the mid-point of a line segment*

**C4.4**
Interpret and obtain the equation of a straight-line graph in the form $y = mx + c$

**E4.4**
*Interpret and obtain the equation of a straight-line graph*

**C4.5     E4.5**
Find the gradient and equation of a straight line parallel to a given line

**E4.6**
*Find the gradient and equation of a straight line perpendicular to a given line*

Elements in purple refer to the Extended curriculum only.

# The French

In the middle of the seventeenth century there were three great French mathematicians, René Descartes, Blaise Pascal and Pierre de Fermat.

René Descartes was a philosopher and a mathematician. His book *The Meditations* asks 'How and what do I know?' His work in mathematics made a link between algebra and geometry. He thought that all nature could be explained in terms of mathematics. Although he was not considered as talented a mathematician as Pascal and Fermat, he has had greater influence on modern thought. The $(x, y)$ coordinates we use are called Cartesian coordinates after Descartes.

*René Descartes (1596–1650)*

Blaise Pascal (1623–1662) was a genius who studied geometry as a child. When he was sixteen, he stated and proved Pascal's Theorem, which relates any six points on any conic section. The theorem is sometimes called the 'Cat's Cradle'. He founded probability theory and made contributions to the invention of calculus. He is best known for Pascal's Triangle.

Pierre de Fermat (1601–1665) was a brilliant mathematician and, along with Descartes, one of the most influential. Fermat invented number theory and worked on calculus. He discovered probability theory with his friend Pascal. It can be argued that Fermat was at least Newton's equal as a mathematician.

Fermat's most famous discovery in number theory is 'Fermat's Last Theorem'. This theorem is derived from Pythagoras' theorem, which states that for a right-angled triangle, $x^2 = y^2 + z^2$ where $x$ is the length of the hypotenuse. Fermat said that if the index (power) was greater than two and $x, y, z$ are all natural numbers, then the equation was never true. (This theorem was only proved in 1995 by the English mathematician Andrew Wiles.)

# 31 Coordinates

On 22 October 1707, four British ships, *The Association* and three others, struck the Gilstone Ledges off the Scilly Isles and more than two thousand men drowned. Why? Because the Admiral had no way of knowing exactly where he was. He needed two coordinates to place his position on the sea. He only had one, his latitude.

The story of how to solve the problem of fixing the second coordinate (longitude) is told in Dava Sobel's book *Longitude*. The British Government offered a prize of £20 000 (millions of pounds at today's prices) to anyone who could solve the problem of how to fix longitude at sea.

## Coordinates

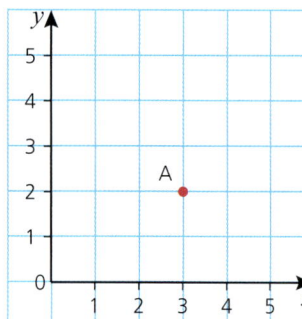

To fix a point in two dimensions (2D), its position is given in relation to a point called the **origin**. Through the origin, axes are drawn perpendicular to each other. The horizontal axis is known as the **x-axis**, and the vertical axis is known as the **y-axis**.

The *x*-axis is numbered from left to right. The *y*-axis is numbered from bottom to top.

The position of point A is given by two coordinates: the *x*-coordinate first, followed by the *y*-coordinate. So the coordinates of point A are (3, 2).

A number line can extend in both directions by extending the *x*- and *y*-axes below zero, as shown in the grid below:

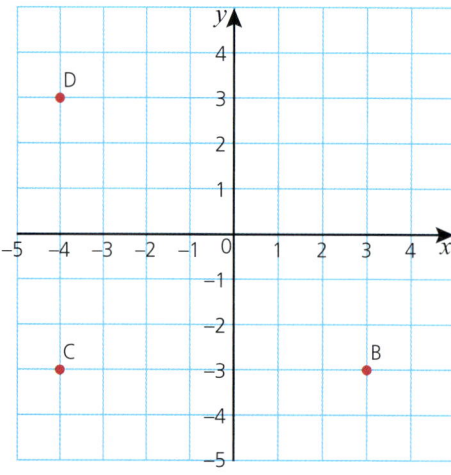

*Coordinates*

Points B, C and D can be described by their coordinates:
Point B is at (3, −3)
Point C is at (−4, −3)
Point D is at (−4, 3)

### Exercise 4.1

1  Draw a pair of axes with both *x* and *y* from −8 to +8.
   Mark each of the following coordinates on your grid:
   a  A = (5, 2)       b  B = (7, 3)        c  C = (2, 4)
   d  D = (−8, 5)      e  E = (−6, −8)      f  F = (3, −7)
   g  G = (7, −3)      h  H = (6, −6)

Draw a separate grid for each of Q.2–4 with *x*- and *y*-axes from −6 to +6.
Plot and join the points in order to name each shape drawn.

2  A = (3, 2)    B = (3, −4)   C = (−2, −4)   D = (−2, 2)
3  E = (1, 3)    F = (4, −5)   G = (−2, −5)
4  H = (−6, 4)   I = (0, −4)   J = (4, −2)    K = (−2, 6)

### Exercise 4.2

Draw a pair of axes with both *x* and *y* from −10 to +10.
1  Plot the points P = (−6, 4), Q = (6, 4) and R = (8, −2).
   a  Plot point S such that PQRS, when drawn, is a parallelogram.
   b  Draw diagonals PR and QS. What are the coordinates of their point of intersection?
   c  What is the area of PQRS?
2  On the same axes, plot point M at (−8, 4) and point N at (4, 4).
   a  Join points MNRS. What shape is formed?
   b  What is the area of MNRS?
   c  Explain your answer to Q.2b.
3  a  On the same axes, plot point J where point J has *y*-coordinate +10 and JRS, when joined, forms an isosceles triangle.
   b  What is the *x*-coordinate of all points on the axis of symmetry of triangle JRS?

### Exercise 4.3

1  a  On a grid with axes numbered from −10 to +10, draw a regular hexagon ABCDEF with centre (0, 0) and coordinate A (0, 8).
   b  Write down the approximate coordinates of points B, C, D, E and F.
2  a  On a similar grid to Q.1, draw an octagon PQRSTUVW which has point P (2, −8), point Q (−6, −8), point R (−7, −5) and a line of symmetry with equation *x* = −2.
      PQ = RS = TU = VW and QR = ST = UV = WP.
   b  List the coordinates of points S, T, U, V and W.
   c  What is the coordinate of the centre of rotational symmetry of the octagon?

277

# 31 COORDINATES

**Exercise 4.4**

1. The points A, B, C and D are not at natural number points on the number line. Point A is at 0.7.

   What are the positions of points B, C and D?

   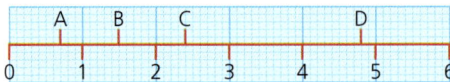

2. On this number line, point E is at 0.4.
   What are the positions of points F, G and H?

   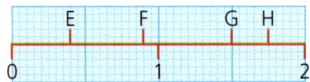

3. What are the positions of points I, J, K, L and M?

   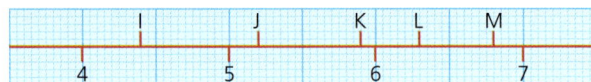

4. Point P is at position 0.4 and point W is at position 9.8.
   What are the positions of points Q, R, S, T, U and V?

**Exercise 4.5**

1. Give the coordinates of points A, B, C and D.

   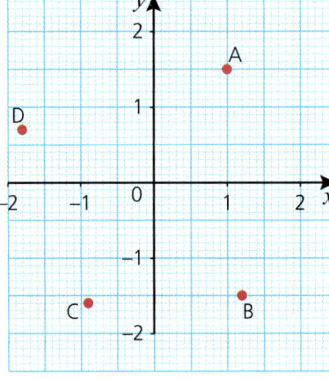

## Coordinates

**2** Give the coordinates of points E, F, G, H.

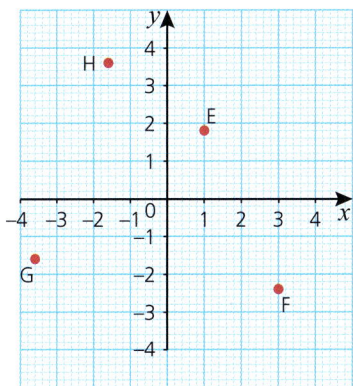

**3** Give the coordinates of points J, K, L and M.

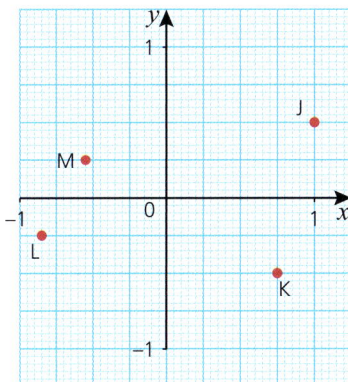

**4** Give the coordinates of points P, Q, R and S.

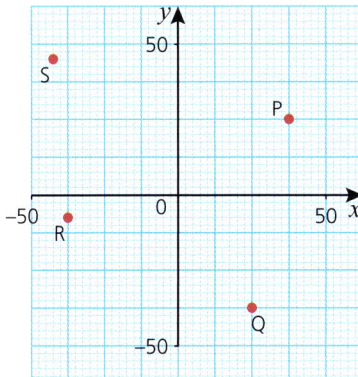

# 32 Line segments

## Calculating the length of a line segment

A line segment is formed when two points are joined by a straight line. To calculate the distance between two points, and therefore the length of the line segment, their coordinates need to be given. Once these are known, Pythagoras' theorem can be used to calculate the distance.

### ➔ Worked example

The coordinates of two points are (1, 3) and (5, 6). Draw a pair of axes, plot the given points and calculate the distance between them.

By dropping a vertical line from the point (5, 6) and drawing a horizontal line from (1, 3), a right-angled triangle is formed. The length of the hypotenuse of the triangle is the length you wish to find.

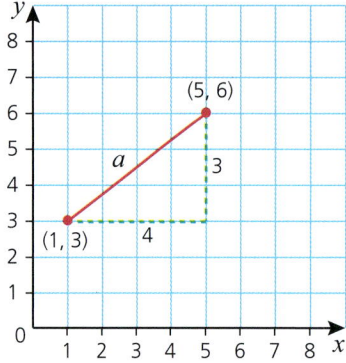

Using Pythagoras' theorem:

$a^2 = 3^2 + 4^2$

$a^2 = 25$

$a = \sqrt{25}$

$a = 5$

The length of the line segment is 5 units.

*You will study Pythagoras' theorem in Chapter 46.*

To find the distance between two points directly from their coordinates, use the following formula:

$$d = \sqrt{(x_1 - x_2)^2 + (y_1 - y_2)^2}$$

*The midpoint of a line segment*

## Worked example

Without plotting the points, calculate the distance between the points (1, 3) and (5, 6).

$$d = \sqrt{(1-5)^2 + (3-6)^2}$$
$$= \sqrt{(-4)^2 + (-3)^2}$$
$$= \sqrt{25}$$
$$= 5$$

The distance between the two points is 5 units.

# The midpoint of a line segment

To find the midpoint of a line segment, use the coordinates of its end points. To find the $x$-coordinate of the midpoint, find the mean of the $x$-coordinates of the end points. Similarly, to find the $y$-coordinate of the midpoint, find the mean of the $y$-coordinates of the end points.

## Worked examples

**a** Find the coordinates of the midpoint of the line segment AB where A is (1, 3) and B is (5, 6).

The $x$-coordinate of the midpoint will be $\frac{1+5}{2} = 3$

The $y$-coordinate of the midpoint will be $\frac{3+6}{2} = 4.5$

So the coordinates of the midpoint are (3, 4.5).

**b** Find the coordinates of the midpoint of a line segment PQ where P is (−2, −5) and Q is (4, 7).

The $x$-coordinate of the midpoint will be $\frac{-2+4}{2} = 1$

The $y$-coordinate of the midpoint will be $\frac{-5+7}{2} = 1$

So the coordinates of the midpoint are (1, 1).

**Exercise 4.6**

1 **i** Plot each of the following pairs of points.
  **ii** Calculate the distance between each pair of points.
  **iii** Find the coordinates of the midpoint of the line segment joining the two points.

  **a** (5, 6) (1, 2)      **b** (6, 4) (3, 1)      **c** (1, 4) (5, 8)
  **d** (0, 0) (4, 8)      **e** (2, 1) (4, 7)      **f** (0, 7) (−3, 1)
  **g** (−3, −3) (−1, 5)   **h** (4, 2) (−4, −2)    **i** (−3, 5) (4, 5)
  **j** (2, 0) (2, 6)      **k** (−4, 3) (4, 5)     **l** (3, 6) (−3, −3)

281

# 32 LINE SEGMENTS

**Exercise 4.6 (cont)**

2  Without plotting the points:
   i  calculate the distance between each of the following pairs of points
   ii find the coordinates of the midpoint of the line segment joining the two points.

   a  (1, 4) (4, 1)
   b  (3, 6) (7, 2)
   c  (2, 6) (6, −2)
   d  (1, 2) (9, −2)
   e  (0, 3) (−3, 6)
   f  (−3, −5) (−5, −1)
   g  (−2, 6) (2, 0)
   h  (2, −3) (8, 1)
   i  (6, 1) (−6, 4)
   j  (−2, 2) (4, −4)
   k  (−5, −3) (6, −3)
   l  (3, 6) (5, −2)

## Gradient of a straight line

The **gradient** of a straight line refers to its 'steepness' or 'slope'. The gradient of a straight line is constant, i.e. it does not change. The gradient can be calculated by considering the coordinates of any two points $(x_1, y_1)$, $(x_2, y_2)$ on the line. It is calculated using the following formula:

$$\text{Gradient} = \frac{\text{vertical distance between the two points}}{\text{horizontal distance between the two points}}$$

By considering the *x*- and *y*-coordinates of the two points, this can be rewritten as:

$$\text{Gradient} = \frac{y_2 - y_1}{x_2 - x_1}$$

### ➡ Worked examples

a  The coordinates of two points on a straight line are (1, 3) and (5, 7). Plot the two points on a pair of axes and calculate the gradient of the line joining them.

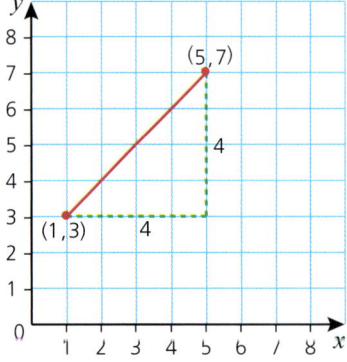

Gradient = $\frac{7-3}{5-1} = \frac{4}{4} = 1$

Note: It does not matter which point you choose to be $(x_1, y_1)$ or $(x_2, y_2)$ as the gradient will be the same. In the example above, reversing the points:

Gradient = $\frac{3-7}{1-5} = \frac{-4}{-4} = 1$

**b** The coordinates of two points on a straight line are (2, 6) and (4, 2). Plot the two points on a pair of axes and calculate the gradient of the line joining them.

If the line was horizontal, there would be no difference in the y-coordinates. This would lead to a numerator of zero. Therefore, the gradient of a horizontal line is zero.

If the line was vertical, there would be no difference in the x-coordinates. This would lead to a division by zero which is undefined. Therefore, the gradient of a vertical line is infinite.

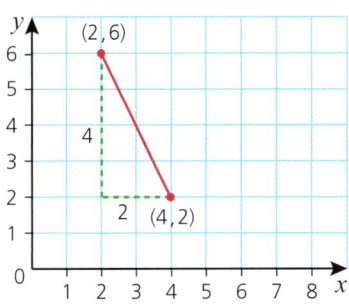

Gradient $= \frac{2-6}{4-2} = \frac{-4}{2} = -2$

---

To check whether or not the sign of the gradient is correct, the following guideline is useful:

A line sloping this way will have a positive gradient.   A line sloping this way will have a negative gradient.

Parallel lines will have the same gradient. Conversely, lines which have the same gradient are parallel. If two lines are parallel to each other, their gradients $m_1$ and $m_2$ are equal.

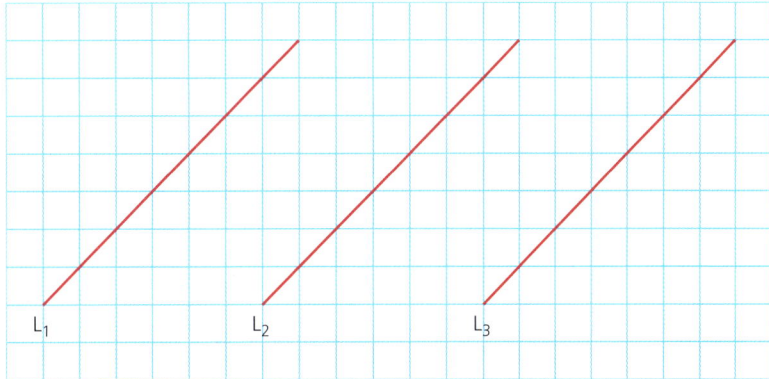

$L_1$, $L_2$, $L_3$ all have the same gradient and so they are parallel.

# 32 LINE SEGMENTS

The $x$-axis and the $y$-axis on a graph intersect at right angles. They are perpendicular to each other. In the graph below, $L_1$ and $L_2$ are perpendicular to each other.

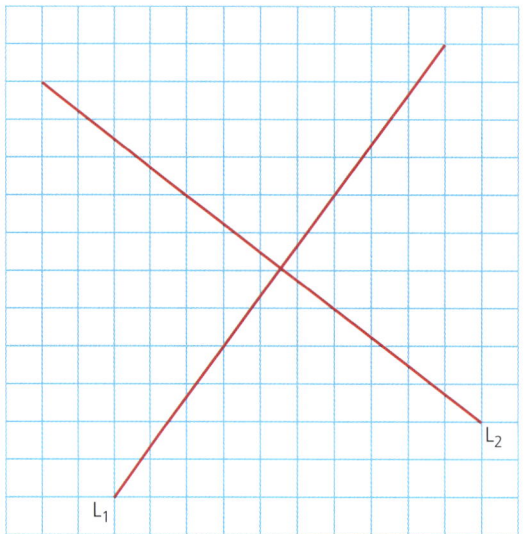

The gradient $m_1$ of line $L_1$ is $\frac{4}{3}$ while the gradient $m_2$ of line $L_2$ is $-\frac{3}{4}$. The product of $m_1 m_2$ gives the result $-1$, i.e. $\frac{4}{3} \times \left(-\frac{3}{4}\right) = -1$.

If two lines are perpendicular to each other, the product of their gradients is $-1$, so $m_1 m_2 = -1$.

Therefore the gradient of one line is the negative reciprocal of the other line, so $m_1 = -\frac{1}{m_2}$.

### Exercise 4.7

Questions 1.ii and 2.ii are relevant to the Extended syllabus only.

1. With the aid of axes if necessary, calculate:
   i the gradient of the line joining the following pairs of points
   ii the gradient of a line perpendicular to this line.
   
   a (5, 6) (1, 2)    b (6, 4) (3, 1)    c (1, 4) (5, 8)
   d (0, 0) (4, 8)    e (2, 1) (4, 7)    f (0, 7) (−3, 1)
   g (−3, −3) (−1, 5) h (4, 2) (−4, −2) i (−3, 5) (4, 5)
   j (2, 0) (2, 6)    k (−4, 3) (4, 5)   l (3, 6) (−3, −3)

2. With the aid of axes if necessary, calculate:
   i the gradient of the line joining the following pairs of points
   ii the gradient of a line perpendicular to this line.
   
   a (1, 4) (4, 1)    b (3, 6) (7, 2)    c (2, 6) (6, −2)
   d (1, 2) (9, −2)   e (0, 3) (−3, 6)   f (−3, −5) (−5, −1)
   g (−2, 6) (2, 0)   h (2, −3) (8, 1)   i (6, 1) (−6, 4)
   j (−2, 2) (4, −4)  k (−5, −3) (6, −3) l (3, 6) (5, −2)

# 33 Equation of a straight line

The coordinates of every point on a straight line all have a common relationship. This relationship, when written algebraically as an equation in terms of $x$ and/or $y$, is known as the equation of the straight line.

## → Worked examples

a  By looking at the coordinates of some of the points on the line below, establish the equation of the straight line.

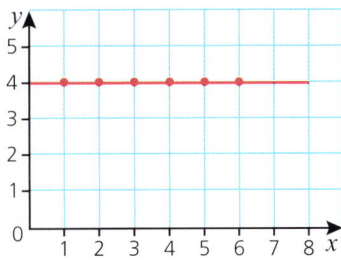

| $x$ | $y$ |
|---|---|
| 1 | 4 |
| 2 | 4 |
| 3 | 4 |
| 4 | 4 |
| 5 | 4 |
| 6 | 4 |

Some of the points on the line have been identified and their coordinates entered in the table above. By looking at the table, it can be seen that the only rule all the points have in common is that $y = 4$.
Hence the equation of the straight line is $y = 4$.

b  By looking at the coordinates of some of the points on the line, establish the equation of the straight line.

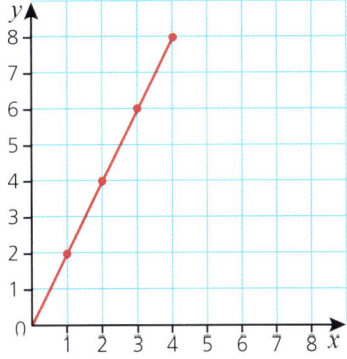

285

# 33 EQUATION OF A STRAIGHT LINE

| x | y |
|---|---|
| 1 | 2 |
| 2 | 4 |
| 3 | 6 |
| 4 | 8 |

Once again, by looking at the table it can be seen that the relationship between the *x*- and *y*-coordinates is that each *y*-coordinate is twice the corresponding *x*-coordinate.

Hence the equation of the straight line is $y = 2x$.

 **Exercise 4.8** For each of the following, identify the coordinates of some of the points on the line and use these to find the equation of the straight line.

1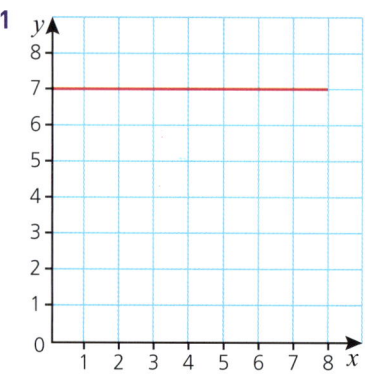

2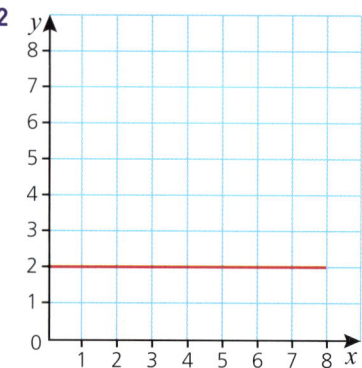

3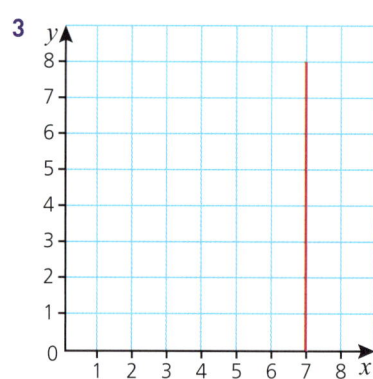

4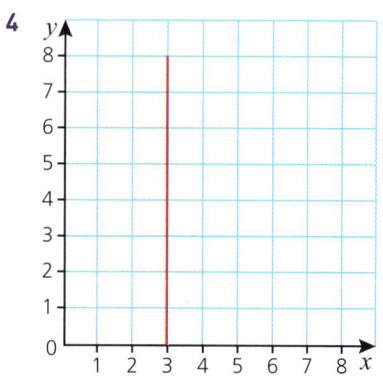

*Equation of a straight line*

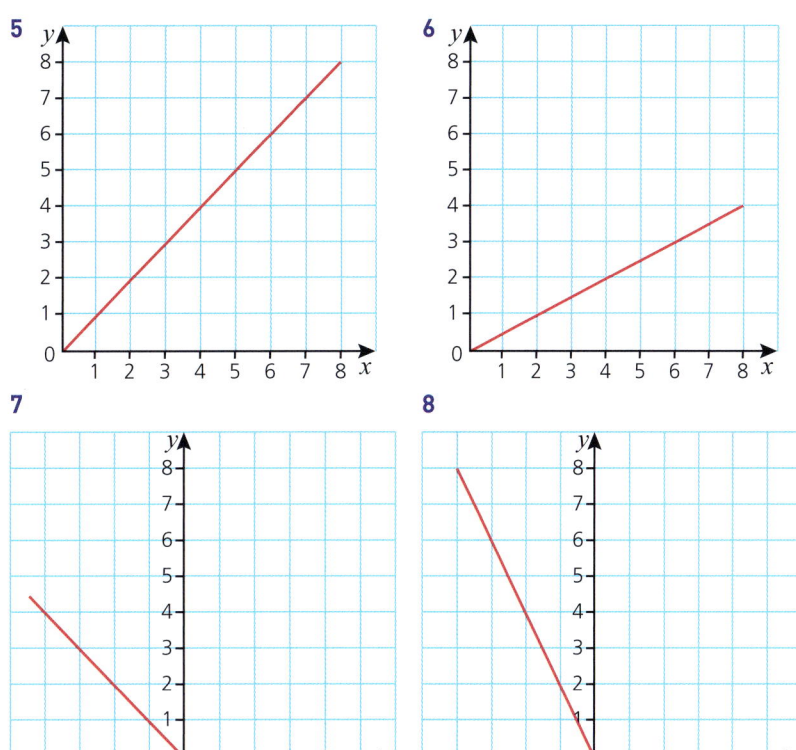

 **Exercise 4.9**

1  For each of the following, identify the coordinates of some of the points on the line and use these to find the equation of the straight line.

a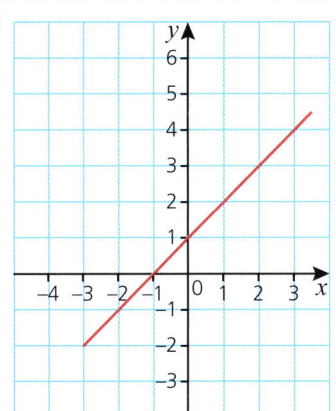

b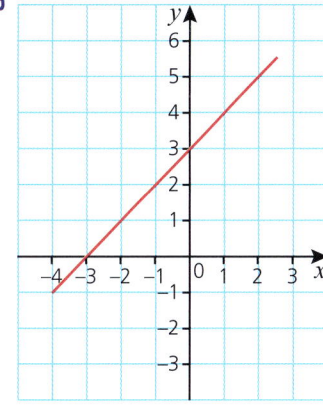

## 33 EQUATION OF A STRAIGHT LINE

**Exercise 4.9 (cont)**

c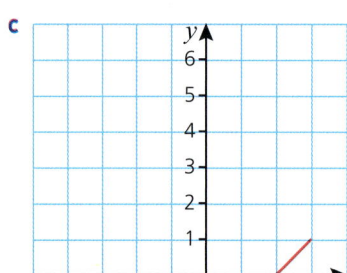

d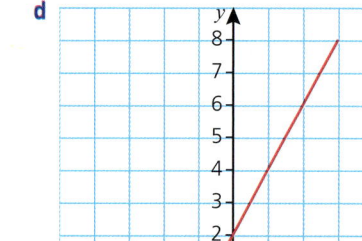

e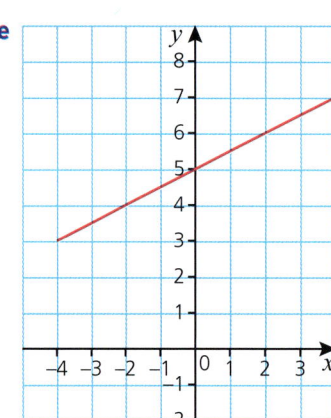

f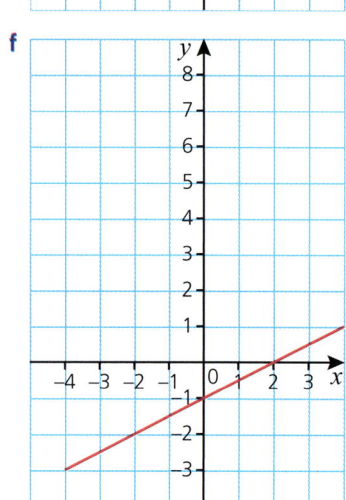

2  For each of the following, identify the coordinates of some of the points on the line and use these to find the equation of the straight line.

*Remember, the equation can be worked out by looking at the relationship between the x- and y- coordinates of points on the line.*

a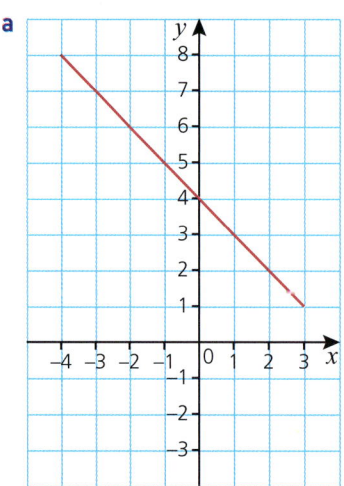

b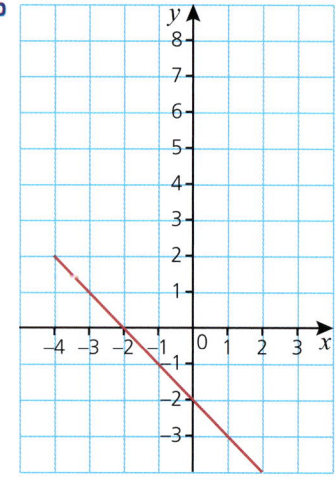

# Equation of a straight line

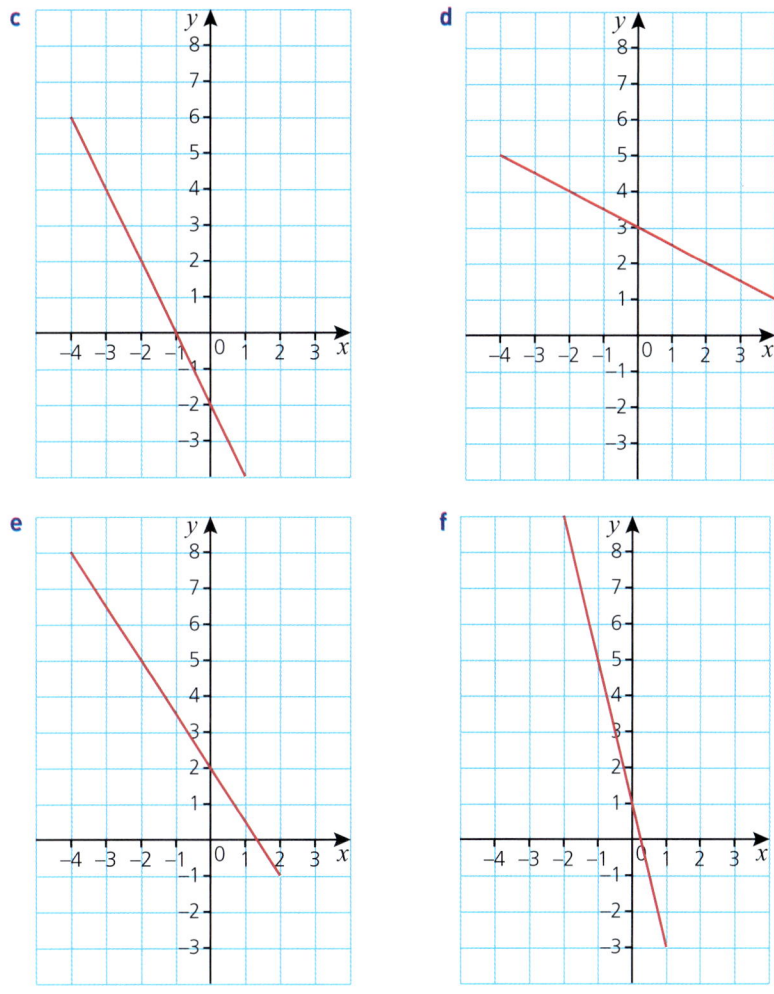

3 a  For each of the graphs in Q.1 and Q.2, calculate the gradient of the straight line.
  b  What do you notice about the gradient of each line and its equation?
  c  What do you notice about the equation of the straight line and where the line intersects the $y$-axis?

4 Copy the diagrams in Q.1. Draw two lines on the diagram parallel to the given line.
  a  Write the equation of these new lines in the form $y = mx + c$.
  b  What do you notice about the equations of these new parallel lines?

5 In Q.2 you found an equation for these lines in the form $y = mx + c$. Change the value of the intercept $c$ and then draw the new line. What do you notice about this new line and the first line?

# 33 EQUATION OF A STRAIGHT LINE

In general the equation of any straight line can be written in the form:
$$y = mx + c$$

where $m$ represents the gradient of the straight line and $c$ the intercept with the $y$-axis. This is shown in the diagram.

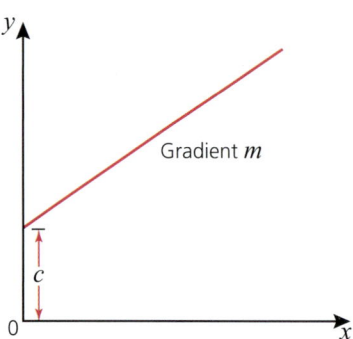

By looking at the equation of a straight line written in the form $y = mx + c$, it is therefore possible to deduce the line's gradient and intercept with the $y$-axis without having to draw it.

## ➡ Worked examples

**a** Find the gradient and $y$-intercept of the following straight lines:

   **i**  $y = 3x - 2$    gradient = 3
                                $y$-intercept = $-2$

   **ii**  $y = -2x + 6$    gradient = $-2$
                                    $y$-intercept = 6

**b** Calculate the gradient and $y$-intercept of the following straight lines:

   **i**  $2y = 4x + 2$

      This needs to be rearranged into **gradient–intercept** form ($y = mx + c$):

      $y = 2x + 1$    gradient = 2
                          $y$-intercept = 1

   **ii**  $y - 2x = -4$

      Rearranging into gradient–intercept form:

      $y = 2x - 4$    gradient = 2
                          $y$-intercept = $-4$

   **iii**  $-4y + 2x = 4$

      Rearranging into gradient–intercept form:

      $y = \frac{1}{2}x - 1$    gradient = $\frac{1}{2}$
                          $y$-intercept = $-1$

*The equation of a line through two points*

iv $\frac{y+3}{4} = -x + 2$

Rearranging into gradient–intercept form:
$y + 3 = -4x + 8$
$y = -4x + 5$     gradient $= -4$
                   $y$-intercept $= 5$

**Exercise 4.10**

For the following linear equations, calculate both the gradient and $y$-intercept.

1  a $y = 2x + 1$     b $y = 3x + 5$     c $y = x - 2$
    d $y = \frac{1}{2}x + 4$     e $y = -3x + 6$     f $y = -\frac{2}{3}x + 1$
    g $y = -x$     h $y = -x - 2$     i $y = -(2x - 2)$

2  a $y - 3x = 1$     b $y + \frac{1}{2}x - 2 = 0$     c $y + 3 = -2x$
    d $y + 2x + 4 = 0$     e $y - \frac{1}{4}x - 6 = 0$     f $-3x + y = 2$
    g $2 + y = x$     h $8x - 6 + y = 0$     i $-(3x + 1) + y = 0$

3  a $2y = 4x - 6$     b $2y = x + 8$     c $\frac{1}{2}y = x - 2$
    d $\frac{1}{4}y = -2x + 3$     e $3y - 6x = 0$     f $\frac{1}{3}y + x = 1$
    g $6y - 6 = 12x$     h $4y - 8 + 2x = 0$     i $2y - (4x - 1) = 0$

4  a $2x - y = 4$     b $x - y + 6 = 0$     c $-2y = 6x + 2$
    d $12 - 3y = 3x$     e $5x - \frac{1}{2}y = 1$     f $-\frac{2}{3}y + 1 = 2x$
    g $9x - 2 = -y$     h $-3x + 7 = -\frac{1}{2}y$     i $-(4x - 3) = -2y$

5  a $\frac{y+2}{4} = \frac{1}{2}x$     b $\frac{y-3}{x} = 2$     c $\frac{y-x}{8} = 0$
    d $\frac{2y - 3x}{2} = 6$     e $\frac{3y - 2}{x} = -3$     f $\frac{\frac{1}{2}y - 1}{x} = -2$
    g $\frac{3x - y}{2} = 6$     h $\frac{6 - 2y}{3} = 2$     i $\frac{-(x + 2y)}{5x} = 1$

6  a $\frac{3x - y}{y} = 2$     b $\frac{-x + 2y}{4} = y + 1$     c $\frac{y - x}{x + y} = 2$
    d $\frac{1}{y} = \frac{1}{x}$     e $\frac{-6(6x + y)}{2} = y + 1$     f $\frac{2x - 3y + 4}{4} = 4$
    g $\frac{y + 1}{x} + \frac{3y - 2}{2x} = -1$     h $\frac{x}{y + 1} + \frac{1}{2y + 2} = -3$     i $\frac{-(x - 3y) - (-x - 2y)}{4 + x - y} = -2$

# The equation of a line through two points

The equation of a straight line can be found once the coordinates of two points on the line are known.

## 33 EQUATION OF A STRAIGHT LINE

### → Worked example

Find the equation of the straight line passing through the points (−3, 3) and (5, 5).

Plotting the two points gives:

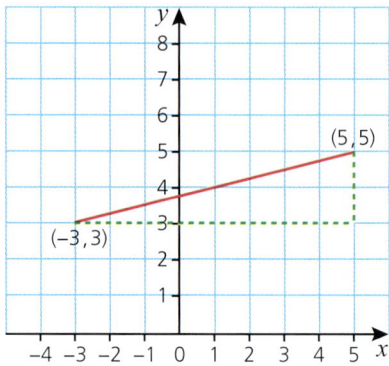

The equation of any straight line can be written in the general form $y = mx + c$. Here you have:

Gradient $= \dfrac{5-3}{5-(-3)} = \dfrac{2}{8} = \dfrac{1}{4}$

The equation of the line now takes the form $y = \dfrac{1}{4}x + c$.

Since the line passes through the two given points, their coordinates must satisfy the equation. So to work out the value of $c$, the $x$ and $y$ coordinates of one of the points are substituted into the equation. Substituting (5, 5) into the equation gives:

$5 = \dfrac{1}{4} \times 5 + c$

$5 = 1\dfrac{1}{4} + c$

Therefore $c = 5 - 1\dfrac{1}{4} = 3\dfrac{3}{4}$

The equation of the straight line passing through (−3, 3) and (5, 5) is:

$y = \dfrac{1}{4}x + 3\dfrac{3}{4}$

---

You have seen that the equation of a straight line takes the form $y = mx + c$. It can, however, also take the form $ax + by = c$. It is possible to write the equation $y = mx + c$ in the form $ax + by = c$ by rearranging the equation.

In the example above, $y = \dfrac{1}{4}x + 3\dfrac{3}{4}$ can firstly be rewritten as:

$y = \dfrac{x}{4} + \dfrac{15}{4}$

Multiplying both sides of the equation by 4 produces the equation $4y = x + 15$.

This can be rearranged to $-x + 4y = 15$, which is the required form, with $a = -1$, $b = 4$ and $c = 15$.

*The equation of a line through two points*

> **Note**
> The letter *c* when used in the form $y = mx + c$ has a specific meaning, i.e. it represents where the line crosses the y-axis. This is not the same for the letter *c* when it appears in the form $ax + by = c$.

## Exercise 4.11

Find the equation of the straight line which passes through each of the following pairs of points. Write your answers in the form:
  i   $y = mx + c$
  ii  $ax + by = c$.

*Questions 1.ii and 2.ii are relevant to the Extended syllabus only.*

1  a  (1, 1)   (4, 7)        b  (1, 4)   (3, 10)
   c  (1, 5)   (2, 7)        d  (0, −4)  (3, −1)
   e  (1, 6)   (2, 10)       f  (0, 4)   (1, 3)
   g  (3, −4)  (10, −18)     h  (0, −1)  (1, −4)
   i  (0, 0)   (10, 5)

2  a  (−5, 3)  (2, 4)        b  (−3, −2) (4, 4)
   c  (−7, −3) (−1, 6)       d  (2, 5)   (1, −4)
   e  (−3, 4)  (5, 0)        f  (6, 4)   (−7, 7)
   g  (−5, 2)  (6, 2)        h  (1, −3)  (−2, 6)
   i  (6, −4)  (6, 6)

## Exercise 4.12

1  Two points A and B have coordinates A(3, 8) and B(9, 5).
   a  Find the equation of the line AB.
   b  Find the equation of the line perpendicular to AB which passes through the point A.
   c  Find the equation of the line perpendicular to AB which passes through the point B.
   d  Comment on any similarity in your answers to parts **b** and **c**.

2  In each of the questions below, the equation of a line $l_1$ and the coordinates of a point P are given.
   i   Find the equation of the line parallel to $l_1$ passing through point P.
   ii  Find the equation of the line perpendicular to $l_1$ passing through point P.
   a  $y = 4x + 1$     P(0, 0)
   b  $y = 2x − 5$     P(2, −2)
   c  $y = −3x − 1$    P(−4, 3)
   d  $4x − 2y = 6$    P(1, 1)

3  Two lines $l_1$ and $l_2$ are shown.

# 33 EQUATION OF A STRAIGHT LINE

$l_1$ passes through (0, 4) and (12, 0).
$l_2$ passes through the origin and is perpendicular to $l_1$.
**a** Find the equation of $l_1$. Give your answer in the form $y = mx + c$
**b** Find the equation of $l_2$.
**c** Work out the coordinates of the point of intersection of the lines $l_1$ and $l_2$.
**d** Work out the ratio of the areas of the triangles A : B in its simplest form.

## Drawing straight line graphs

To draw a straight line graph, only two points need to be known. Once these have been plotted the line can be drawn between them and extended if necessary at both ends. It is important to check that your line is correct by taking a point from your graph and ensuring it satisfies the original equation.

### ➡ Worked examples

**a** Plot the line $y = x + 3$.

To identify two points simply choose two values of $x$. Substitute these into the equation and calculate their corresponding $y$-values.

When $x = 0$, $y = 3$.

When $x = 4$, $y = 7$.

Therefore two of the points on the line are (0, 3) and (4, 7).

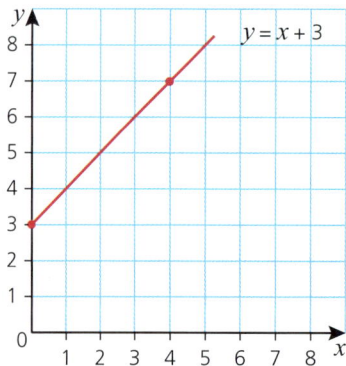

The straight line $y = x + 3$ is plotted as shown.

Check using a third point, e.g. (1, 4).

When $x = 1$, $y = x + 3 = 4$, so (1, 4) satisfies the equation of the line.

### Drawing straight line graphs

**b** Plot the line $y = -2x + 4$.

When $x = 2$, $y = 0$.

When $x = -1$, $y = 6$.

The coordinates of two points on the line are $(2, 0)$ and $(-1, 6)$, and the line is plotted as shown.

Check using the point $(0, 4)$.

When $x = 0$, $y = -2x + 4 = 4$, so $(0, 4)$ satisfies the equation of the line.

Note that, in questions of this sort, it is often easier to rearrange the equation into gradient–intercept form first.

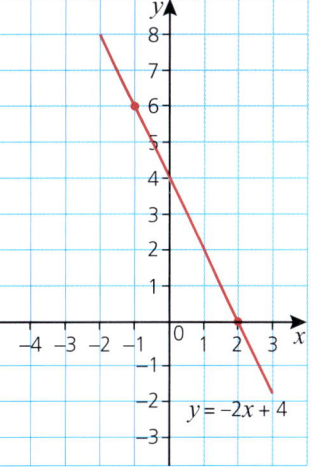

## Exercise 4.13

**1** Plot the following straight lines:
- **a** $y = 2x + 3$
- **b** $y = x - 4$
- **c** $y = 3x - 2$
- **d** $y = -2x$
- **e** $y = -x - 1$
- **f** $y = -\frac{1}{2}x - 4$
- **g** $-y = 3x - 3$
- **h** $2y = 4x - 2$
- **i** $y - 4 = 3x$

**2** Plot the following straight lines:
- **a** $-2x + y = 4$
- **b** $-4x + 2y = 12$
- **c** $3y = 6x - 3$
- **d** $2x = x + 1$
- **e** $3y - 6x = 9$
- **f** $2y + x = 8$
- **g** $x + y + 2 = 0$
- **h** $3x + 2y - 4 = 0$
- **i** $4 = 4y - 2x$

**3** Plot the following straight lines:
- **a** $\frac{x+y}{2} = 1$
- **b** $x + \frac{y}{2} = 1$
- **c** $\frac{x}{3} + \frac{y}{2} = 1$
- **d** $y + \frac{x}{2} = 3$
- **e** $\frac{y}{5} + \frac{x}{3} = 0$
- **f** $\frac{-(2x+y)}{4} = 1$
- **g** $\frac{y-(x-y)}{3x} = -1$
- **h** $\frac{y}{2x+3} - \frac{1}{2} = 0$
- **i** $-2(x+y) + 4 = -y$

# Investigations, modelling and ICT 4

## Investigation: Plane trails

In an aircraft show, planes often fly with a coloured smoke trail. Depending on the formation of the planes, the trails can intersect in different ways.

In the diagram below, the three smoke trails do not cross as they are parallel.

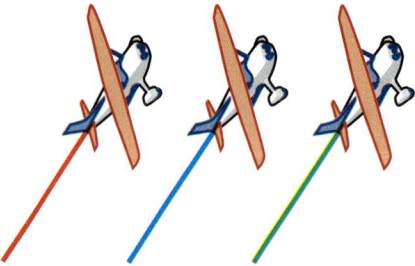

In the following diagram, there are two crossing points.

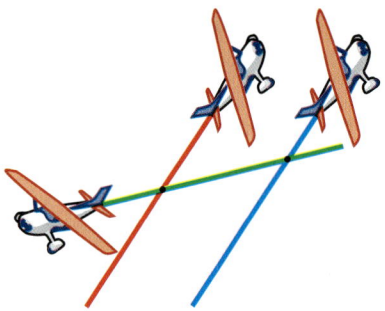

By flying differently, the three planes can produce trails that cross at three points.

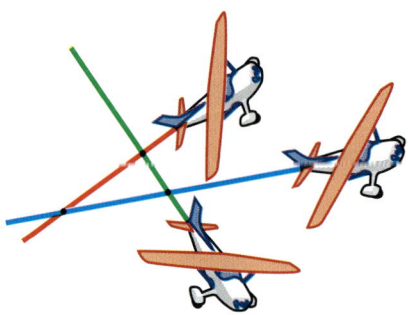

1. Investigate the connection between the maximum number of crossing points and the number of planes.
2. Record the results of your investigation in an ordered table.
3. Write an algebraic rule linking the number of planes ($p$) and the maximum number of crossing points ($n$).

# Investigation: Hidden treasure

A television show sets up a puzzle for its contestants to try to solve. Some buried treasure is hidden on a 'treasure island'. The treasure is hidden in a treasure chest. Each contestant stands by one of the treasure chests.

The treasure is hidden according to the following rule:

It is not hidden in chest 1.

Chest 2 is left empty for the time being.

It is not hidden in chest 3.

Chest 4 is left empty for the time being.

It is not hidden in chest 5.

The pattern of crossing out the first chest and then alternate chests is continued until only one chest is left. This will involve going round the circle several times, continuing the pattern. The treasure is hidden in the last chest left.

The diagrams below show how the last chest is chosen:

# INVESTIGATIONS, MODELLING AND ICT 4

After one round, chests 1, 3, 5, 7, 9 and 11 have been discounted.

After the second round, chests 2, 6 and 10 have also been discounted.

After the third round, chests 4 and 12 have also been discounted. This leaves only chest 8.

The treasure is therefore hidden in chest 8.

Unfortunately for participants, the number of contestants changes each time, as does the number of chests.

1 Investigate which treasure chest you would choose if there are:
  a 4 contestants
  b 5 contestants
  c 8 contestants
  d 9 contestants
  e 15 contestants.
2 Investigate the winning treasure chest for other numbers of contestants and enter your results in an ordered table.
3 State any patterns you notice in your table of results.
4 Use your patterns to predict the winning chest for 31, 32 and 33 contestants.
5 Write a rule linking the winning chest $x$ and the number of contestants $n$.

# Modelling: Stretching a spring

A spring is attached to a clamp stand as shown below.

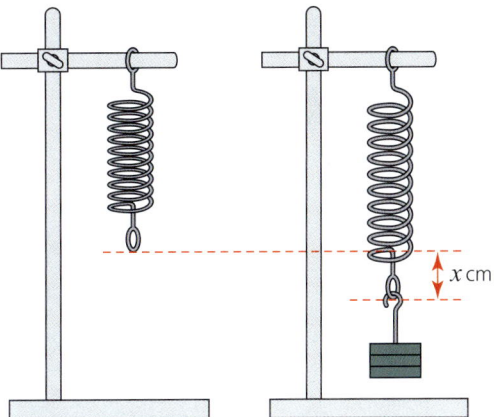

Different weights are attached to the end of the spring. The mass (*m*) in grams is noted, as is the amount by which the spring stretches (*x*) in cm.

The data collected is shown in the table below:

| Mass (g) | 50 | 100 | 150 | 200 | 250 | 300 | 350 | 400 | 450 | 500 |
|---|---|---|---|---|---|---|---|---|---|---|
| Extension (cm) | 3.1 | 6.3 | 9.5 | 12.8 | 15.4 | 18.9 | 21.7 | 25.0 | 28.2 | 31.2 |

1. Plot a graph of mass against extension.
2. Describe the approximate relationship between the mass and the extension.
3. Draw a line of best fit through the data.
4. Find the equation of the line of best fit.
5. Use your model to predict what the length of the spring would be for a weight of 275 g.
6. Explain why it is unlikely that the model would be useful to find the extension if a weight of 5 kg was added to the spring.
7. What range of values for the mass is the model valid for?

**INVESTIGATIONS, MODELLING AND ICT 4**

## ICT Activity

Your graphic display calculator is able to graph inequalities and shade the appropriate region. The examples below show some screenshots taken from a graphic display calculator.

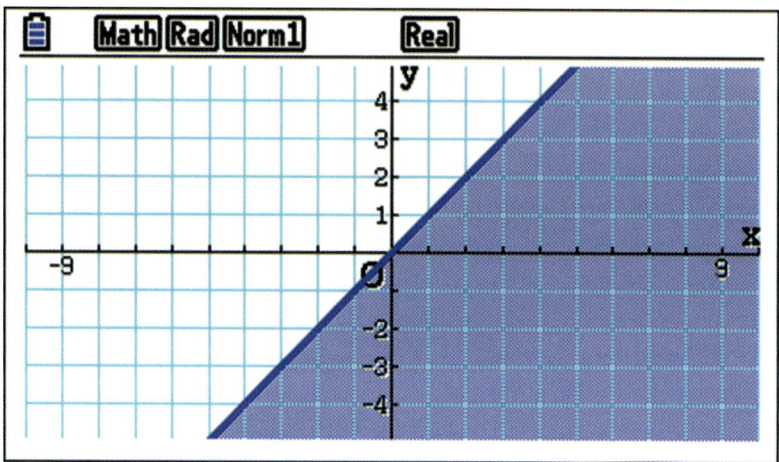

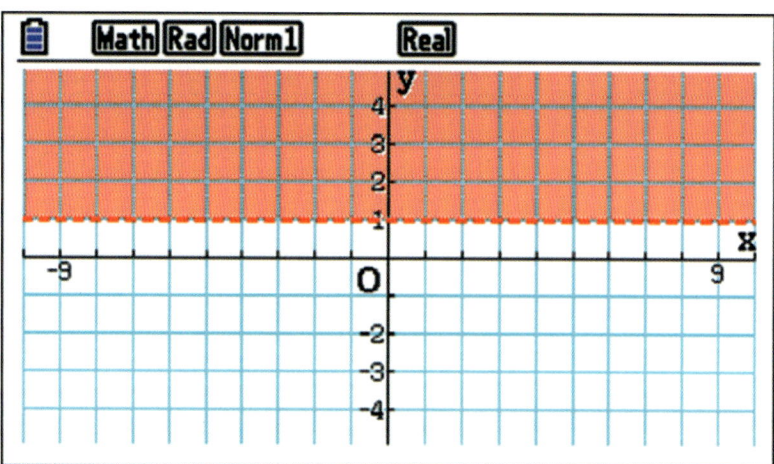

Investigate how your calculator can graph linear inequalities.

# Student assessment 4

## Student assessment 1

1. Sketch the following graphs on the same pair of axes, labelling each clearly.
   - a  $x = 3$
   - b  $y = -2$
   - c  $y = -3x$
   - d  $y = \frac{x}{4} + 4$

2. For each of the following linear equations:
   - i  Find the gradient and $y$-intercept
   - ii plot the graph.
   - a  $y = 2x + 3$
   - b  $y = 4 - x$
   - c  $2x - y = 3$
   - d  $-3x + 2y = 5$

3. Find the equation of the straight line which passes through the following pairs of points:
   - a  $(-2, -9)\ (5, 5)$
   - b  $(1, -1)\ (-1, 7)$

4. The coordinates of the end points of two line segments are given below. Calculate the length of each of the lines.
   - a  $(2, 6)\ (-2, 3)$
   - b  $(-10, -10)\ (0, 14)$

5. The coordinates of two points are given as $P(1, 12)$ and $Q(5, 2)$.
   - a  Find the equation of the line passing through P and Q.
   - b  Find the equation of the perpendicular bisector of PQ.

# TOPIC 5

## Geometry

### Contents

Chapter 34  Geometrical vocabulary (C5.1, E5.1)
Chapter 35  Symmetry (C5.4, E5.4)
Chapter 36  Measuring and drawing angles and bearings (C5.2, E5.2)
Chapter 37  Angle properties (C5.5, E5.5)
Chapter 38  Similarity (C5.3, E5.3)
Chapter 39  Properties of circles (C5.6, C5.7, E5.6, E5.7)

### Learning objectives

**C5.1      E5.1**
Use and interpret the geometrical terms: point; vertex; line; **plane**; parallel; perpendicular; **perpendicular bisector**; bearing; right angle, acute, obtuse and reflex angles; interior and exterior angles; similar and congruent; scale factor

Use and interpret the vocabulary of triangles, special quadrilaterals, polygons, simple solids **and solids**

Use and interpret the vocabulary of a circle

**C5.2      E5.2**
1  Measure and draw lines and angles
2  Use and interpret three-figure bearings

**C5.3      E5.3**
1  Calculate lengths of similar shapes
2  **Use the relationships between lengths and areas of similar shapes and lengths, surface areas and volumes of similar solids**
3  **Solve problems and give simple explanations involving similarity**

**C5.4      E5.4**
1  Recognise line symmetry and order of rotational symmetry in two dimensions
2  **Recognise symmetry properties of prisms, cylinders, pyramids and cones**

**C5.5      E5.5**
1  Calculate unknown angles and give explanations using the following geometrical properties:
   - sum of angles at a point = 360°
   - sum of angles at a point on a straight line = 180°
   - vertically opposite angles are equal
   - angle sum of a triangle = 180° and angle sum of a quadrilateral = 360°
2  Calculate unknown angles and give geometric explanations for angles formed within parallel lines:
   - corresponding angles are equal
   - alternate angles are equal
   - co-interior (supplementary) angles sum to 180°
3  Know and use angle properties of regular polygons **and irregular polygons**

**C5.6      E5.6**

Calculate unknown angles and give explanations using the following geometrical properties of circles:
- angle in a semicircle = 90°
- angle between tangent and radius = 90°
- **angle at the centre is twice the angle at the circumference**
- **angles in the same segment are equal**
- **opposite angles of a cyclic quadrilateral sum to 180° (supplementary)**
- **alternate segment theorem**

**C5.7      E5.7**

Use the following symmetry properties of circles:
- **equal chords are equidistant from the centre**
- **the perpendicular bisector of a chord passes through the centre**
- tangents from an external point are equal in length

Elements in purple refer to the Extended curriculum only.

# The Greeks

Many of the great Greek mathematicians came from the Greek Islands, from cities such as Ephesus or Miletus (which are in present day Turkey) or from Alexandria in Egypt. This introduction briefly mentions some of the Greek mathematicians of 'The Golden Age'. You may wish to find out more about them.

Thales of Alexandria invented the 365-day calendar and predicted the dates of eclipses of the Sun and the Moon.

Pythagoras of Samos founded a school of mathematicians and worked with geometry. His successor as leader was Theano, the first woman to hold a major role in mathematics.

Archimedes (287–212BC)

Eudoxus of Asia Minor (Turkey) worked with irrational numbers such as pi and discovered the formula for the volume of a cone.

Euclid of Alexandria formed what would now be called a university department. His book became the set text in schools and universities for 2000 years.

Apollonius of Perga (Turkey) worked on, and gave names to, the parabola, the hyperbola and the ellipse.

Archimedes is accepted today as one of the greatest mathematicians of all time. However, he was so far ahead of his time that his influence on his contemporaries was limited by their lack of understanding.

# 34 Geometrical vocabulary

## Angles and lines

Different types of angle have different names:

**acute angles** lie between 0° and 90°

**right angles** are exactly 90°

**obtuse angles** lie between 90° and 180°

**reflex angles** lie between 180° and 360°

To find the shortest distance between two **points**, you measure the length of the **straight line** which joins them.

Two lines which meet at right angles are **perpendicular** to each other.

So in the diagram below, CD is **perpendicular** to AB, and AB is perpendicular to CD.

As CD is perpendicular to AB, the distance CD is also the shortest distance from point D to the line AB.

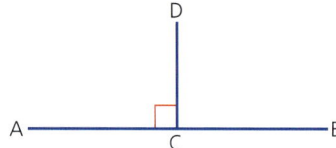

*As CD is perpendicular to AB, the distance CD is also the shortest distance of point D to the line AB.*

If the lines AD and BD are drawn to form a triangle, the line CD can be called the **height** or **altitude** of the triangle ABD.

**Parallel** lines are straight lines which can be continued to infinity in either direction without meeting.

Railway lines are an example of parallel lines. Parallel lines are marked with arrows, as shown:

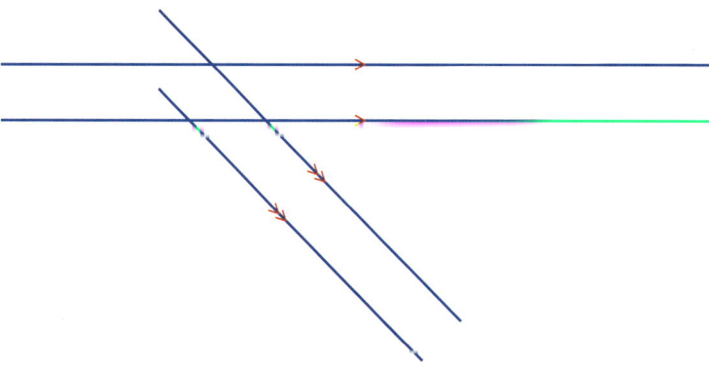

# Triangles

Triangles can be described in terms of their sides or their angles, or both.

An **acute-angled** triangle has all its angles less than 90°.

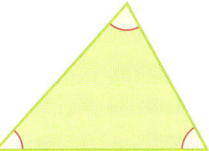

A **right-angled** triangle has an angle of 90°.

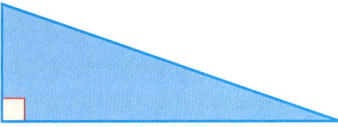

An **obtuse-angled** triangle has one angle greater than 90°.

An **isosceles** triangle has two sides of equal length, and the angles opposite the equal sides are equal.

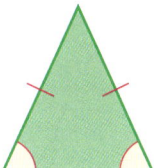

An **equilateral** triangle has three sides of equal length and three equal angles.

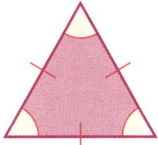

A **scalene** triangle has three sides of different lengths and all three angles are different.

# 34 GEOMETRICAL VOCABULARY

## Congruent triangles

*You will not be expected to prove that two shapes are congruent.*

**Congruent** triangles are **identical**. They have corresponding sides of the same length, and corresponding angles which are equal.

## Similar triangles

If the angles of two triangles are the same, then their corresponding sides will also be in proportion to each other. When this is the case, the triangles are said to be **similar**.

In the diagram below, triangle ABC is similar to triangle XYZ. Similar shapes are covered in more detail in Chapter 38 in this topic.

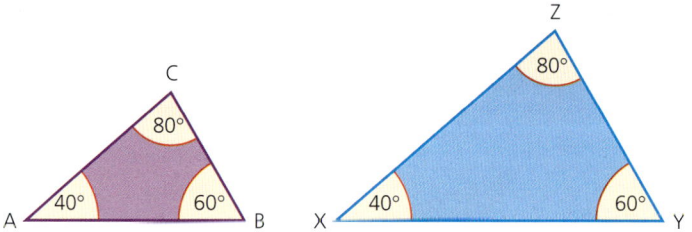

## Circles

*A tangent to a circle is a straight line which just touches the circumference of the circle at one point.*

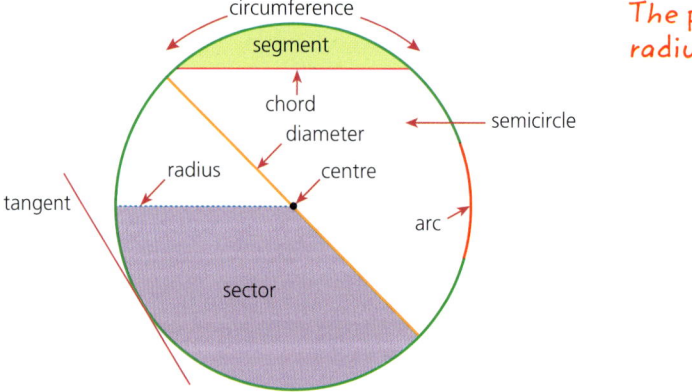

*The plural of radius is radii.*

The circumference is the perimeter all the way around the circle. An arc is part of the circumference. If an arc is less than half the circumference it is called a **minor arc**. If the arc is more than half the circumference it is called a **major arc**.

306

# Quadrilaterals

A **quadrilateral** is a plane shape consisting of four angles and four sides. There are several types of quadrilateral. The main ones, and their properties, are described below.

A corner of a 2D shape is called a **vertex**. The plural of vertex is **vertices**.

Two pairs of parallel sides.

All sides are equal.

All angles are equal.

Diagonals intersect at right angles.

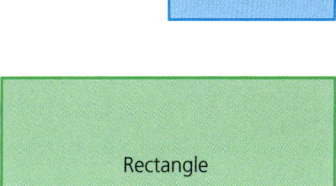

Two pairs of parallel sides.

Opposite sides are equal.

All angles are equal.

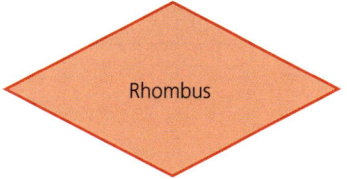

Two pairs of parallel sides.

All sides are equal.

Opposite angles are equal.

Diagonals bisect each other at right angles.

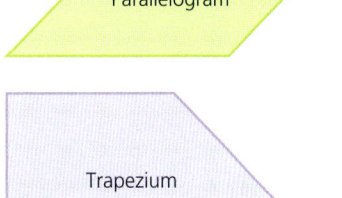

Two pairs of parallel sides.

Opposite sides are equal.

Opposite angles are equal.

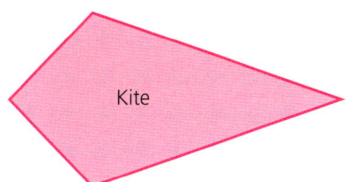

One pair of parallel sides.

An isosceles trapezium has one pair of parallel sides and the other pair of sides are equal in length.

Two pairs of equal sides.

One pair of equal angles.

Diagonals intersect at right angles.

# 34 GEOMETRICAL VOCABULARY

**Exercise 5.1**

**1** Copy and complete the following table. The first line has been started for you.

|  | Rectangle | Square | Parallelogram | Kite | Rhombus | Equilateral triangle |
|---|---|---|---|---|---|---|
| Opposite sides equal in length | Yes | Yes |  |  |  |  |
| All sides equal in length |  |  |  |  |  |  |
| All angles right angles |  |  |  |  |  |  |
| Both pairs of opposite sides parallel |  |  |  |  |  |  |
| Diagonals equal in length |  |  |  |  |  |  |
| Diagonals intersect at right angles |  |  |  |  |  |  |
| All angles equal |  |  |  |  |  |  |

## Polygons

Any closed figure made up of straight lines is called a **polygon**.

If the sides are the same length and the interior angles are equal, the figure is called a **regular polygon**.

*If the sides of a polygon are not all the same length and the interior angles not all the same size, the polygon is an irregular polygon.*

The names of the common polygons are:

| 3 sides | **tri**angle |
| 4 sides | **quad**rilateral |
| 5 sides | **penta**gon |
| 6 sides | **hexa**gon |
| 7 sides | **hepta**gon |
| 8 sides | **octa**gon |
| 9 sides | **nona**gon |
| 10 sides | **deca**gon |
| 12 sides | **dodeca**gon |

Two polygons are said to be **similar** if

**a** their angles are the same

**b** corresponding sides are in proportion.

# Nets

The diagram below is the **net** of a cube. It shows the faces of the cube opened out into a two-dimensional plan. The net of a three-dimensional shape can be folded up to make that shape.

**Exercise 5.2**  Draw the following on squared paper:
1 Two other possible nets of a cube
2 The net of a cuboid (rectangular prism)
3 The net of a triangular prism
4 The net of a cylinder
5 The net of a square-based pyramid
6 The net of a tetrahedron

*A tetrahedron is a pyramid with four faces and each face is a triangle.*

# 35 Symmetry

## Line symmetry

A **line of symmetry** divides a two-dimensional (flat) shape into two congruent (identical) shapes.

e.g.

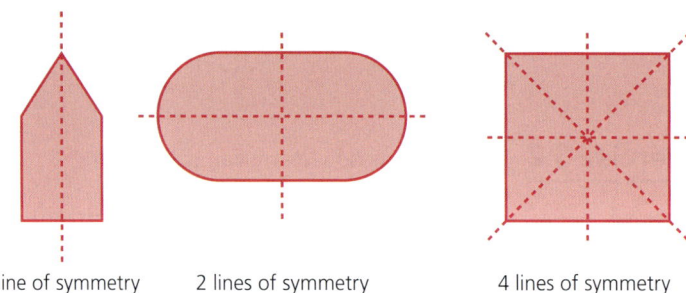

1 line of symmetry     2 lines of symmetry     4 lines of symmetry

**Exercise 5.3**

1. Draw the following shapes and, where possible, show all their lines of symmetry:
   - a square
   - b rectangle
   - c equilateral triangle
   - d isosceles triangle
   - e kite
   - f regular hexagon
   - g regular octagon
   - h regular pentagon
   - i isosceles trapezium
   - j circle

2. Copy the shapes below and, where possible, show all their lines of symmetry:

   a

   b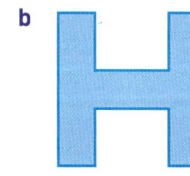

   c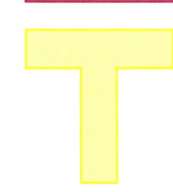

   d

   e

   f

*Line symmetry*

**g**  **h**

**3** Copy the shapes below and complete them so that the **bold** line becomes a line of symmetry:

**a** **b**

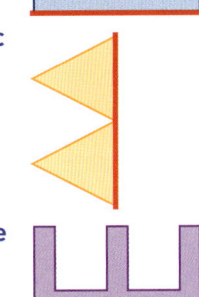 

**c** **d**

**e**  **f**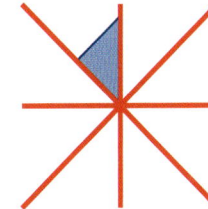

**4** Copy the shapes below and complete them so that the **bold** lines become lines of symmetry:

**a** **b**

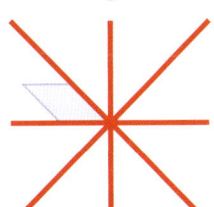 

**c** **d**

# 35 SYMMETRY

## Rotational symmetry

A two-dimensional shape has **rotational symmetry** if, when rotated about a central point, it looks the same as its starting position. The number of times it looks the same during a complete revolution is called the **order of rotational symmetry**. For example:

rotational symmetry of order 2     rotational symmetry of order 4

**Exercise 5.4**

1  Draw the following shapes. Identify the centre of rotation, and state the order of rotational symmetry:
   a square
   b equilateral triangle
   c regular pentagon
   d parallelogram
   e rectangle
   f rhombus
   g regular hexagon
   h regular octagon
   i circle

2  Copy the shapes below. Indicate the centre of rotation, and state the order of rotational symmetry:

   a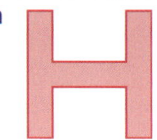
   b
   c
   d
   e

3  Copy and complete the following table of the symmetry properties for the given regular polygons. It has been started for you.

| Regular polygon | Number of lines of symmetry | Order of rotational symmetry |
|---|---|---|
| Equilateral triangle | 3 | |
| Square | | |
| Pentagon | | |
| Hexagon | | 6 |

## Symmetry in three dimensions

| Regular polygon | Number of lines of symmetry | Order of rotational symmetry |
|---|---|---|
| Heptagon | | |
| Octagon | | |
| Nonagon | | |
| Decagon | | |

# Symmetry in three dimensions

A plane of symmetry divides a three-dimensional (solid) shape into two congruent solid shapes.

e.g.

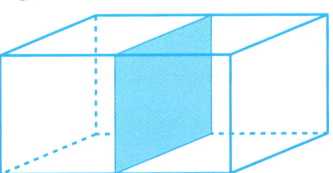 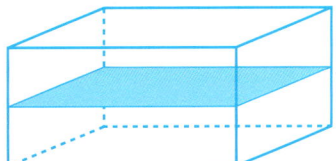

A cuboid has at least three planes of symmetry, two of which are shown above.

A three-dimensional shape has **rotational symmetry** if, when rotated about a central **axis**, it looks the same at certain intervals.

e.g.

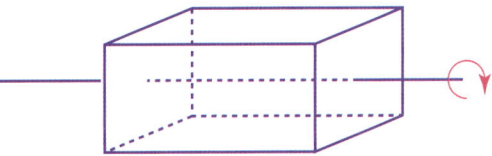

The cuboid has rotational symmetry of order 2 about the axis shown.

**Exercise 5.5**

1   Draw each of the solid shapes below twice, then:
    i   on each drawing of the shape, draw a different plane of symmetry
    ii  state how many planes of symmetry the shape has in total.

a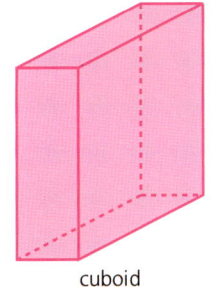
cuboid

b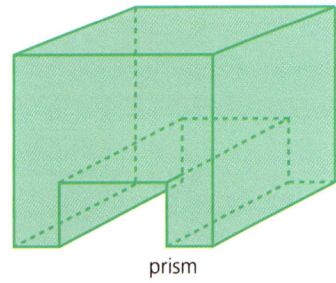
prism

# 35 SYMMETRY

**Exercise 5.5 (cont)**

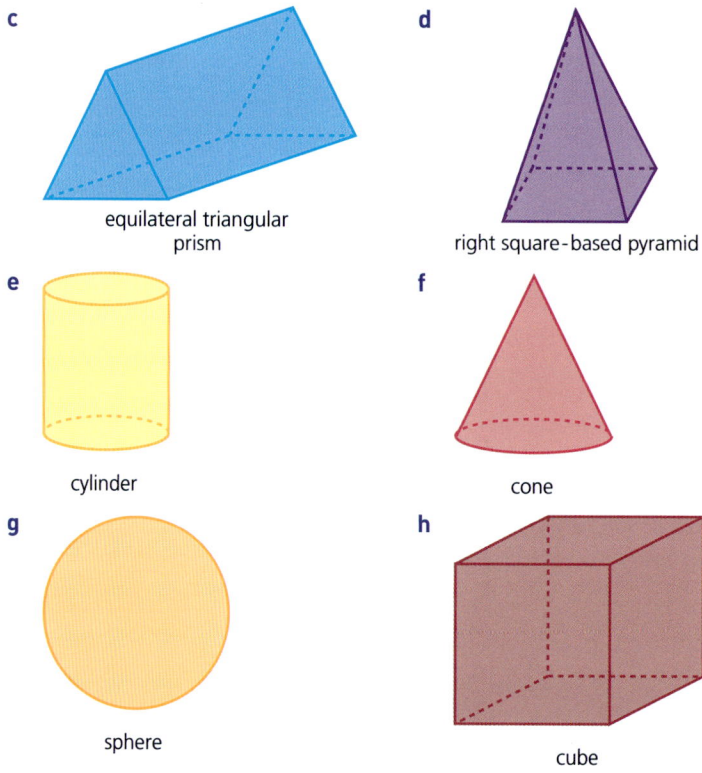

c equilateral triangular prism
d right square-based pyramid
e cylinder
f cone
g sphere
h cube

2 For each of the shapes shown below, determine the order of rotational symmetry about the axis shown.

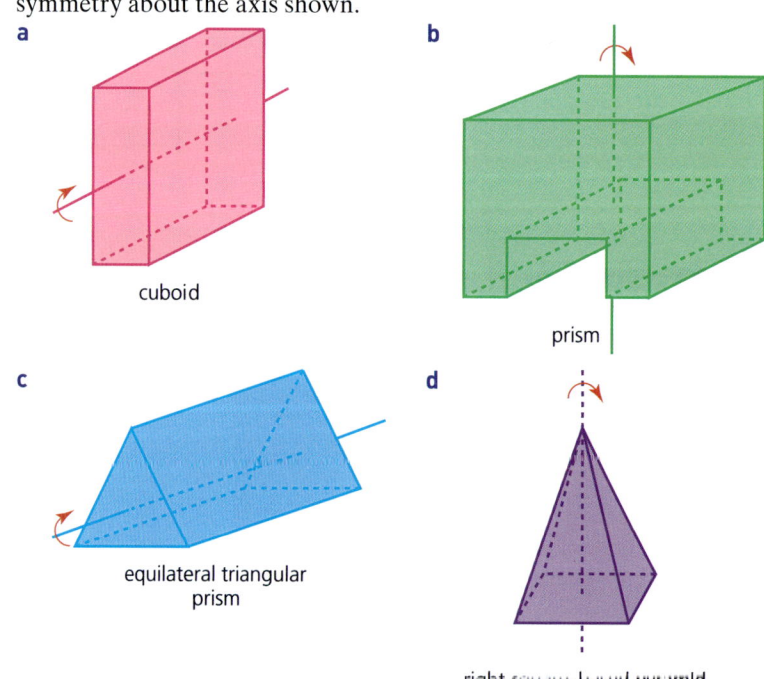

a cuboid
b prism
c equilateral triangular prism
d right square-based pyramid

*Symmetry in three dimensions*

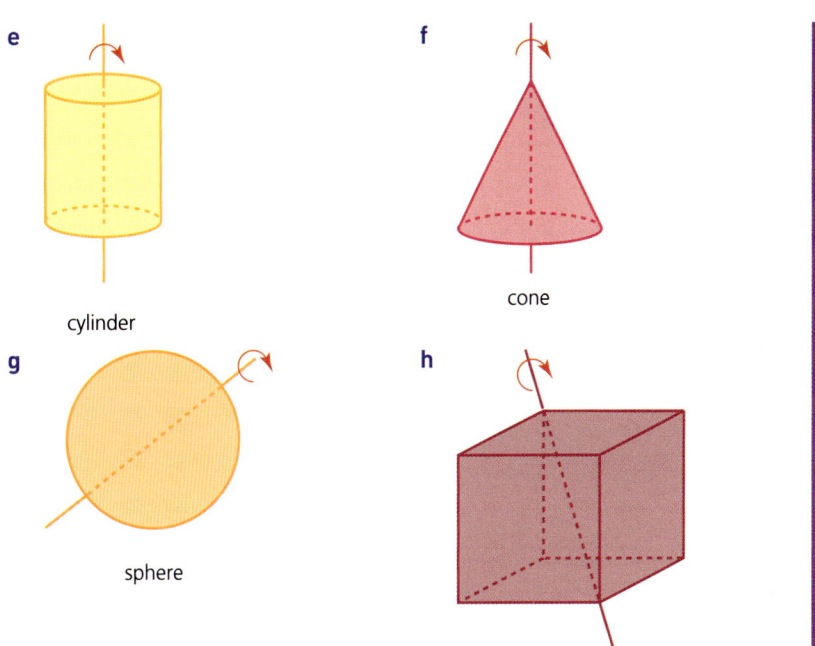

# 36 Measuring and drawing angles and bearings

## Angles

An angle is a measure of turn. When drawn, it can be measured using either a protractor or an angle measurer. The units of turn are degrees (°). Measuring with a protractor needs care, as there are two scales marked on it – an inner one and an outer one.

### → Worked examples

a   Measure the angle drawn (below):
- Place the protractor over the angle so that the cross lies on the point where the two arms meet.
- Align the 0° with one of the arms.
- Decide which scale is appropriate. In this case, it is the inner scale as it starts at 0°.
- Measure the angle using the inner scale.

The angle is 45°.

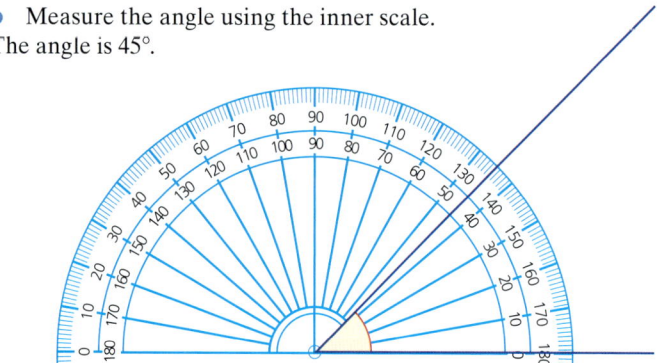

*All straight lines should be drawn using a ruler.*

b   Draw an angle of 110°.
- Start by drawing a straight line.
- Place the protractor on the line so that the cross is on one of the end points of the line. Ensure that the line is aligned with the 0° on the protractor.
- Decide which scale to use. In this case, it is the outer scale as it starts at 0°.
- Mark where the protractor reads 110°.

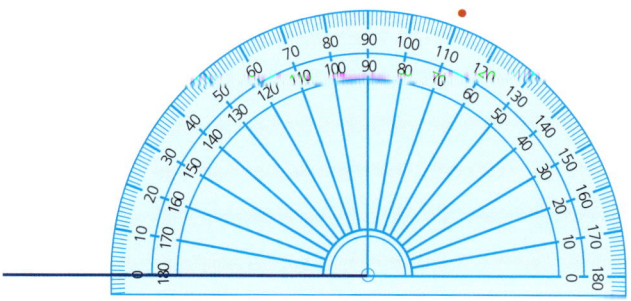

316

*Angles*

- Join the mark made to the end point of the original line.

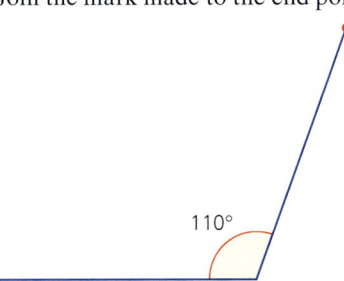

**Exercise 5.6**  1  Measure each of the following angles:

a 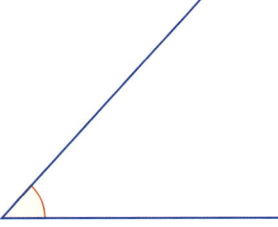  b 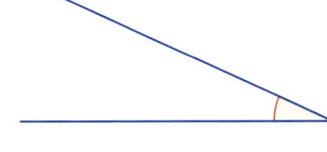  c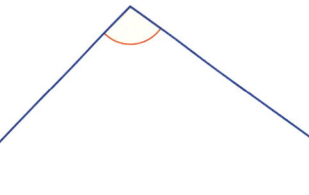

d e 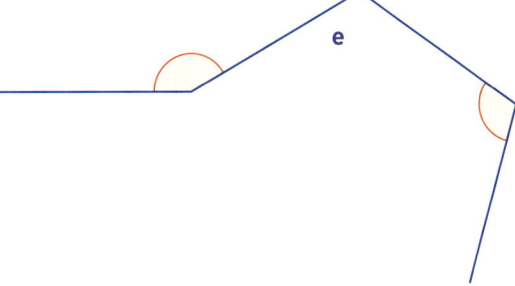  f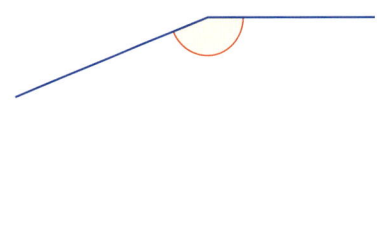

2  Measure each of the following angles:

a 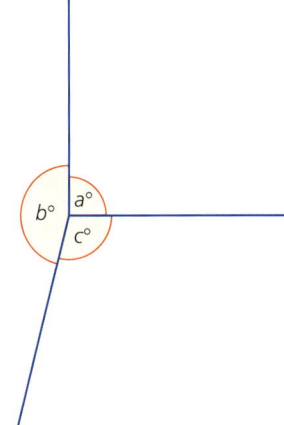  b 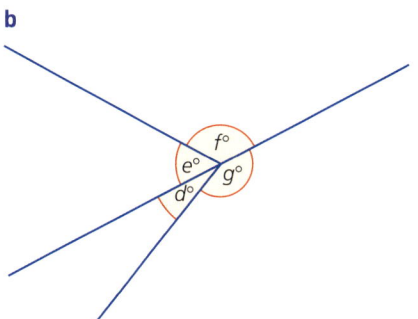  c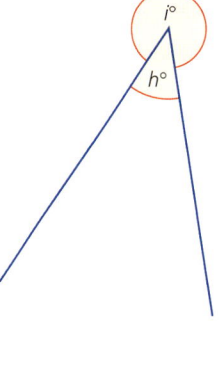

# 36 MEASURING AND DRAWING ANGLES AND BEARINGS

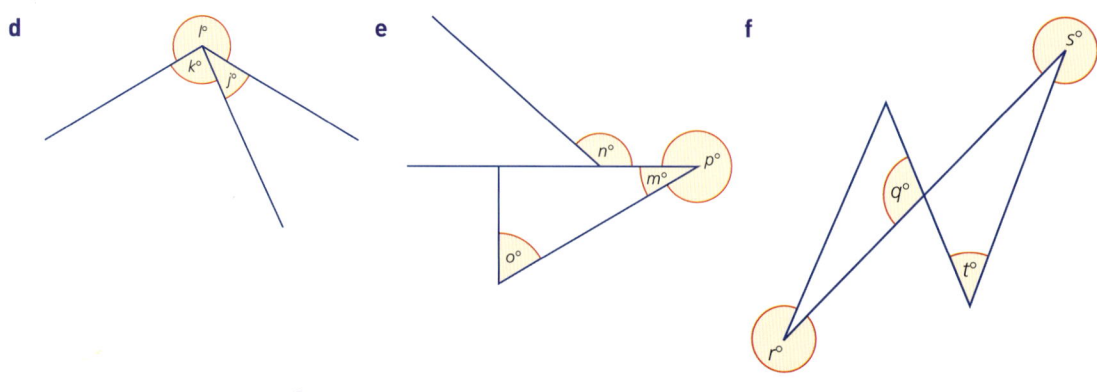

3  Draw angles of the following sizes:
   a  20°       b  45°       c  90°
   d  120°      e  157°      f  172°
   g  14°       h  205°      i  311°
   j  283°      k  198°      l  352°

## Bearings

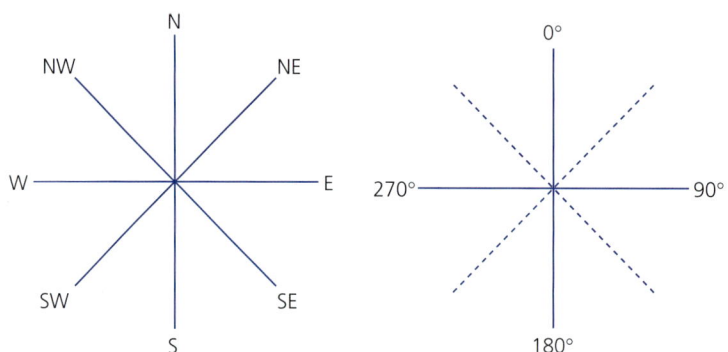

In the days when sailing ships travelled the oceans of the world, compass **bearings**, such as those in the diagram above left, were used for navigation.

As the need for more accurate direction arose, extra points were added to N, S, E, W, NE, SE, SW and NW. Midway between north and north east was north north east, and midway between north east and east was east north east, and so on. This gave sixteen points of the compass. This was later extended even further, eventually to sixty-four points.

As the speed of travel increased, a new system was required. The new system was the **three-figure bearing** system. North was given the bearing zero. 360° in a clockwise direction was one full rotation.

# Bearings

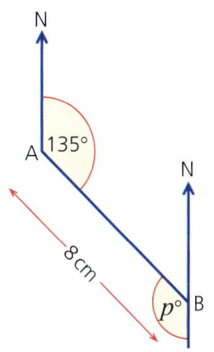

## Back bearings

The bearing of B from A is 135° and the distance from A to B is 8 cm, as shown in the diagram. The bearing of A from B is called the **back bearing**.

Since the two north lines are parallel:

$P = 135°$ (alternate angles), so the back bearing is $(180 + 135)°$.

That is, 315°.

*(There a number of methods of solving this type of problem.)*

*You will cover alternate angles in the next chapter.*

### → Worked example

The bearing of B from A is 245°.

What is the bearing of A from B?

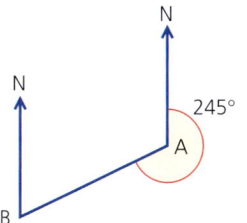 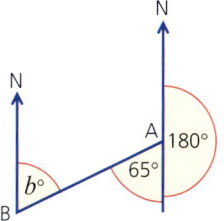

Since the two north lines are parallel:

$B = 65°$ (alternate angles), so the bearing is $(245 - 180)°$. That is, 065°.

## Exercise 5.7

1. Draw a diagram to show the following compass bearings and journey. Use a scale of 1 cm : 1 km. North can be taken to be a line vertically up the page. Start at point A. Travel a distance of 7 km on a bearing of 135° to point B. From B, travel 12 km on a bearing of 250° to point C. Measure the distance and bearing of A from C.
2. Given the following bearings of point B from point A, draw diagrams and use them to calculate the bearings of A from B.
   a  bearing 163°
   b  bearing 214°
3. Given the following bearings of point D from point C, draw diagrams and use them to calculate the bearings of C from D.
   a  bearing 300°
   b  bearing 282°

# 37 Angle properties

There are many angle relationships in geometry, the most common of which are explained below.

## Angles at a point

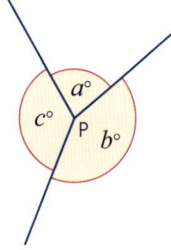

The diagram shows that if a person standing at P turns through each of the angles $a°$, $b°$ and $c°$ in turn, then the total amount they have rotated would be 360° (a complete turn). Therefore:

$a° + b° + c° = 360°$

**Angles about a point add up to 360°.**

## Angles on a straight line

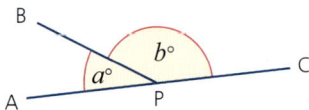

The points APC lie on a straight line. A person standing at point P, initially facing point A, turns through an angle $a°$ to face point B and then turns a further angle $b°$ to face point C. The person has turned through half a complete turn and therefore rotated through 180°. Therefore $a° + b° = 180°$. This can be summarised as:

**Angles on a straight line, about a point, add up to 180°.**

### ➜ Worked example

Calculate the value of $x$ in the diagram:

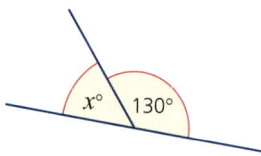

The sum of all the angles at a point on a straight line is 180°. Therefore:

$x° + 130° = 180°$

$x° = 180° - 130°$

Therefore angle $x$ is 50°.

## Angles on a straight line

 **Exercise 5.8**   **1** Calculate the size of angle $x$ in each of the following:

a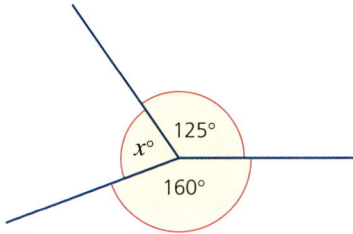

b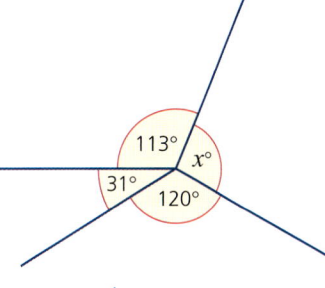

c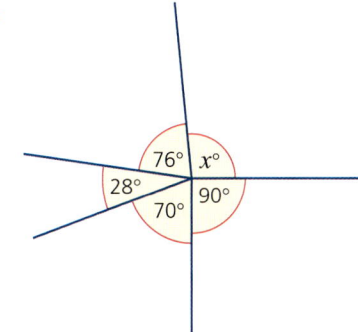

d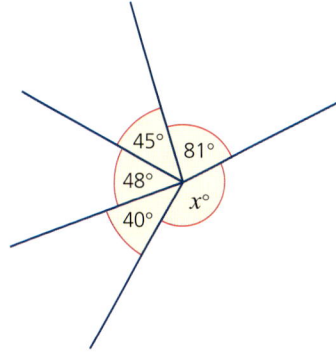

**2** Calculate the size of angle $y$ in each of the following:

a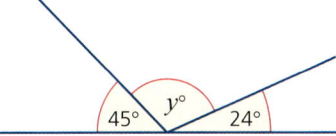

b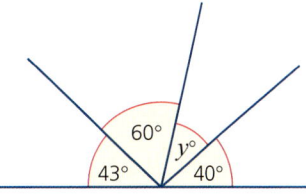

c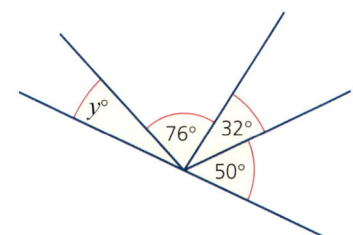

d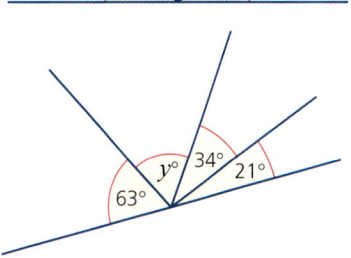

# 37 ANGLE PROPERTIES

**Exercise 5.8 (cont)**  3  Calculate the size of angle *p* in each of the following:

a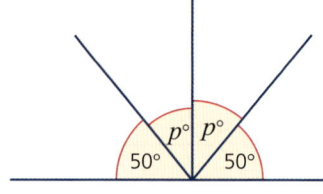

b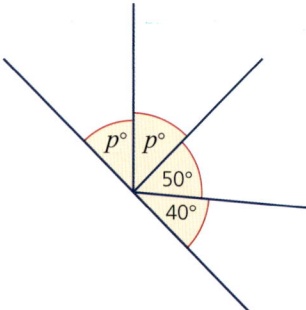

c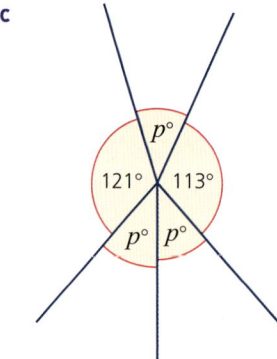

d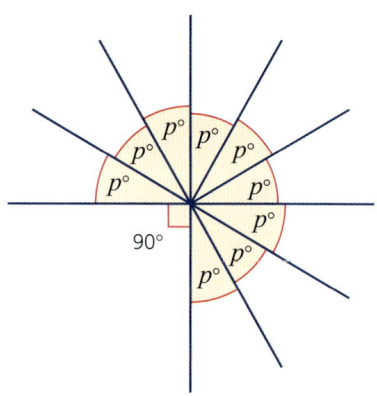

## Angles formed within parallel lines

**Exercise 5.9**  1  Draw a similar diagram to the one shown below. Measure carefully each of the labelled angles and write them down.

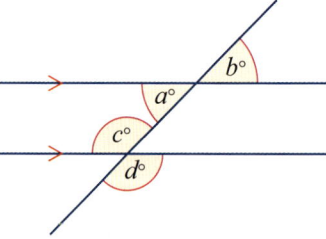

2  Draw a similar diagram to the one shown below. Measure carefully each of the labelled angles and write them down.

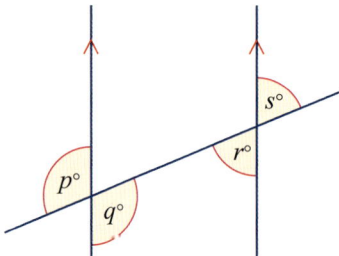

*Angles formed within parallel lines*

**3** Draw a similar diagram to the one shown below. Measure carefully each of the labelled angles and write them down.

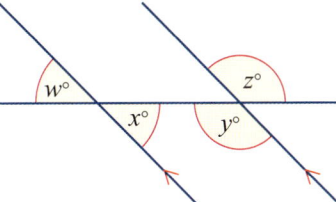

**4** Write down what you have noticed about the angles you measured in Q.1–3.

When two straight lines cross, it is found that the angles opposite each other are the same size. They are known as **vertically opposite angles**. By using the fact that angles at a point on a straight line add up to 180°, it can be shown why vertically opposite angles must always be equal in size.

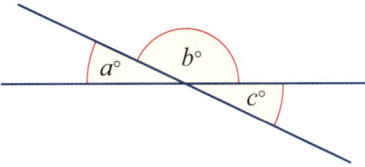

$a° + b° = 180°$

$c° + b° = 180°$

Therefore, $a$ is equal to $c$.

## Exercise 5.10

**1** Draw a similar diagram to the one shown below. Measure carefully each of the labelled angles and write them down.

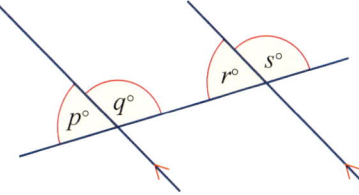

**2** Draw a similar diagram to the one shown below. Measure carefully each of the labelled angles and write them down.

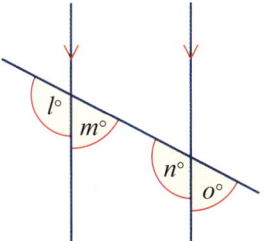

# 37 ANGLE PROPERTIES

**Exercise 5.10 (cont)**

3  Draw a similar diagram to the one shown below. Measure carefully each of the labelled angles and write them down.

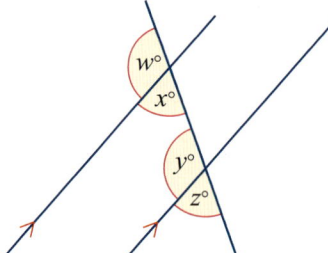

4  Write down what you have noticed about the angles you measured in Q.1–3.

When a line intersects two parallel lines, as in the diagram below, it is found that certain angles are the same size.

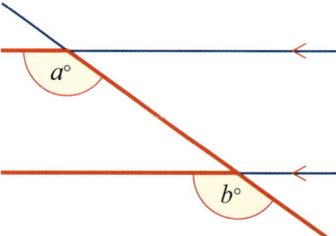

The angles $a$ and $b$ are equal and are known as **corresponding angles**. Corresponding angles can be found by looking for an 'F' formation in a diagram.

A line intersecting two parallel lines also produces another pair of equal angles known as **alternate angles**. These can be shown to be equal by using the fact that both vertically opposite and corresponding angles are equal.

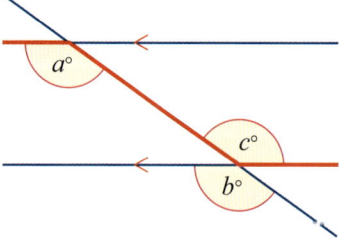

In the diagram above, $a = b$ (corresponding angles). But $b = c$ (vertically opposite). So $a = c$.

Angles $a$ and $c$ are alternate angles. These can be found by looking for a 'Z' formation in a diagram.

*Angles formed within parallel lines*

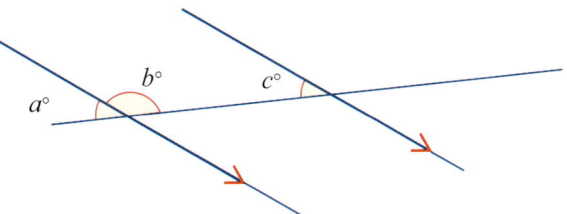

As $a° = c°$ (corresponding angles) and $a° + b° = 180°$ (angles on a straight line add up to 180°) then $b° + c° = 180°$.

$b$ and $c$ are **co-interior** angles as they face each other between parallel lines. Co-interior angles therefore add up to 180°.

*Pairs of angles which add up to 180° are also called supplementary angles.*

**Exercise 5.11**  In each of the following questions, some of the angles are given.

Find, giving reasons, the size of the other labelled angles.

1
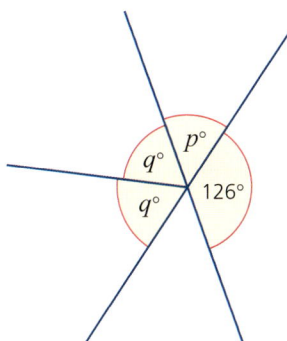

2
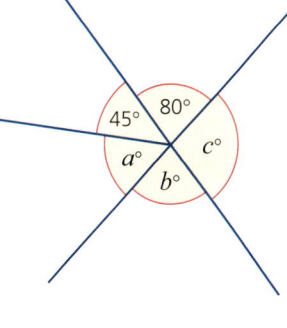

3
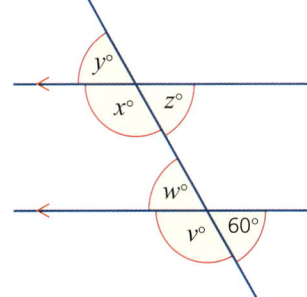

4
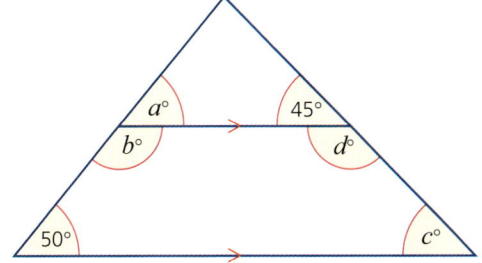

325

# 37 ANGLE PROPERTIES

**Exercise 5.11 (cont)**

5

6

7

8

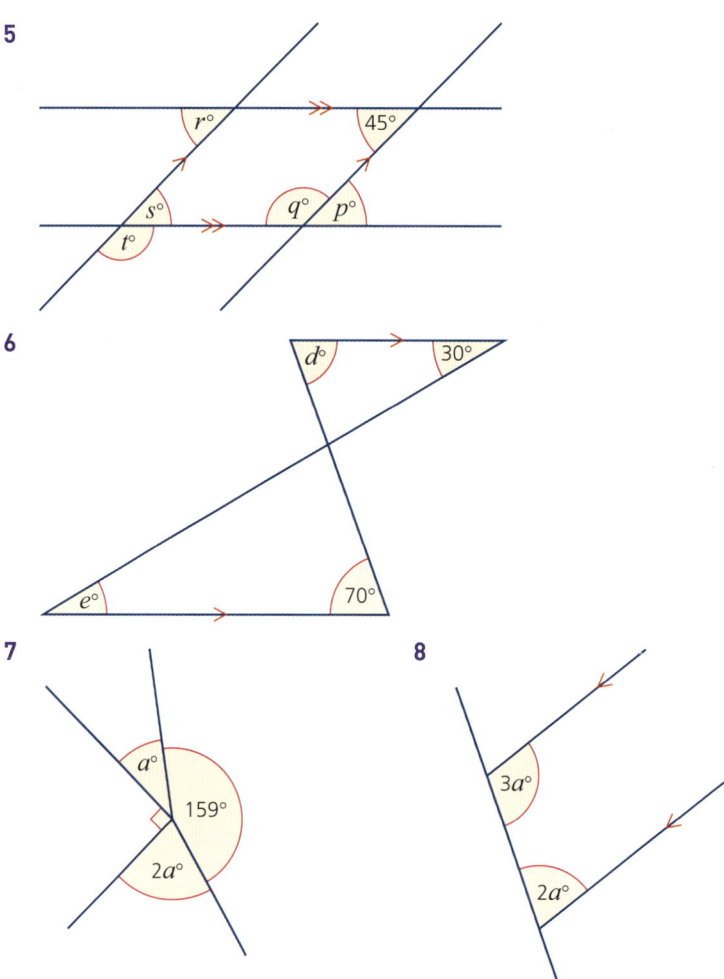

## Properties of polygons

A polygon is a closed, two-dimensional shape formed by straight lines, with at least three sides. Examples of polygons include triangles, quadrilaterals and hexagons. The pattern below shows a number of different polygons **tessellating**; that is, fitting together with no gaps or overlaps.

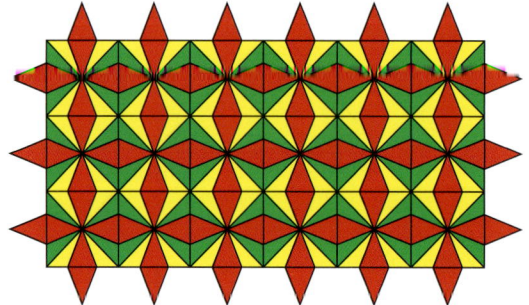

# Angle properties of polygons

The polygons in the pattern above tessellate because they have particular properties.

The properties of several triangles and quadrilaterals are listed on pages 305 and 307.

**Exercise 5.12**

1. Identify as many different polygons as you can in the tessellating pattern on the previous page.
2. Name each of the polygons drawn below. Give reasons for each answer.

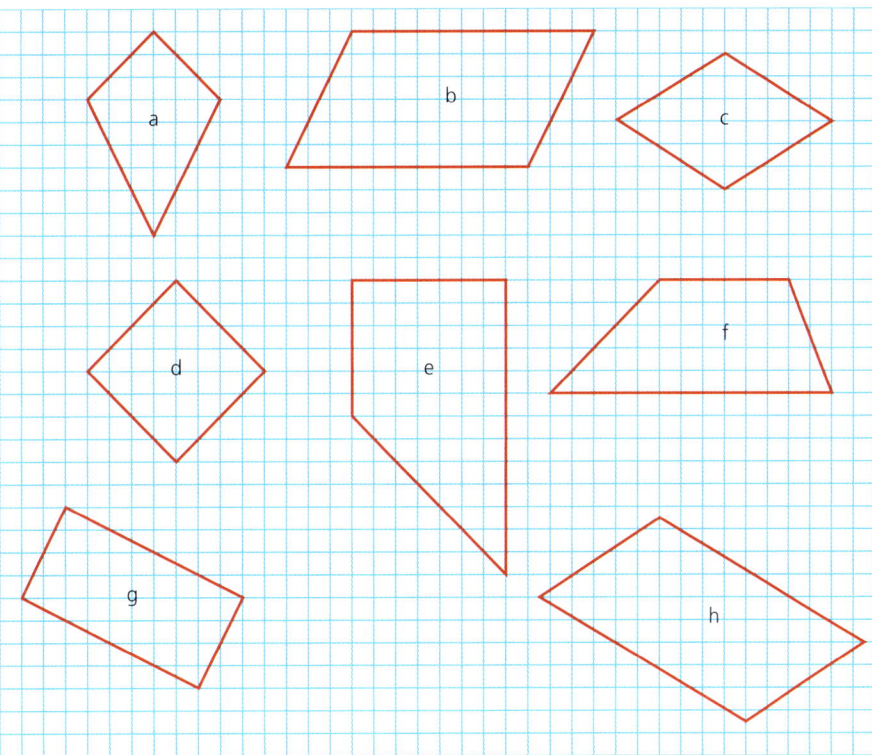

## Angle properties of polygons

In the triangle shown, the **interior angles** are labelled $a$, $b$ and $c$, while the **exterior angles** are labelled $d$, $e$ and $f$.

Imagine a person standing at one of the **vertices** (corners) and walking along the edges of the triangle in a clockwise direction until they are at the start again. At each vertex they would have turned through an angle equivalent to the exterior angle at that point. This shows that, during the complete journey, they would have turned through an angle equivalent to one complete turn, i.e. 360°.

# 37 ANGLE PROPERTIES

Therefore, $d° + e° + f° = 360°$.

It is also true that $a° + d° = 180°$ (angles on a straight line), $b° + e° = 180°$ and $c° + f° = 180°$.

Therefore, $a° + b° + c° + d° + e° + f° = 540°$

$a° + b° + c° + 360° = 540°$

$a° + b° + c° = 180°$

These findings lead us to two more important rules:
1  **The exterior angles of a triangle (indeed of any polygon) add up to 360°.**
2  **The interior angles of a triangle add up to 180°.**

By looking at the triangle again, it can now be stated that:

$a° + d° = 180°$

and also     $a° + b° + c° = 180°$

Therefore    $d° = b° + c°$

**The exterior angle of a triangle is equal to the sum of the opposite two interior angles.**

  **Exercise 5.13**   1  For each of the triangles below, use the information given to calculate the size of angle $x$:

a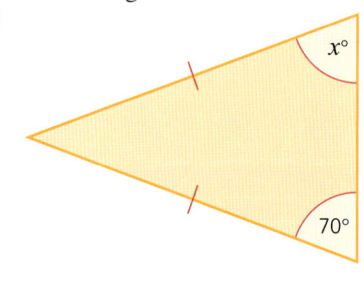

b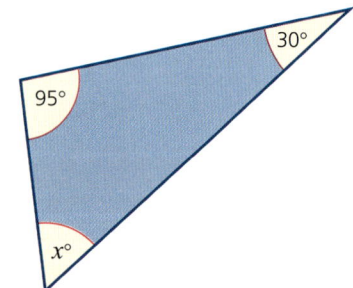

c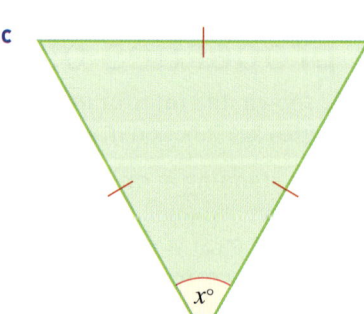

d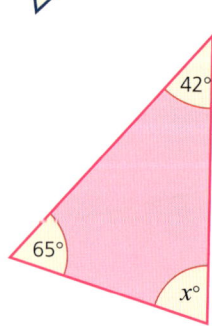

*Angle properties of polygons*

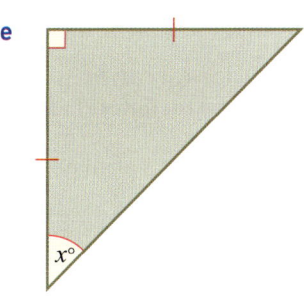
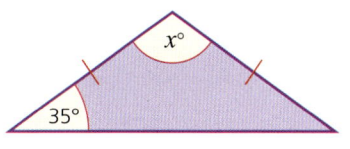

**2** In each of the diagrams below, calculate the size of the labelled angles:

**a**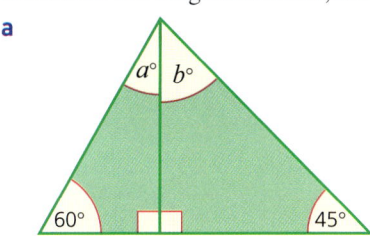

**b**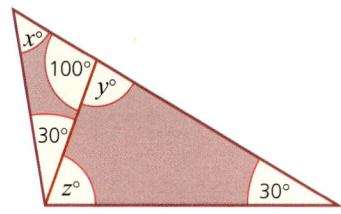

**c**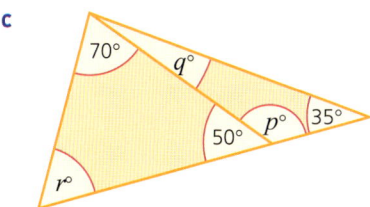

**d**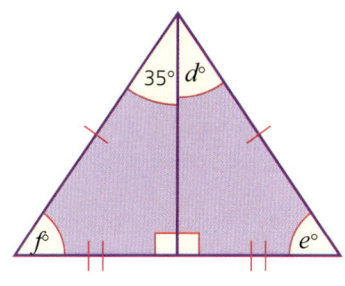

**e**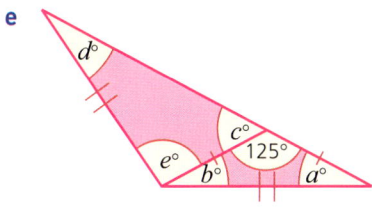

**f**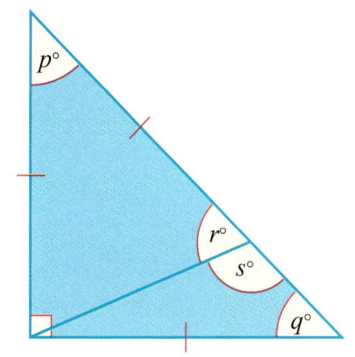

In the quadrilaterals below, a straight line is drawn from one of the vertices to the opposite vertex. The result is to split each quadrilateral into two triangles.

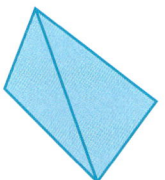  

# 37 ANGLE PROPERTIES

You already know that the sum of the angles in a triangle is 180°.

Therefore, as a quadrilateral can be split into two triangles, **the sum of the four angles of any quadrilateral must be 360°.**

### Exercise 5.14
In each of the diagrams below, calculate the size of the lettered angles.

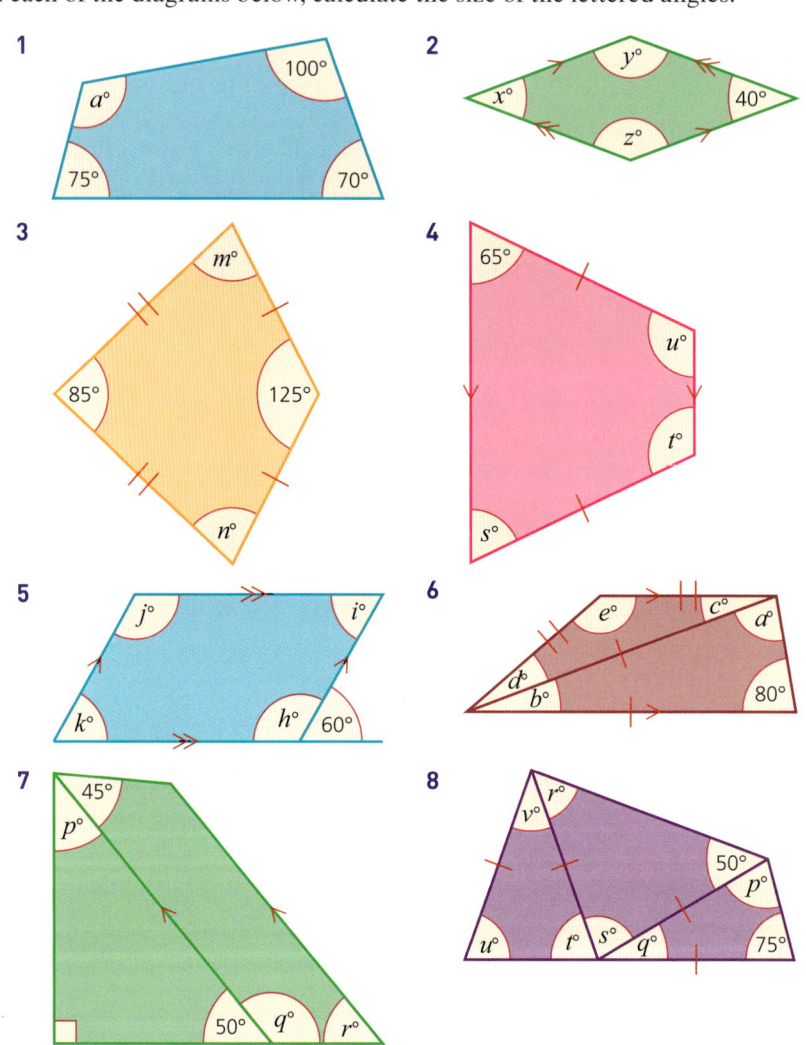

You already know that a polygon is a closed two-dimensional shape, bounded by straight lines. A **regular polygon** is distinctive in that all its sides are of equal length and all its angles are of equal size. Below are some examples of regular polygons.

*The sum of the interior angles of a polygon*

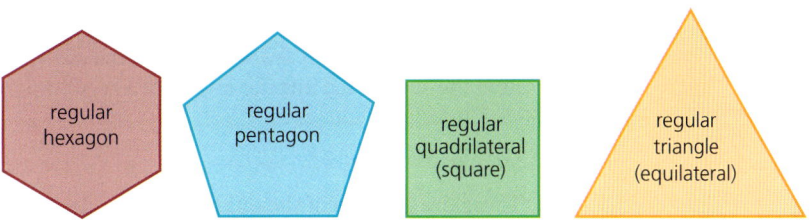

The name of each polygon is derived from the number of angles it contains. The following list identifies some of these polygons.

| | | |
|---|---|---|
| 3 angles | = | **tri**angle |
| 4 angles | = | **quad**rilateral (tetragon) |
| 5 angles | = | **pent**agon |
| 6 angles | = | **hex**agon |
| 7 angles | = | **hept**agon (septagon) |
| 8 angles | = | **oct**agon |
| 9 angles | = | **non**agon |
| 10 angles | = | **dec**agon |
| 12 angles | = | **dodec**agon |

# The sum of the interior angles of a polygon

In the polygons below, a straight line is drawn from each vertex to vertex A.

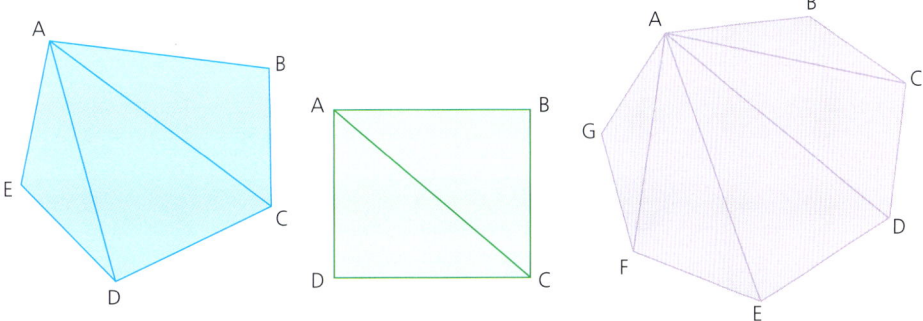

(Note: the above shapes are **irregular** polygons since their sides are not of equal length.)

As can be seen, the number of triangles is always two less than the number of sides the polygon has, i.e. if there are $n$ sides, then there will be $(n-2)$ triangles.

Since the angles of a triangle add up to 180°, the sum of the interior angles of a polygon is therefore $180(n-2)$ degrees.

331

# 37 ANGLE PROPERTIES

## → Worked example

Find the sum of the interior angles of a regular pentagon and hence the size of each interior angle.

For a pentagon, $n = 5$.

Therefore the sum of the interior angles $= 180(5 - 2)°$

$$= 180 \times 3$$
$$= 540°$$

For a regular pentagon the interior angles are of equal size.

Therefore each angle is $\frac{540}{5} = 108°$.

## The sum of the exterior angles of a polygon

The angles marked $a$, $b$, $c$, $d$, $e$ and $f$ represent the exterior angles of the regular hexagon drawn. As you have already found, the sum of the exterior angles of any polygon is 360°.

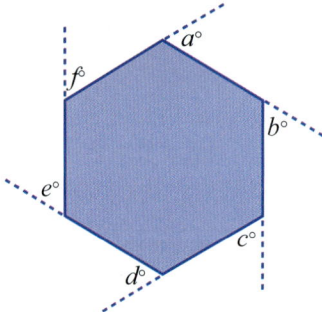

If the polygon is regular and has $n$ sides, then each exterior angle is $\frac{360°}{n}$.

## → Worked example

a  Find the size of an exterior angle of a regular nine-sided polygon.

$$\frac{360}{9} = 40°$$

b  Calculate the number of sides a regular polygon has if each exterior angle is 15°.

$$n = \frac{360}{15}$$
$$= 24$$

The polygon has 24 sides.

## The sum of the exterior angles of a polygon

**Exercise 5.15**

1. Find the sum of the interior angles of the following polygons:
   a. a hexagon    b. a nine-sided polygon    c. a seven-sided polygon
2. Find the value of each interior angle of the following regular polygons:
   a. an octagon
   b. a square
   c. a decagon
   d. a 12-sided polygon
3. Find the size of each exterior angle of the following regular polygons:
   a. a pentagon    b. a 12-sided polygon    c. a seven-sided polygon
4. The exterior angles of regular polygons are given below. In each case, calculate the number of sides the polygon has.
   a. 20°        b. 36°        c. 10°
   d. 45°        e. 18°        f. 3°
5. The interior angles of regular polygons are given below. In each case, calculate the number of sides the polygon has.
   a. 108°       b. 150°       c. 162°
   d. 156°       e. 171°       f. 179°
6. Calculate the number of sides a regular polygon has if an interior angle is five times the size of an exterior angle.
7. Copy and complete the table below for regular polygons:

| Number of sides | Name | Sum of exterior angles | Size of an exterior angle | Sum of interior angles | Size of an interior angle |
|---|---|---|---|---|---|
| 3 | | | | | |
| 4 | | | | | |
| 5 | | | | | |
| 6 | | | | | |
| 7 | | | | | |
| 8 | | | | | |
| 9 | | | | | |
| 10 | | | | | |
| 12 | | | | | |

8. An irregular polygon has six sides.
   Its interior angles are $x°, 2x°, 3x°, 3x°, 4x°$ and $174°$.
   a. Calculate the size of each of the five unknown interior angles.
   b. Calculate the size of each of the corresponding six exterior angles.
9. The diagram shows an irregular pentagon ABCDE.
   Although the sides are all the same length, the angles are not all the same size.
   Calculate the size of angle CAB.

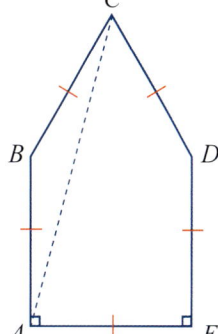

# 38 Similarity

## Similar shapes

Two polygons are said to be **similar** if a) they are equiangular and b) the corresponding sides are in proportion.

For triangles, being equiangular implies that corresponding sides are in proportion. The converse is also true.

In the diagrams below, $\triangle ABC$ and $\triangle PQR$ are similar.

For similar figures the ratios of the lengths of the sides are the same and represent the **scale factor**, i.e.

$\frac{p}{a} = \frac{q}{b} = \frac{r}{c} = k$ (where $k$ is the scale factor of enlargement)

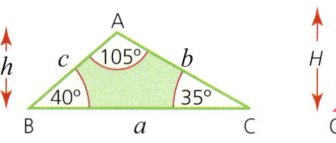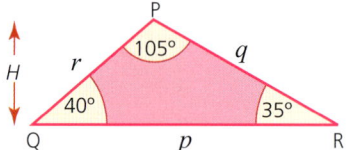

The heights of similar triangles are proportional also:

$\frac{H}{h} = \frac{p}{a} = \frac{q}{b} = \frac{r}{c} = k$

The ratio of the areas of similar triangles (the **area factor**) is equal to the square of the scale factor.

$\frac{\text{Area of } \triangle PQR}{\text{Area of } \triangle ABC} = \frac{\frac{1}{2} H \times p}{\frac{1}{2} h \times a} = \frac{H}{h} \times \frac{p}{a} = k \times k = k^2$

**Exercise 5.16**

1 a Explain why the two triangles below are similar.

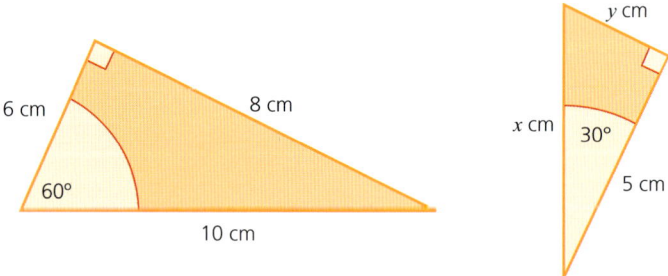

b Calculate the scale factor which reduces the larger triangle to the smaller one.
c Calculate the value of $x$ and the value of $y$.

*Similar shapes*

**2** Match up the triangles that are similar to each other.

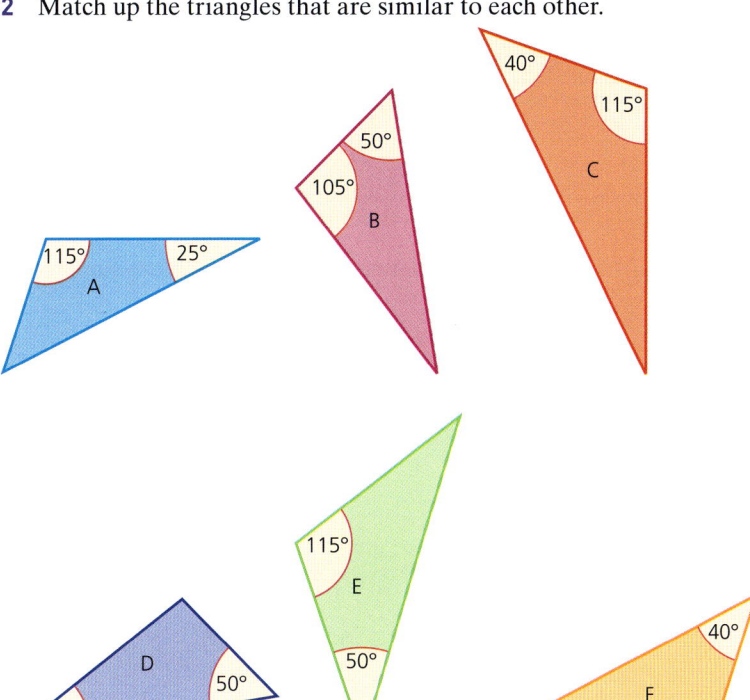

**3** The triangles below are similar.

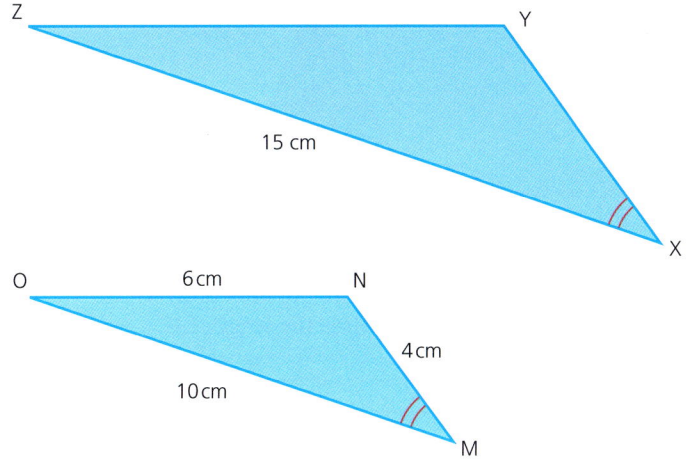

**a** Calculate the length XY.
**b** Calculate the length YZ.

# 38 SIMILARITY

**Exercise 5.16 (cont)**

4  In the triangle below, calculate the lengths of sides $p$, $q$ and $r$.

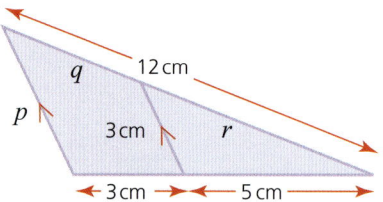

5  The triangles PQR and LMN are similar.

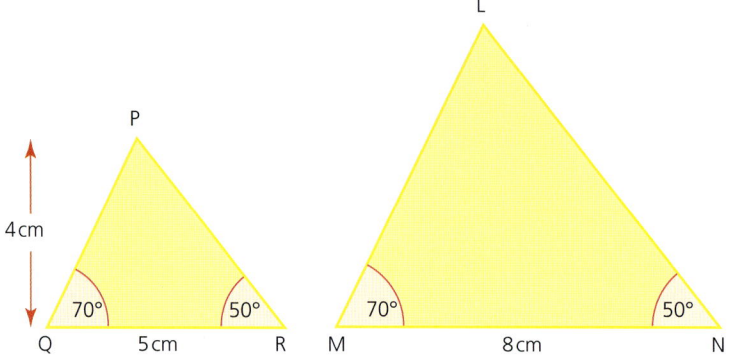

Calculate:
a  the area of △PQR
b  the scale factor of enlargement
c  the area of △LMN.

6  The triangle ADE shown below has an area of $12\,\text{cm}^2$.

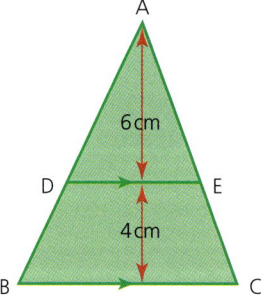

a  Calculate the area of △ABC.
b  Calculate the length BC.

Similar shapes

7 The following parallelograms are similar.

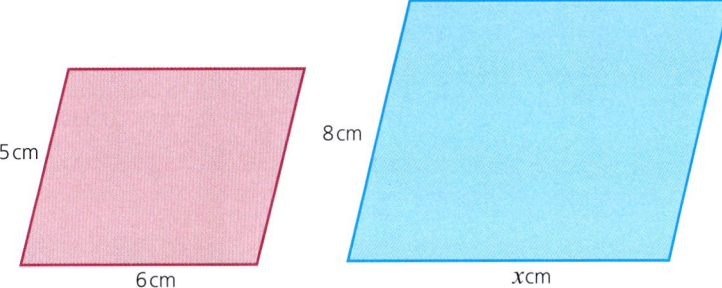

Calculate the length of the side marked $x$.

8 The diagram below shows two rhombuses.

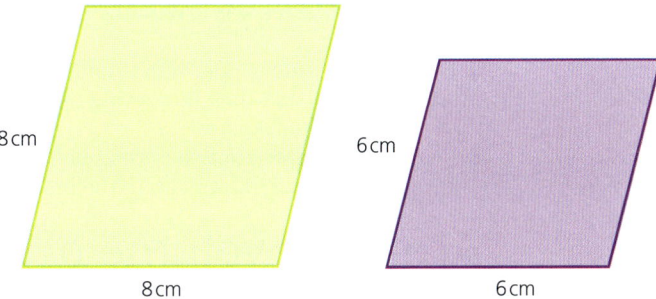

Explain, giving reasons, whether the two rhombuses are definitely similar.

9 The diagram below shows a trapezium within a trapezium. Explain, giving reasons, whether the two trapeziums are definitely similar.

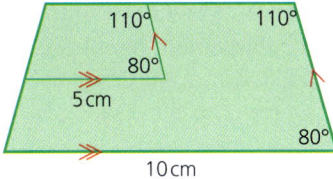

## Exercise 5.17

1 In the hexagons shown, hexagon B is an enlargement of hexagon A by a scale factor of 2.5.

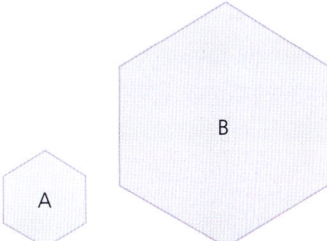

If the area of A is $8\,cm^2$, calculate the area of B.

2 P and Q are two regular pentagons. Q is an enlargement of P by a scale factor of 3. If the area of pentagon Q is $90\,cm^2$, calculate the area of P.

# 38 SIMILARITY

**Exercise 5.17 (cont)**

3 Below is a row of four triangles A, B, C and D. Each is an enlargement of the previous one by a scale factor of 1.5.
   a If the area of C is 202.5 cm², calculate the area of:
      i  triangle D       ii  triangle B       iii  triangle A.
   b If the triangles were to continue in this sequence, which letter triangle would be the first to have an area greater than 15 000 cm²?

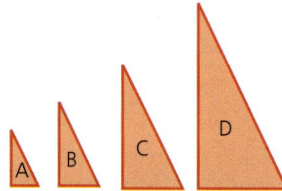

4 A square is enlarged by increasing the length of its sides by 10%. If the length of its sides was originally 6 cm, calculate the area of the enlarged square.

5 A square of side length 4 cm is enlarged by increasing the lengths of its sides by 25% and then increasing them by a further 50%. Calculate the area of the final square.

6 An equilateral triangle has an area of 25 cm². If the lengths of its sides are reduced by 15%, calculate the area of the reduced triangle.

## Area and volume of similar shapes

Earlier in the topic you found the following relationship between the scale factor and the area factor of enlargement:

Area factor = (scale factor)²

A similar relationship can be stated for volumes of similar shapes:

Consider a cuboid $c$ with dimensions $l$, $w$ and $h$ as shown:

Volume of cuboid $c$ is $l \times w \times h$

A cuboid $C$ is an enlargement of cuboid $c$ by a scale factor $k$. Its dimensions are therefore as shown:

Volume of cuboid $C$ is $kl \times kw \times kh = k^3 lwh$

The ratio of the volume of cuboid $C$ to cuboid $c$ is $\frac{k^3 lwh}{lwh} = k^3$

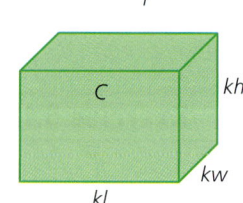

The volume of the enlarged cuboid is therefore a multiple of the original volume by a factor of $k^3$.

Volume factor = (scale factor)³

**Exercise 5.18**

1 A cuboid has dimensions as shown:

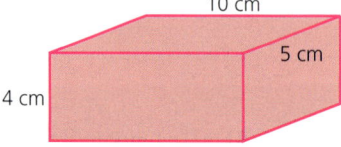

If the cuboid is enlarged by a scale factor of 2.5, calculate:
   a the total surface area of the original cuboid
   b the total surface area of the enlarged cuboid

c the volume of the original cuboid
   d the volume of the enlarged cuboid.
2 A cube has side length 3 cm.
   a Calculate its total surface area.
   b If the cube is enlarged and has a total surface area of 486 cm², calculate the scale factor of enlargement.
   c Calculate the volume of the enlarged cube.
3 Two cubes P and Q are of different sizes. If $n$ is the ratio of their corresponding sides, write in terms of $n$:
   a the ratio of their surface areas
   b the ratio of their volumes.
4 The cuboids A and B shown below are similar.

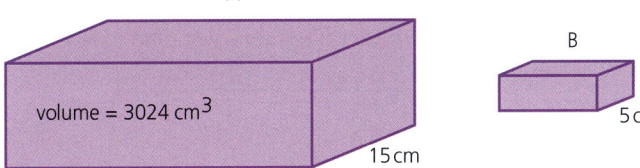

Calculate the volume of cuboid B.

5 Two similar troughs X and Y are shown below.

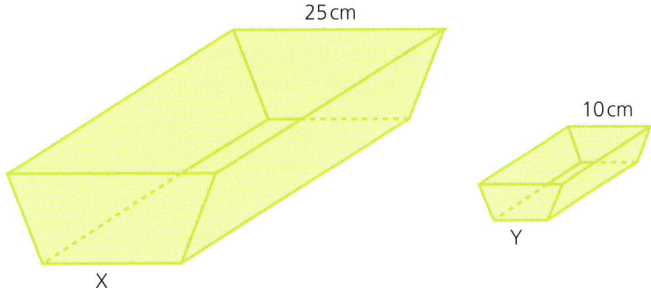

If the capacity of X is 10 litres, calculate the capacity of Y.

## Exercise 5.19

1 The two cylinders L and M shown below are similar.

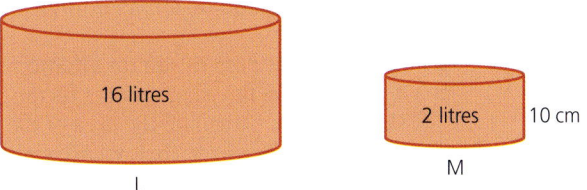

If the height of cylinder M is 10 cm, calculate the height of cylinder L.

2 A square-based pyramid is cut into two shapes by a cut running parallel to the base and made halfway up as shown.
   a Calculate the ratio of the volume of the smaller pyramid to that of the original one.
   b Calculate the ratio of the volume of the small pyramid to that of the remaining larger solid.

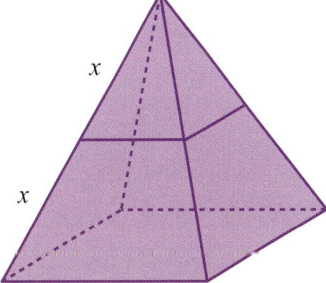

## 38 SIMILARITY

**Exercise 5.19 (cont)**

3  The two cones A and B shown below are similar. Cone B is an enlargement of A by a scale factor of 4.

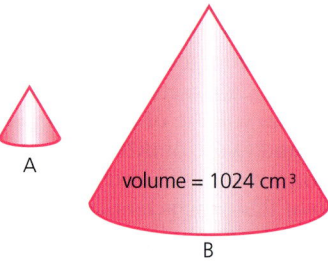

If the volume of cone B is 1024 cm³, calculate the volume of cone A.

4  A cylinder is split in two. One part is shaded, the other is not.

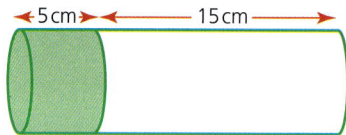

  a  Stating your reasons clearly, decide whether the two cylinders shown above are similar.
  b  What is the ratio of the curved surface area of the shaded cylinder to that of the unshaded cylinder?

5  The diagram (below) shows a triangle.

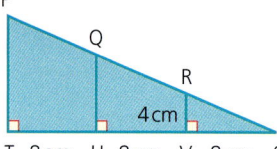

  a  Calculate the area of $\triangle RSV$.
  b  Calculate the area of $\triangle QSU$.
  c  Calculate the area of $\triangle PST$.

6  The area of an island on a map is 30 cm². The scale used on the map is 1 : 100 000.
  a  Calculate the area in square kilometres of the real island.
  b  An airport on the island is on a rectangular piece of land measuring 3 km by 2 km. Calculate the area of the airport on the map in cm².

7  The two blocks X and Y (below) are similar.
  The total surface area of block Y is four times that of block X.

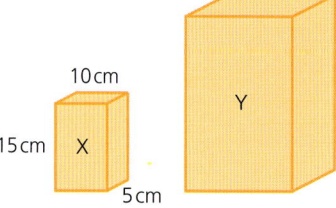

Calculate:
  a  the dimensions of block Y
  b  the mass of block X if block Y has a mass of 800 g.

# 39 Properties of circles

You will already be familiar with the terms used to describe aspects of the circle shown in the diagram.

*The circumference is the perimeter all the way around the circle. An arc is part of the circumference. If an arc is less than half the circumference it is called a **minor arc**. If the arc is more than half the circumference it is called a **major arc**.*

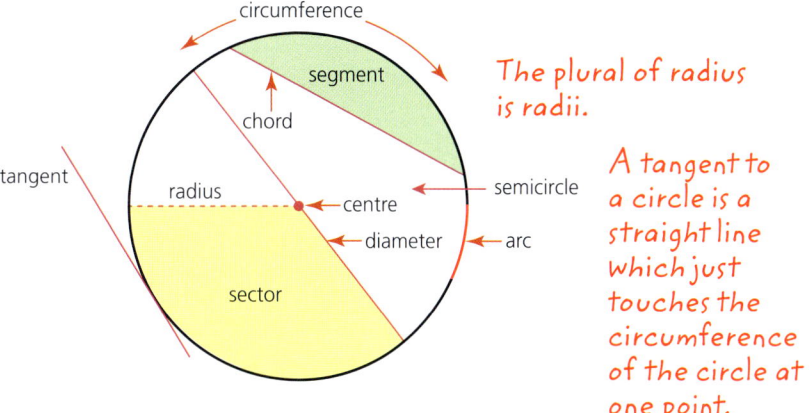

*The plural of radius is radii.*

*A tangent to a circle is a straight line which just touches the circumference of the circle at one point.*

## The angle in a semicircle

If AB represents the diameter of the circle, then angle ACB is 90°.

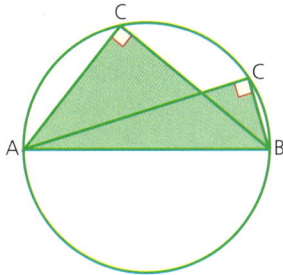

*This is a special case of the theorem on p.344 where the angle at the centre of a circle is twice the angle at the circumference on the same arc.*

 **Exercise 5.20**

In each of the following diagrams, O marks the centre of the circle. Calculate the value of $x$ in each case. Give reasons for your answers.

1  
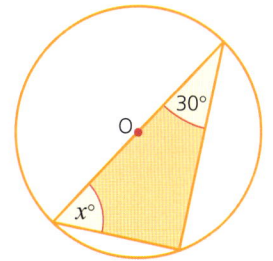

2  
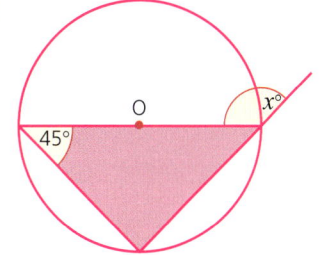

341

# 39 PROPERTIES OF CIRCLES

**Exercise 5.20 (cont)**

3, 4, 5, 6 [diagrams with circles showing angles: 110°, 58°, 20° and x° to find]

## The angle between a tangent and a radius of a circle

A tangent to a circle is a straight line or plane that touches the circumference of the circle at one point only, but does not cross it at that point.

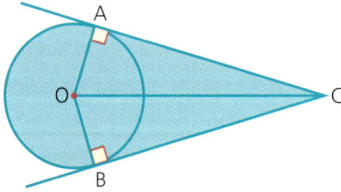

The angle between a tangent at a point and the radius to the same point on the circle is a right angle.

Triangles OAC and OBC are congruent as angle OAC and angle OBC are right angles, OA = OB because they are both radii and OC is common to both triangles.

*If two triangles are congruent, it means that they are exactly the same shape and size.*

*The angle between a tangent and a radius of a circle*

**Exercise 5.21** In each of the following diagrams, O marks the centre of the circle. Calculate the value of *x* in each case.

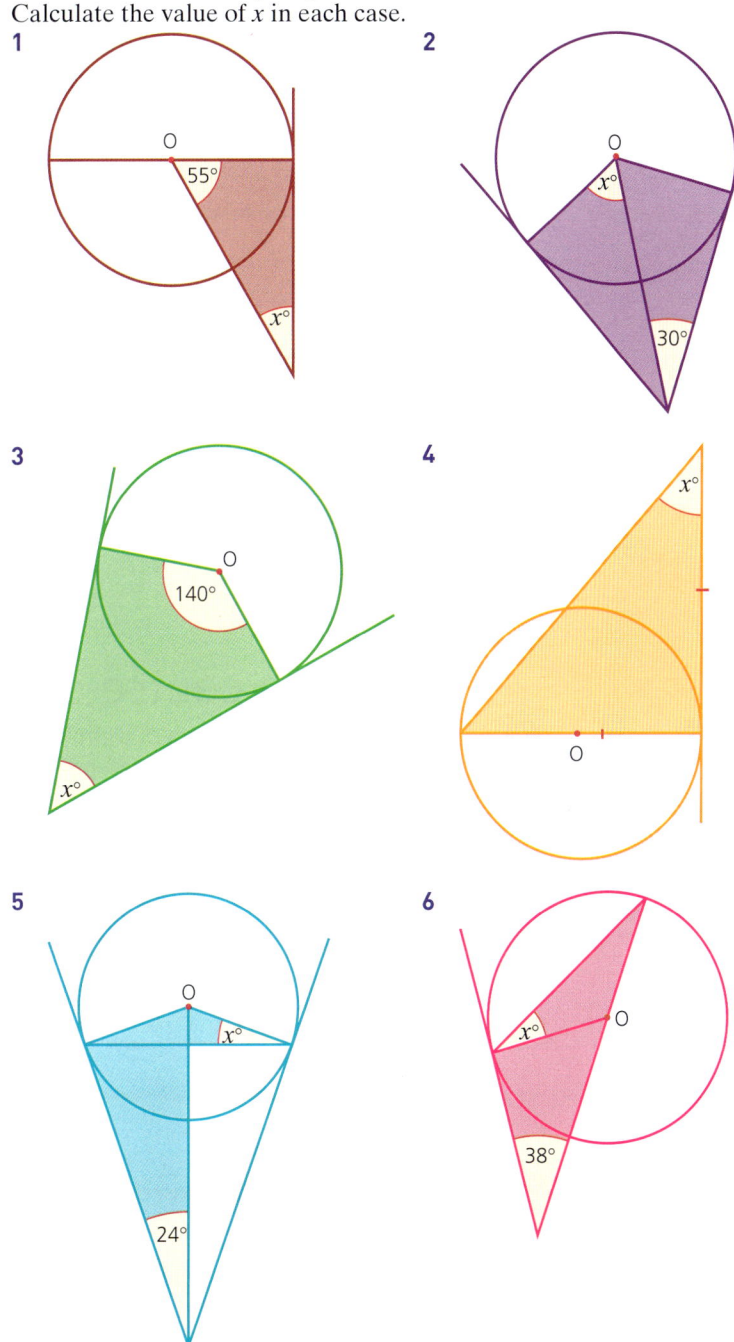

# 39 PROPERTIES OF CIRCLES

**Exercise 5.21 (cont)**

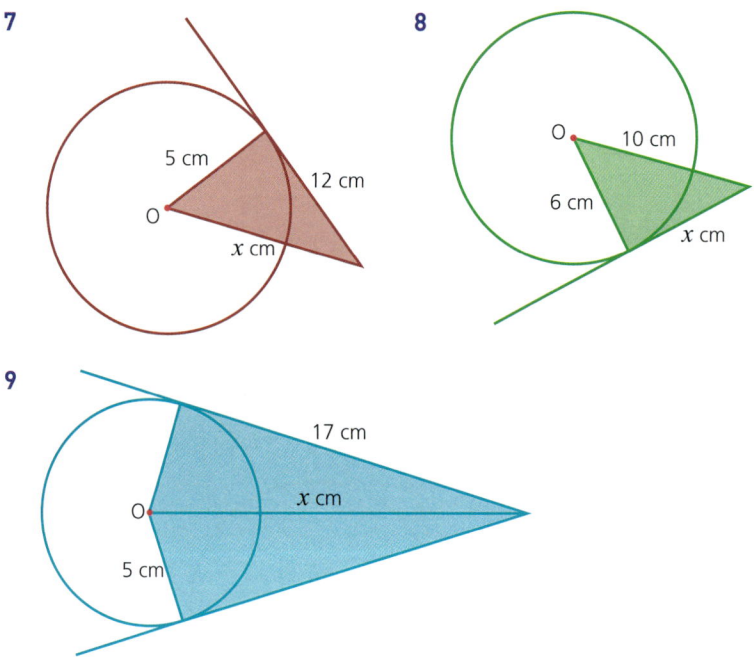

## Angle at the centre of a circle

The angle on an arc at the centre of a circle is twice the size of the angle made by the same arc at the circumference of the circle.

Both diagrams below illustrate this theorem.

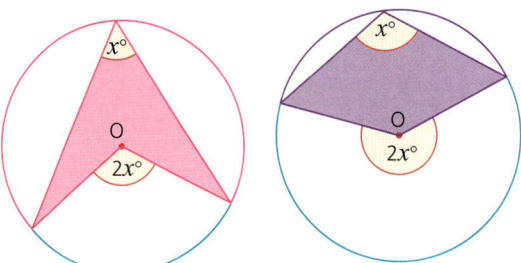

*The angle in a semicircle theorem on p. 341 is a specific case of this theorem. The angle at the centre is 180° and the angle in the semicircle on is 90°.*

This can be proved as follows:

The two triangles shown are isosceles as two of their sides are radii of the circle.

$2x + w = 180°$ (angles of a triangle sum to 180°)

$2y + z = 180°$ (angles of a triangle sum to 180°)

Therefore $w = 180 - 2x$ and $z = 180 - 2y$

However $a + w + z = 360°$ (angles around a point sum to 360°)

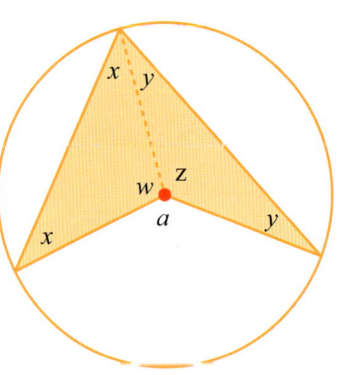

Therefore $a + (180 - 2x) + (180 - 2y) = 360$
$a + 360 - 2x - 2y = 360$
so $a = 2x + 2y$
$a = 2(x + y)$

As can be seen from the diagram, $x+y$ is also the angle at the circumference, therefore **the angle at the centre is twice the angle at the circumference on the same arc.**

### Exercise 5.22

In each of the following diagrams, O marks the centre of the circle. Calculate the size of the lettered angles.

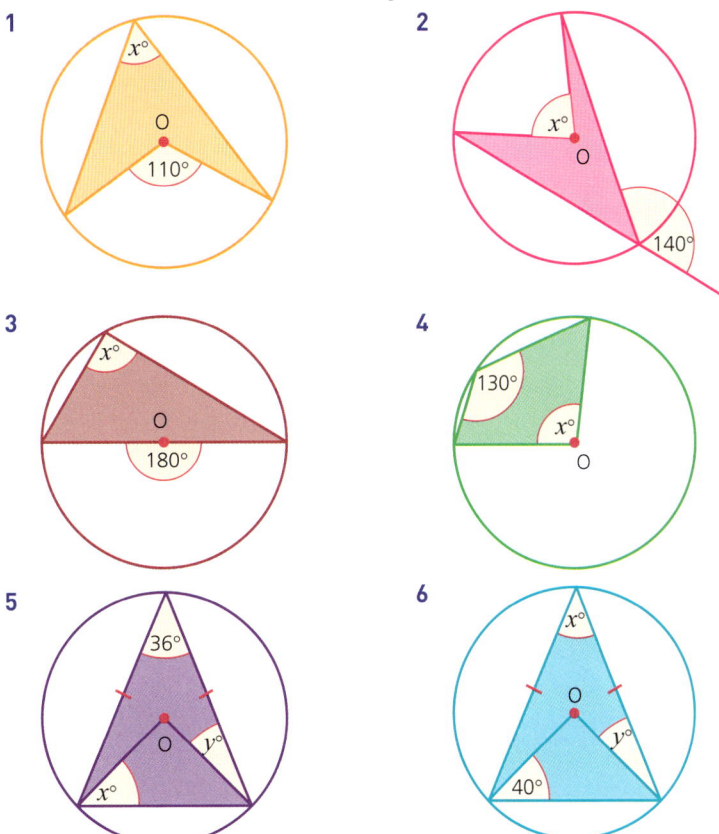

## Angles in the same segment

Angles in the same segment of a circle are equal.

This can be explained by using the theorem that the angle on the arc at the centre is twice the angle on the circumference. Looking at the diagram (right), if the angle at the centre is $2x°$, then each of the angles at the circumference must be equal to $x°$.

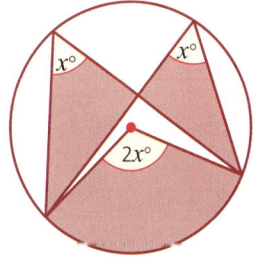

# 39 PROPERTIES OF CIRCLES

**Exercise 5.23** Calculate the lettered angles in the following diagrams:

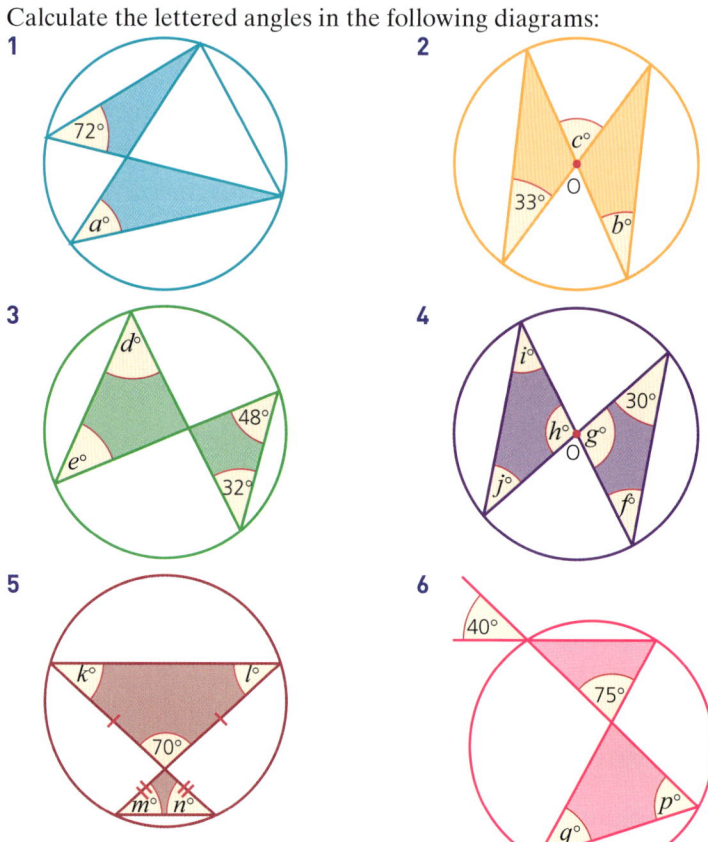

## Opposite angles of a cyclic quadrilateral

A **cyclic quadrilateral** is a quadrilateral whose vertices all lie on the circumference of a circle:

If the angle at B is $x°$ then the angle at the centre on the same arc must be $2x°$. If the angle at D is $y°$ then the angle at the centre on the same arc must be $2y°$.

But $2x + 2y = 360°$ (angles around a point sum to 360°) i.e. $2(x + y) = 360°$

Therefore $x + y = 180°$

So **the opposite angles of the cyclic quadrilateral x and y sum to 180°**.

## Angles in opposite segments

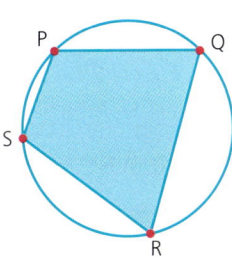
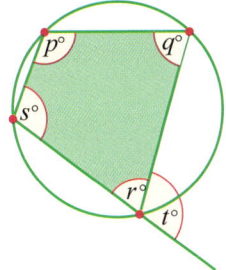

Points P, Q, R and S all lie on the circumference of the circle above. They are called **concyclic points**. Joining the points P, Q, R and S produces a cyclic quadrilateral.

The opposite angles are **supplementary**, i.e. they add up to 180°.

Since $p° + r° = 180°$ (supplementary angles) and $r° + t° = 180°$ (angles on a straight line), it follows that $p° = t°$.

Therefore the exterior angle of a cyclic quadrilateral is equal to the interior opposite angle.

## Alternate segment theorem

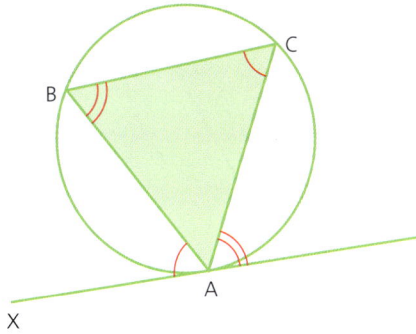

XY is a tangent to the circle at point A. AB is a chord of the circle.

The alternate segment theorem states that the angle between a tangent and a chord at their point of contact is equal to the angle in the alternate segment.

i.e. angle BAX = angle ACB; similarly, angle CAY = angle ABC

This can be proved as follows:

Point O is the centre of the circle. Drawing radii to points A, B and C creates isosceles triangles ∆AOC, ∆BOC and ∆AOB.

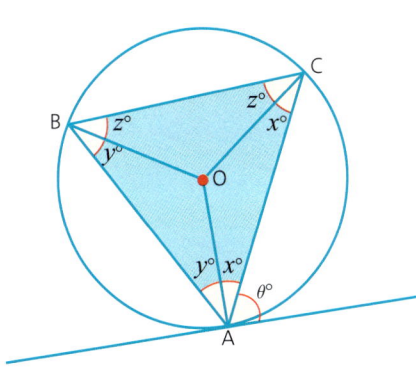

The angle $\theta$ is the angle between the chord AC and the tangent to the circle at A.

$\theta = 90 - x$, as the angle between a radius and a tangent to a circle at a point is a right angle.

## 39 PROPERTIES OF CIRCLES

In △ABC, $(x + y) + (y + z) + (x + z) = 180°$

Therefore $\quad 2x + 2y + 2z = 180°$

$\quad\quad\quad\quad\quad\quad 2(x + y + z) = 180°$

$\quad\quad\quad\quad\quad\quad x + y + z = 90$

If $x + y + z = 90$ then $y + z = 90 - x$

As angle ABC = $y + z$, then angle ABC = $90 - x$

As $\theta = 90 - x$, then angle ABC = $\theta$

**Exercise 5.24** Calculate the size of the lettered angles in each of the following:

1

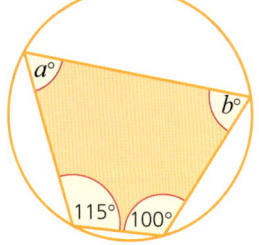

2

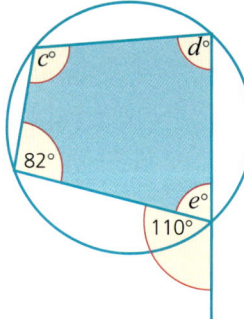

3

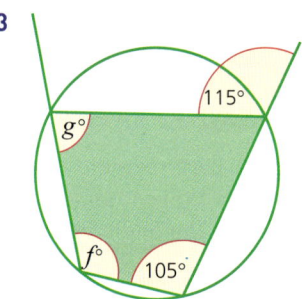

4

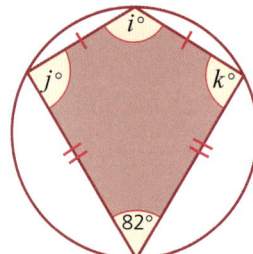

5

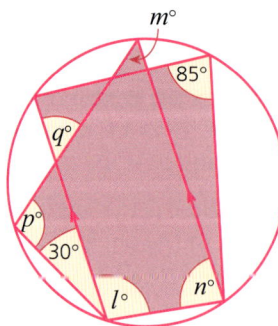

6

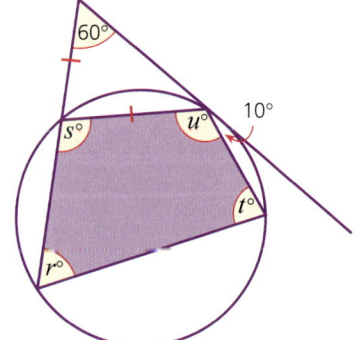

**7**

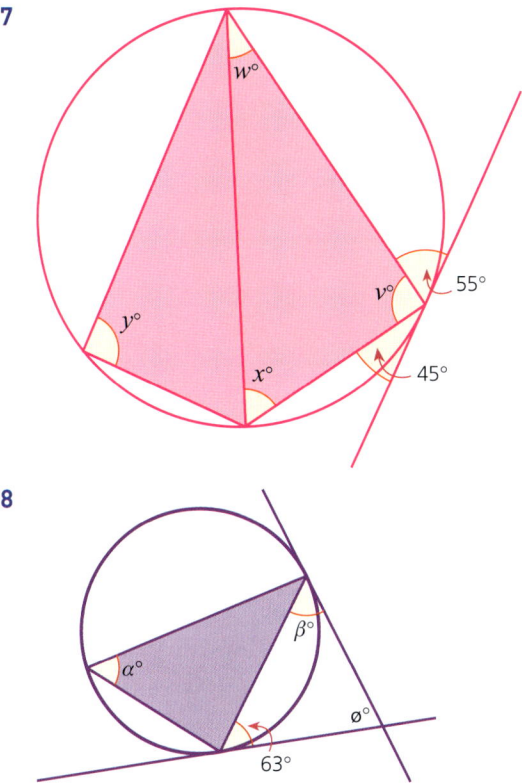

**8**

# Equal chords and perpendicular bisectors

If chords AB and XY are of equal length, then, since OA, OB, OX and OY are radii, the triangles OAB and OXY are congruent isosceles triangles. It follows that:
- the section of a line of symmetry OM through △OAB is the same length as the section of a line of symmetry ON through △OXY
- OM and ON are **perpendicular bisectors** of AB and XY respectively.

*The perpendicular bisector of a line segment AB is the line that intersects the line segment AB at a right angle and cuts it into two equal parts.*

# 39 PROPERTIES OF CIRCLES

**Exercise 5.25**

1  In the diagram below, O is the centre of the circle, PQ and RS are chords of equal length and M and N are their respective midpoints.

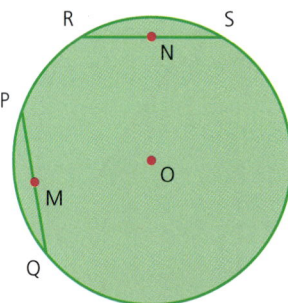

   a  What kind of triangle is ΔPOQ?
   b  Describe the line ON in relation to RS.
   c  If angle POQ is 80°, calculate angle OQP.
   d  Calculate angle ORS.

2  In the diagram below, O is the centre of the circle. AB and CD are equal chords and the points R and S are their midpoints respectively.

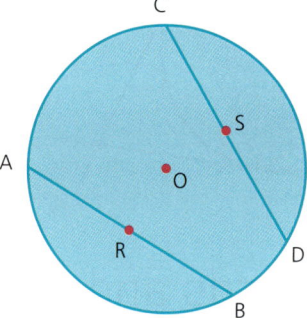

Identify which of these statements are true and which are false, giving reasons for your answers.
   a  angle COD = 2 × angle AOR
   b  OR = OS
   c  If angle ROB is 60°, then ΔAOB is equilateral.
   d  OR and OS are perpendicular bisectors of AB and CD respectively.

3  Using the diagram (right) identify which of the following statements are true and which are false, giving reasons for your answers.
   a  If ΔVOW and ΔTOU are isosceles triangles, then T, U, V and W would all lie on the circumference of a circle with its centre at O.
   b  If ΔVOW and ΔTOU are congruent isosceles triangles, then T, U, V and W would all lie on the circumference of a circle with its centre at O.

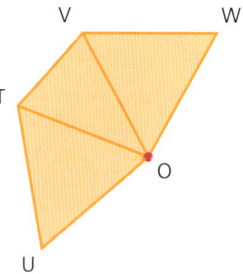

# Tangents from an external point

Triangles OAC and OBC are congruent since angle OAC and angle OBC are right angles, OA = OB because they are both radii, and OC is common to both triangles. Hence AC = BC.

Therefore, tangents being drawn to the same circle from an external point are equal in length.

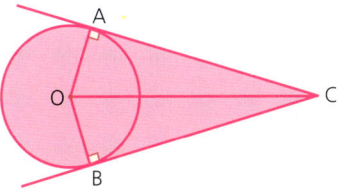

**Exercise 5.26**

1  Copy each of the diagrams below and calculate the size of the angle marked $x°$ in each case. Assume that the lines drawn from points on the circumference are tangents.

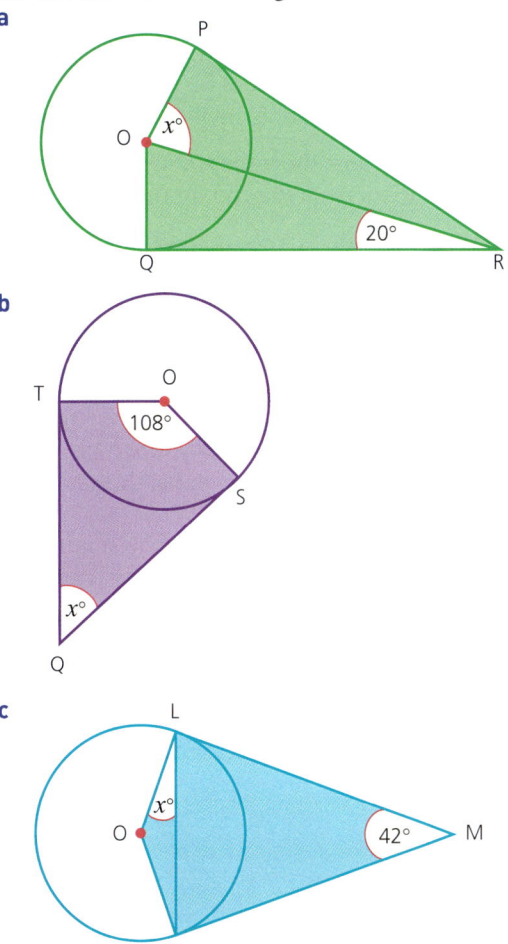

# Investigations, modelling and ICT 5

## Investigation: Towers of Hanoi

This investigation is based on an old Vietnamese legend. The legend is as follows:

At the beginning of time a temple was created by the gods. Inside the temple stood three giant rods. On one of these rods, 64 gold discs, all of different diameters, were stacked in descending order of size, i.e. the largest at the bottom rising to the smallest at the top. Priests at the temple were responsible for moving the discs onto the remaining two rods until all 64 discs were stacked in the same order on one of the other rods. When this task was completed, time would cease and the world would come to an end.

The discs, however, could only be moved according to certain rules. These were:
- Only one disc could be moved at a time.
- A disc could only be placed on top of a larger one.

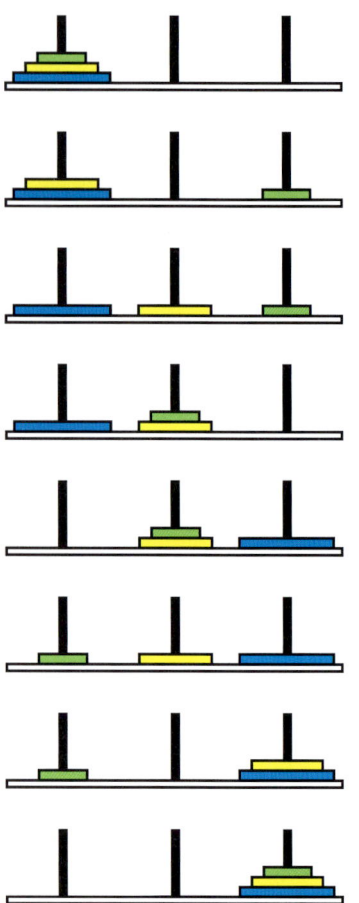

The diagrams show the smallest number of moves required to transfer three discs from the rod on the left to the rod on the right.

With three discs, the smallest number of moves is seven.

1. What is the smallest number of moves needed for two discs?
2. What is the smallest number of moves needed for four discs?
3. Investigate the smallest number of moves needed to move different numbers of discs.
4. Display the results of your investigation in an ordered table.
5. Describe any patterns you see in your results.
6. Predict, from your results, the smallest number of moves needed to move 10 discs.
7. Determine a formula for the smallest number of moves for $n$ discs.
8. Assume the priests have been transferring the discs at the rate of one per second and assume the Earth is approximately 4.54 billion years old ($4.54 \times 10^9$ years).
   According to the legend, is the world coming to an end soon? Explain your answer with relevant calculations.

## Investigation: Fountain borders

The Alhambra Palace in Granada, Spain, has many fountains, which pour water into pools. Many of the pools are surrounded by beautiful ceramic tiles. This investigation looks at the number of square tiles needed to surround a particular shape of pool.

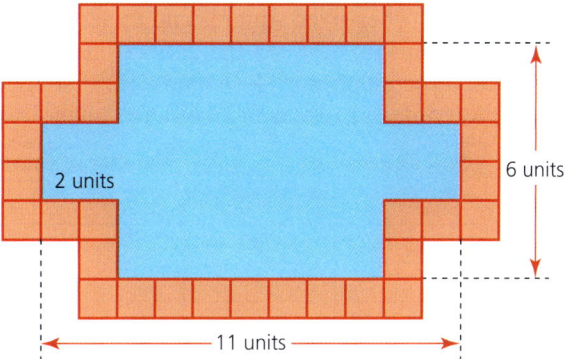

The diagram above shows a rectangular pool, 11 × 6 units, in which a square of dimension 2 × 2 units is taken from each corner.

The total number of unit square tiles needed to surround the pool is 38.

The shape of the pools can be generalised as shown below:

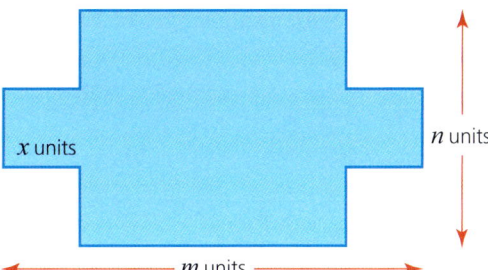

1. Investigate the number of unit square tiles needed for different-sized pools. Record your results in an ordered table.
2. From your results, write an algebraic rule in terms of $m$, $n$ and $x$ (if necessary) for the number of tiles $T$ needed to surround a pool.
3. Explain, in words and using diagrams, why your rule works.

## Investigation: Tiled walls

Many cultures have used tiles to decorate buildings. Putting tiles on a wall takes skill. These days, to make sure that each tile is in the correct position, spacers are used between the tiles.

# INVESTIGATIONS, MODELLING AND ICT 5

You can see from the diagram that there are + shaped and T shaped spacers.

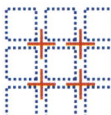 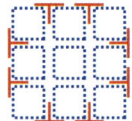

1. Draw other-sized squares and rectangles and investigate the relationship between the dimensions of the shape (length and width) and the number of + shaped and T shaped spacers.
2. Record your results in an ordered table similar to the one shown here.

| Wall dimensions | | Number of spacers | |
|---|---|---|---|
| Length | Width | + | T |
| 3 | 3 | 4 | 8 |
| 2 | 3 | | |
| | | | |

3. Write an algebraic rule for the number of + shaped spacers $c$ in a rectangle $l$ tiles long by $w$ tiles wide.
4. Write an algebraic rule for the number of T shaped spacers $t$ in a rectangle $l$ tiles long by $w$ tiles wide.

## Modelling: Gift wrap

A net of a cube with a side length of 10 cm is shown.

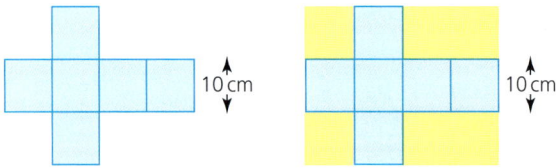

In order to cover the assembled cube with wrapping paper, the net is laid on top of a rectangular sheet of paper as shown.
1. What is the area of the sheet of paper?
2. What is the area of the rectangular sheet of paper if the assembled cube has a side length of $x$ cm?

A different net can also be assembled to form a cube. The shape of the rectangle needed to cover it is also different, as shown.

3. Calculate the rectangular area of paper needed for a cube of side length $x$ cm assembled from this type of net.

## Modelling: Gift wrap

A cuboid has dimensions $p \times p \times r$.

**4** Copy and complete the diagram to show a possible net of the cuboid.

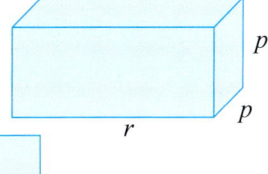

**5** Show that the area ($A$) of a rectangular sheet of paper needed to wrap the cuboid is given by the formula $A = 6p(p + r)$.

**6** Given that $p \leq r \leq 4p$, calculate, in terms of $p$, the maximum and minimum area of paper needed.

**7** Draw axes similar to those below, with $p$ on the $x$-axis and $A$ on the $y$-axis.

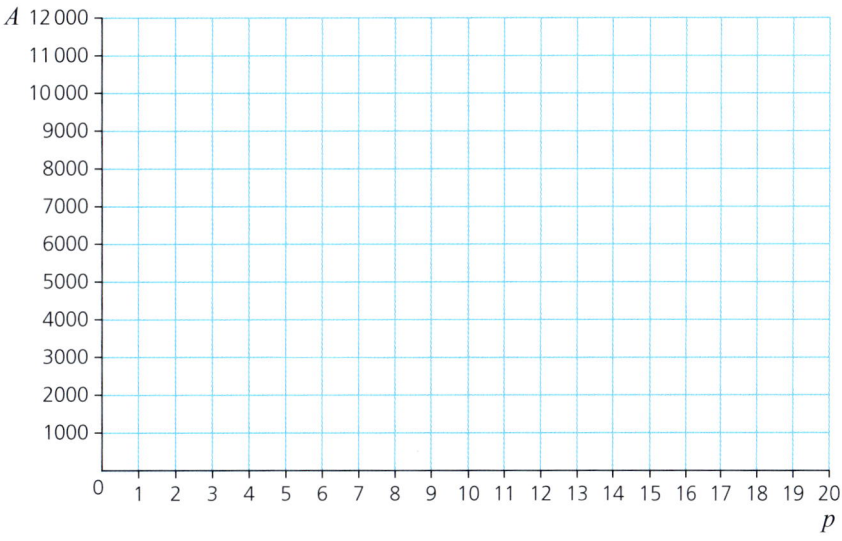

  **a** Using your graphic display calculator to help, sketch a graph of the function for the minimum area on the axes.
  **b** Using your graphic display calculator again to help, sketch on the same axes the graph of the function for the maximum area.

**8** Explain, in the context of the problem, the significance of the region between the two graphs.

**9** A wrapping paper manufacturer makes rectangular sheets of paper for these cuboids, with an area in the range $2000\,\text{cm}^2 \leq A \leq 4000\,\text{cm}^2$.

Using the graphs on your graphic display calculator, identify a value of $p$ that satisfies this area range. State the value of $p$ and the area ($A$) of paper used.

**10** Use your answer to Q.9 above to deduce the dimension ($r$) of the cuboid.

# INVESTIGATIONS, MODELLING AND ICT 5

## ICT Activity 1

In this activity you should use a dynamic geometry package to demonstrate that for the triangle shown:

$$\frac{AB}{ED} = \frac{AC}{EC} = \frac{BC}{DC}$$

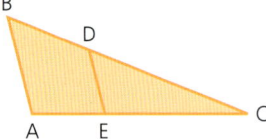

1. **a** Using the geometry package, construct the triangle ABC.
   **b** Construct the line segment ED such that it is parallel to AB.
   **c** Using a 'measurement' tool, measure each of the lengths AB, AC, BC, ED, EC and DC.
   **d** Using a 'calculator' tool, calculate the ratio $\frac{AB}{ED}, \frac{AC}{EC}, \frac{BC}{DC}$
2. Comment on your answers to Q.1d.
3. **a** Grab vertex B and move it to a new position. What happens to the ratios you calculated in Q.1d?
   **b** Grab the vertices A and C in turn and move them to new positions. What happens to the ratios? Explain why this happens.
4. Grab point D and move it to a new position along the side BC. Explain, giving reasons, what happens to the ratios.

## ICT Activity 2

Using a geometry package, demonstrate the following angle properties of a circle:

**i** The angle on an arc at the centre of a circle is twice the size of the angle made by the same arc at the circumference of the circle.
The diagram below demonstrates the construction that needs to be formed:

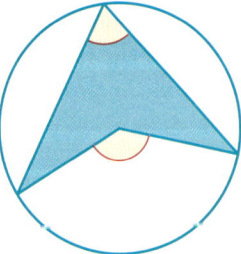

**ii** The angles in the same segment of a circle are equal.
**iii** The exterior angle of a cyclic quadrilateral is equal to the interior opposite angle.

# Student assessments 5

## Student assessment 1

1. A climber gets to the top of Mont Blanc. He can see in the distance a number of ski resorts. He uses his map to find the bearing and distance of the resorts, and records them as shown below.

   Val d'Isère 30 km       bearing 082°
   Les Arcs 40 km          bearing 135°
   La Plagne 45 km         bearing 205°
   Méribel 35 km           bearing 320°

   Choose an appropriate scale and draw a diagram to show the position of each resort. What are the distance and bearing of the following?
   a  Val d'Isère from La Plagne
   b  Méribel from Les Arcs

2. A coastal radar station picks up a distress call from a ship. It is 50 km away on a bearing of 345°. The radar station contacts a lifeboat at sea which is 20 km away on a bearing of 220°.
   Make a scale drawing and use it to find the distance and bearing of the ship from the lifeboat.

3. An aircraft is seen on radar at airport A. The aircraft is 210 km away from the airport on a bearing of 065°. The aircraft is diverted to airport B, which is 130 km away from A on a bearing of 215°. Choose an appropriate scale and make a scale drawing to find the distance and bearing of airport B from the aircraft.

## Student assessment 2

1. Cones M and N are similar.

   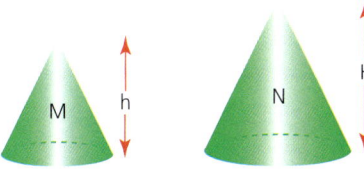

   a  Write the ratio of their surface areas in the form, area of M : area of N.
   b  Write the ratio of their volumes in the form, volume of M : volume of N.

2. Calculate the values of $x$, $y$ and $z$ in the triangle below.

   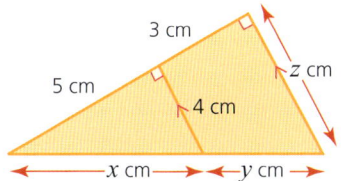

3. The tins A and B shown below are similar. The capacity of tin B is three times that of tin A. If the label on tin A has an area of 75 cm², calculate the area of the label on tin B.

4. The cube shown below is enlarged by a scale factor of 2.5.

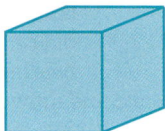

4 cm

   a. Calculate the volume of the enlarged cube.
   b. Calculate the surface area of the enlarged cube.

5. The two troughs X and Y shown below are similar.

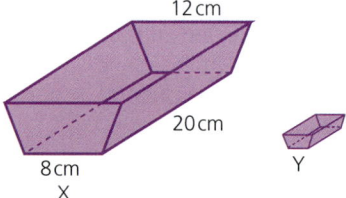

The scale factor of enlargement from Y to X is 4. If the capacity of trough X is 1200 cm³, calculate the capacity of trough Y.

6. The rectangular floor plan of a house measures 8 cm by 6 cm. If the scale of the plan is 1:50, calculate:
   a. the dimensions of the actual floor
   b. the area of the actual floor in m².

7. The volume of the cylinder shown below is 400 cm³.

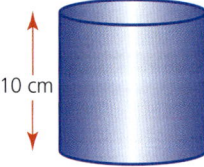

10 cm

Calculate the volume of a similar cylinder formed by enlarging the one shown by a scale factor of 2.

# Student assessments 5

NB: Diagrams are not drawn to scale.

## Student assessment 3

1. For the diagrams below, calculate the size of the labelled angles.

   a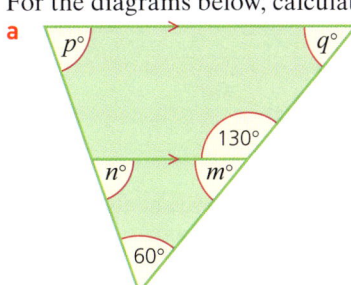
   
   b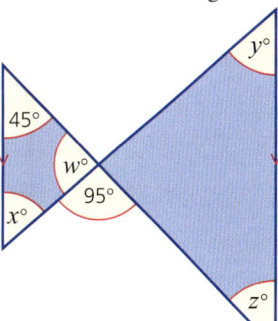
   
   c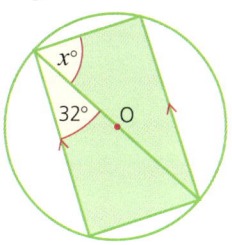

   In Q.2–5, O marks the centre of the circle. Calculate the size of the angle marked $x$ in each case.

2.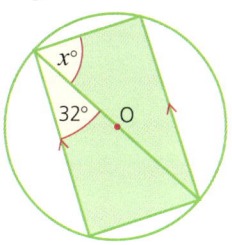

3.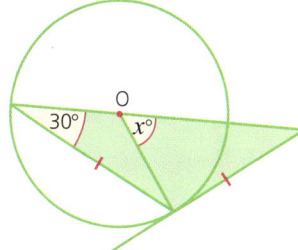

4.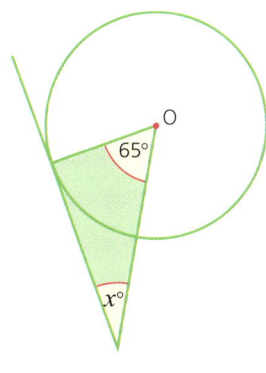

5.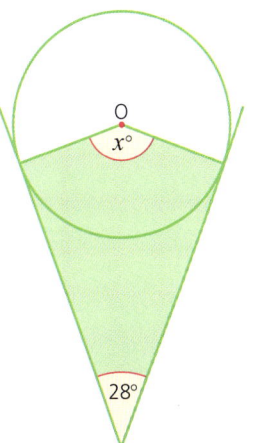

# STUDENT ASSESSMENTS 5

**6** If OA is a radius of the circle and PB the tangent to the circle at A, calculate angle ABO.

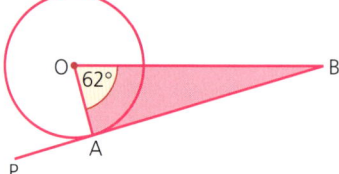

In Q.7–10, O marks the centre of the circle. Calculate the size of the angle marked $x$ in each case.

**7**

**8**

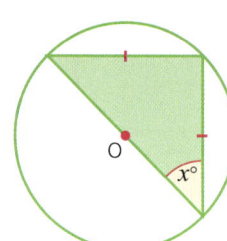

**9**

**10**

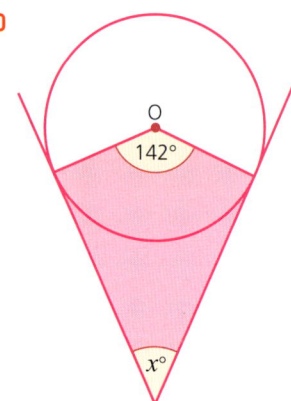

**11**

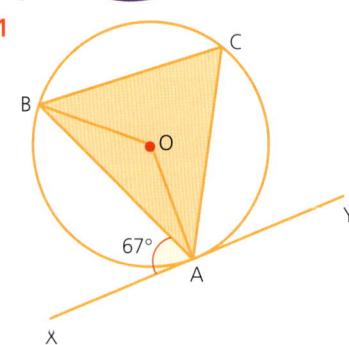

In the diagram above, XY is a tangent to the circle at A. O is the centre of the circle. Calculate each of the following angles, explaining each of your answers.
  **a**  angle ACB
  **b**  the reflex angle AOB
  **c**  angle ABO

## Student assessment 4

1   In the following diagrams, O is the centre of the circle. Identify which angles are:
    i   supplementary angles
    ii  right angles
    iii equal.

    a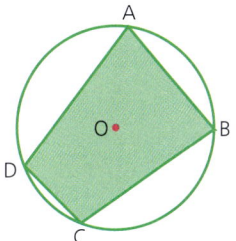

    b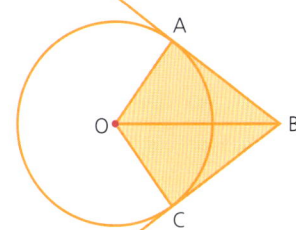

    c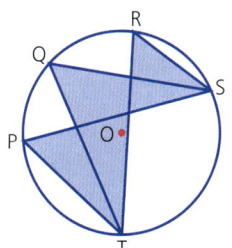

    d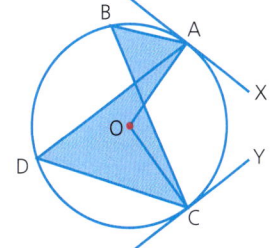

2   If angle AOC is 72°, calculate angle ABC.

    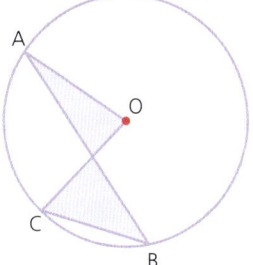

3   If angle AOB = 130°, calculate angle ABC, angle OAB and angle CAO.

    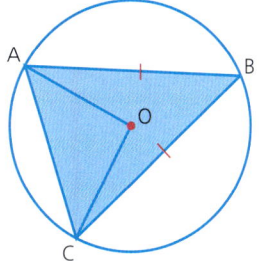

4   Show that ABCD is a cyclic quadrilateral.

    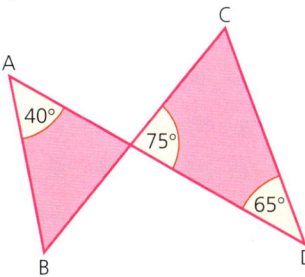

5   Calculate $f$ and $g$.

    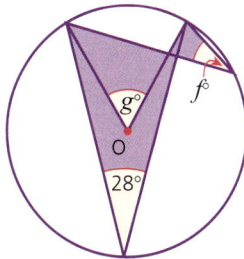

# STUDENT ASSESSMENTS 5

**6** If $y = 22.5$, calculate the value of $x$.

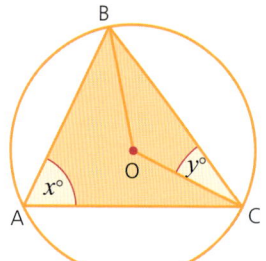

**7**

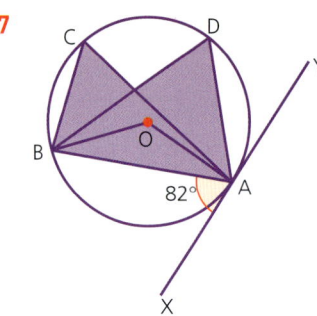

In the diagram above, XY is a tangent to the circle at A. O is the centre of the circle.

Calculate each of the following angles, explaining each of your answers.
**a** angle ACB
**b** angle ABO

## Student assessment 5

**1** Given that PQ and QR are both tangents to the circle, calculate the size of the angle marked $x°$.

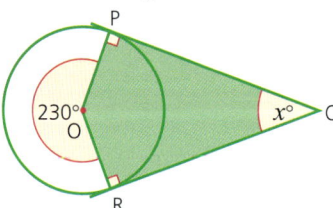

2. In the diagram, OM and ON are perpendicular bisectors of AB and XY respectively. OM = ON.

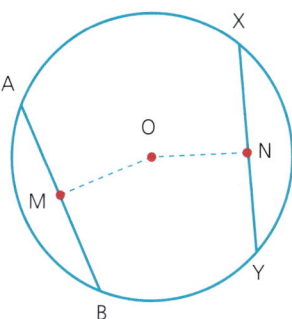

Prove that AB and XY are chords of equal length.

3.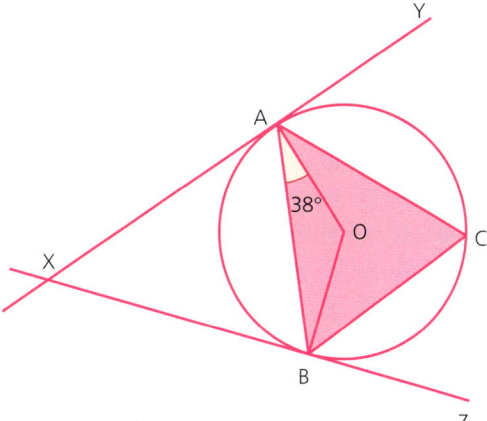

In the diagram, XY and XZ are both tangents to the circle at points A and B respectively. O is the centre of the circle.
If angle OAB = 38°, calculate the following angles, showing clearly your working and explaining your answers.
  a  angle ACB
  b  angle AXB

# TOPIC 6

## Mensuration

### Contents

Chapter 40  Measures (C6.1, E6.1)
Chapter 41  Perimeter and area of simple plane shapes (C6.2, E6.2)
Chapter 42  Circumference and area of a circle (C6.3, E6.3)
Chapter 43  Arc length and area of a sector (C6.3, E6.3)
Chapter 44  Area and volume of further plane shapes and prisms (C6.2, C6.4, E6.2, E6.4)
Chapter 45  Surface area and volume of other solids (C6.3, C6.4, C6.5, E6.3, E6.4, E6.5)

### Learning objectives

**C6.1       E6.1**
Use metric units of mass, length, area, volume and capacity in practical situations and convert quantities into larger or smaller units

**C6.2       E6.2**
Carry out calculations involving the perimeter and area of a rectangle, triangle and trapezium

**C6.3       E6.3**
Carry out calculations involving the circumference and area of a circle

Carry out calculations involving arc length and sector area as fractions of the circumference and area of a circle

**C6.4       E6.4**
Carry out calculations and solve problems involving the surface area and volume of a
- cuboid
- prism
- cylinder
- sphere
- pyramid
- cone

**C6.5       E6.5**
1 Carry out calculations and solve problems involving perimeters and areas of
   - compound shapes
   - parts of shapes
2 Carry out calculations and solve problems involving surface areas and volumes of
   - compound solids
   - parts of solids

Elements in purple refer to the Extended curriculum only.

# The British

Isaac Newton was born in Lincolnshire in 1642 and was probably the greatest scientist and mathematician ever to have lived. He was 22 and on leave from Cambridge University when he began mathematical work on optics, dynamics, thermodynamics, acoustics and astronomy. He studied gravitation and the idea that white light is a mixture of all the rainbow's colours. He also designed the first reflecting telescope, the first reflecting microscope and the sextant.

*Isaac Newton (1642–1727)*

Newton is widely regarded as the 'Father of Calculus'. He discovered what is now called the Fundamental Theorem of Calculus, i.e. that integration and differentiation are each other's inverse operation. He applied calculus to solve many problems including finding areas, tangents, the lengths of curves and the maxima and minima of functions.

In 1687, Newton published *Philosophiae Naturalis Principia Mathematica*, one of the greatest scientific books ever written. The movement of the planets was not understood before Newton's Laws of Motion and the Law of Universal Gravitation. The idea that the Earth rotated about the Sun was introduced in ancient Greece, but Newton explained why this happens.

# 40 Measures

## Metric units

The metric system uses a variety of units for length, mass and capacity.
- The common units of length are: kilometre (km), metre (m), centimetre (cm) and millimetre (mm).
- The common units of mass are: tonne (t), kilogram (kg), gram (g) and milligram (mg).
- The common units of capacity are: litre (L or l) and millilitre (ml).

Note: 'centi' comes from the Latin *centum* meaning hundred (a centimetre is one hundredth of a metre);
'milli' comes from the Latin *mille* meaning thousand (a millimetre is one thousandth of a metre);
'kilo' comes from the Greek *khilloi* meaning thousand (a kilometre is one thousand metres).
It may be useful to have some practical experience of estimating lengths, volumes and capacities before starting the following exercises.

**Exercise 6.1**

Copy and complete the sentences below:
1. a There are ... centimetres in one metre.
   b There are ... millimetres in one metre.
   c One metre is one ... of a kilometre.
   d One thousandth of a litre is one ... .
2. Which of the units below would best be used to measure the following?
   mm, cm, m, km, mg, g, kg, ml, litres
   a your height
   b the length of your finger
   c the mass of a shoe
   d the amount of liquid in a cup
   e the height of a van
   f the mass of a ship
   g the capacity of a swimming pool
   h the length of a highway
   i the capacity of the petrol tank of a car

# Converting from one unit to another

## Length

1 km = 1000 m

Therefore 1 m = $\frac{1}{1000}$ km

1 m = 1000 mm

Therefore 1 mm = $\frac{1}{1000}$ m

1 m = 100 cm

Therefore 1 cm = $\frac{1}{100}$ m

1 cm = 10 mm

Therefore 1 mm = $\frac{1}{10}$ cm

### ➜ Worked examples

a  Change 5.8 km into m.
   Since 1 km = 1000 m, 5.8 km is 5.8 × 1000 m
   5.8 km = 5800 m

b  Change 4700 mm to m.
   Since 1 m is 1000 mm,
   4700 mm is 4700 ÷ 1000 m
   4700 mm = 4.7 m

c  Convert 2.3 km into cm.
   2.3 km is 2.3 × 1000 m = 2300 m
   2300 m is 2300 × 100 cm
   2.3 km = 230 000 cm

### Exercise 6.2

1  Put in the missing unit to make the following statements correct:
   a  300 ... = 30 cm
   b  6000 mm = 6 ...
   c  3.2 m = 3200 ...
   d  4.2 ... = 4200 mm
   e  2.5 km = 2500 ...

2  Convert the following to millimetres:
   a  8.5 cm
   b  23 cm
   c  0.83 m
   d  0.05 m
   e  0.0004 m

3  Convert the following to metres:
   a  560 cm
   b  6.4 km
   c  96 cm
   d  0.004 km
   e  12 mm

4  Convert the following to kilometres:
   a  1150 m
   b  250 000 m
   c  500 m
   d  70 m
   e  8 m

## Mass

1 tonne is 1000 kg

Therefore 1 kg = $\frac{1}{1000}$ tonne

1 kilogram is 1000 g

Therefore 1 g = $\frac{1}{1000}$ kg

1 g is 1000 mg

Therefore 1 mg = $\frac{1}{1000}$ g

*1 litre of water has a mass of 1 kilogram.*

# 40 MEASURES

## → Worked examples

**a** Convert 8300 kg to tonnes.
Since 1000 kg = 1 tonne, 8300 kg is 8300 ÷ 1000 tonnes
8300 kg = 8.3 tonnes

**b** Convert 2.5 g to mg.
Since 1 g is 1000 mg, 2.5 g is 2.5 × 1000 mg
2.5 g = 2500 mg

### Exercise 6.3

*Capacity and volume of a container are slightly different. Capacity refers to the space inside the container, whereas its volume refers to the amount of space the container occupies.*

**1** Convert the following:
   **a** 3.8 g to mg           **b** 28 500 kg to tonnes
   **c** 4.28 tonnes to kg    **d** 320 mg to g
   **e** 0.5 tonnes to kg

## Capacity

1 litre is the capacity of a 10 cm × 10 cm × 10 cm cube
1 litre is 1000 millilitres which is equivalent to 1000 cm³
Therefore $1\,\text{ml} = \frac{1}{1000}$ litre = 1 cm³
1 centilitre is 10 millilitres which is equivalent to 10 cm³
Therefore 100 cl = 1 litre and $1\,\text{cl} = \frac{1}{100}$ litre = 10 cm³

### Exercise 6.4

**1** Calculate the following and give the totals in ml:
   **a** 3 litres + 1500 ml      **b** 0.88 litre + 650 ml
   **c** 0.75 litre + 6300 ml    **d** 45 cl + 0.55 litre

**2** Calculate the following and give the totals in litres:
   **a** 0.75 litre + 450 ml      **b** 850 ml + 49 cl
   **c** 0.6 litre + 0.8 litre      **d** 8 cl + 62 cl + 0.7 litre

## Area and volume conversions

Converting between units for area and volume is not as straightforward as converting between units for length.
The diagram below shows a square of side length 1 m.

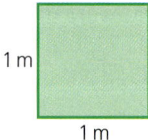

Area of the square = 1 m²
However, if the lengths of the sides are written in cm, each of the sides are 100 cm.
Area of the square = 100 × 100 = 10 000 cm²
Therefore an area of 1 m² = 10 000 cm².

# Area and volume conversions

Similarly, a square of side length 1 cm is the same as a square of side length 10 mm. Therefore an area of 1 cm² is equivalent to an area of 100 mm².

The diagram below shows a cube of side length 1 m.

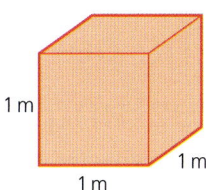

Volume of the cube = 1 m³

Once again, if the lengths of the sides are written in cm, each of the sides is 100 cm.

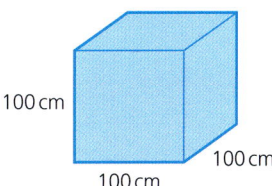

Volume of the cube = $100 \times 100 \times 100 = 1\,000\,000$ cm³
Therefore a volume of 1 m³ = 1 000 000 cm³.
Similarly, a cube of side length 1 cm is the same as a cube of side length 10 mm.
Therefore a volume of 1 cm³ is equivalent to a volume of 1000 mm³.

## Exercise 6.5

1  Convert the following areas:
   a  10 m² to cm²
   b  2 m² to mm²
   c  5 km² to m²
   d  3.2 km² to m²
   e  8.3 cm² to mm²

2  Convert the following areas:
   a  500 cm² to m²
   b  15 000 mm² to cm²
   c  1000 m² to km²
   d  40 000 mm² to m²
   e  2 500 000 cm² to km²

3  Convert the following volumes and capacities:
   a  2.5 m³ to cm³
   b  3.4 cm³ to mm³
   c  2 km³ to m³
   d  0.2 m³ to litres
   e  0.03 m³ to litres

4  Convert the following volumes:
   a  150 000 cm³ to m³
   b  24 000 mm³ to cm³
   c  850 000 m³ to km³
   d  300 mm³ to cm³
   e  15 cm³ to m³

# 41 Perimeter and area of simple plane shapes

## The perimeter and area of a rectangle

The **perimeter** of a shape is the distance around the outside of the shape. Perimeter can be measured in mm, cm, m, km, etc.

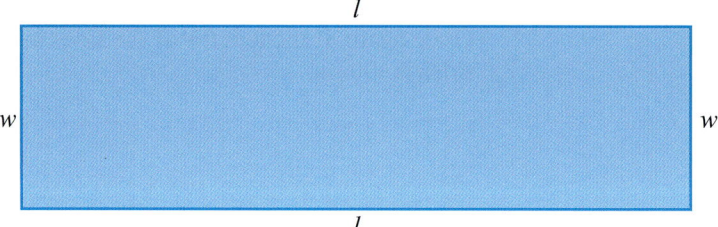

The perimeter of the rectangle of length $l$ and width $w$ above is therefore:

Perimeter = $l + w + l + w$

This can be rearranged to give:

Perimeter = $2l + 2w$

This in turn can be factorised to give:

Perimeter = $2(l + w)$

The **area** of a shape is the amount of surface that it covers. Area is measured in $mm^2$, $cm^2$, $m^2$, $km^2$, etc.
The area $A$ of the rectangle above is given by the formula: $A = lw$

###  Worked example

Calculate the width of a rectangle of area $200\,cm^2$ and length $25\,cm$.
$A = lw$
$200 = 25w$
$w = 8$
So the width is $8\,cm$.

# The area of a triangle

Rectangle ABCD has a triangle CDE drawn inside it.

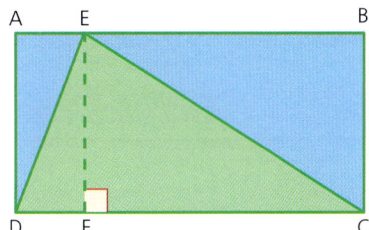

Point E is said to be a **vertex** of the triangle.
EF is the **height** or **altitude** of the triangle.
CD is the **length** of the rectangle, but is called the **base** of the triangle.
It can be seen from the diagram that triangle DEF is half the area of the rectangle AEFD.
Also, triangle CFE is half the area of rectangle EBCF.
It follows that **triangle CDE is half the area of rectangle ABCD**.

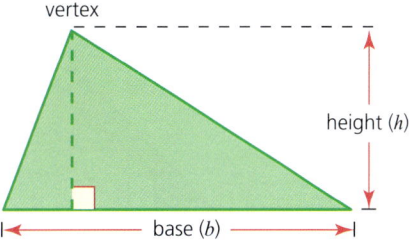

Area of a triangle $A = \frac{1}{2}bh$, where $b$ is the base length and $h$ is the height.
Note: it does not matter which side is called the base, but the height must be measured at right angles from the base to the opposite vertex.

## Exercise 6.6

1 Calculate the areas of the triangles below:

a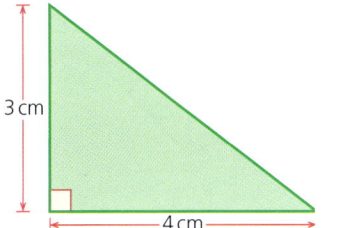

b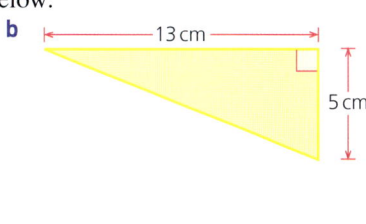

# 41 PERIMETER AND AREA OF SIMPLE PLANE SHAPES

**Exercise 6.6 (cont)**

c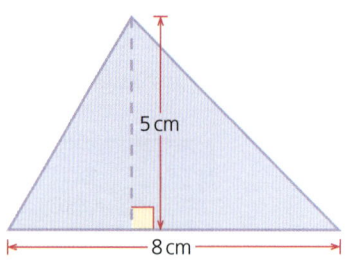

d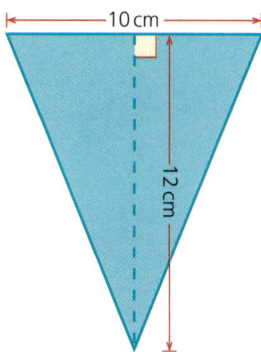

e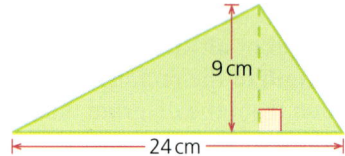

f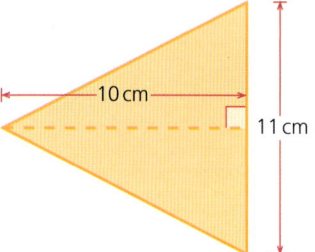

2   Calculate the areas of the shapes below:

a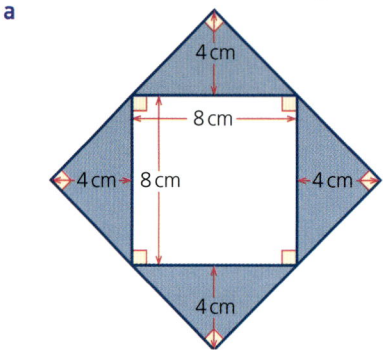

b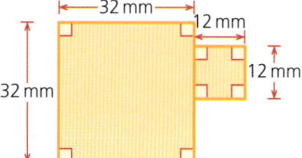

c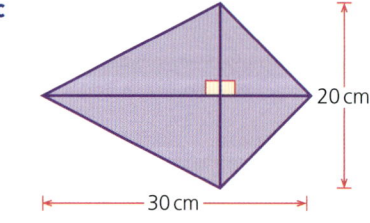

d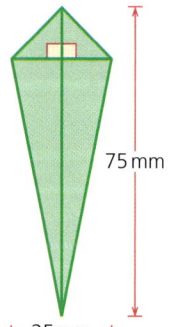

# 42 Circumference and area of a circle

*NB: Diagrams are not drawn to scale.*

All circles are similar shapes. As a result, the ratio of their circumference to diameter is constant. That is:

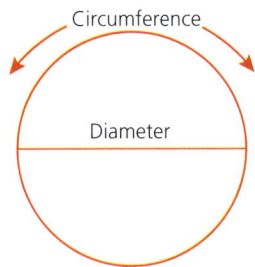

$\frac{\text{Circumference}}{\text{Diameter}} = \text{constant}$

The constant is π (pi), which is 3.14 to 3 s.f.

Therefore $\frac{C}{D} = \pi$

But the diameter is 2 × radius, so the above equation can be written as $\frac{C}{2r} = \pi$.

So the equation for the circumference of a circle is $C = 2\pi r$.

The formula for the area $A$ of a circle can be explained as follows.
The diagram below shows a circle divided into four sectors. The sectors have then been rearranged and assembled as shown.

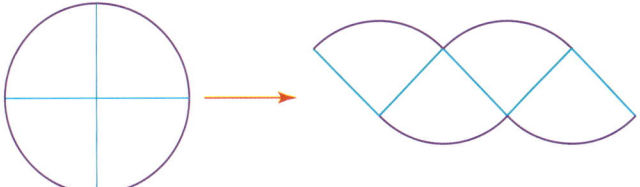

The total length of the curved edges is the same as the circumference of the circle.
If the circle is divided into eight sectors and each assembled as before, the diagram is:

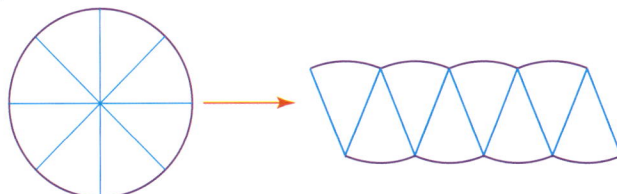

As the number of sectors increases, the assembled shape begins to look more and more like a rectangle, as shown below with 32 sectors.

# 42 CIRCUMFERENCE AND AREA OF A CIRCLE

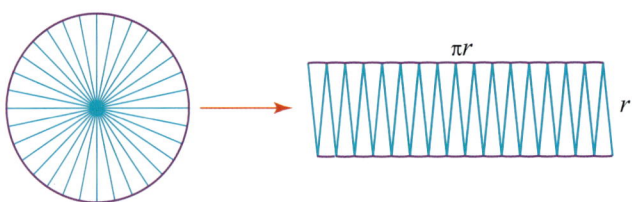

The top and bottom edges of the 'rectangle' are still equivalent to the circumference of the circle, i.e. $2\pi r$.

The top edge is therefore half the circumference, i.e. $\pi r$.

The height of the 'rectangle' is nearly equivalent to the radius of the circle.

With an infinite number of sectors, the circle can be rearranged to form a rectangle with a width $\pi r$ and a height $r$.

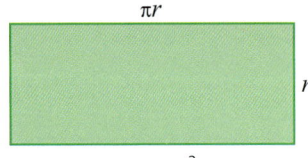

Area = $\pi r^2$

So the equation for the area of a circle is $A = \pi r^2$.

## → Worked examples

**a** Calculate the circumference of this circle, giving your answer to 3 s.f.

$C = 2\pi r$

$= 2\pi \times 3 = 18.8$

The circumference is 18.8 cm.

Note: the answer 18.8 cm is only correct to 3 s.f. and therefore only an approximation. An **exact** answer involves leaving the answer in terms of pi.

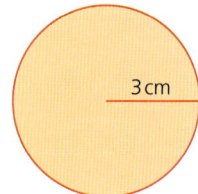

i.e. $C = 2\pi r$

$= 2\pi \times 3$

$= 6\pi$ cm

**b** If the circumference of this circle is 12 cm, calculate the radius, giving your answer:

  **i** to 3 s.f.
  **ii** in terms of $\pi$

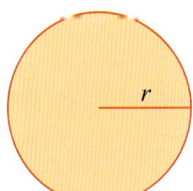

**i** $C = 2\pi r$

$r = \dfrac{C}{2\pi}$

$r = \dfrac{12}{2\pi}$

$= 1.91$

The radius is 1.91 cm (3 s.f.).

ii  $r = \frac{C}{2\pi} = \frac{12}{2\pi}$

$= \frac{6}{\pi}$ cm

c  Calculate the area of this circle, giving your answer:
i   to 3 s.f.
ii  in exact form

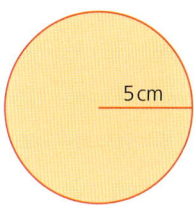
5 cm

i   $A = \pi r^2$
$= \pi \times 5^2 = 78.5$
The area is $78.5\,\text{cm}^2$ (3 s.f.).

ii  $A = \pi r^2$
$= \pi \times 5^2$
$= 25\pi\,\text{cm}^2$

d  If the area of a circle is $34\,\text{cm}^2$, calculate the radius, giving your answer:
i   to 3 s.f.
ii  in terms of $\pi$

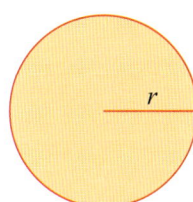
$r$

i   $A = \pi r^2$
$r = \sqrt{\frac{A}{\pi}}$
$r = \sqrt{\frac{34}{\pi}} = 3.29$
The radius is $3.29\,\text{cm}$ (3 s.f.).

ii  $r = \sqrt{\frac{A}{\pi}} = \sqrt{\frac{34}{\pi}}$ cm

## Exercise 6.7

1  Calculate the circumference of each circle, giving your answer to 3 s.f.

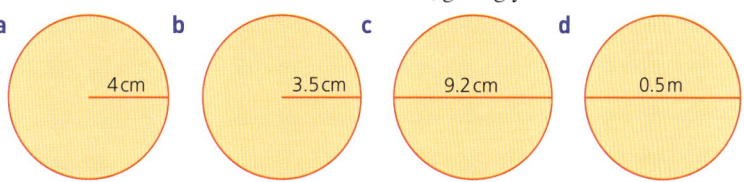
a  4 cm    b  3.5 cm    c  9.2 cm    d  0.5 m

2  Calculate the area of each of the circles in Q.1. Give your answers to 3 s.f.
3  Calculate the radius of a circle when the circumference is:
  a  15 cm    b  $\pi$ cm    c  4 m    d  8 mm
4  Calculate the diameter of a circle when the area is:
  a  $16\,\text{cm}^2$    b  $9\pi\,\text{cm}^2$    c  $8.2\,\text{m}^2$    d  $14.6\,\text{mm}^2$

# 42 CIRCUMFERENCE AND AREA OF A CIRCLE

## Exercise 6.8

1  The wheel of a car has an outer radius of 25 cm. Calculate:
   a  how far the car has travelled after one complete turn of the wheel
   b  how many times the wheel turns for a journey of 1 km.

2  If the wheel of a bicycle has a diameter of 60 cm, calculate how far a cyclist will have travelled after the wheel has rotated 100 times.

3  A circular ring has a cross-section as shown (right). If the outer radius is 22 mm and the inner radius 20 mm, calculate the cross-sectional area of the ring. Give your answer in terms of $\pi$.

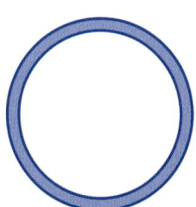

4  Four circles are drawn in a line and enclosed by a rectangle, as shown. If the radius of each circle is 3 cm, calculate the unshaded area within the rectangle, giving your answer in exact form.

5  A garden is made up of a rectangular patch of grass and two semicircular vegetable patches. If the dimensions of the rectangular patch are 16 m (length) and 8 m (width), calculate in exact form:

   a  the perimeter of the garden
   b  the total area of the garden.

# 43 Arc length and area of a sector

*NB: Diagrams are not drawn to scale.*

## Arc length

An **arc** is part of the circumference of a circle between two radii. The minor arc is the smaller of the two arcs formed, whilst the major arc is the larger of the two.

Its length is proportional to the size of the angle $\theta$ between the two radii. The length of the arc as a fraction of the circumference of the whole circle is therefore equal to the fraction that $\theta$ is of 360°.

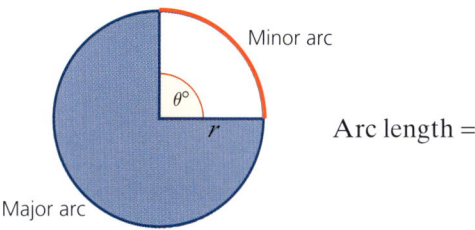

Arc length = $\frac{\theta}{360} \times 2\pi r$

## ➔ Worked examples

a   Find the length of the minor arc in the circle (right).

   i   Give your answer to 3 s.f.
       Arc length = $\frac{80}{360} \times 2 \times \pi \times 6$
       = 8.38 cm (3 s.f.)

   ii  Give your answer in terms of $\pi$.
       Arc length = $\frac{80}{360} \times 2 \times \pi \times 6$
       = $\frac{8}{3} \pi$ cm

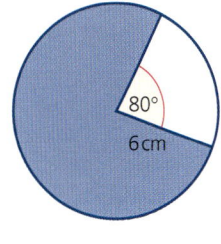

b   In this circle, the length of the minor arc is 12.4 cm and the radius is 7 cm.

   i   Calculate the angle $\theta°$.
       Arc length = $\frac{\theta}{360} \times 2\pi r$
       $12.4 = \frac{\theta}{360} \times 2 \times \pi \times 7$
       $\frac{12.4 \times 360}{2 \times \pi \times 7} = \theta$
       $\theta = 101.5°$ (1 d.p.)

   ii  Calculate the length of the major arc.
       $C = 2\pi r = 2 \times \pi \times 7 = 44.0$ cm (3 s.f.)
       Major arc = circumference − minor arc
       = (44.0 − 12.4) cm
       = 31.6 cm

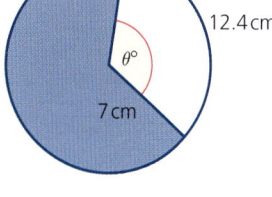

> **Note**
> Students following the Core syllabus are only expected to work with sectors where the angle is a factor of 360°.

# 43 ARC LENGTH AND AREA OF A SECTOR

**Exercise 6.9**

1  For each of the following, give the length of the arc to 3 s.f. O is the centre of the circle.

   a  (sector, 45°, radius 8 cm)
   b  (sector, 8°, radius 15 cm)
   c  (sector, 120°, radius 6 cm)
   d  (sector, 270°, radius 5 cm)

2  A **sector** is the region of a circle enclosed by two radii and an arc. In the questions below the radius $r$ and arc length $a$ are given in each case.
   i   Calculate the angle $\theta$ in each case.
   ii  State whether the arc is a major or minor arc.
   
   a  $r = 14$ cm    $a = 8$ cm
   b  $r = 4$ cm     $a = 16$ cm
   c  $r = 7.5$ cm   $a = 7.5$ cm
   d  $r = 6.8$ cm   $a = 13.6$ cm

3  Calculate the radius $r$ for each of the following sectors. The angle $\theta$ and arc length $a$ are given in each case.
   a  $\theta = 75°$   $a = 16$ cm
   b  $\theta = 300°$  $a = 24$ cm
   c  $\theta = 20°$   $a = 6.5$ cm
   d  $\theta = 243°$  $a = 17$ cm

**Exercise 6.10**

1  Calculate the perimeters of these shapes. Give your answers in exact form.

   a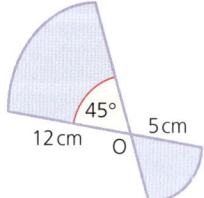
   b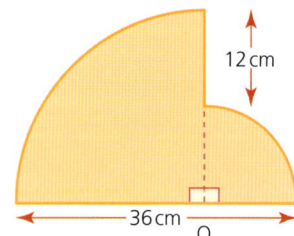

2. A shape is made from two sectors arranged in such a way that they share the same centre. The radius of the smaller sector is 7 cm and the radius of the larger sector is 10 cm. If the angle at the centre of the smaller sector is 30° and the arc length of the larger sector is 12 cm, calculate:
   a. the arc length of the smaller sector
   b. the total perimeter of the two sectors
   c. the angle at the centre of the larger sector.

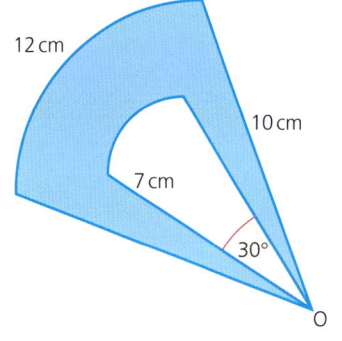

3. For the diagram on the right, calculate:
   a. the radius of the smaller sector
   b. the perimeter of the shape
   c. the angle $\theta°$.

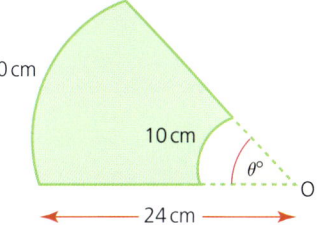

# The area of a sector

A **sector** is the region of a circle enclosed by two radii and an arc. Its area is proportional to the size of the angle $\theta°$ between the two radii. The area of the sector as a fraction of the area of the whole circle is therefore equal to the fraction that $\theta°$ is of 360°.

$$\text{Area of sector} = \frac{\theta}{360} \times \pi r^2$$

As before, the circle is divided into a minor and major sector.

## ➡ Worked examples

a. Calculate the area of the sector (right), giving your answer
   i. to 3 s.f.     ii. in terms of $\pi$

   i. Area $= \frac{\theta}{360} \times \pi r^2$
      $= \frac{45}{360} \times \pi \times 12^2 = 56.5 \text{ cm}^2$

   ii. Area $= \frac{45}{360} \times \pi \times 12^2$
       $= 18\pi \text{ cm}^2$

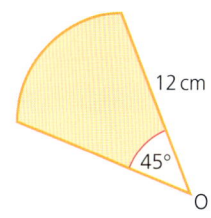

# 43 ARC LENGTH AND AREA OF A SECTOR

**b** Calculate the radius of the sector, giving your answer to 3 s.f.

$$\text{Area} = \frac{\theta}{360} \times \pi r^2$$

$$50 = \frac{30}{360} \times \pi \times r^2$$

$$\frac{50 \times 360}{30\pi} = r^2$$

$$r = 13.8$$

The radius is 13.8 cm.

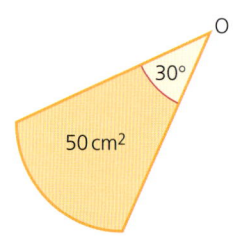

**Exercise 6.11**

1 Calculate the area of each of the following sectors, using the values of the angles $\theta$ and radius $r$ in each case.
  a $\theta = 60°$    $r = 8$ cm
  b $\theta = 120°$   $r = 14$ cm
  c $\theta = 2°$     $r = 18$ cm
  d $\theta = 320°$   $r = 4$ cm

2 Calculate the radius for each of the following sectors, using the values of the angle $\theta$ and the area $A$ in each case.
  a $\theta = 40°$    $A = 120$ cm$^2$
  b $\theta = 12°$    $A = 42$ cm$^2$
  c $\theta = 150°$   $A = 4$ cm$^2$
  d $\theta = 300°$   $A = 400$ cm$^2$

3 Calculate the value of the angle $\theta$, to the nearest degree, of each of the following sectors, using the values of $A$ and $r$ in each case.
  a $r = 12$ cm    $A = 60$ cm$^2$
  b $r = 26$ cm    $A = 0.02$ m$^2$
  c $r = 0.32$ m   $A = 180$ cm$^2$
  d $r = 38$ mm    $A = 16$ cm$^2$

4 A circle of radius 7 cm is divided into two sectors. If the minor arc length is 8.6 cm, calculate the area of the major sector.

5 A circle has a radius of $r$ cm.
  If the area of the major sector : minor sector is $4:1$, calculate the ratio of the major arc : minor arc

**Exercise 6.12**

1 A rotating sprinkler is placed in one corner of a garden as shown. If it has a reach of 8 m and rotates through an angle of 30°, calculate the area of garden not being watered. Give your answer in terms of $\pi$.

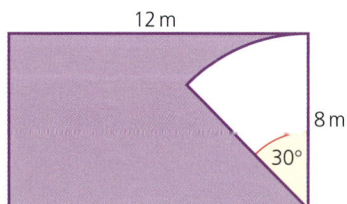

## The area of a sector

2  Two sectors AOB and COD share the same centre O. The area of AOB is three times the area of COD. Calculate:
   a  the area of sector AOB
   b  the area of sector COD
   c  the radius $r$ cm of sector COD.

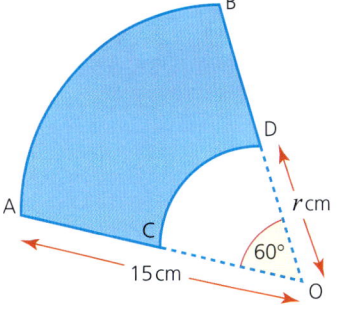

3  A circular cake is cut. One of the slices is shown. Calculate:

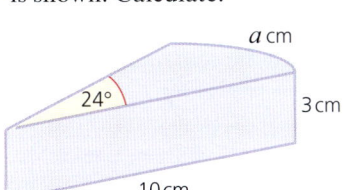

   a  the length $a$ cm of the arc
   b  the total surface area of all the sides of the slice.

4  The diagram shows a plan view of four tiles in the shape of sectors placed in the bottom of a box. C is the midpoint of the arc AB and intersects the chord AB at point D. If the length ADB is 8 cm and the length OB is 10 cm, calculate:

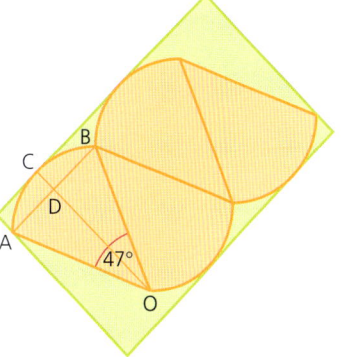

   a  the length OD
   b  the length CD
   c  the area of the sector AOB
   d  the length and width of the box
   e  the area of the base of the box not covered by the tiles.

381

# 44 Area and volume of further plane shapes and prisms

## The area of parallelograms and trapeziums

A **parallelogram** can be rearranged to form a rectangle as shown below:

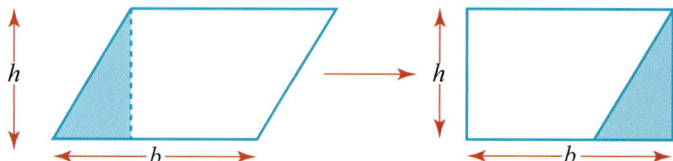

Therefore, area of parallelogram = base length × perpendicular height.

A **trapezium** can be visualised as being split into two triangles as shown:

Area of triangle A = $\frac{1}{2} \times a \times h$

Area of triangle B = $\frac{1}{2} \times b \times h$

Area of the trapezium = area of triangle A + area of triangle B

$$= \tfrac{1}{2}ah + \tfrac{1}{2}bh$$
$$= \tfrac{1}{2}h(a + b)$$

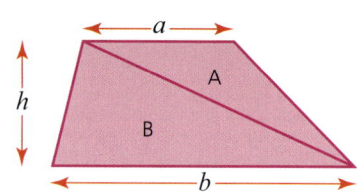

### ➔ Worked examples

a  Calculate the area of the parallelogram shown below:

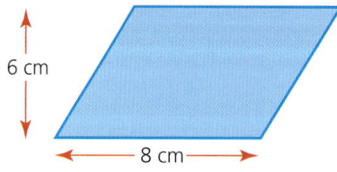

Area = base length × perpendicular height
= 8 × 6
= 48 cm²

*The area of parallelograms and trapeziums*

**b** Calculate the shaded area in the shape shown:

Area of rectangle = $12 \times 8 = 96 \, cm^2$

Area of trapezium = $\frac{1}{2} \times 5(3 + 5)$

$= 2.5 \times 8 = 20 \, cm^2$

Shaded area = $96 - 20 = 76 \, cm^2$

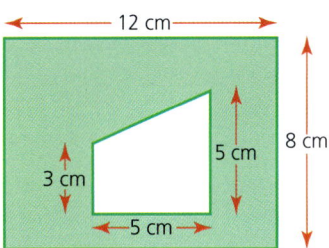

**Exercise 6.13** Find the area of each of the following shapes:

**a**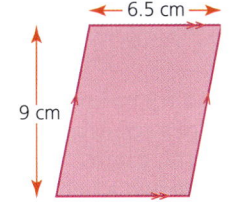

**b**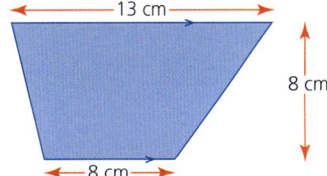

**c**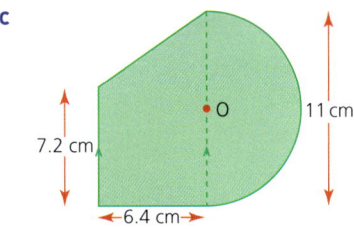

**d**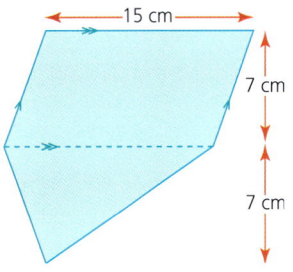

**Exercise 6.14**

**1** Calculate $a$.

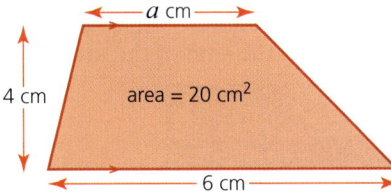

**2** If the areas of this trapezium and parallelogram are equal, calculate $x$.

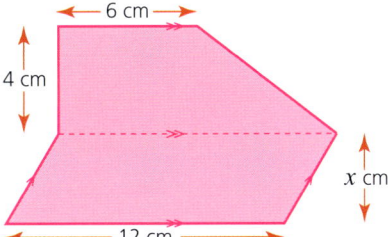

# 44 AREA AND VOLUME OF FURTHER PLANE SHAPES AND PRISMS

**Exercise 6.14 (cont)**

3 The end view of a house is as shown in the diagram.

If the door has a width of 0.75 m and a height of 2 m and the circular window has a diameter of 0.8 m, calculate the area of brickwork.

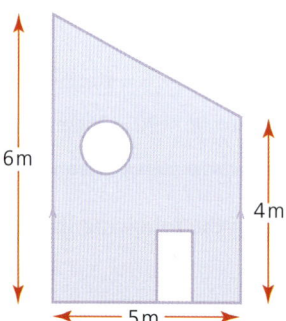

4 A garden in the shape of a trapezium is split into three parts: two flower beds in the shape of a triangle and a parallelogram and a section of grass in the shape of a trapezium.
The area of the grass is two and a half times the total area of flower beds.
Calculate:
a the area of each flower bed
b the area of grass
c the value of $x$.

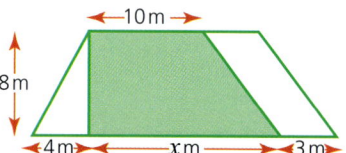

## The surface area of a cuboid and cylinder

To calculate the surface area of a **cuboid**, start by looking at its individual faces. These are either squares or rectangles. The surface area of a cuboid is the sum of the areas of its faces.

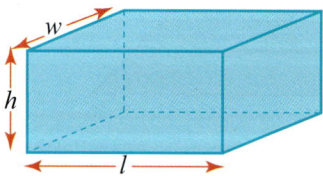

Area of top = $wl$    Area of bottom = $wl$

Area of front = $lh$    Area of back = $lh$

Area of one side = $wh$    Area of other side = $wh$

Total surface area

$= 2wl + 2lh + 2wh$

$= 2(wl + lh + wh)$

## The surface area of a cuboid and cylinder

For the surface area of a **cylinder**, it is best to visualise the net of the solid: it is made up of one rectangular piece and two circular pieces.

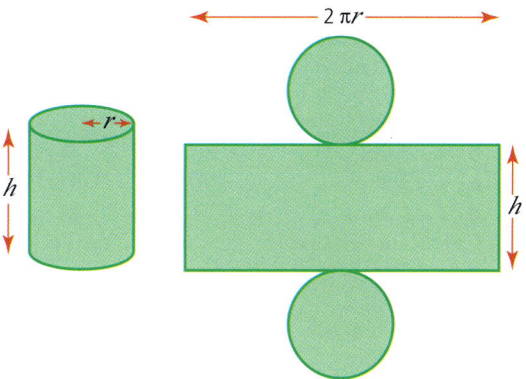

Area of circular pieces = $2 \times \pi r^2$
Area of rectangular piece = $2\pi r \times h$
Total surface area = $2\pi r^2 + 2\pi rh$
       = $2\pi r(r + h)$

### → Worked examples

a  Calculate the surface area of the cuboid shown below:

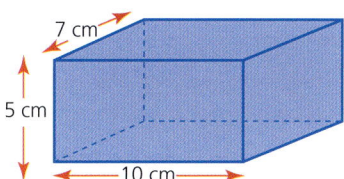

Total area of top and bottom = $2 \times 7 \times 10 = 140 \, cm^2$
Total area of front and back = $2 \times 5 \times 10 = 100 \, cm^2$
Total area of both sides = $2 \times 5 \times 7 = 70 \, cm^2$
Total surface area = $310 \, cm^2$

b  If the height of a cylinder is 7 cm and the radius of its circular top is 3 cm, calculate its surface area.

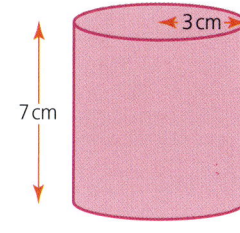

Total surface area = $2\pi r(r + h)$
       = $2\pi \times 3 \times (3 + 7)$
       = $6\pi \times 10$
       = $60\pi$
       = $188 \, cm^2$ (3 s.f.)

The total surface area is $188 \, cm^2$.

*Leaving the answer as $60\pi$ is called leaving the answer in terms of $\pi$ or giving the answer in exact form.*

# 44 AREA AND VOLUME OF FURTHER PLANE SHAPES AND PRISMS

**Exercise 6.15**

1. Calculate the surface area of each of the following cuboids if:
   a  $l = 12$ cm      $w = 10$ cm     $h = 5$ cm
   b  $l = 4$ cm       $w = 6$ cm      $h = 8$ cm
   c  $l = 4.2$ cm     $w = 7.1$ cm    $h = 3.9$ cm
   d  $l = 5.2$ cm     $w = 2.1$ cm    $h = 0.8$ cm

2. Calculate the height of each of the following cuboids if:
   a  $l = 5$ cm       $w = 6$ cm      surface area = 104 cm$^2$
   b  $l = 2$ cm       $w = 8$ cm      surface area = 112 cm$^2$
   c  $l = 3.5$ cm     $w = 4$ cm      surface area = 118 cm$^2$
   d  $l = 4.2$ cm     $w = 10$ cm     surface area = 226 cm$^2$

3. Calculate the surface area of each of the following cylinders if:
   a  $r = 2$ cm       $h = 6$ cm
   b  $r = 4$ cm       $h = 7$ cm
   c  $r = 3.5$ cm     $h = 9.2$ cm
   d  $r = 0.8$ cm     $h = 4.3$ cm

4. Calculate the height of each of the following cylinders. Give your answers to 1 d.p.
   a  $r = 2.0$ cm     surface area = 40 cm$^2$
   b  $r = 3.5$ cm     surface area = 88 cm$^2$
   c  $r = 5.5$ cm     surface area = 250 cm$^2$
   d  $r = 3.0$ cm     surface area = 189 cm$^2$

**Exercise 6.16**

1. Two cubes are placed next to each other.

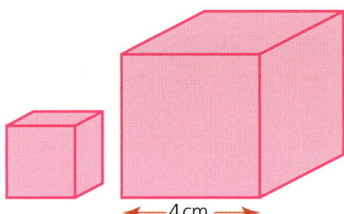

   The length of each of the edges of the larger cube is 4 cm. If the ratio of their surface areas is 1 : 4, calculate:
   a  the surface area of the small cube
   b  the length of an edge of the small cube.

2. A cube and a cylinder have the same surface area. If the cube has an edge length of 6 cm and the cylinder a radius of 2 cm, calculate:
   a  the surface area of the cube
   b  the height of the cylinder. Give your answer:
      i   to 1 d.p.
      ii  in terms of $\pi$

3. The two cylinders have the same surface area. The shorter of the two has a radius of 3 cm and a height of 2 cm, and the taller cylinder has a radius of 1 cm.

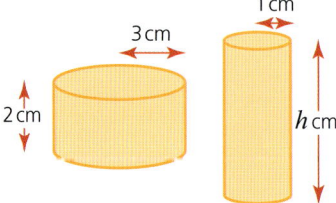

Calculate:
a the surface area of one of the cylinders
b the height *h* of the taller cylinder.
4 Two cuboids have the same surface area. The dimensions of one of them are: length = 3 cm, width = 4 cm and height = 2 cm. Calculate the height of the other cuboid if its length is 1 cm and its width is 4 cm.

# The volume of prisms

A prism is any three-dimensional object which has a constant cross-sectional area.

Below are a few examples of some of the more common types of prisms:

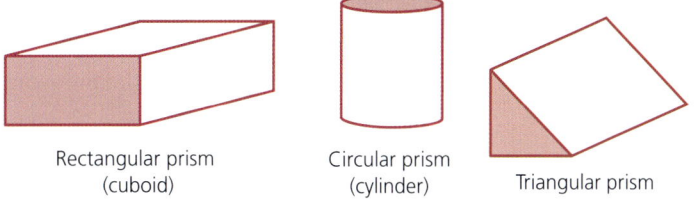

Rectangular prism (cuboid)   Circular prism (cylinder)   Triangular prism

When each of the shapes is cut parallel to the shaded face, the cross-section is constant and the shape is therefore classified as a prism.

**Volume of a prism = area of cross-section × length**

## → Worked examples

a Calculate the volume of the cylinder shown in the diagram:

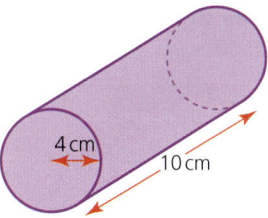

Volume = cross-sectional area × length = $\pi \times 4^2 \times 10$

Volume = 503 cm³ (3 s.f.)

b Calculate the volume of the L-shaped prism shown in the diagram:

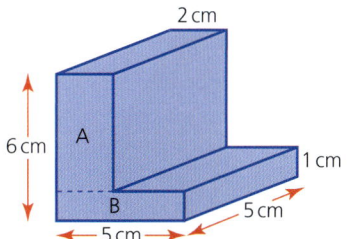

# 44 AREA AND VOLUME OF FURTHER PLANE SHAPES AND PRISMS

The cross-sectional area can be split into two rectangles:
Area of rectangle A = 5 × 2 = 10 cm²
Area of rectangle B = 5 × 1 = 5 cm²
Total cross-sectional area = (10 cm² + 5 cm²) = 15 cm²
Volume of prism = 15 × 5 = 75 cm³

### Exercise 6.17

1  Calculate the volume of each of the following cuboids:
   a  Width 2 cm        Length 3 cm         Height 4 cm
   b  Width 6 cm        Length 1 cm         Height 3 cm
   c  Width 6 cm        Length 23 mm        Height 2 cm
   d  Width 42 mm       Length 3 cm         Height 0.007 m

2  Calculate the volume of each of the following cylinders:
   a  Radius 4 cm        Height 9 cm
   b  Radius 3.5 cm      Height 7.2 cm
   c  Radius 25 mm       Height 10 cm
   d  Radius 0.3 cm      Height 17 mm

3  Calculate the volume of each of the following triangular prisms:
   a  Base length 6 cm
      Perpendicular height 3 cm
      Length 12 cm
   b  Base length 4 cm
      Perpendicular height 7 cm
      Length 10 cm
   c  Base length 5 cm
      Perpendicular height 24 mm
      Length 7 cm
   d  Base length 62 mm
      Perpendicular height 2 cm
      Length 0.01 m

4  Calculate the volume of each of the following prisms. All dimensions are given in centimetres.

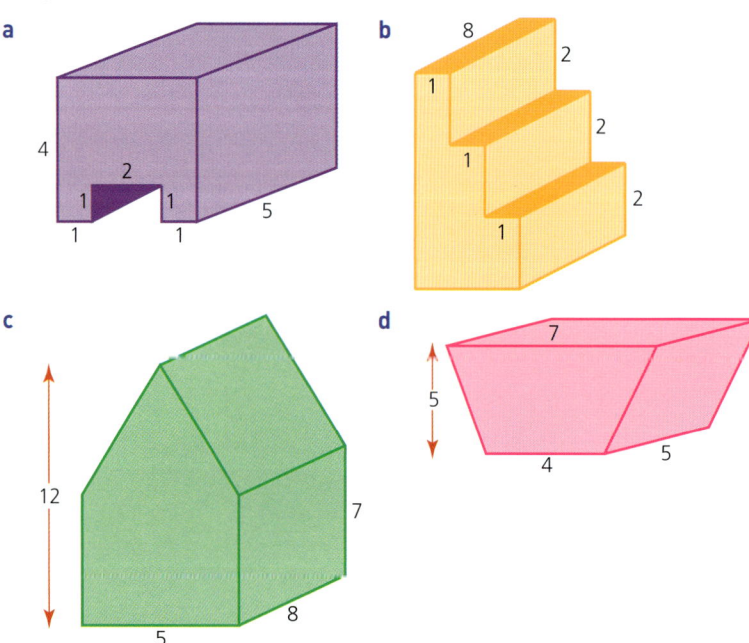

## The volume of prisms

**Exercise 6.18**

1  The diagram shows a plan view of a cylinder inside a box the shape of a cube. If the radius of the cylinder is 8 cm, calculate:
   a  the height of the cube
   b  the volume of the cube
   c  the volume of the cylinder
   d  the percentage volume of the cube not occupied by the cylinder.

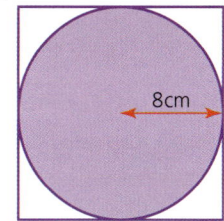

2  A chocolate bar is made in the shape of a triangular prism. The triangular face of the prism is equilateral and has an edge length of 4 cm and a perpendicular height of 3.5 cm. The manufacturer also sells these in special packs of six bars arranged as a hexagonal prism.
   If the prisms are 20 cm long, calculate:
   a  the cross-sectional area of the pack
   b  the volume of the pack.

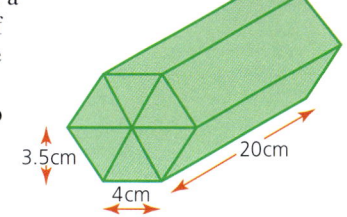

3  A cuboid and a cylinder have the same volume. The radius and height of the cylinder are 2.5 cm and 8 cm respectively. If the length and width of the cuboid are each 5 cm, calculate its height to 1 d.p.

4  A section of steel pipe is shown in the diagram.

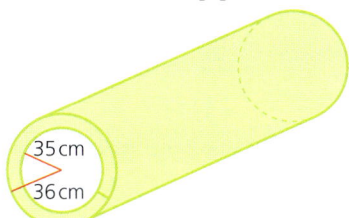

The inner radius is 35 cm and the outer radius 36 cm. Calculate the volume of steel used in making the pipe if it has a length of 130 m.

5  The diagram below shows the net of a right-angled triangular prism.

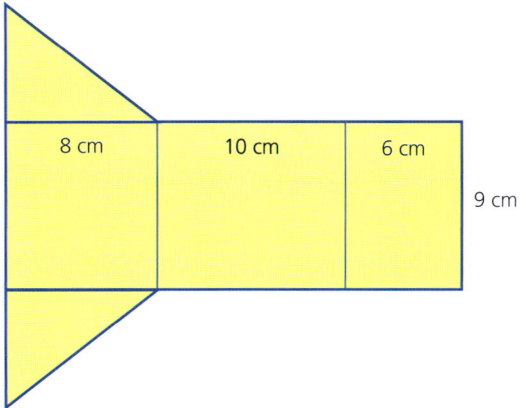

Calculate:
a  the surface area of the prism
b  the volume of the prism.

# 45 Surface area and volume of other solids

*The formula for the volume of a sphere will be given to you in any assessment.*

## Volume of a sphere

**Volume of sphere** $= \frac{4}{3}\pi r^3$

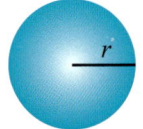

### → Worked examples

**a** Calculate the volume of the sphere, giving your answer

   **i** to 3 s.f.            **ii** in terms of $\pi$

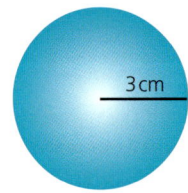

  **i** Volume of sphere $= \frac{4}{3}\pi r^3$

$= \frac{4}{3} \times \pi \times 3^3$

$= 113.1$

The volume is $113\,\text{cm}^3$ (3 s.f.).

  **ii** Volume of sphere $= \frac{4}{3}\pi \times 3^3$

$= 36\pi\,\text{cm}^3$

**b** Given that the volume of a sphere is $150\,\text{cm}^3$, calculate its radius to 1 d.p.

$V = \frac{4}{3}\pi r^3$

$r^3 = \frac{3V}{4\pi}$

$r^3 = \frac{3 \times 150}{4 \times \pi}$

$r = \sqrt[3]{35.8} = 3.3$

The radius is $3.3\,\text{cm}$ (1 d.p.)

## Volume of a sphere

**Exercise 6.19**

1. Calculate the volume of each of the following spheres. The radius is given in each case.
   a  6 cm    b  9.5 cm    c  8.2 cm    d  0.7 cm
2. Calculate the radius of each of the following spheres. Give your answers in centimetres and to 1 d.p. The volume is given in each case.
   a  130 cm³    b  720 cm³    c  0.2 m³    d  1000 mm³

**Exercise 6.20**

1. Given that sphere B has twice the volume of sphere A, calculate the radius of sphere B. Give your answer to 1 d.p.

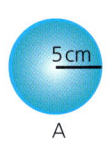
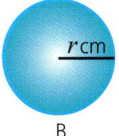

2. Calculate the volume of material used to make the hemispherical bowl shown, if the inner radius of the bowl is 5 cm and its outer radius 5.5 cm. Give your answer in terms of π.

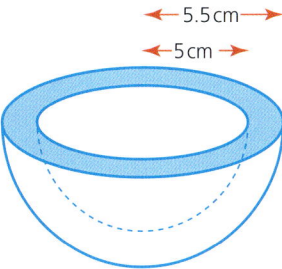

3. The volume of the material used to make the sphere and hemispherical bowl below are the same. Given that the radius of the sphere is 7 cm and the inner radius of the bowl is 10 cm, calculate, to 1 d.p., the outer radius $r$ cm of the bowl.

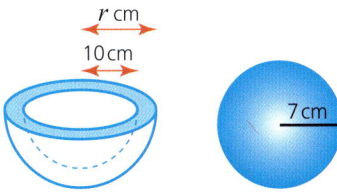

4. A ball is placed inside a box into which it will fit tightly. If the radius of the ball is 10 cm, calculate the percentage volume of the box not occupied by the ball.

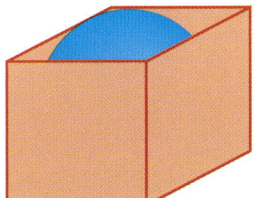

# 45 SURFACE AREA AND VOLUME OF OTHER SOLIDS

**Exercise 6.20 (cont)**

5  A steel ball is melted down to make eight smaller identical balls. If the radius of the original steel ball was 20 cm, calculate to the nearest millimetre the radius of each of the smaller balls.

6  A steel ball of volume 600 cm³ is melted down and made into three smaller balls, A, B and C. If the volumes of A, B and C are in the ratio 7 : 5 : 3, calculate to 1 d.p. the radius of each of A, B and C.

7  The cylinder and sphere shown have the same radius and the same height. Calculate the ratio of their volumes, giving your answer in the form, volume of cylinder : volume of sphere.

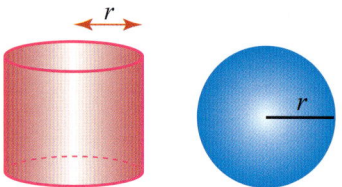

## The surface area of a sphere

**Surface area of sphere = $4\pi r^2$**

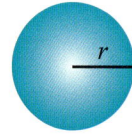

*The formula for the surface area of a sphere will be given to you in any assessment.*

**Exercise 6.21**

1  Calculate the surface area of each of the following spheres. The radius is given in each case.
   a  6 cm      b  4.5 cm      c  12.25 cm      d  $\frac{1}{\sqrt{\pi}}$ cm

2  Calculate the radius of each of the following spheres. The surface area is given in each case.
   a  50 cm²      b  16.5 cm²      c  120 mm²      d  $\pi$ cm²

3  Sphere A has a radius of 8 cm and sphere B has a radius of 16 cm. Calculate the ratio of their surface areas in the form $1 : n$

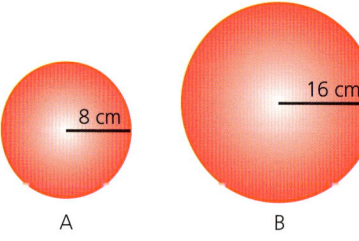

# The volume of a pyramid

*A hemisphere is half of a sphere. Note this content is relevant to the Extended syllabus only.*

4  A **hemisphere** of diameter 10 cm is attached to a cylinder of equal diameter as shown.

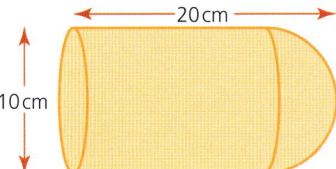

If the total length of the shape is 20 cm, calculate the surface area of the whole shape.

5  A sphere and a cylinder both have the same surface area and the same height of 16 cm.

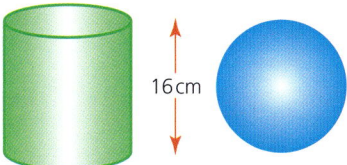

Calculate the radius of the cylinder.

## The volume of a pyramid

A pyramid is a three-dimensional shape in which each of its faces must be plane. A pyramid has a polygon for its base and the other faces are triangles with a common vertex, known as the **apex**. Its individual name is taken from the shape of the base.

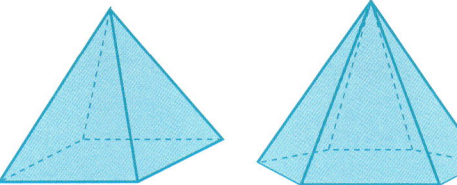

Square-based pyramid   Hexagonal-based pyramid

To derive the formula for the volume of a pyramid requires mathematics at a higher level than covered in this book. However, below are diagrams to show how an approximate value for the volume of a pyramid is derived.

393

# 45 SURFACE AREA AND VOLUME OF OTHER SOLIDS

Consider first a cube of side length 1 cm.

Its volume is 1 cm³.

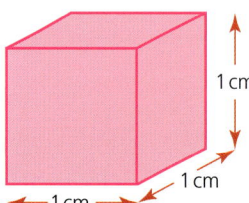

Now consider a step pyramid of two layers made of cubes of side length $\frac{1}{2}$ cm.

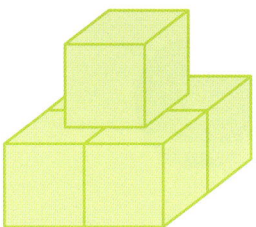

Volume of top layer $\left(\frac{1}{2}\right)^3 = \frac{1}{8}$ cm³

Volume of second layer $\frac{1}{2} \times 1 \times 1 = \frac{1}{2}$ cm³

Total volume $= \frac{1}{2} + \frac{1}{8} = \frac{5}{8}$ cm³ $= 0.625$ cm³

Now consider a step pyramid of four layers, made of cubes of side length $\frac{1}{4}$ cm.

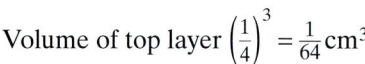

Volume of top layer $\left(\frac{1}{4}\right)^3 = \frac{1}{64}$ cm³

Volume of second layer $= \frac{1}{4} \times \frac{2}{4} \times \frac{2}{4} = \frac{4}{64}$ cm³

Volume of third layer $= \frac{1}{4} \times \frac{3}{4} \times \frac{3}{4} = \frac{9}{64}$ cm³

Volume of bottom layer $= \frac{1}{4} \times \frac{4}{4} \times \frac{4}{4} = \frac{16}{64}$ cm³

So the total volume $= \frac{1}{64} + \frac{4}{64} + \frac{9}{64} + \frac{16}{64} = \frac{30}{64}$ cm³ $\approx 0.469$ cm³

A step pyramid of 10 layers, made of cubes of side length $\frac{1}{10}$ cm can be shown to have a total volume of $\frac{77}{200}$ cm³ $= 0.385$ cm³

In fact as the number of layers increases, the total volume for a step cube of total height 1 unit and base length and width of 1 unit gets closer and closer to $\frac{1}{3}$ units³.

## The volume of a pyramid

This can be shown by dividing a cube, of side length $x$, by drawing a line from each vertex to its diagonally opposite vertex:

The cube has been divided into six congruent square-based pyramids.

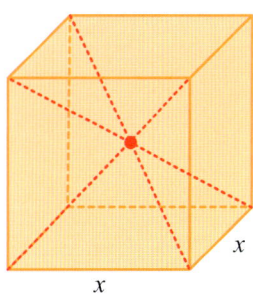

Volume of cube = $x^3$ units$^3$

Therefore volume of each pyramid = $\frac{1}{6}x^3$ units$^3$

$= \frac{1}{3}x^2 \times \frac{1}{2}x$

i.e the volume of each pyramid is $\frac{1}{3} \times$ area of its base $\times$ its perpendicular height. This can be applied to the example below.

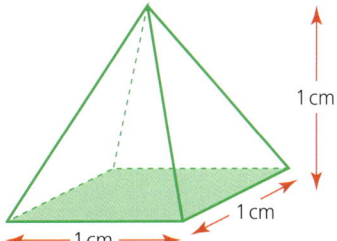

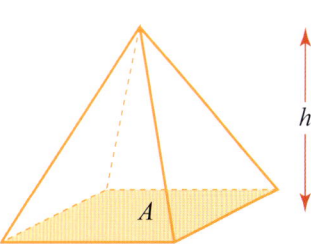

Volume = $\frac{1}{3} \times 1 \times 1 \times 1 = \frac{1}{3}$ cm$^3$

In general, for any pyramid:

Volume = $\frac{1}{3} \times$ area of base $\times$ perpendicular height

**Volume = $\frac{1}{3}Ah$**

*The formula for the volume of a pyramid will be given to you in any assessment.*

### → Worked examples

a  A rectangular-based pyramid has a perpendicular height of 5 cm and base dimensions as shown. Calculate the volume of the pyramid.

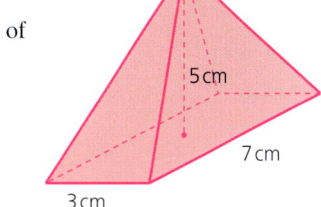

Volume = $\frac{1}{3} \times$ base area $\times$ height

$= \frac{1}{3} \times 3 \times 7 \times 5$

$= 35$

The volume is 35 cm$^3$.

b  The pyramid shown below has a volume of 60 cm$^3$. Calculate its perpendicular height $h$ in cm.

Volume = $\frac{1}{3} \times$ base area $\times$ height

Height = $\frac{3 \times \text{volume}}{\text{base area}}$

$h = \frac{3 \times 60}{\frac{1}{2} \times 8 \times 5}$

$h = 9$

The height is 9 cm.

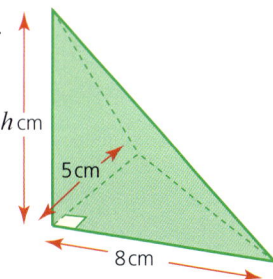

# 45 SURFACE AREA AND VOLUME OF OTHER SOLIDS

 **Exercise 6.22**   Find the volume of each of the following pyramids:

1

2

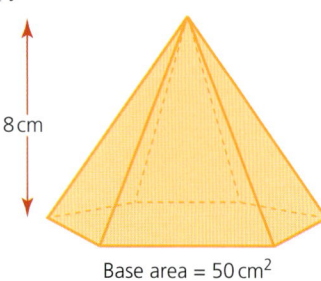

Base area = 50 cm²

3

4

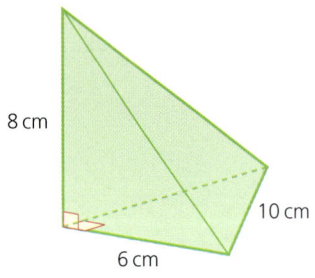

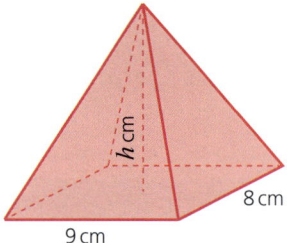

**Exercise 6.23**

1 Calculate the perpendicular height $h$ in cm for the pyramid, given that it has a volume of 168 cm³.

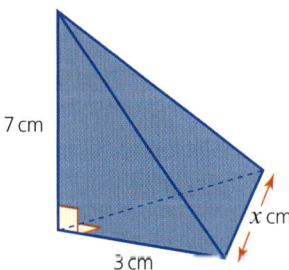

2 Calculate the length of the edge marked $x$ cm, given that the volume of the pyramid is 14 cm³.

*The volume of a cone*

3 The top of a square-based pyramid (below) is cut off.

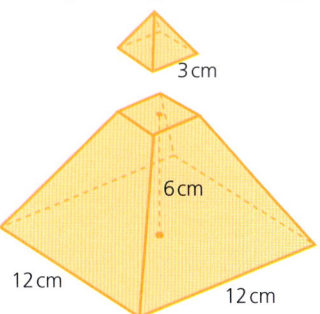

The cut is made parallel to the base. If the base of the smaller pyramid has a side length of 3 cm and the vertical height of the remaining solid pyramid is 6 cm, calculate:
  a  the height of the original pyramid
  b  the volume of the original pyramid
  c  the volume of the frustum.

*When the top of a pyramid or cone is cut off parallel to the base, the remaining bottom part is known as a frustum.*

4 The top of a triangular-based pyramid (tetrahedron) is cut off. The cut is made parallel to the base. If the vertical height of the top is 6 cm, calculate:
  a  the height of the frustum (bottom piece)
  b  the volume of the small pyramid
  c  the volume of the original pyramid.

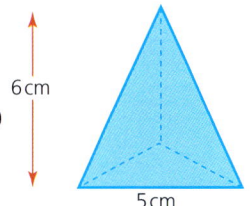

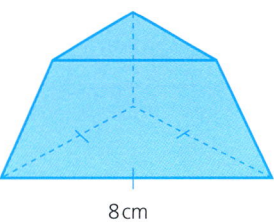

## The volume of a cone

A cone is a pyramid with a circular base. The formula for its volume is therefore the same as for any other pyramid.

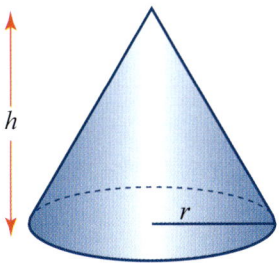

Volume = $\frac{1}{3}$ × base area × height

= $\frac{1}{3} \times \pi r^2 h$

*The formula for the volume of a cone will be given to you in any assessment.*

397

# 45 SURFACE AREA AND VOLUME OF OTHER SOLIDS

## Worked examples

a Calculate the volume of the cone.

$\text{Volume} = \frac{1}{3}\pi r^2 h$

$= \frac{1}{3} \times \pi \times 4^2 \times 8$

$= 134 \ (3 \text{ s.f.})$

The volume is $134 \text{ cm}^3$.

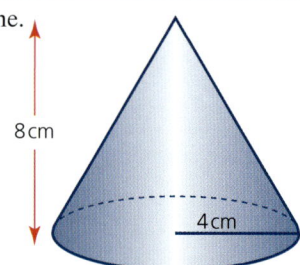

b The sector below is assembled to form a cone as shown:

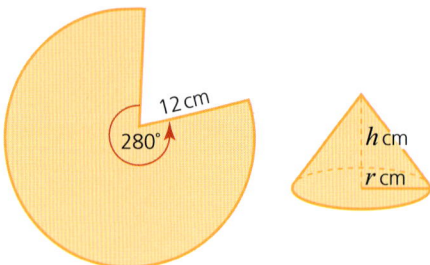

i Calculate, in terms of $\pi$, the base circumference of the cone.
The base circumference of the cone is equal to the arc length of the sector.
The radius of the sector is equal to the slant height of the cone (i.e. 12 cm).

Sector arc length $= \frac{\theta}{360} \times 2\pi r$

$= \frac{280}{360} \times 2\pi \times 12 = \frac{56}{3}\pi$

So the base circumference is $\frac{56}{3}\pi$ cm.

ii Calculate, as a fraction, the base radius of the cone.
The base of a cone is circular. Therefore:

$C = 2\pi r$

$r = \frac{C}{2\pi}$

$= \frac{\frac{56}{3}\pi}{2\pi} = \frac{56}{6} = \frac{28}{3}$

So the radius is $\frac{28}{3}$ cm.

*The volume of a cone*

**You will learn about Pythagoras' theorem in Chapter 46.**

iii Calculate the exact vertical height of the cone.

The vertical height of the cone can be calculated using Pythagoras' theorem on the right-angled triangle enclosed by the base radius, vertical height and the sloping face, as shown below.

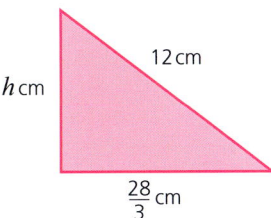

Note that the length of the sloping face is equal to the radius of the sector.

$$12^2 = h^2 + \left(\frac{28}{3}\right)^2$$

$$h^2 = 12^2 - \left(\frac{28}{3}\right)^2$$

$$h^2 = \frac{512}{9}$$

$$h = \frac{\sqrt{512}}{3} = \frac{16\sqrt{2}}{3}$$

Therefore the vertical height is $\frac{16\sqrt{2}}{3}$ cm.

iv Calculate the volume of the cone, leaving your answer both in terms of $\pi$ and to 3 s.f.

$$\text{Volume} = \tfrac{1}{3} \times \pi r^2 h$$

$$= \tfrac{1}{3} \pi \times \left(\frac{28}{3}\right)^2 \times \frac{16\sqrt{2}}{3}$$

$$= \frac{12544\sqrt{2}}{81} \pi \text{ cm}^3$$

$$= 688 \text{ cm}^3 \text{ (3 s.f.)}$$

In the examples above, the previous answer was used to calculate the next stage of the question. By using exact values each time, you avoid introducing rounding errors into the calculation.

**Exercise 6.24**

1 Calculate the volume of each of the following cones. Use the values for the base radius $r$ and the vertical height $h$ given in each case.
   a  $r = 3$ cm      $h = 6$ cm
   b  $r = 6$ cm      $h = 7$ cm
   c  $r = 8$ mm      $h = 2$ cm
   d  $r = 6$ cm      $h = 44$ mm

2 Calculate the base radius of each of the following cones. Use the values for the volume $V$ and the vertical height $h$ given in each case.
   a  $V = 600$ cm³     $h = 12$ cm
   b  $V = 225$ cm³     $h = 18$ mm
   c  $V = 1400$ mm³    $h = 2$ cm
   d  $V = 0.04$ m³     $h = 145$ mm

# 45 SURFACE AREA AND VOLUME OF OTHER SOLIDS

**Exercise 6.24 (cont)**

*You will need to use Pythagoras' theorem to calculate the vertical heights.*

3 The base circumference $C$ and the length of the sloping face $l$ is given for each of the following cones. Calculate in each case:
  i the base radius
  ii the vertical height
  iii the volume.
  Give all answers to 3 s.f.
  a  $C = 50$ cm     $l = 15$ cm
  b  $C = 100$ cm    $l = 18$ cm
  c  $C = 0.4$ m     $l = 75$ mm
  d  $C = 240$ mm    $l = 6$ cm

**Exercise 6.25**

1 The two cones A and B shown below have the same volume. Using the dimensions shown and given that the base circumference of cone B is 60 cm, calculate the height $h$ in cm.

*You will need to use Pythagoras' theorem to answer these questions.*

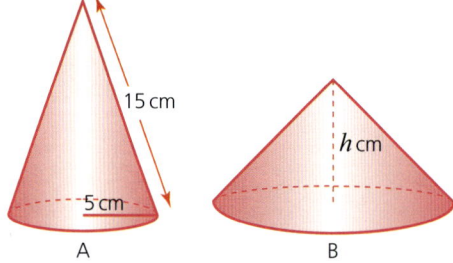

2 The sector shown is assembled to form a cone. Calculate the following, giving your answers to parts a, b and c in exact form.

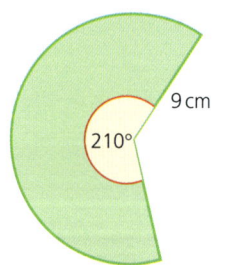

  a  the base circumference of the cone
  b  the base radius of the cone
  c  the vertical height of the cone
  d  the volume of the cone.
      Give your answer correct to 3 s.f.

3 A cone is placed inside a cuboid as shown.

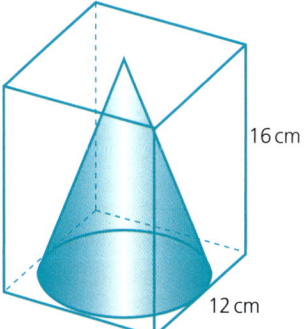

If the base diameter of the cone is 12 cm and the height of the cuboid is 16 cm, calculate the volume of the cuboid not occupied by the cone.

*The volume of a cone*

4 Two similar sectors (below) are assembled into cones. Calculate the ratio of their volumes.

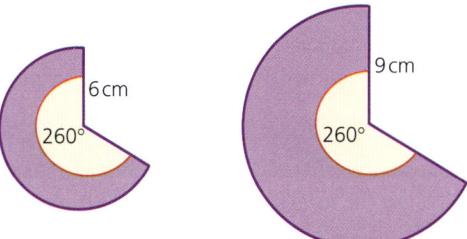

**Exercise 6.26**

1 An ice cream consists of a hemisphere and a cone. Calculate, in exact form, its total volume.

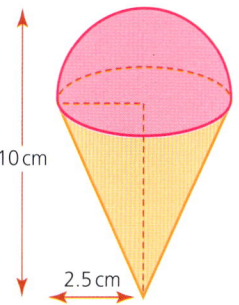

2 A cone is placed on top of a cylinder. Using the dimensions given, calculate the total volume of the shape.

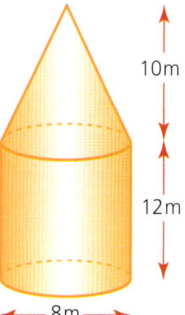

3 Two identical frustums are placed end to end as shown:

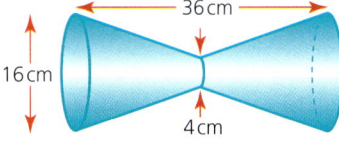

Calculate the total volume of the shape.

4 Two cones A and B are placed in either end of a cylindrical tube as shown.

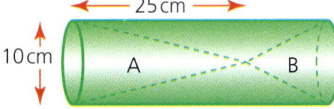

Given that the volumes of A and B are in the ratio 2:1, calculate:
a the volume of cone A
b the height of cone B
c the volume of the cylinder.

401

# 45 SURFACE AREA AND VOLUME OF OTHER SOLIDS

## The surface area of a cone

The surface area of a cone comprises the area of the circular base and the area of the curved face. The area of the curved face is equal to the area of the sector from which it is formed.

### ➡ Worked example

Calculate the total surface area of the cone shown:

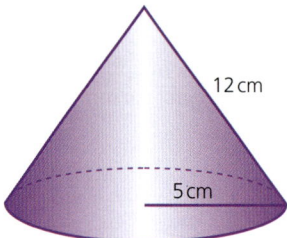

Surface area of base = $\pi r^2$
$= 25\pi \, cm^2$

The curved surface area can best be visualised if drawn as a sector as shown in the diagram:
The radius of the sector is equivalent to the slant height of the cone. The arc length of the sector is equivalent to the base circumference of the cone i.e. $10\pi$ cm.

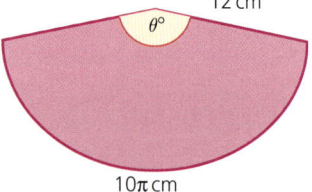

$$\frac{\theta}{360} = \frac{10\pi}{24\pi}$$

Therefore $\theta = 150$

Area of sector = $\frac{150}{360} \times \pi \times 12^2 = 60\pi \, cm^2$

Total surface area = $60\pi + 25\pi = 85\pi = 267$ (3 s.f.)

The total surface area is $267 \, cm^2$.

---

*The formula for the curved surface area of a cone, i.e. the area of the sector, can be calculated using the formula shown here. This formula will be given to you in any assessment.*

In the worked example, the area of the sector was calculated to be $60\pi \, cm^2$. This is therefore also the area of the curved surface of the cone.

The curved surface of a cone can also be calculated using the formula **Area = $\pi r l$**, where $r$ represents the radius of the circular base and $l$ the slant length of the cone.

The formula can be proved as follows.
Consider a cone with a slant length $l$ and a base radius $r$.

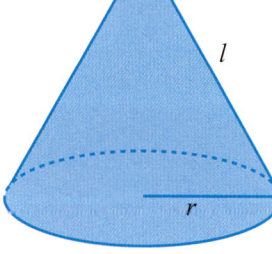

## The surface area of a cone

which can be cut and opened to form the sector shown:

The fraction that the area of the sector is to that of the whole circle is the same as the fraction of its arc length to that of the circumference, i.e. $\frac{2\pi r}{2\pi l}$

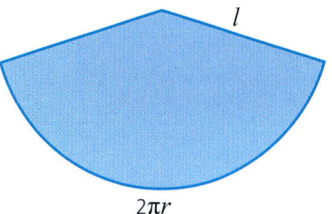

Therefore area of sector = $\frac{2\pi r}{2\pi l} \times \pi l^2 = \pi r l$

As the curved surface area of cone is the same as area of sector, its area is = $\pi r l$

In the worked example above, the curved surface area = $\pi \times 5 \times 12$
$= 60\pi \text{cm}^2$

### Exercise 6.27

1  Calculate the surface area of the following cones (you may use the formula to work out the area of the curved surface of each cone):

a 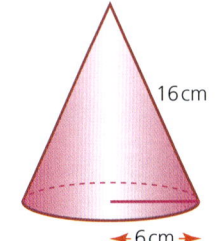   b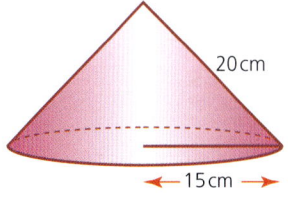

2  Two cones with the same base radius are stuck together as shown. Calculate the surface area of the shape.

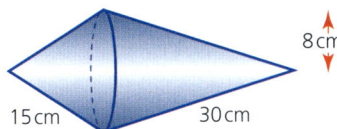

3  Two cones have the same total surface area.

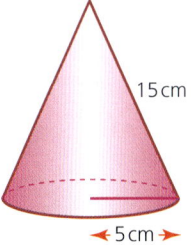   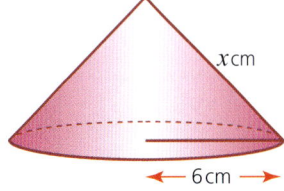

Calculate:
a  the total surface area of each cone
b  the value of $x$.

# Investigations, modelling and ICT 6

## Investigation: Tennis balls

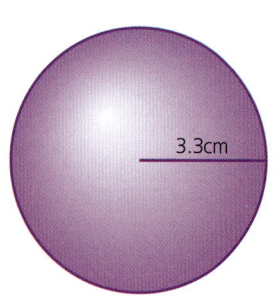

Tennis balls are spherical and have a radius of 3.3 cm.
A manufacturer wishes to make a cuboidal container with a lid to hold 12 tennis balls. The container is to be made of cardboard. The manufacturer wishes to use as little cardboard as possible.

1. Sketch some of the different containers that the manufacturer might consider.
2. For each container, calculate the total area of cardboard used and therefore decide on the most economical design.

## Modelling: Metal trays

A rectangular sheet of metal measures $30 \times 40$ cm.

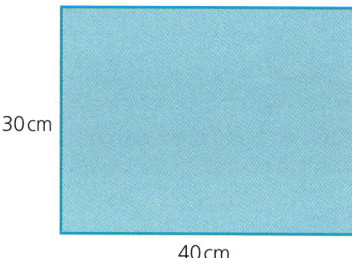

The sheet has squares of equal size cut from each corner. It is then folded to form a metal tray as shown:

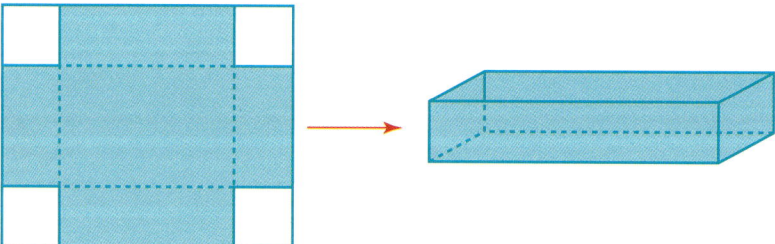

1  a  Calculate the length, width and height of the tray if a square of side length 1 cm is cut from each corner of the sheet of metal.
   b  Calculate the volume of this tray.
2  a  Calculate the length, width and height of the tray if a square of side length 2 cm is cut from each corner of the sheet of metal.
   b  Calculate the volume of this tray.
3  A square of side length $x$ cm is cut from each corner.
   a  What is the maximum length of $x$?
   b  Show that the volume ($V$) of the tray can be modelled by the formula $V = 4x^3 - 140x^2 + 1200x$.
4  Using your graphic display calculator to help:
   a  Sketch the graph of the function $V = 4x^3 - 140x^2 + 1200x$ for the valid domain of $x$.
   b  Find the maximum value of $V$ and the value of $x$ (to 1 d.p.) at which it occurs.
5  What range of values of $x$ is your model valid for?

# ICT Activity

In this topic you will have seen that it is possible to construct a cone from a sector. The dimensions of the cone are dependent on the dimensions of the sector. In this activity, you will be using a spreadsheet to investigate the maximum possible volume of a cone constructed from a sector of fixed radius.

Circles of radius 10 cm are cut from paper and used to construct cones. Different-sized sectors are cut from the circles and then arranged to form a cone. For example:

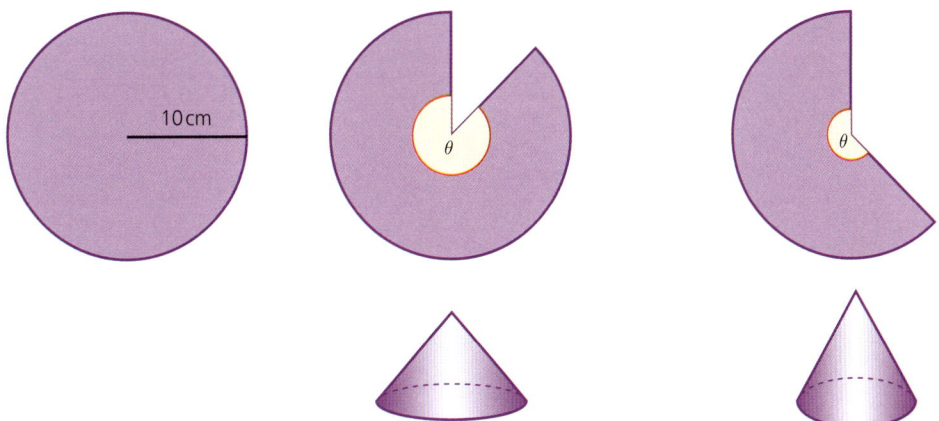

# INVESTIGATIONS, MODELLING AND ICT 6

1. Using a spreadsheet similar to the one below, calculate the maximum possible volume for a cone constructed from one of these circles.

| | A | B | C | D | E | F |
|---|---|---|---|---|---|---|
| 1 | Angle of sector ($\theta$) | Sector arc length (cm) | Base circumference of cone (cm) | Base radius of cone (cm) | Vertical height of cone (cm) | Volume of cone (cm$^3$) |
| 2 | 5 | 0.873 | 0.873 | 0.139 | 9.999 | 0.202 |
| 3 | 10 | 1.745 | 1.745 | 0.278 | 9.996 | 0.808 |
| 4 | 15 | 2.618 | 2.618 | 0.417 | 9.991 | 1.816 |
| 5 | 20 | | | | | |
| 6 | 25 | | | | | |
| 7 | 30 | | | | | |
| 8 | Continue to 355° | Enter formulae here to calculate the results for each column | | | | |

2. Plot a graph to show how the volume changes as $\theta$ increases. Comment on your graph.

# Student assessments 6

## Student assessment 1

1. Calculate the area of the shape below.

   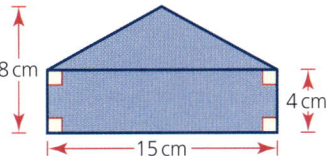

2. Calculate the circumference and area of each of the following circles. Give your answers to 3 s.f.

   a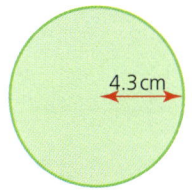
   
   b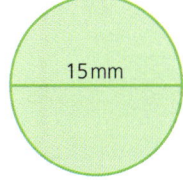

3. A semicircular shape is cut out of the side of a rectangle as shown. Calculate the shaded area, giving your answer in exact form.

   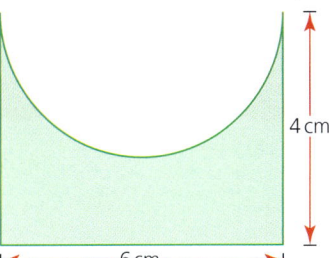

4. For the diagram (right), calculate the area of:
   a the semicircle
   b the trapezium
   c the whole shape.

   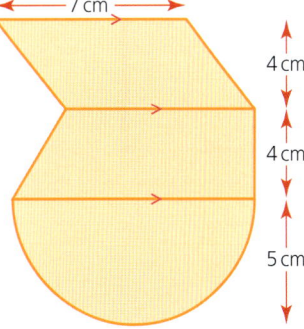

407

# STUDENT ASSESSMENTS 6

5  A cylindrical tube has an inner diameter of 6 cm, an outer diameter of 7 cm and a length of 15 cm. Calculate the following to 3 s.f.:

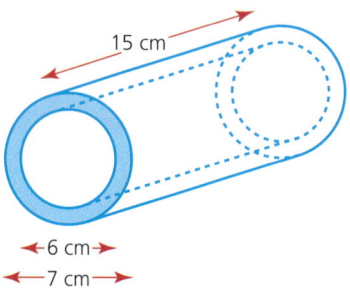

a  the surface area of the shaded end
b  the inside surface area of the tube
c  the total surface area of the tube.

6  Calculate the volume of each of the following cylinders. Give your answers in terms of π.

a 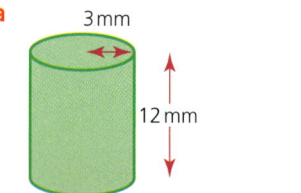  b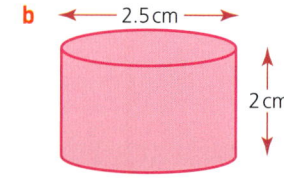

## Student assessment 2

1  Calculate the area of this shape:

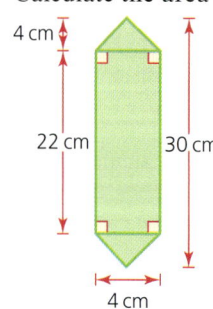

*Student assessments 6*

**2** A rectangle of length 32 cm and width 20 cm has a semicircle cut out of two of its sides as shown (below). Calculate the shaded area to 3 s.f.

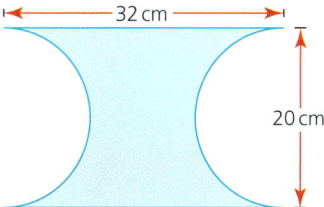

**3** Calculate the area of:

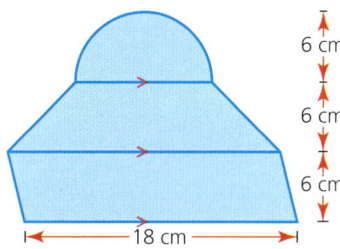

 **a** the semicircle
 **b** the parallelogram
 **c** the whole shape.

**4** A prism in the shape of a hollowed-out cuboid has dimensions as shown. If the end is square, calculate the volume of the prism.

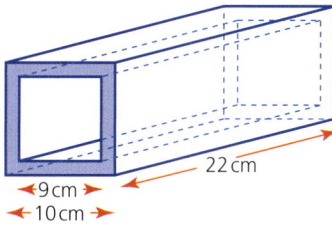

**5** Calculate the surface area of each of the following cylinders. Give your answers in terms of π.

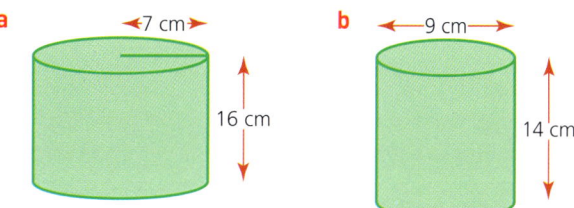

409

## Student assessment 3

1. Calculate the arc length of each of the following sectors. The angle $\theta$ and radius $r$ are given in each case.
   a. $\theta = 45°$      $r = 15\,\text{cm}$
   b. $\theta = 150°$     $r = 13.5\,\text{cm}$

2. Calculate the angle $\theta$ in each of the following sectors. The radius $r$ and arc length $a$ are given in each case.
   a. $r = 20\,\text{mm}$     $a = 95\,\text{mm}$
   b. $r = 9\,\text{cm}$      $a = 9\,\text{mm}$

3. Calculate the area of the sector shown below, giving your answer in exact form.

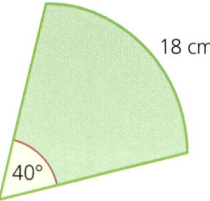

4. A sphere has a radius of 6.5 cm. Calculate to 3 s.f.:
   a. its total surface area
   b. its volume.

5. A pyramid with a base the shape of a regular hexagon is shown. If the length of each of its sloping edges is 24 cm, calculate:

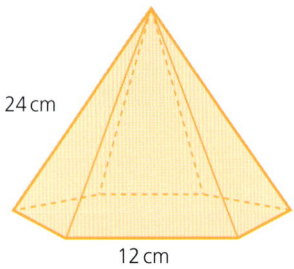

   a. its total surface area
   b. its volume.

## Student assessment 4

1. Calculate the arc length of the following sectors. The angle θ and radius $r$ are given in each case.
   a  θ = 255°      $r$ = 40 cm
   b  θ = 240°      $r$ = 16.3 mm
2. Calculate the angle θ in each of the following sectors. The radius $r$ and arc length $a$ are given in each case.
   a  $r$ = 40 cm    $a$ = 100 cm
   b  $r$ = 20 cm    $a$ = 10 mm
3. Calculate the area of the sector shown below:

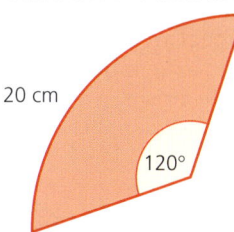

4. A hemisphere has a radius of 8 cm. Calculate to 1 d.p.:
   a  its total surface area      b  its volume.
5. A cone has its top cut as shown. Calculate the volume of the frustum.

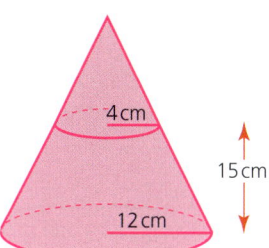

## Student assessment 5

1  The prism has a cross-sectional area in the shape of a sector.

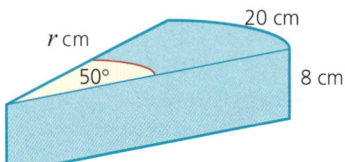

Calculate:
a the radius $r$ cm
b the cross-sectional area of the prism
c the total surface area of the prism
d the volume of the prism.

2  The cone and sphere shown (below) have the same volume.

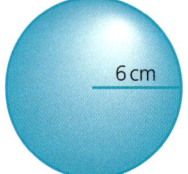 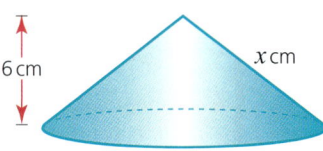

If the radius of the sphere and the height of the cone are both 6 cm, calculate each of the following. Give your answers in exact form.
a the volume of the sphere
b the base radius of the cone

3  The top of a cone is cut off and a hole, the shape of a cylinder, is drilled out of the remaining base.
Calculate:
a the height of the original cone
b the volume of the original cone
c the volume of the solid frustum
d the volume of the cylindrical hole
e the volume of the remaining cone.

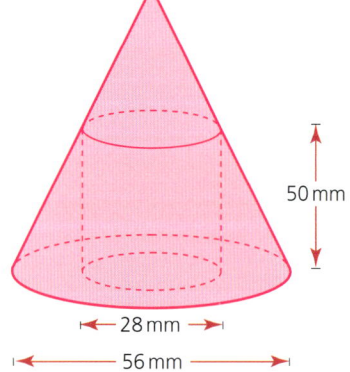

**4** A metal object is made from a hemisphere and a cone, both of base radius 12 cm. The height of the object when upright is 36 cm.

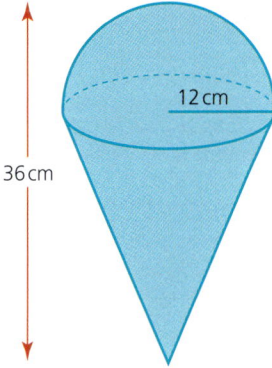

Calculate:
- **a** the volume of the hemisphere
- **b** the volume of the cone
- **c** the curved surface area of the hemisphere
- **d** the total surface area of the object.

# TOPIC 7

# Trigonometry

## Contents

Chapter 46 Pythagoras' theorem (C7.1, C7.2, E7.1, E7.2)
Chapter 47 Sine, cosine and tangent ratios (C7.2, E7.2)
Chapter 48 Special angles and their trigonometric ratios (E3.1, E7.3, E7.4)
Chapter 49 The sine and cosine rules (E7.5)
Chapter 50 Applications of trigonometry (E7.2, E7.6)
Chapter 51 Trigonometric graphs, properties and transformations (E3.1, E3.6, E7.4)

## Learning objectives

**E3.1**
Recognise the following function types from the shape of their graphs:

trigonometric $f(x) = a \sin(bx); a \cos(bx); \tan x$

**E3.6**
Describe and identify transformations to a graph of $y = f(x)$

When $y = f(x) + k$,
$y = f(x + k)$

**C7.1**      **E7.1**
Know and use Pythagoras' theorem

**C7.2**      **E7.2**
1   Know and use the sine, cosine and tangent ratios for acute angles in calculations involving sides and angles of a right-angled triangle
2   Solve problems in two dimensions using Pythagoras' theorem and trigonometry
3   Know that the perpendicular distance from a point to a line is the shortest distance to the line
4   Carry out calculations involving angles of elevation and depression

**E7.3**
Know the exact values of:

$\sin x$ and $\cos x$ for $x = 0°, 30°, 45°, 60°$ and $90°$

$\tan x$ for $x = 0°, 30°, 45°, 60°$

**E7.4**
Know the exact values of:

$\sin x$ and $\cos x$ for $x = 0°, 30°, 45°, 60°$ and $90°$

$\tan x$ for $x = 0°, 30°, 45°, 60°$

**E7.5**
1   Use the sine and cosine rules in calculations involving lengths and angles for any triangle
2   Use the formula area of triangle = $\frac{1}{2}ab \sin C$

**E7.6**
Carry out calculations and solve problems in three dimensions using Pythagoras' theorem and trigonometry, including calculating the angle between a line and a plane

Elements in purple refer to the Extended curriculum only.

# The Swiss

## Leonhard Euler

Euler, like Newton before him, was the greatest mathematician of his generation. He studied all areas of mathematics and continued to work hard after he had gone blind.

As a young man, Euler discovered and proved:

the sum of the infinite series $\sum \left(\dfrac{1}{n^2}\right) = \dfrac{\pi^2}{6}$

i.e. $\dfrac{1}{1^2} + \dfrac{1}{2^2} + \dfrac{1}{3^2} + \ldots + \dfrac{1}{n^2} = \dfrac{\pi^2}{6}$

This brought him to the attention of other mathematicians.

Leonhard Euler (1707–1783)

Euler made discoveries in many areas of mathematics, especially calculus and trigonometry. He also developed the ideas of Newton and Leibniz.

Euler worked on graph theory and functions and was the first to prove several theorems in geometry. He studied relationships between a triangle's height, midpoint and circumscribing and inscribing circles, and he also discovered an expression for the volume of a tetrahedron (a triangular pyramid) in terms of its sides.

He also worked on number theory and found the largest prime number known at the time.

Some of the most important constant symbols in mathematics, π, e and i (the square root of –1), were introduced by Euler.

## The Bernoulli family

The Bernoullis were a family of Swiss merchants who were friends of Euler. The two brothers, Johann and Jacob, were very gifted mathematicians and scientists, as were their children and grandchildren. They made discoveries in calculus, trigonometry and probability theory in mathematics. In science, they worked on astronomy, magnetism, mechanics, thermodynamics and more.

Unfortunately, many members of the Bernoulli family were not pleasant people. The older members of the family were jealous of each other's successes and often stole the work of their sons and grandsons and pretended that it was their own.

# 46 Pythagoras' theorem

Pythagoras' theorem states the relationship between the lengths of the three sides of a right-angled triangle.

Pythagoras' theorem states that:

$$a^2 = b^2 + c^2$$

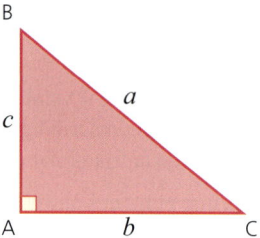

There are many proofs of Pythagoras' theorem. One of them is shown below.

Consider four congruent (identical) right-angled triangles:

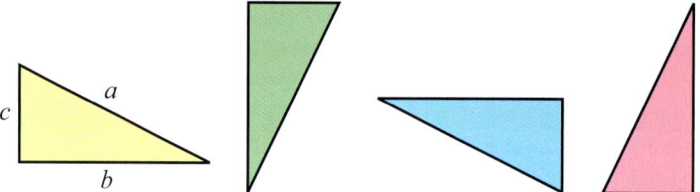

Three are rotations of the first triangle by 90°, 180° and 270° clockwise respectively.

Each triangle has an area equal to $\frac{bc}{2}$.

The triangles can be joined together to form a square of side length $a$ as shown:

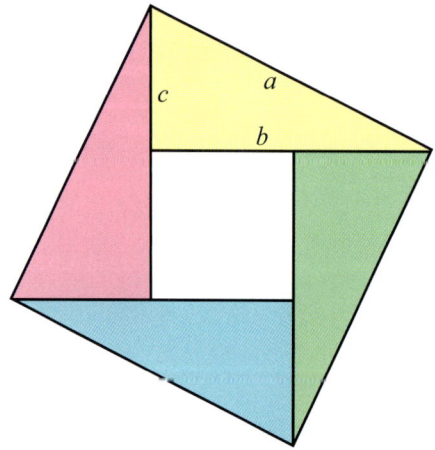

# Pythagoras' theorem

The square has a hole at its centre of side length $b - c$.

Therefore, the area of the centre square is
$$(b - c)^2 = b^2 - 2bc + c^2$$

The area of the four triangles is $4 \times \frac{bc}{2} = 2bc$

The area of the large square is $a^2$

Therefore
$$a^2 = (b - c)^2 + 2bc$$
$$a^2 = b^2 - 2bc + c^2 + 2bc$$
$$a^2 = b^2 + c^2$$

hence proving Pythagoras' theorem.

## → Worked examples

**a** Calculate the length of the side BC.

Using Pythagoras:
$$a^2 = b^2 + c^2$$
$$a^2 = 8^2 + 6^2$$
$$a^2 = 64 + 36 = 100$$
$$a = \sqrt{100}$$
$$a = 10$$
$$BC = 10\,\text{m}$$

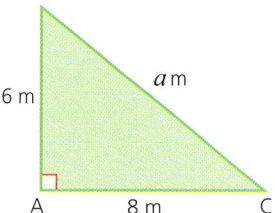

**b** Calculate the length of the side AC.

Using Pythagoras:
$$a^2 = b^2 + c^2$$
$$a^2 - c^2 = b^2$$
$$b^2 = 144 - 25 = 119$$
$$b = \sqrt{119}$$
$$b = 10.9 \text{ (3 s.f.)}$$
$$AC = 10.9\,\text{m} \text{ (3 s.f.)}$$

*Leaving the answer as $\sqrt{119}$ is in **exact** form.*

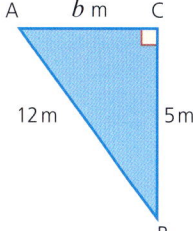

The converse of Pythagoras' theorem can also be used to show whether or not a triangle is right-angled.

In the triangle ABC, the lengths of the three sides are given, but it is not indicated whether any of the angles are a right angle. It is important not to assume that a triangle is right-angled just because it may look like it is.

# 46 PYTHAGORAS' THEOREM

If triangle ABC is right-angled, then it will satisfy Pythagoras' theorem.

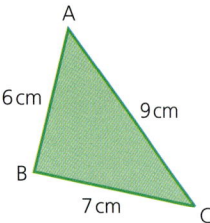

That is $(AC)^2 = (AB)^2 + (BC)^2$

$$9^2 = 6^2 + 7^2$$
$$81 = 36 + 49$$
$$81 = 85$$

$(AC)^2 = 9^2 = 81$

$(AB)^2 + (BC)^2 = 6^2 + 7^2 = 85$

As $81 \neq 85$, the triangle cannot be right-angled.

**Exercise 7.1** In each of the diagrams in Q.1 and Q.2, use Pythagoras' theorem to calculate the length of the marked side.

**1 a** 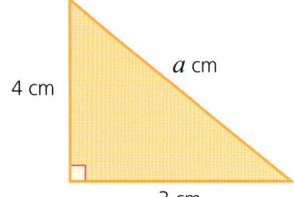    **b**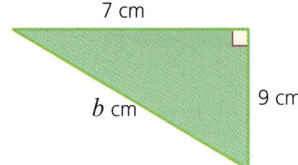

**c** 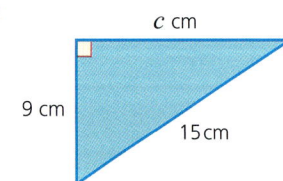    **d**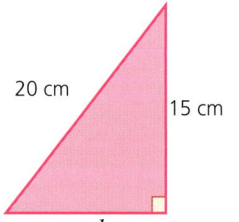

**2 a** 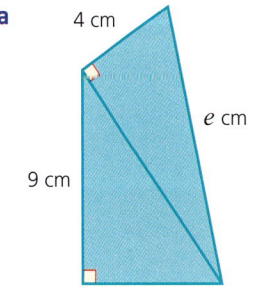    **b**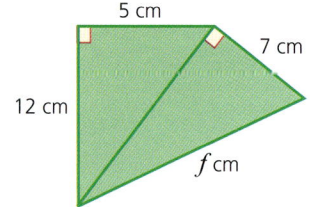

# Pythagoras' theorem

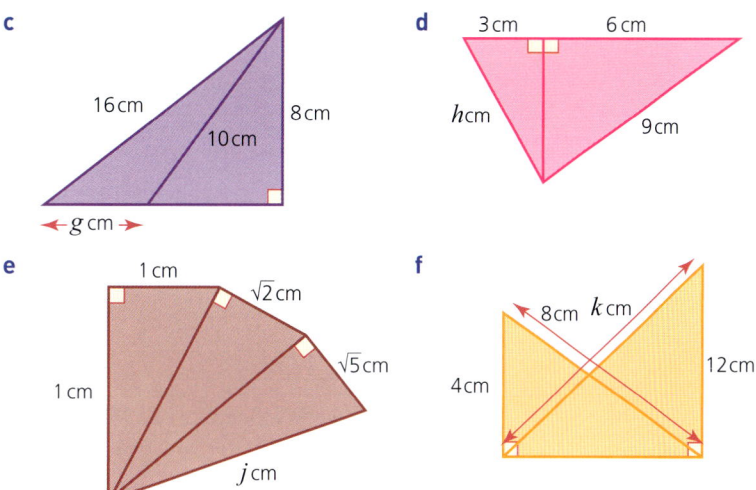

3  By applying Pythagoras' theorem, decide which of the following triangles are right-angled.

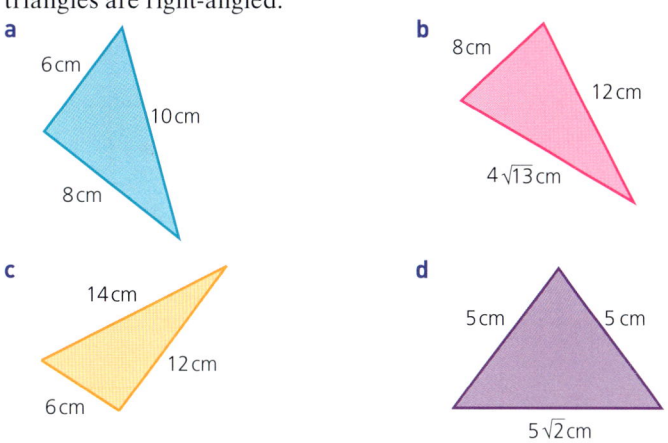

4  Villages A, B and C lie on the edge of the Namib Desert. Village A is 30 km due north of village C. Village B is 65 km due east of A. Calculate the shortest distance between villages C and B, giving your answer to the nearest 0.1 km.

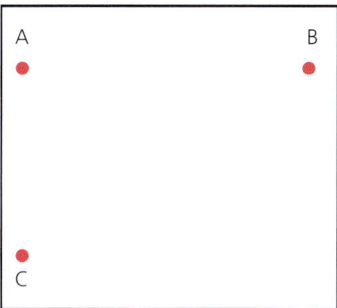

## 46 PYTHAGORAS' THEOREM

**Exercise 7.1 (cont)**

5  Town X is 54 km due west of town Y. The shortest distance between town Y and town Z is 86 km. If town Z is due south of X, calculate the distance between X and Z, giving your answer to the nearest kilometre.

6  Village B is on a bearing of 135° and at a distance of 40 km from village A. Village C is on a bearing of 225° and a distance of 62 km from village A.

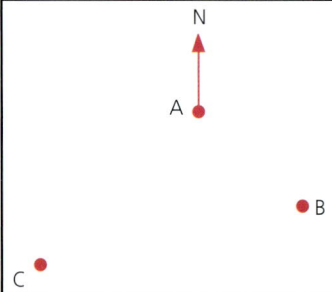

  a  Show that triangle ABC is right-angled.
  b  Calculate the distance from B to C, giving your answer to the nearest 0.1 km.

7  Two boats set off from X at the same time. Boat A sets off on a bearing of 325° and with a velocity of 14 km/h. Boat B sets off on a bearing of 235° with a velocity of 18 km/h.

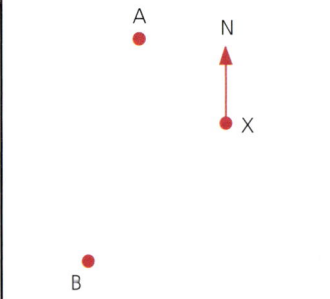

Calculate the distance between the boats after they have been travelling for 2.5 hours. Give your answer to the nearest metre.

**8** A boat sets off on a trip from S. It heads towards B, a point 6 km away and due north. At B, it changes direction and heads towards point C, also 6 km away and due east of B. At C, it changes direction once again and heads on a bearing of 135° towards D, which is 13 km from C.

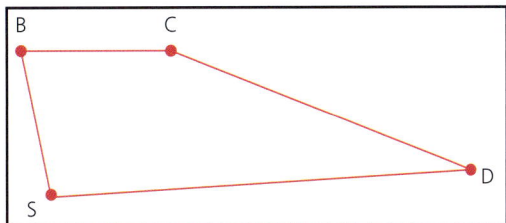

  **a** Calculate the distance between S and C to the nearest 0.1 km.
  **b** Calculate the distance the boat will have to travel if it is to return to S from D.

**9** Two trees are standing on flat ground.

The height of the smaller tree is 7 m. The distance between the top of the smaller tree and the base of the taller tree is 15 m.

The distance between the top of the taller tree and the base of the smaller tree is 20 m.

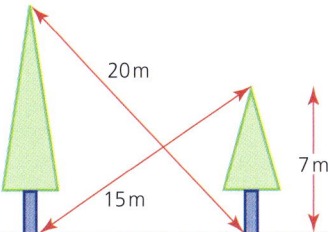

  **a** Calculate the horizontal distance between the two trees.
  **b** Calculate the height of the taller tree.

## Exercise 7.2

**1** The diagram below shows a scale model of a garage. Its width is 5 cm, its length 10 cm and the height of its walls 6 cm.

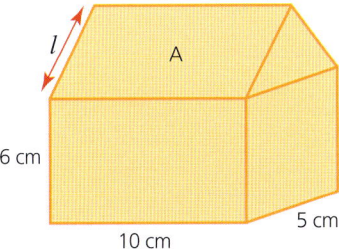

  **a** If the width of the real garage is 4 m, calculate:
   **i** the length of the real garage
   **ii** the real height of the garage wall.
  **b** If the apex of the roof of the real garage is 2 m above the top of the walls, use Pythagoras' theorem to find the real slant length $l$.
  **c** What is the actual area of the roof section marked A?

# 46 PYTHAGORAS' THEOREM

**Exercise 7.2 (cont)**

2  The triangles ABC and XYZ below are similar.

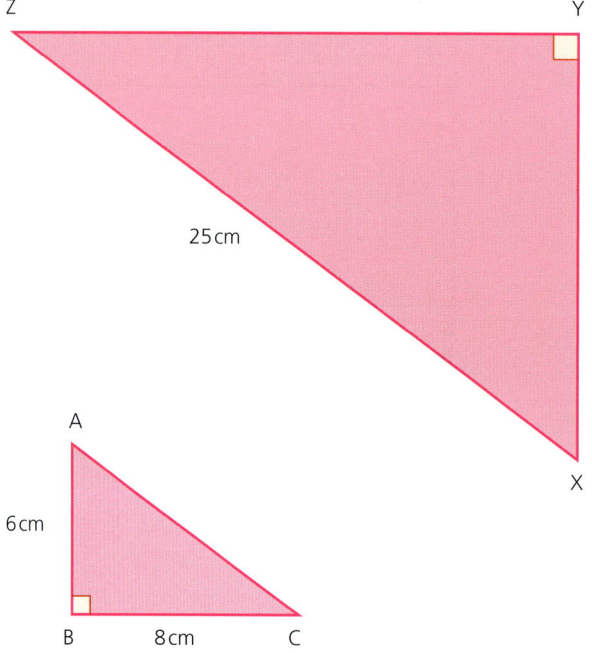

  a  Using Pythagoras' theorem calculate the length of AC.
  b  Calculate the scale factor of enlargement.
  c  Calculate the area of △XYZ.

3  Copy each of the diagrams below and calculate the length of the side marked $y$ cm in each case. Assume that the lines drawn from points on the circumference are tangents.

  a
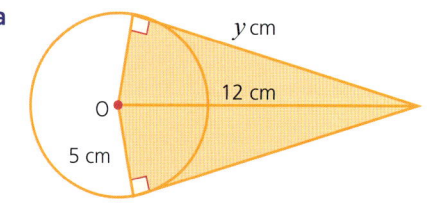

  b
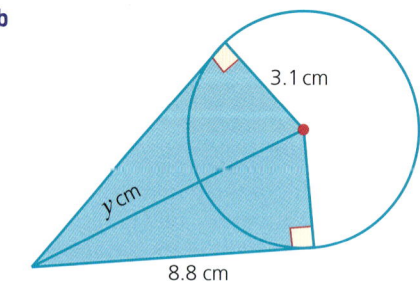

# 47 Sine, cosine and tangent ratios

There are three basic trigonometric ratios: sine, cosine and tangent.

Each of these relates an angle of a right-angled triangle to a ratio of the lengths of two of its sides.

The sides of the triangle have names, two of which are dependent on their position in relation to a specific angle. The longest side (always opposite the right angle) is called the **hypotenuse**. The side opposite the angle is called the **opposite** side and the side next to the angle is called the **adjacent** side.

Note that, when the chosen angle is at A, the sides labelled opposite and adjacent change as shown:

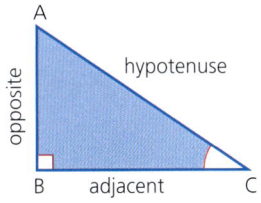 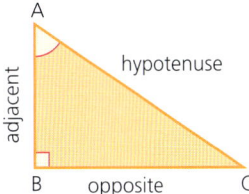

## Tangent

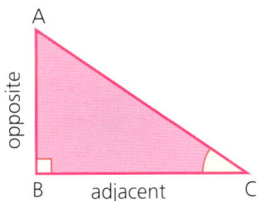

$$\tan C = \frac{\text{length of opposite side}}{\text{length of adjacent side}}$$

### ➡ Worked examples

a  Calculate the size of angle BAC in the following triangles:

i
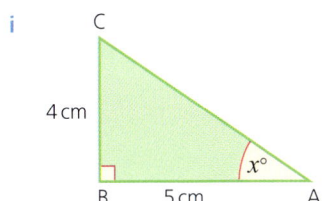

# 47 SINE, COSINE AND TANGENT RATIOS

$$\tan x = \frac{\text{opposite}}{\text{adjacent}} = \frac{4}{5}$$

$$x = \tan^{-1}\left(\frac{4}{5}\right)$$

$$x = 38.7 \text{ (1 d.p.)}$$

angle BAC = 38.7° (1 d.p.)

ii

[Triangle with C at top, right angle at B, angle x° at A; CB = 8 cm, BA = 3 cm]

$$\tan x° = \frac{8}{3}$$

$$x = \tan^{-1}\left(\frac{8}{3}\right)$$

$$x = 69.4 \text{ (1 d.p.)}$$

angle BAC = 69.4° (1 d.p.)

**b** Calculate the length of the opposite side QR.

$$\tan 42° = \frac{p}{6}$$

$$6 \times \tan 42° = p$$

$$p = 5.40 \text{ (3 s.f.)}$$

QR = 5.40 cm (3 s.f.)

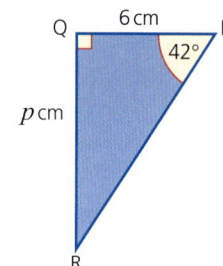

**c** Calculate the length of the adjacent side XY.

$$\tan 35° = \frac{6}{z}$$

$$z \times \tan 35° = 6$$

$$z = \frac{6}{\tan 35°}$$

$$z = 8.57 \text{ (3 s.f.)}$$

XY = 8.57 cm (3 s.f.)

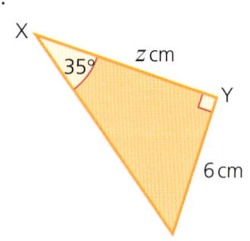

# Tangent

**Exercise 7.3** Calculate the length of the side marked $x$ cm in each of the diagrams in Q.1 and 2. Give your answers to 3 s.f.

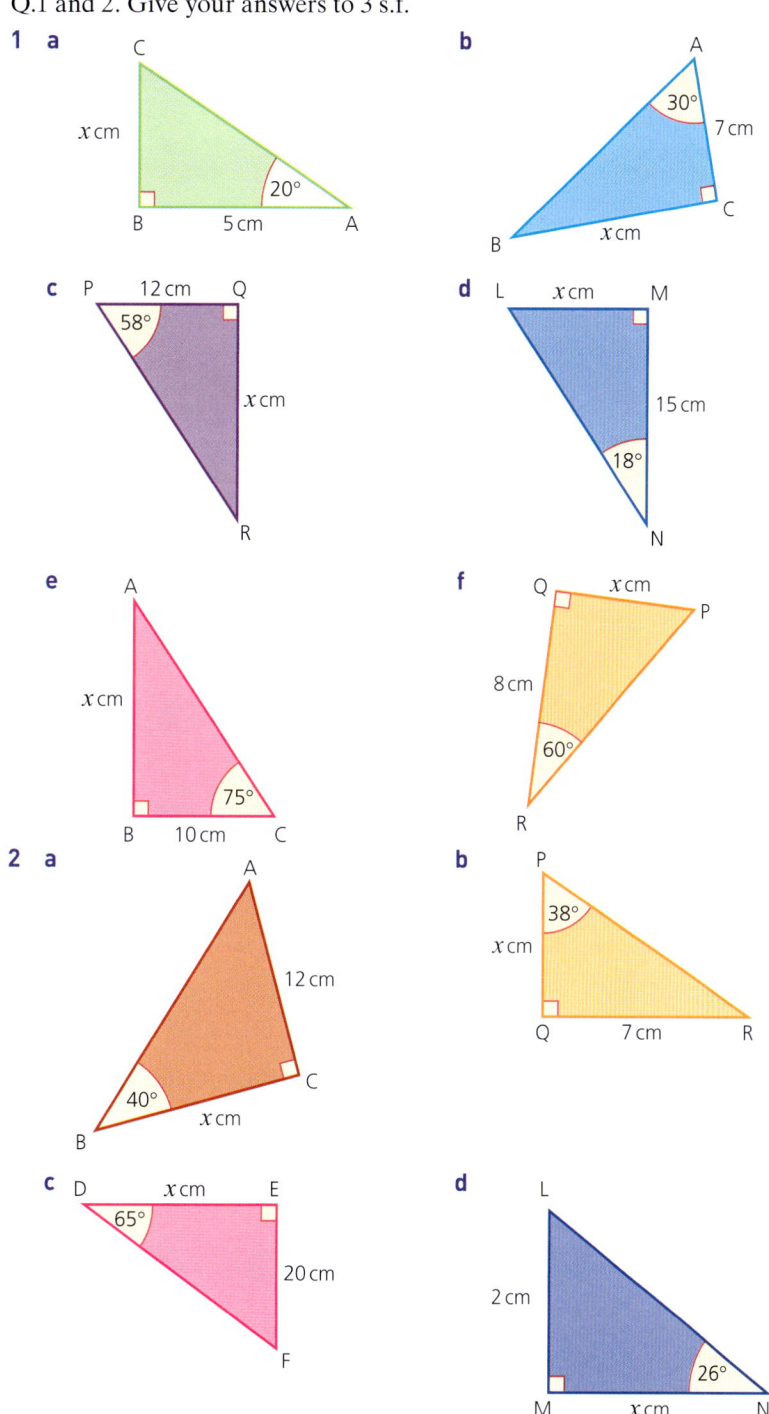

# 47 SINE, COSINE AND TANGENT RATIOS

## Exercise 7.3 (cont)

e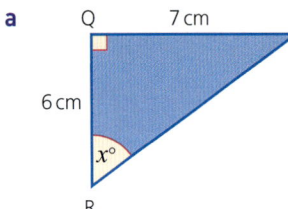

f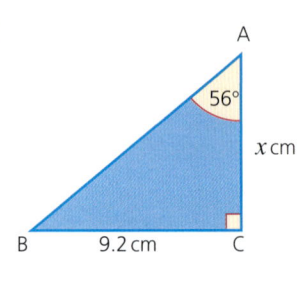

**3** Calculate the size of the marked angle $x°$ in each of the following diagrams. Give your answers to 1 d.p.

a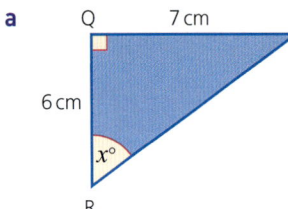

b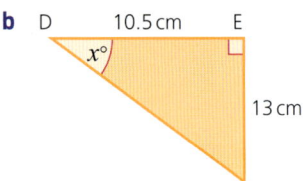

c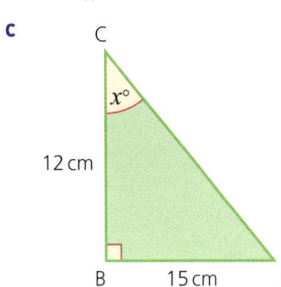

d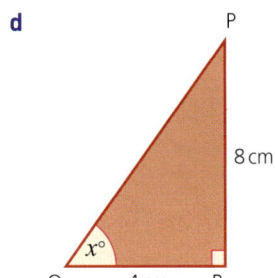

e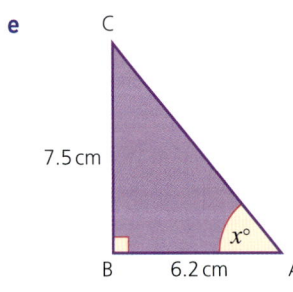

f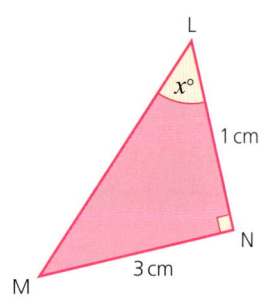

## Sine

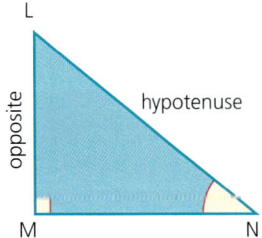

$$\sin N = \frac{\text{length of opposite side}}{\text{length of hypotenuse}}$$

# Sine

sin⁻¹ means the angle whose sin is.

> **Worked examples**

a Calculate the size of angle BAC.

$\sin x = \dfrac{\text{opposite}}{\text{hypotenuse}} = \dfrac{7}{12}$

$x = \sin^{-1}\left(\dfrac{7}{12}\right)$

$x = 35.7$ (1 d.p.)

angle BAC = 35.7° (1 d.p.)

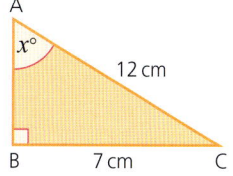

b Calculate the length of the hypotenuse PR.

$\sin 18° = \dfrac{11}{q}$

$q \times \sin 18° = 11$

$q = \dfrac{11}{\sin 18°}$

$q = 35.6$ (3 s.f.)

PR = 35.6 cm (3 s.f.)

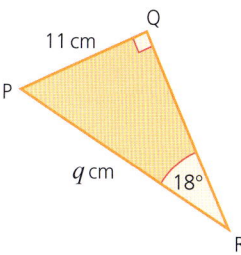

## Exercise 7.4

1 Calculate the length of the marked side in each of the following diagrams. Give your answers to 3 s.f.

a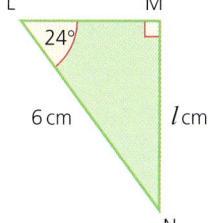

b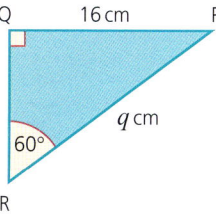

c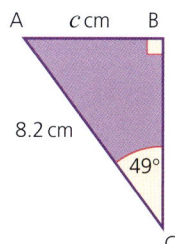

d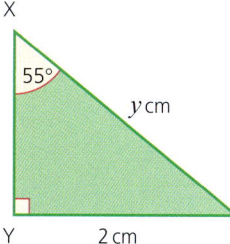

e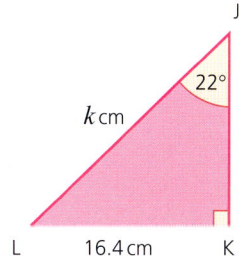

f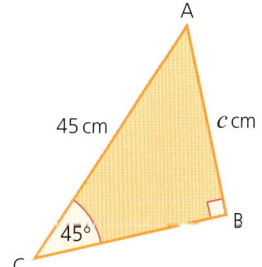

# 47 SINE, COSINE AND TANGENT RATIOS

**Exercise 7.4 (cont)**

2  Calculate the size of the angle marked $x$ in each of the following diagrams. Give your answers to 1 d.p.

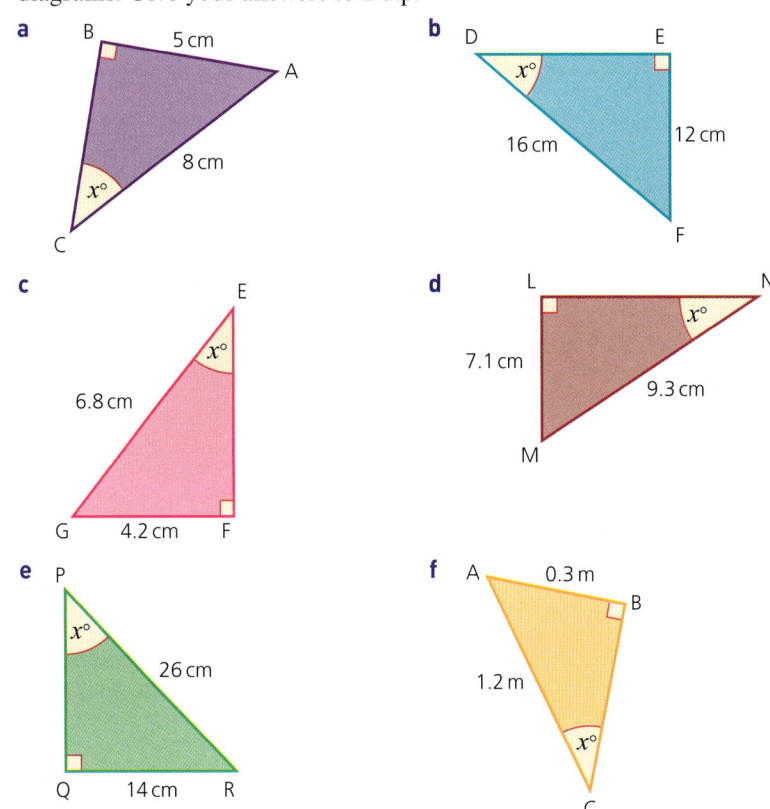

## Cosine

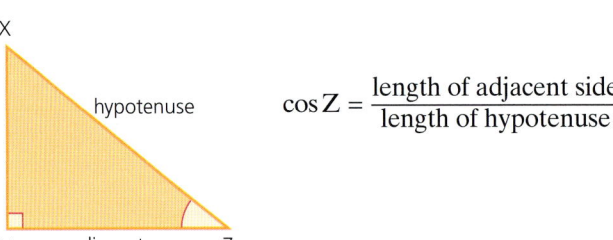

$$\cos Z = \frac{\text{length of adjacent side}}{\text{length of hypotenuse}}$$

# Cosine

## → Worked examples

**a** Calculate the length XY.

$$\cos 62° = \frac{\text{adjacent}}{\text{hypotenuse}} = \frac{z}{20}$$

$$z = 20 \times \cos 62°$$

$$z = 9.39 \text{ (3 s.f.)}$$

$$XY = 9.39 \text{ cm (3 s.f.)}$$

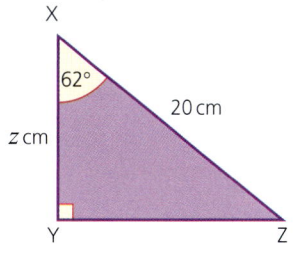

**b** Calculate the size of angle ABC.

$$\cos x = \frac{5.3}{12}$$

$$x = \cos^{-1}\left(\frac{5.3}{12}\right)$$

$$x = 63.8 \text{ (1 d.p.)}$$

angle ABC = 63.8° (1 d.p.)

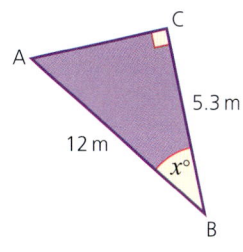

## Exercise 7.5

Calculate either the marked side or angle in each of the following diagrams. Give your answers to 1 d.p.

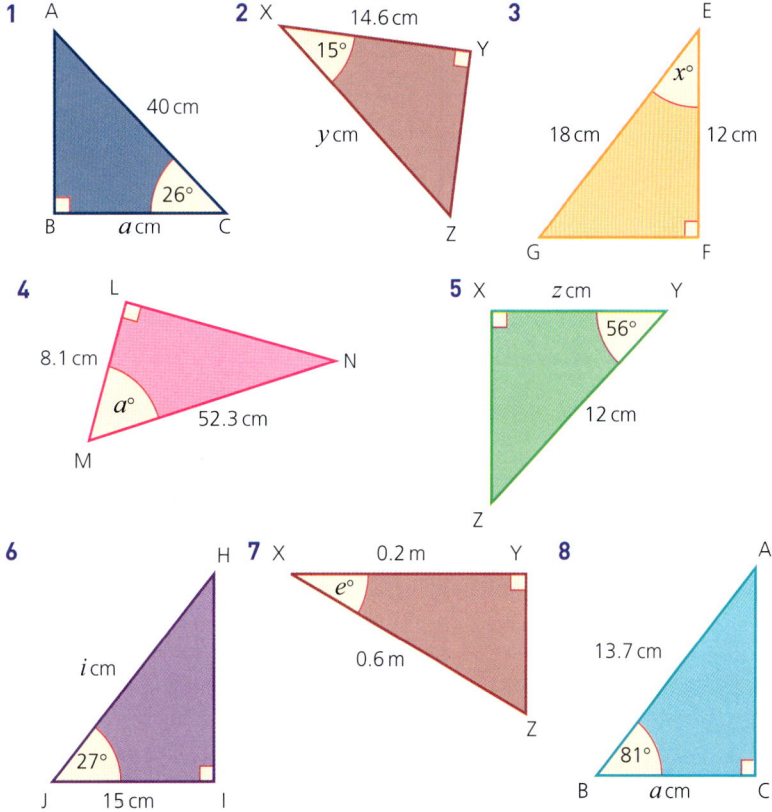

# 47 SINE, COSINE AND TANGENT RATIOS

**Exercise 7.6**

1. By using Pythagoras' theorem, trigonometry or both, calculate the marked value in each of the following diagrams. In each case give your answer to 1 d.p.

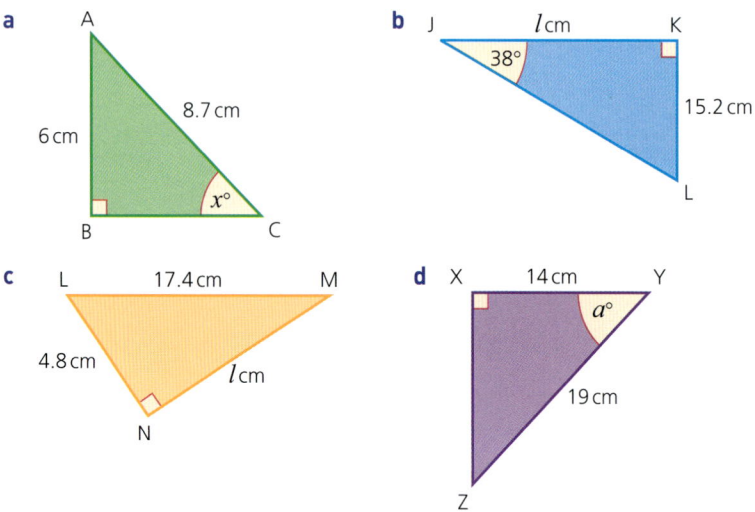

2. A sailing boat sets off from a point X and heads towards Y, a point 17 km north. At point Y it changes direction and heads towards point Z, a point 12 km away on a bearing of 090°. Once at Z the crew want to sail back to X.

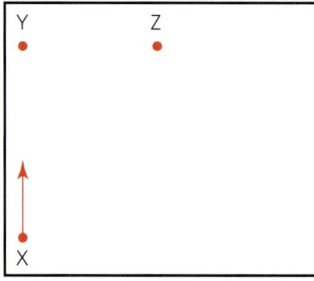

Calculate:
   a  the distance ZX
   b  the bearing of X from Z.

3. An aeroplane sets off from G on a bearing of 024° towards H, a point 250 km away. At H, it changes course and heads towards J on a bearing of 055° and a distance of 180 km away.
   a  How far is H to the north of G?
   b  How far is H to the east of G?
   c  How far is J to the north of H?
   d  How far is J to the east of H?
   e  What is the shortest distance between G and J?
   f  What is the bearing of G from J?

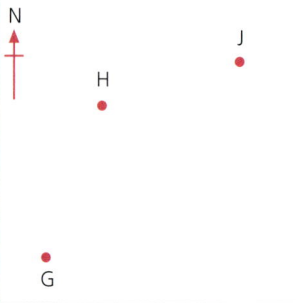

## Cosine

4  Two trees are standing on flat ground. The angle of elevation of their tops from a point X on the ground is 40°. If the horizontal distance between X and the small tree is 8 m and the distance between the tops of the two trees is 20 m, calculate:
   a  the height of the small tree
   b  the height of the tall tree
   c  the horizontal distance between the trees.

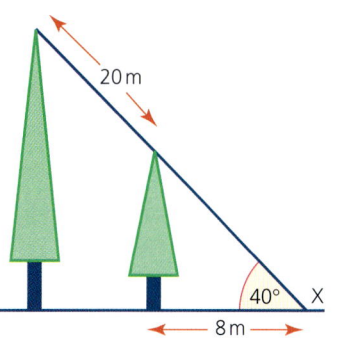

5  PQRS is a quadrilateral. The sides RS and QR are the same length. The sides QP and RS are parallel.
   Calculate:
   a  angle SQR
   b  angle PSQ
   c  length PQ
   d  length PS
   e  the area of PQRS.

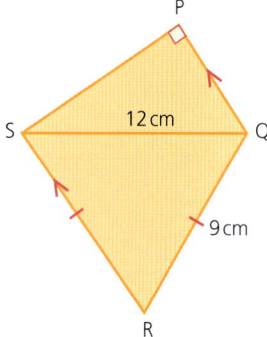

# 48 Special angles and their trigonometric ratios

So far, most of the angles you have worked with have required the use of a calculator in order to calculate their sine, cosine or tangent. However, some angles produce exact values and a calculator is both unnecessary and unhelpful when exact solutions are required.

There are a number of angles which have 'neat' trigonometric ratios, for example 0°, 30°, 45°, 60° and 90°. Their trigonometric ratios are derived below.

Consider the right-angled isosceles triangle ABC.
Let the perpendicular sides AC and BC each have a length of 1 unit.

As △ABC is isosceles, angle ABC = angle CAB = 45°.

Using Pythagoras' theorem, AB can also be calculated:

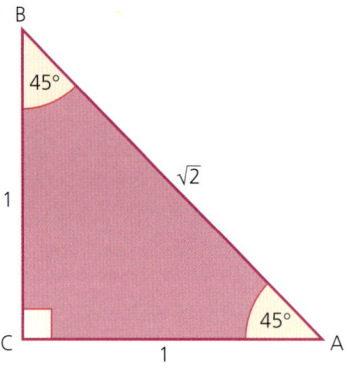

$(AB)^2 = (AC)^2 + (BC)^2$

$(AB)^2 = 1^2 + 1^2 = 2$

$AB = \sqrt{2}$

From the triangle, it can be deduced that $\sin 45° = \frac{1}{\sqrt{2}}$.

When rationalised, this can be written as $\sin 45° = \frac{\sqrt{2}}{2}$.

Similarly, $\cos 45° = \frac{1}{\sqrt{2}} = \frac{\sqrt{2}}{2}$

Therefore $\sin 45° = \cos 45°$

$\tan 45° = \frac{1}{1} = 1$

Consider also the equilateral triangle XYZ in which each of its sides has a length of 2 units.

## Special angles and their trigonometric ratios

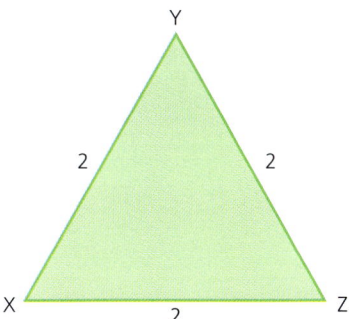

If a vertical line is dropped from the vertex Y until it meets the base XZ at P, the triangle is bisected. Consider now the right-angled triangle XYP.

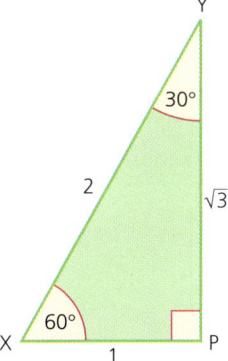

angle XYP = 30° as it is half of angle XYZ.

XP = 1 unit length as it is half of XZ.

The length YP can be calculated using Pythagoras' theorem:

$$(XY)^2 = (XP)^2 + (YP)^2$$
$$(YP)^2 = (XY)^2 - (XP)^2 = 2^2 - 1^2 = 3$$
$$YP = \sqrt{3}$$

Therefore from this triangle the following trigonometric ratios can be deduced:

$\sin 30° = \dfrac{1}{2}$  $\quad \cos 30° = \dfrac{\sqrt{3}}{2}$  $\quad \tan 30° = \dfrac{1}{\sqrt{3}} = \dfrac{\sqrt{3}}{3}$

$\sin 60° = \dfrac{\sqrt{3}}{2}$  $\quad \cos 60° = \dfrac{1}{2}$  $\quad \tan 60° = \dfrac{\sqrt{3}}{1} = \sqrt{3}$

# 48 SPECIAL ANGLES AND THEIR TRIGONOMETRIC RATIOS

These results and those obtained from the trigonometric graphs shown on the next page are summarised in the table below:

| Angle ($\theta$) | sin ($\theta$) | cos ($\theta$) | tan ($\theta$) |
|---|---|---|---|
| 0° | 0 | 1 | 0 |
| 30° | $\frac{1}{2}$ | $\frac{\sqrt{3}}{2}$ | $\frac{1}{\sqrt{3}} = \frac{\sqrt{3}}{3}$ |
| 45° | $\frac{1}{\sqrt{2}} = \frac{\sqrt{2}}{2}$ | $\frac{1}{\sqrt{2}} = \frac{\sqrt{2}}{2}$ | 1 |
| 60° | $\frac{\sqrt{3}}{2}$ | $\frac{1}{2}$ | $\sqrt{3}$ |
| 90° | 1 | 0 | – |

There are other angles which have the same trigonometric ratios as those shown in the table. The following section explains why, using a unit circle, i.e. a circle with a radius of 1 unit.

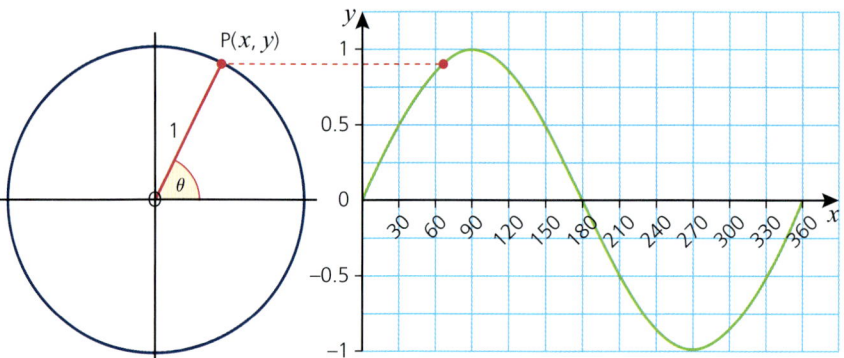

In the diagram above, P is a point on the circumference of a circle with centre at O and a radius of 1 unit. P has coordinates $(x, y)$. As the position of P changes, then so does the angle $\theta$.

$\sin \theta = \frac{y}{1} = y$ i.e. the sine of the angle $\theta$ is represented by the $y$-coordinate of P.

The graph therefore shows the different values of $\sin \theta$ as $\theta$ varies. A more complete diagram is shown below. Note that the angle $\theta$ is measured anticlockwise from the positive $x$-axis.

## Special angles and their trigonometric ratios

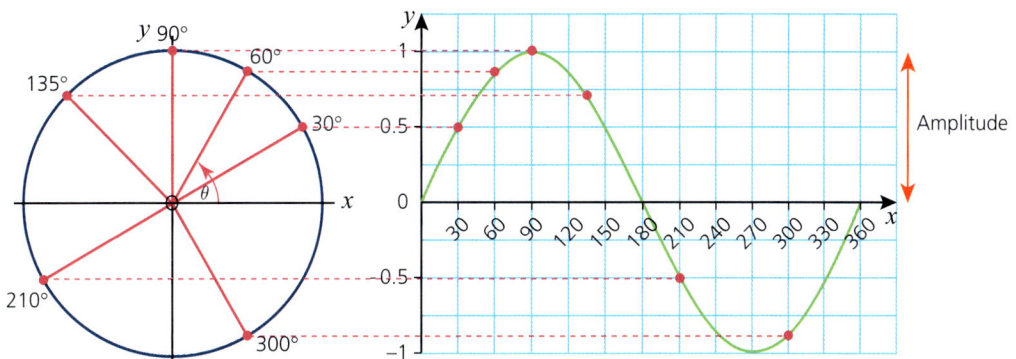

The graph of $y = \sin x$ has:
- a period of 360° (i.e. it repeats itself every 360°)
- a maximum value of +1
- a minimum value of −1
- an amplitude of 1
- symmetry, e.g. $\sin \theta = \sin (180 - \theta)$.

Similar diagrams and graphs can be constructed for $\cos \theta$ and $\tan \theta$.

From the unit circle, it can be deduced that $\cos \theta = \frac{x}{1} = x$,

i.e. the cosine of the angle $\theta$ is represented by the $x$-coordinate of P. Since $\cos \theta = x$, to be able to compare the graphs, the circle should be rotated through 90° as shown.

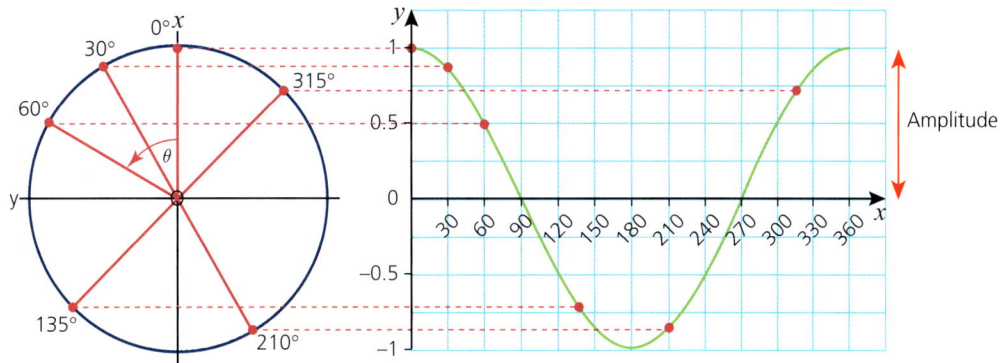

The properties of the cosine curve are similar to those of the sine curve. It has:
- a period of 360°
- a maximum value of +1
- a minimum value of −1
- an amplitude of 1
- symmetry, e.g. $\cos \theta = \cos (360 - \theta)$

The cosine curve is a translation of the sine curve of −90°, i.e. $\cos \theta = \sin (\theta + 90)$.

435

# 48 SPECIAL ANGLES AND THEIR TRIGONOMETRIC RATIOS

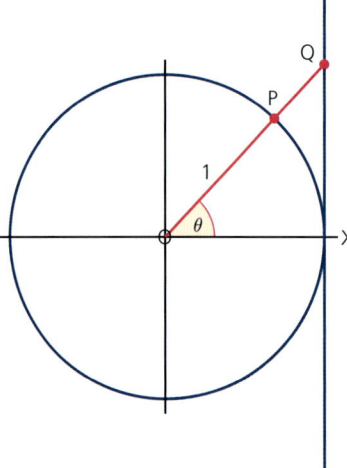

From the unit circle it can be deduced that $\tan \theta = \frac{y}{x}$.

In order to compare all the graphs, a tangent to the unit circle is drawn at (1, 0). OP is extended to meet the tangent at Q as shown.

As OX = 1 (radius of the unit circle), $\tan \theta = \frac{QX}{OX} = QX$.

i.e. $\tan \theta$ is equal to the $y$-coordinate of Q.

The graph of $\tan \theta$ is therefore shown below:

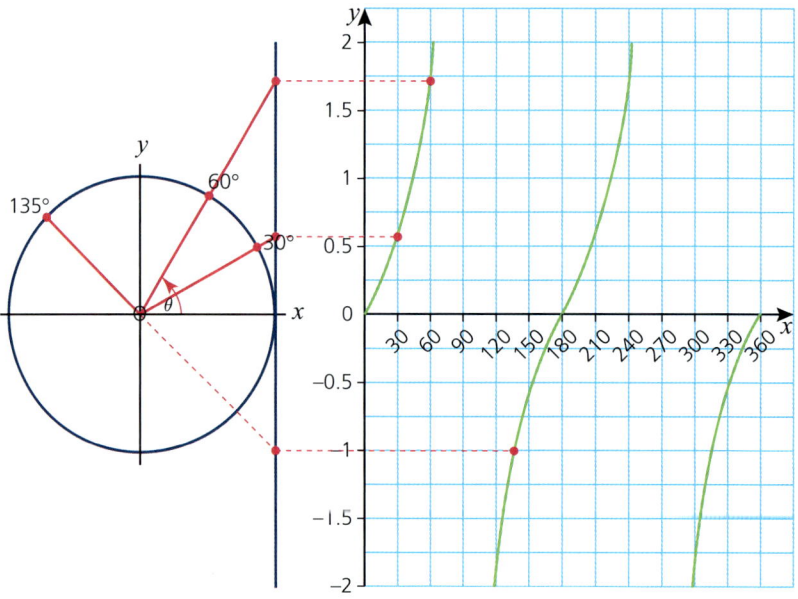

The graph of $\tan \theta$ has:
- a period of 180°
- no maximum or minimum value
- symmetry
- asymptotes at 90° and 270°

## Worked examples

**a** sin 30° = 0.5. Which other angle between 0° and 360° has a sine of 0.5?

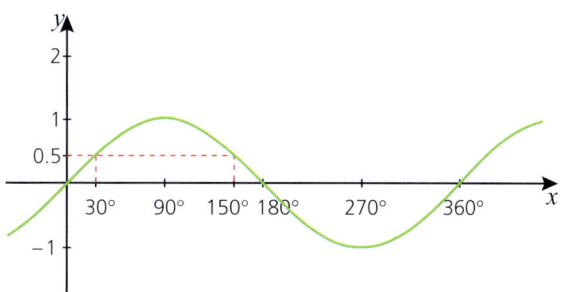

From the graph above it can be seen that sin 150° = 0.5.
Also sin $x$ = sin (180° − $x$); therefore sin 30° = sin (180° − 30) = sin 150°.

**b** cos 60° = 0.5. Which other angle between 0° and 360° has a cosine of 0.5?

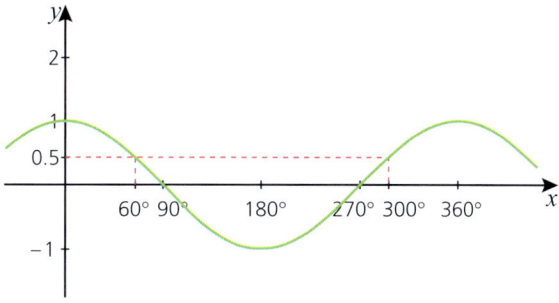

From the graph above it can be seen that cos 300° = 0.5.
Also cos $x$ = cos (360° − $x$); therefore cos 60° = cos (360° − 60) = cos 300°

**c** The cosine of which angle between 0° and 180° is equal to the negative of cos 50°?

cos 50° has the same magnitude but different sign to cos 130° because of the symmetrical properties of the cosine curve.

Therefore cos 130° = −cos 50°

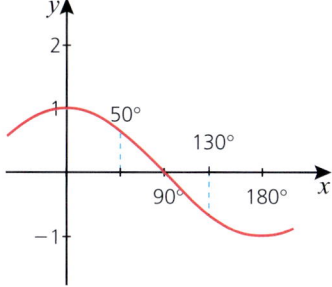

# 48 SPECIAL ANGLES AND THEIR TRIGONOMETRIC RATIOS

**Exercise 7.7**

1  Write each of the following in terms of the sine of another angle between 0° and 360°:
   a  sin 60°
   b  sin 80°
   c  sin 115°
   d  sin 200°
   e  sin 300°
   f  sin 265°

2  Write each of the following in terms of the sine of another angle between 0° and 360°:
   a  sin 35°
   b  sin 50°
   c  sin 30°
   d  sin 248°
   e  sin 304°
   f  sin 327°

3  Find the two angles between 0° and 360° which have the following sine. Give each angle to the nearest degree.
   a  0.33
   b  0.99
   c  0.09
   d  $-\frac{1}{2}$
   e  $-\frac{\sqrt{3}}{2}$
   f  $-\frac{1}{\sqrt{2}}$

4  Find the two angles between 0° and 360° which have the following sine. Give each angle to the nearest degree.
   a  0.94
   b  0.16
   c  0.80
   d  −0.56
   e  −0.28
   f  −0.33

**Exercise 7.8**

1  Write each of the following in terms of the cosine of another angle between 0° and 360°:
   a  cos 20°
   b  cos 85°
   c  cos 32°
   d  cos 95°
   e  cos 147°
   f  cos 106°

2  Write each of the following in terms of the cosine of another angle between 0° and 360°:
   a  cos 98°
   b  cos 144°
   c  cos 160°
   d  cos 183°
   e  cos 211°
   f  cos 234°

3  Write each of the following in terms of the cosine of another angle between 0° and 180°:
   a  −cos 100°
   b  −cos 90°
   c  −cos 110°
   d  −cos 45°
   e  −cos 122°
   f  −cos 25°

4  The cosine of which acute angle has the same value as:
   a  −cos 100°
   b  −cos 105°
   c  −cos 120°
   d  −cos 98°
   e  −cos 92°
   f  −cos 110°?

5  Explain with reference to a right-angled triangle why the tangent of 90° is undefined.

# Solving trigonometric equations

Knowledge of the graphs of trigonometric functions enables you to solve trigonometric equations.

## ➜ Worked examples

*This ratio is one of the special angles covered earlier. Therefore a calculator is not really needed to work out the size of A.*

a  Angle A is an obtuse angle. If $\sin A = \frac{\sqrt{3}}{2}$ calculate the size of A.

If $\sin A = \frac{\sqrt{3}}{2}$

$A = \sin^{-1} \frac{\sqrt{3}}{2}$

$A = 60°$

However, the question states that A is an obtuse angle (i.e. $90° < A < 180°$). Therefore $A \neq 60°$.

Because of the symmetry properties of the sine curve, it can be deduced that $\sin 60° = \sin 120°$ as shown.

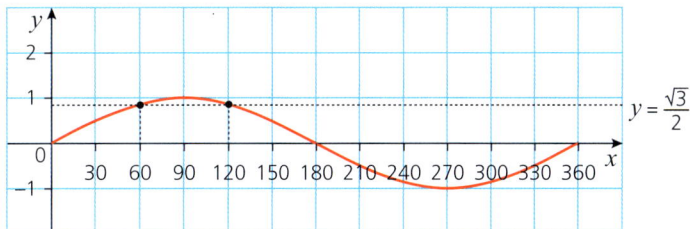

*This ratio is also of a special angle. A calculator is therefore not needed.*

b  If $\tan \theta = \frac{1}{\sqrt{3}}$, calculate the possible values for $\theta$ in the range $0° \leq \theta \leq 360°$.

$\tan \theta = \frac{1}{\sqrt{3}}$

$\theta = \tan^{-1} \frac{1}{\sqrt{3}} = 30°$

But the graph of $y = \tan \theta$ has a period of $180°$. Therefore another solution in the range would be $30° + 180° = 210°$.

Therefore $\theta = 30°$ or $210°$.

# 48 SPECIAL ANGLES AND THEIR TRIGONOMETRIC RATIOS

**Exercise 7.9**

1. Solve each of the following equations, giving all solutions in the range $0° \leq \theta \leq 360°$.
   a. $\sin\theta = \frac{1}{4}$
   b. $\cos\theta = \frac{1}{\sqrt{2}}$
   c. $\sin\theta = -\frac{1}{2}$
   d. $\tan\theta = -\sqrt{3}$
   e. $5\cos\theta + 1 = 2$
   f. $\frac{1}{2}\tan\theta + 2 = 1$

2. In the triangle, $\tan\theta = \frac{3}{4}$.

   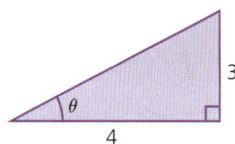

   Find, without a calculator, the value of:
   a. $\sin\theta$
   b. $\cos\theta$

3. In the triangle, $\sin\theta = \frac{3}{10}$.

   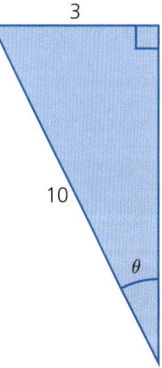

   Find, without a calculator, the exact value of:
   a. $\cos\theta$
   b. $\tan\theta$

4. By using the triangle as an aid, explain why the solution to the equation $\sin\theta = \cos\theta$ occurs when $\theta = 45°$.

   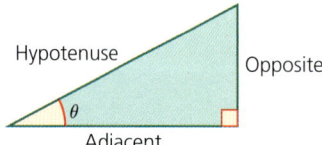

# 49 The sine and cosine rules

With right-angled triangles you can use the basic trigonometric ratios of sine, cosine and tangent. The **sine rule** is a relationship which can be used with triangles which are not right-angled.

The sine rule can be derived as follows:

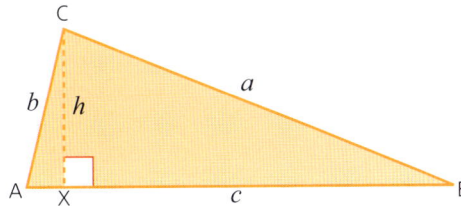

In triangle ACX, $\sin A = \frac{h}{b}$, therefore $h = b \sin A$.

In triangle BCX, $\sin B = \frac{h}{a}$, therefore $h = a \sin B$.

As the height $h$ is common to both triangles, it can be deduced that:

$b \sin A = a \sin B$.

This can be rearranged to: $\frac{a}{\sin A} = \frac{b}{\sin B}$.

Similarly, when a perpendicular line is drawn from A to meet side BC at Y, another equation is formed.

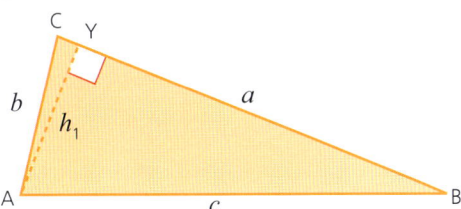

In triangle ACY, $\sin C = \frac{h_1}{b}$, therefore $h_1 = b \sin C$.

In triangle BAY, $\sin B = \frac{h_1}{c}$, therefore $h_1 = c \sin B$.

As the height $h_1$ is common to both triangles, it can be deduced that:

$b \sin C = c \sin B$.

This can be rearranged to: $\frac{b}{\sin B} = \frac{c}{\sin C}$.

Both equations can be combined to form $\frac{a}{\sin A} = \frac{b}{\sin B} = \frac{c}{\sin C}$.

This is the sine rule.

The reciprocal of each fraction can be taken, resulting in another form of the sine rule: $\frac{\sin A}{a} = \frac{\sin B}{b} = \frac{\sin C}{c}$.

*The sine rule formula will be provided in any assessment.*

# 49 THE SINE AND COSINE RULES

The sine rule proved above also works for obtuse-angled triangles as shown.

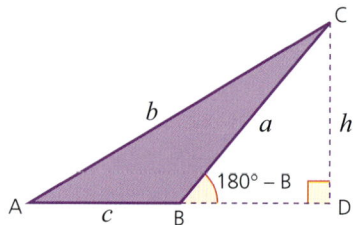

Consider the obtuse-angled triangle ABC.

The height of triangle ABC is $h$.

From $\triangle$ACD, $\sin A = \frac{h}{b}$; therefore $h = b \sin A$. (Equation 1)

From $\triangle$BCD, $\sin(180 - B) = \frac{h}{a}$; therefore $h = a \sin(180 - B)$. (Equation 2)

However, $\sin(180 - B) = \sin B$. This can be seen from the graph of the sine curve below:

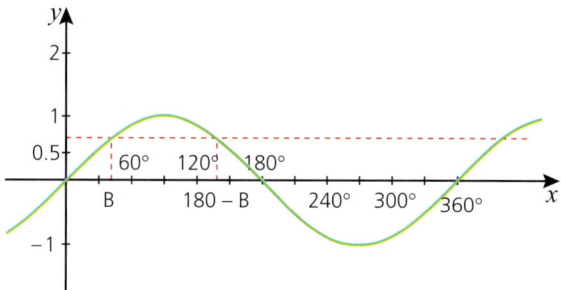

So equation 2 above can be rewritten as $h = a \sin B$. (Equation 3)

Combining equations 1 and 3 gives the equation $b \sin A = a \sin B$.

Rearranging this gives the sine rule $\frac{a}{\sin A} = \frac{b}{\sin B}$.

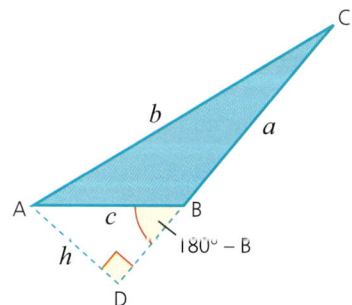

By considering the triangle as shown and using a similar proof as above, the sine rule $\frac{b}{\sin B} = \frac{c}{\sin C}$ can be derived.

Combining the two results produces $\frac{a}{\sin A} = \frac{b}{\sin B} = \frac{c}{\sin C}$ as before.

## Worked examples

**a** Calculate the length of side BC.

Using the sine rule:

$$\frac{a}{\sin A} = \frac{b}{\sin B}$$

$$\frac{a}{\sin 40°} = \frac{6}{\sin 30°}$$

$$a = \frac{6 \times \sin 40°}{\sin 30°}$$

BC = 7.71 cm (3 s.f.)

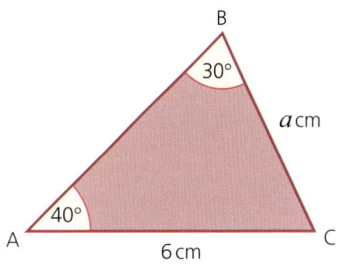

**b** Calculate the size of angle C.
Using the sine rule:

$$\frac{\sin A}{a} = \frac{\sin C}{c}$$

$$\sin C = \frac{6.5 \times \sin 60°}{6} = 0.94$$

Note that the reciprocal of both sides has been used.
This makes the subsequent rearrangement easier.

$C = \sin^{-1}(0.94)$

$C = 69.8°$ (1 d.p.)

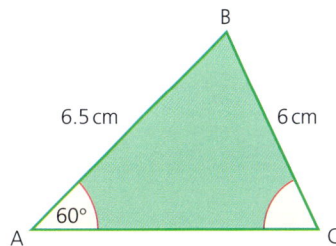

In the worked example part b above, $C = 69.8°$ is not the only possible solution from the information given as the question does not state that the angle is acute and often diagrams are not drawn to scale.

The triangle $ABC_1$ is as shown in the worked example part b above and therefore angle $AC_1B = 69.8°$ as calculated.

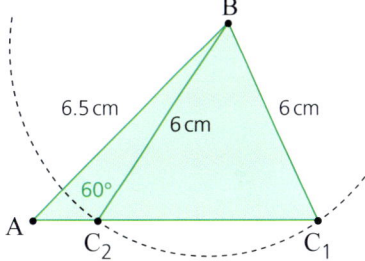

However, if a circle of radius 6 cm and centre at B is drawn, then it can be seen to intersect the base of the triangle in another place, $C_2$.

Therefore, triangle $ABC_2$ is another possible triangle.

But angle $AC_2B$ is not the same size as angle $AC_1B$.

It can be calculated in the same way as before, however using your knowledge of the sine curve, it can be calculated that angle $AC_2B$ is

$180° - 69.8° = 110.2°$

This is known as the **ambiguous case** of the sine rule as there are two possible answers.

# 49 THE SINE AND COSINE RULES

**Exercise 7.10**

1 Calculate the length of the side marked $x$ in each of the following. Give your answers to 3 s.f.

a
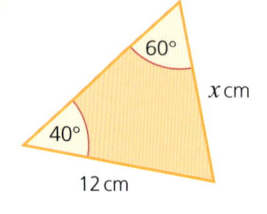

b
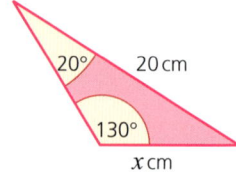

c
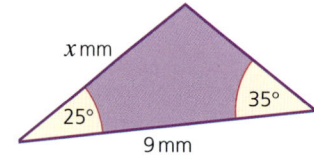

d
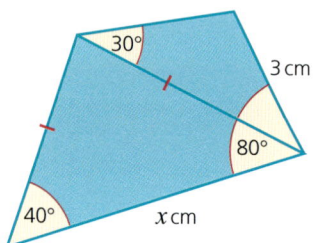

2 Calculate the size of the angle marked $\theta°$ in each of the following. If two possible values for the angle are possible, give both values. Give your answers to 1 d.p.

a
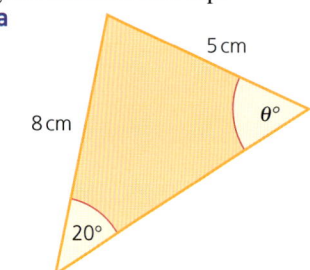

b
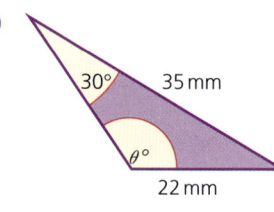

c
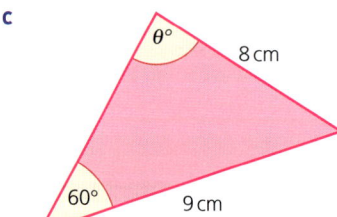

d
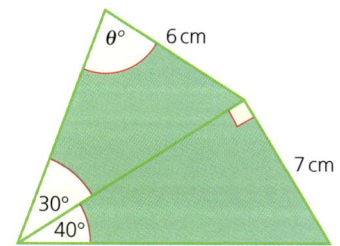

3 $\triangle ABC$ has the following dimensions:
AC = 10 cm, AB = 8 cm and angle ACB = 20°.
  a  Calculate the two possible values for angle CBA.
  b  Sketch and label the two possible shapes for $\triangle ABC$.

4 $\triangle PQR$ has the following dimensions:
PQ = 6 cm, PR = 4 cm and angle PQR = 40°.
  a  Calculate the two possible values for angle QRP.
  b  Sketch and label the two possible shapes for $\triangle PQR$.

# The cosine rule

The **cosine rule** is another relationship which can be used with triangles which are not right-angled.

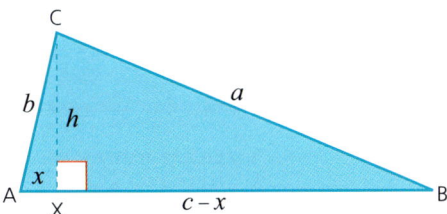

Using Pythagoras' theorem, two equations can be constructed.

For triangle ACX:

$b^2 = x^2 + h^2$, which can be rearranged to give $h^2 = b^2 - x^2$

For triangle BCX:

$a^2 = h^2 + (c - x)^2$, which can be rearranged to give $h^2 = a^2 - (c - x)^2$

As $h^2$ is common to both equations, it can be deduced that:

$a^2 - (c - x)^2 = b^2 - x^2$

$\Rightarrow \quad a^2 = b^2 - x^2 + (c - x)^2$
$\Rightarrow \quad a^2 = b^2 - x^2 + (c^2 - 2cx + x^2)$
$\Rightarrow \quad a^2 = b^2 + c^2 - 2cx$

But from triangle ACX, $\cos A = \frac{x}{b}$; therefore $x = b \cos A$.

Substituting $x = b \cos A$ into $a^2 = b^2 + c^2 - 2cx$ gives:

$$a^2 = b^2 + c^2 - 2bc \cos A$$

*The cosine rule formula will be provided in any assessment.*

This is the cosine rule.

If angle A is required the formula can be rearranged to give:

$\cos A = \dfrac{b^2 + c^2 - a^2}{2bc}$.

A similar proof can be applied if angle A is obtuse.

## → Worked examples

a Calculate the length of the side BC.

Using the cosine rule:

$a^2 = b^2 + c^2 - 2bc \cos A$
$a^2 = 9^2 + 7^2 - (2 \times 9 \times 7 \times \cos 50°)$
$\quad = 81 + 49 - (126 \times \cos 50°) = 49.0$
$a = \sqrt{49.0}$
$a = 7$

BC = 7 cm

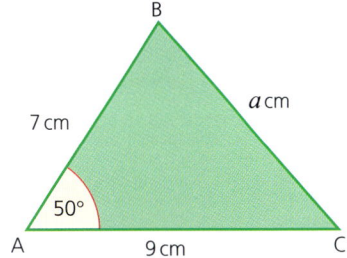

# 49 THE SINE AND COSINE RULES

**b** Calculate the size of angle A.
Using the cosine rule:
$a^2 = b^2 + c^2 - 2bc \cos A$
Rearranging the equation gives:

$$\cos A = \frac{b^2 + c^2 - a^2}{2bc}$$

$$\cos A = \frac{15^2 + 12^2 - 20^2}{2 \times 15 \times 12} = -0.086$$

$A = \cos^{-1}(-0.086)$
$A = 94.9°$ (1 d.p.)

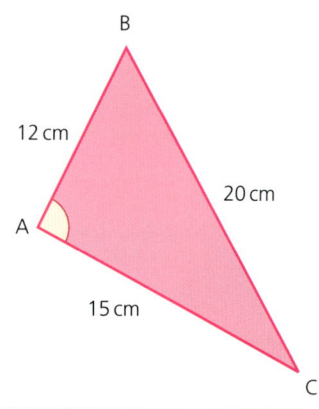

### Exercise 7.11

**1** Calculate the length of the side marked $x$ in each of the following. Give your answers to 3 s.f.

**a**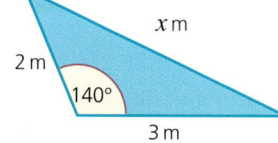

**b**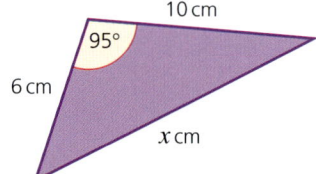

**c**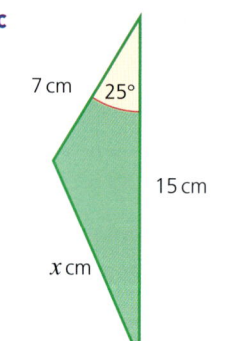

**d**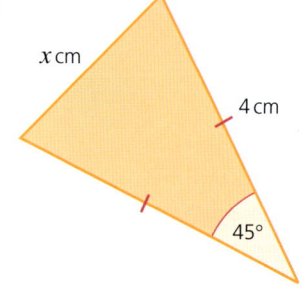

**e**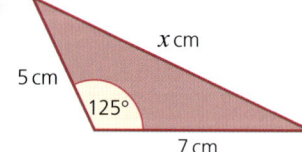

## The cosine rule

**2** Calculate the angle marked $\theta°$ in each of the following. Give your answers to 1 d.p.

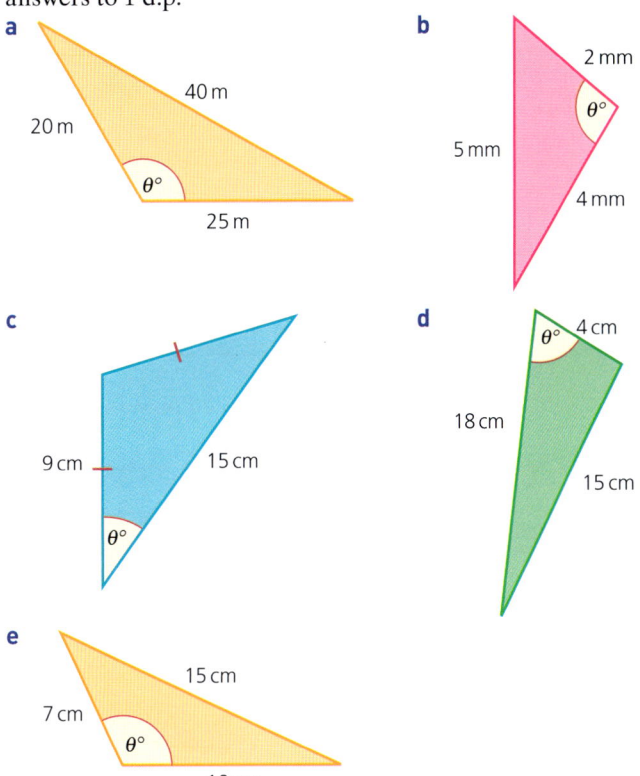

**3** The circle shown has a radius of 8 cm. Two points A and B lie on its circumference.
The angle AOB is 100°.

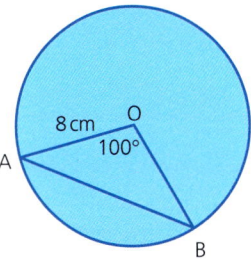

  **a** Calculate the length of the chord AB.
  **b** Calculate the shortest distance between the chord AB and the centre of the circle O.

# 49 THE SINE AND COSINE RULES

**Exercise 7.12**

1. Four players, W, X, Y and Z, are on a rugby pitch. The diagram shows a plan view of their relative positions.

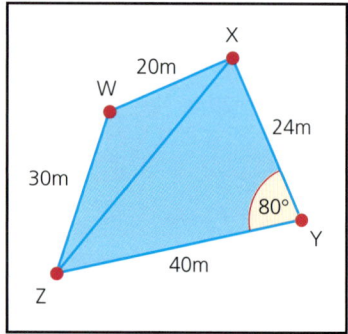

   Calculate:
   a  the distance between players X and Z
   b  angle ZWX
   c  angle WZX
   d  angle YZX
   e  the distance between players W and Y.

2. Three yachts, A, B and C, are racing off the 'Cape'. Their relative positions are shown below.

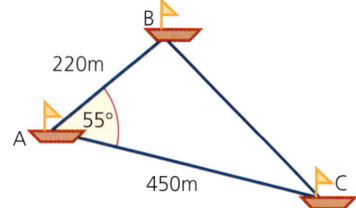

   Calculate the distance between B and C to the nearest 10 m.

3. There are two trees standing on one side of a river bank. On the opposite side is a boy standing at X.

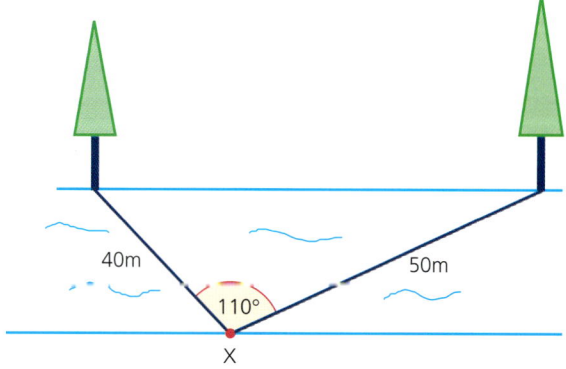

   Using the information given, calculate the distance between the two trees.

# The area of a triangle

The area of a triangle can be calculated without the need for knowing its height.

Area = $\frac{1}{2}bh$

Also: $\sin C = \frac{h}{a}$

Rearranging gives: $h = a \sin C$

Therefore **area = $\frac{1}{2} ab \sin C$**

*The area of a triangle formula will be provided in any assessment.*

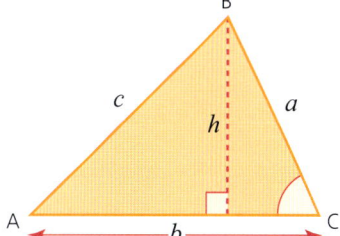

## Exercise 7.13

1  Calculate the area of the following triangles. Give your answers to 3 s.f.

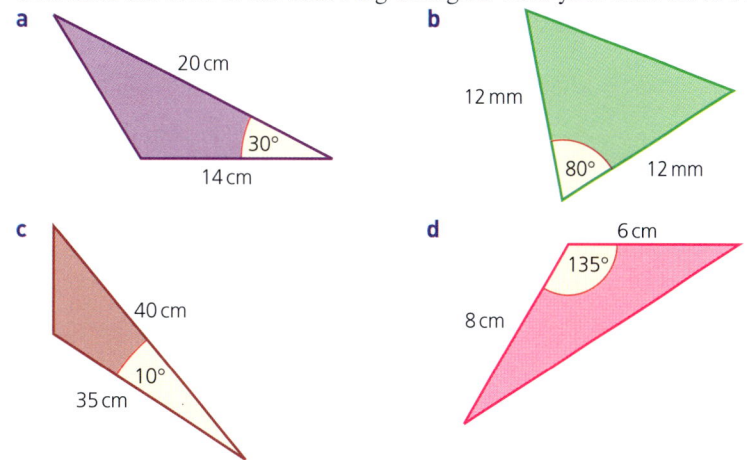

2  Calculate the value of $x$ in each of the following. Give your answers correct to 1 d.p.

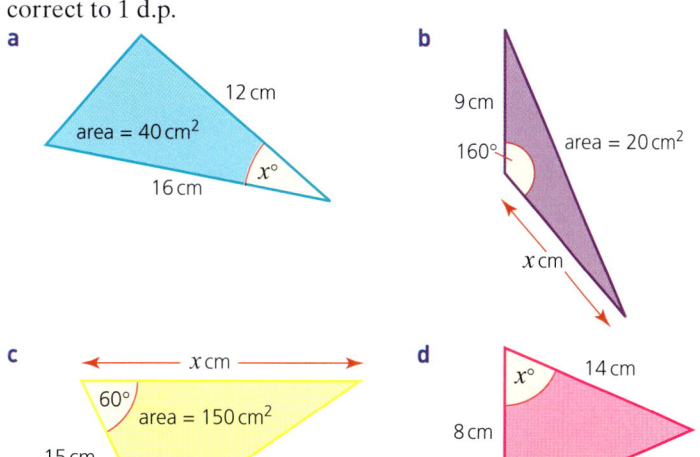

# 49 THE SINE AND COSINE RULES

**Exercise 7.13 (cont)**

3  ABCD is a school playing field. The following lengths are known:
   OA = 83 m, OB = 122 m, OC = 106 m, OD = 78 m

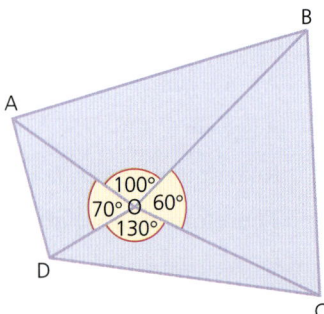

Calculate the area of the school playing field to the nearest 100 m².

4  The roof of a garage has a slanting length of 3 m and makes an angle of 120° at its vertex. The height of the garage wall is 4 m and its depth is 9 m.
   Calculate:
   a  the cross-sectional area of the roof
   b  the volume occupied by the whole garage.

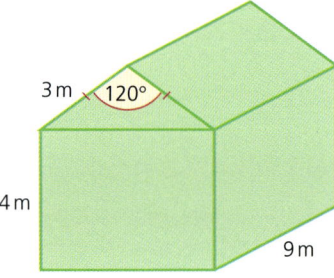

# 50 Applications of trigonometry

## Angles of elevation and depression

The **angle of elevation** is the angle above the horizontal through which a line of view is raised. The **angle of depression** is the angle below the horizontal through which a line of view is lowered.

### Worked examples

**a** The base of a tower is 60 m away from a point X on the ground. If the angle of elevation of the top of the tower from X is 40°, calculate the height of the tower.
Give your answer to the nearest metre.

$\tan 40° = \dfrac{h}{60}$

$h = 60 \times \tan 40° = 50.3$

The height is 50 m.

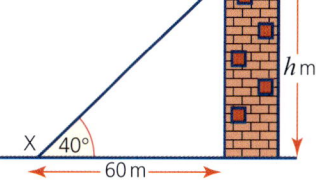

**b** An aeroplane receives a signal from a point X on the ground 6 km away. If the angle of depression of point X from the aeroplane is 30°, calculate the height at which the plane is flying.

Give your answer to the nearest 0.1 km.

$\sin 30° = \dfrac{h}{6}$

$h = 6 \times \sin 30° = 3.0$

The height is 3.0 km.

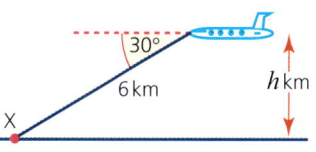

### Exercise 7.14

1  A and B are two villages. If the horizontal distance between them is 12 km, and the vertical distance between them is 2 km, calculate:

  **a**  the shortest distance between the two villages
  **b**  the angle of elevation of B from A.

2  X and Y are two towns. If the horizontal distance between them is 10 km and the angle of depression of Y from X is 7°, calculate:
  **a**  the shortest distance between the two towns
  **b**  the vertical height between the two towns.

# 50 APPLICATIONS OF TRIGONOMETRY

**Exercise 7.14 (cont)**

3  A girl standing on a hill at A, overlooking a lake, can see a small boat at a point B on the lake. If the girl is at a height of 50 m above B and at a horizontal distance of 120 m away from B, calculate:

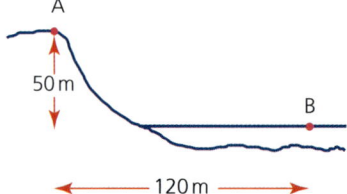

   a  the angle of depression of the boat from the girl
   b  the shortest distance between the girl and the boat.

4  Two hot air balloons are 1 km apart in the air. If the angle of elevation of the higher from the lower balloon is 20°, calculate the following, giving your answers to the nearest metre:

   a  the vertical height between the two balloons
   b  the horizontal distance between the two balloons.

5  A boy X can be seen by two of his friends Y and Z, who are swimming in the sea. If the angle of elevation of X from Y is 23° and from Z is 32°, and the height of X above Y and Z is 40 m, calculate:

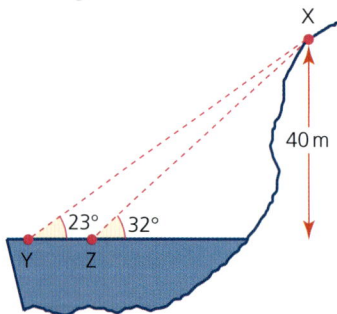

   a  the horizontal distance between X and Z
   b  the horizontal distance between Y and Z.

   Note: XYZ is a vertical plane.

6  A plane is flying at an altitude of 6 km directly over the line AB. It spots two boats, A and B, on the sea.

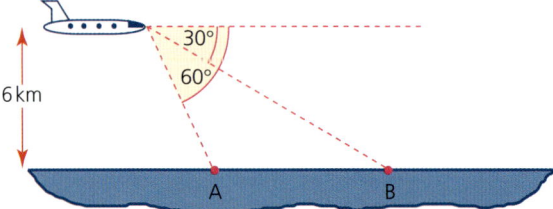

   If the angles of depression of A and B from the plane are 60° and 30° respectively, calculate the horizontal distance between A and B.

## Angles of elevation and depression

**7** A plane is flying at a constant altitude over the sea directly over the line XY. It can see two boats X and Y which are 4 km apart.

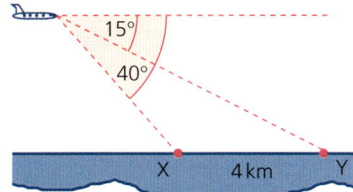

If the angles of depression of X and Y from the plane are 40° and 15° respectively, calculate:
a the horizontal distance between Y and the plane
b the altitude at which the plane is flying.

**8** Two planes are flying directly above each other. A person standing at P can see both of them. The horizontal distance between the two planes and the person is 2 km.

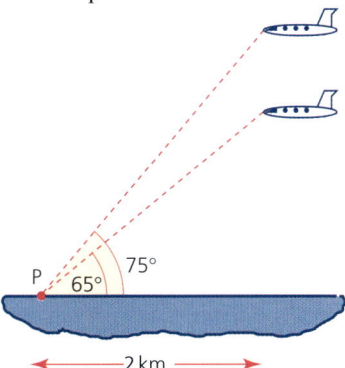

If the angles of elevation of the planes from the person are 65° and 75°, calculate:
a the altitude at which the higher plane is flying
b the vertical distance between the two planes.

**9** Three villagers A, B and C can see each other across a valley. The horizontal distance between A and B is 8 km, and the horizontal distance between B and C is 12 km. The angle of depression of B from A is 20° and the angle of elevation of C from B is 30°.
Note: A, B and C are in the same vertical plane.

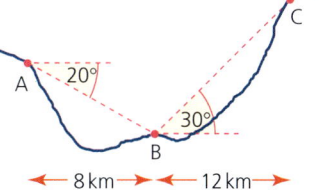

Calculate, giving all answers to 1 d.p.:
a the vertical height between A and B
b the vertical height between B and C
c the angle of elevation of C from A
d the shortest distance between A and C.

453

# 50 APPLICATIONS OF TRIGONOMETRY

**Exercise 7.14 (cont)**

10  Using binoculars, three people P, Q and R can see each other across a valley. The horizontal distance between P and Q is 6.8 km and the horizontal distance between Q and R is 10 km. If the shortest distance between P and Q is 7 km and the angle of depression of Q from R is 15°, calculate, giving appropriate answers:

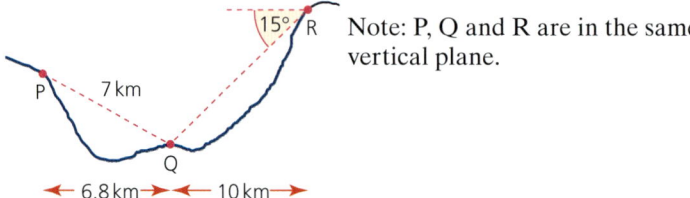

Note: P, Q and R are in the same vertical plane.

a   the vertical height between Q and R
b   the vertical height between P and R
c   the angle of elevation of R from P
d   the shortest distance between P and R.

11  Two people A and B are standing either side of a transmission mast. A is 130 m away from the mast and B is 200 m away.

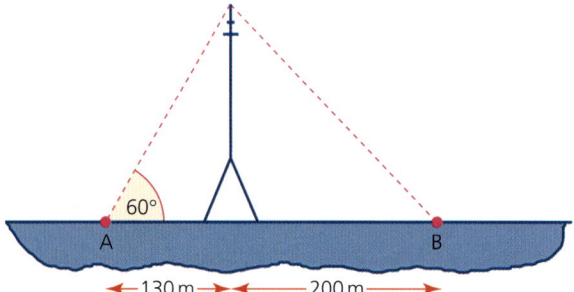

If the angle of elevation of the top of the mast from A is 60°, calculate:
a   the height of the mast to the nearest metre
b   the angle of elevation of the top of the mast from B.

12  Three boats X, Y and Z are shown below.

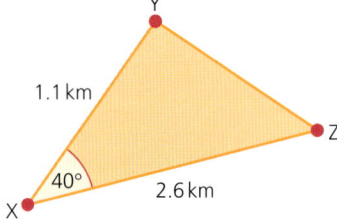

Find the distance between boats Y and Z, giving your answer to the nearest 100 m.

13  Three hot air balloons P, Q and R travelling in the same vertical plane are shown.

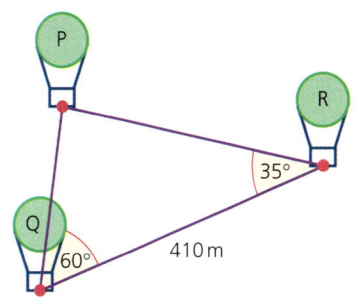

If angle PQR = 60°, angle PRQ = 35° and the distance between balloons Q and R is 410 m, calculate:
   a  the distance between balloons P and R, to the nearest 10 m
   b  the distance between balloons P and Q, to the nearest 10 m.

14  A triangle AOB lies inside a circle. Vertices A and B lie on the circumference of the circle, O at its centre.

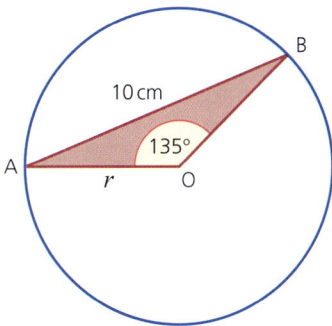

If the angle AOB = 135° and the chord AB = 10 cm, calculate the length of the radius $r$.

15  The diagram shows two people at X and Y standing 5 km apart on a shore. A large cruise ship is at Z.
The bearing of Y from X is 095°.
The bearing of Z from X is 050° and the bearing of Y from Z is 175°.
   a  Calculate the angle XZY.
   b  Calculate the distance between X and Z. Give your answer to the nearest metre.
   c  Calculate the distance between Y and Z. Give your answer to the nearest metre.

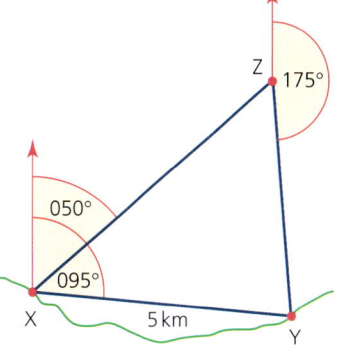

# 50 APPLICATIONS OF TRIGONOMETRY

## Trigonometry in three dimensions

### ➡ Worked examples

The diagram shows a cube of edge length 3 cm.

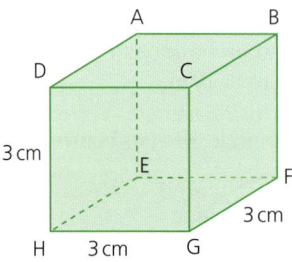

i  Calculate the length EG.

Triangle EHG is right-angled. Use Pythagoras' theorem to calculate the length EG:

$(EG)^2 = (EH)^2 + (HG)^2$

$(EG)^2 = 3^2 + 3^2 = 18$

EG $= \sqrt{18}$ cm

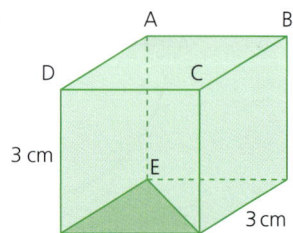
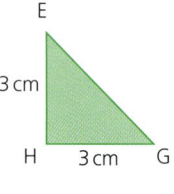

ii  Calculate the length AG.

Triangle AEG is right-angled. Use Pythagoras' theorem to calculate the length AG:

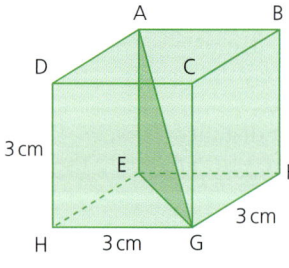
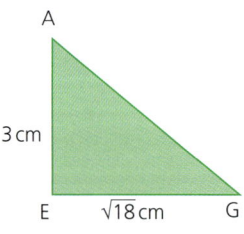

$(AG)^2 = (AE)^2 + (EG)^2$
$(AG)^2 = 3^2 + (\sqrt{18})^2$
$(AG)^2 = 9 + 18$
AG $= \sqrt{27}$ cm $= 3\sqrt{3}$ cm

iii  Calculate the angle EGA.

To calculate angle EGA, use the triangle EGA:

$\tan G = \dfrac{3}{\sqrt{18}}$

G – 35.3° (1 d.p.)

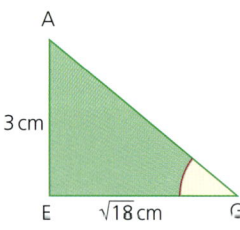

## Exercise 7.15

**1**
  **a** Calculate the length HF.
  **b** Calculate the length HB.
  **c** Calculate the angle BHG.

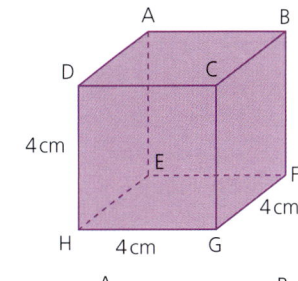

**2**
  **a** Calculate the length CA.
  **b** Calculate the length CE.
  **c** Calculate the angle ACE.

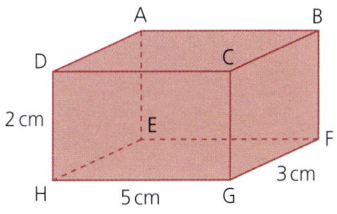

**3**
  **a** Calculate the length EG.
  **b** Calculate the length AG.
  **c** Calculate the angle AGE.

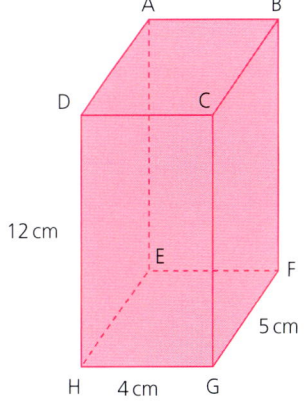

**4**
  **a** Calculate the angle BCE.
  **b** Calculate the angle GFH.

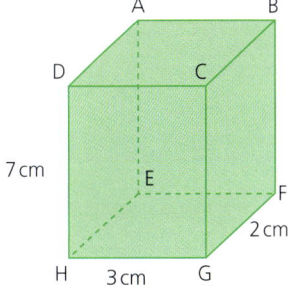

**5** The diagram shows a rectangular-based pyramid where A is vertically above X.
  **a** **i** Calculate the length DB.
     **ii** Calculate the angle DAX.
  **b** **i** Calculate the angle CED.
     **ii** Calculate the angle DBA.

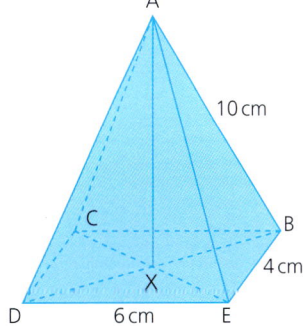

# 50 APPLICATIONS OF TRIGONOMETRY

**Exercise 7.15 (cont)**

6 The diagram shows a rectangular-based pyramid where A is vertically above X.
  a i Calculate the length CE.
    ii Calculate the angle CAX.
  b i Calculate the angle BDE.
    ii Calculate the angle ADB.

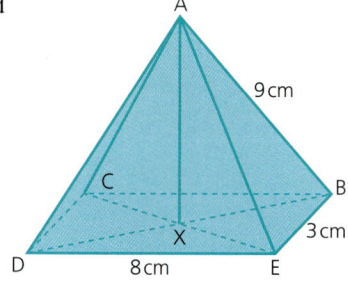

7 The diagram shows the net of a square-based pyramid, with dimensions as shown. Calculate, giving your answers to 3 s.f.
  a the total surface area of the pyramid
  b the total volume of the pyramid.

8 In this cone, angle YXZ = 60°.
  a Calculate the length XY.
  b Calculate the length YZ.
  c Calculate the circumference of the base.

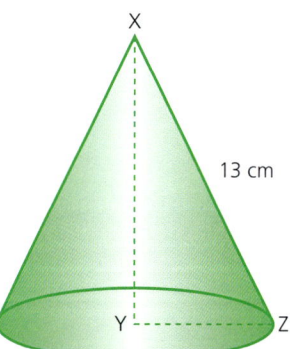

9 In this cone, angle XZY = 40°.
  a Calculate the length XZ.
  b Calculate the length XY.

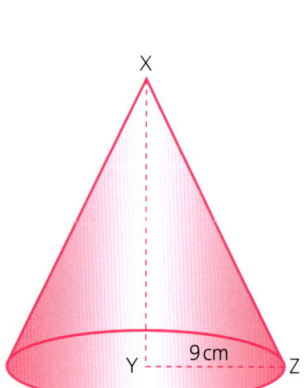

*Trigonometry in three dimensions*

**10** One corner of this cuboid has been sliced off along the plane QTU. WU = 4 cm.
  **a** Calculate the length of the three sides of the triangle QTU.
  **b** Calculate the three angles P, Q and T in triangle PQT.
  **c** Calculate the area of triangle PQT.

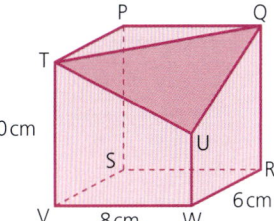

**11** Calculate the surface area of a regular tetrahedron with edge length 2 cm.

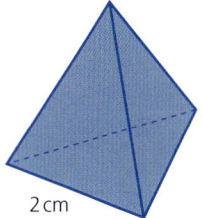

**12** The rectangular-based pyramid shown (below) has a sloping edge length of 12 cm. Calculate its surface area.

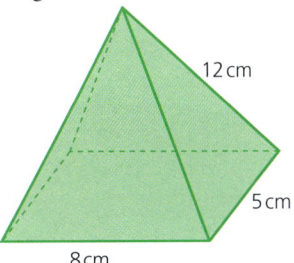

**13** Two square-based pyramids are glued together as shown. Given that all the triangular faces are identical, calculate the surface area of the whole shape.

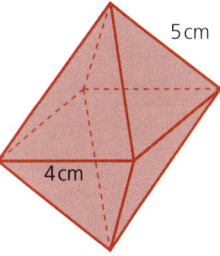

**14** Calculate the surface area of the frustum of the square-based pyramid (below). Assume that all the sloping faces are identical.

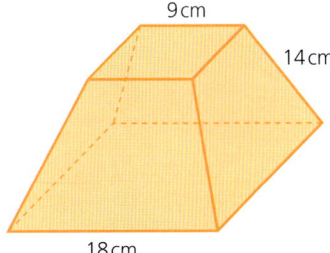

# 50 APPLICATIONS OF TRIGONOMETRY

**Exercise 7.15 (cont)**

15 The two pyramids shown below have the same surface area.

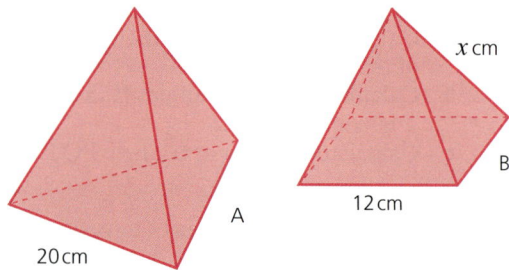

Calculate:
a the surface area of the regular tetrahedron
b the area of one of the triangular faces on the square-based pyramid
c the value of $x$.

## The angle between a line and a plane

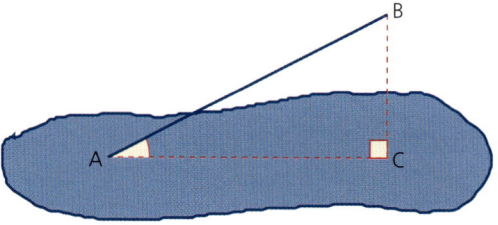

To calculate the size of the angle between the line AB and the shaded **plane**, drop a perpendicular from B. It meets the shaded plane at C. Then join AC.

The angle between the lines AB and AC represents the angle between the line AB and the shaded plane.

The line AC is the projection of the line AB on the shaded plane.

*A plane is a flat two dimensional surface.*

### ➔ Worked examples

i Calculate the length EC.
First use Pythagoras' theorem to calculate the length EG:

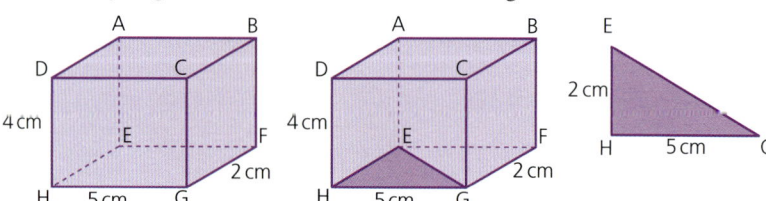

$(EG)^2 = (EH)^2 + (HG)^2$
$(EG)^2 = 2^2 + 5^2$
$(EG)^2 = 29$
$EG = \sqrt{29}$ cm

Now use Pythagoras' theorem to calculate EC:

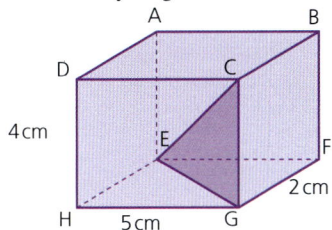 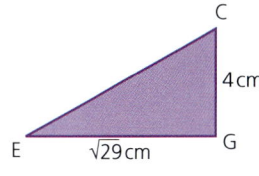

$(EC)^2 = (EG)^2 + (CG)^2$
$(EC)^2 = (\sqrt{29})^2 + 4^2$
$(EC)^2 = 29 + 16$
$EC = \sqrt{45}$ cm $= 3\sqrt{5}$ cm

ii  Calculate the angle between the line CE and the plane ADHE.

To calculate the angle between the line CE and the plane ADHE, use the right-angled triangle CED and calculate the angle CED:

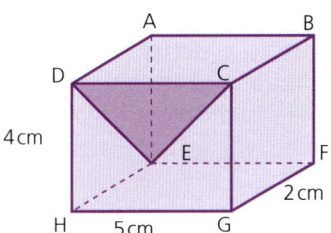 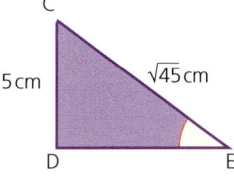

$\sin E = \dfrac{CD}{CE}$

$\sin E = \dfrac{5}{3\sqrt{5}}$

$E = \sin^{-1}\left(\dfrac{5}{3\sqrt{5}}\right)$

$E = 48.2°$ (1 d.p.)

**Exercise 7.16**

1  Name the projection of each line onto the given plane:
 a  TR onto RSWV
 b  TR onto PQUT
 c  SU onto PQRS
 d  SU onto TUVW
 e  PV onto QRVU
 f  PV onto RSWV

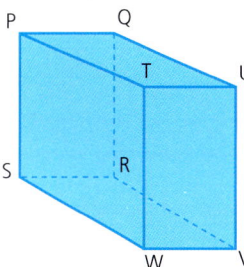

# 50 APPLICATIONS OF TRIGONOMETRY

**Exercise 7.16 (cont)**

2 Name the projection of each line onto the given plane:
   a KM onto IJNM
   b KM onto JKON
   c KM onto HIML
   d IO onto HLOK
   e IO onto JKON
   f IO onto LMNO

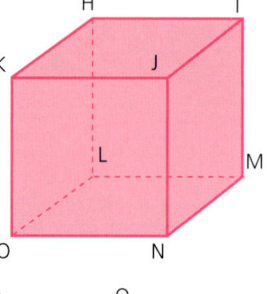

3 Name the angle between the given line and plane:
   a PT and PQRS
   b PU and PQRS
   c SV and PSWT
   d RT and TUVW
   e SU and QRVU
   f PV and PSWT

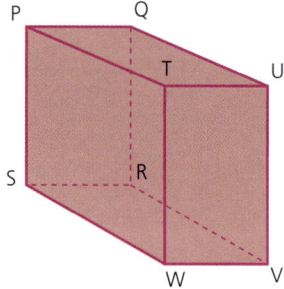

4 a Calculate the length BH.
  b Calculate the angle between the line BH and the plane EFGH.

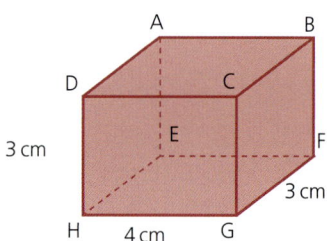

5 a Calculate the length AG.
  b Calculate the angle between the line AG and the plane EFGH.
  c Calculate the angle between the line AG and the plane ADHE.

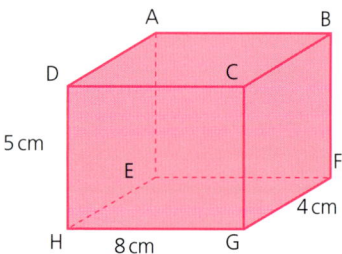

6 The diagram shows a rectangular-based pyramid where A is vertically above X.
  a Calculate the length BD.
  b Calculate the angle between AB and CBED.

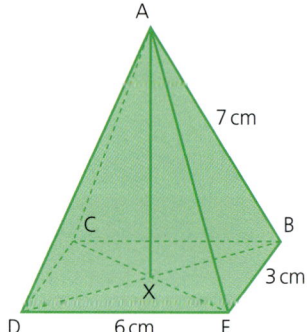

*The angle between a line and a plane*

7 The diagram shows a rectangular-based pyramid where U is vertically above X.

   a Calculate the length WY.
   b Calculate the length UX.
   c Calculate the angle between UX and UZY.

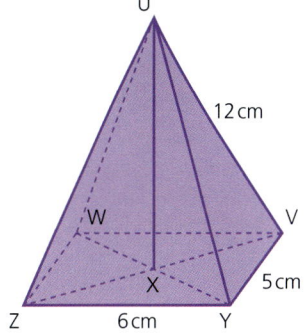

8 ABCD and EFGH are square faces lying parallel to each other.

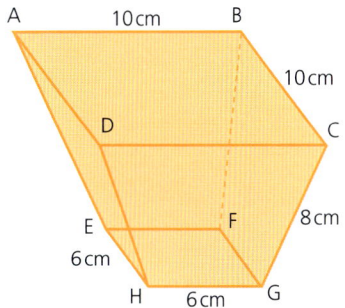

Calculate:
   a the length DB
   b the length HF
   c the vertical height of the object
   d the angle DH makes with the plane ABCD.

9 ABCD and EFGH are square faces lying parallel to each other.
Calculate:
   a the length AC
   b the length EG
   c the vertical height of the object
   d the angle CG makes with the plane EFGH.

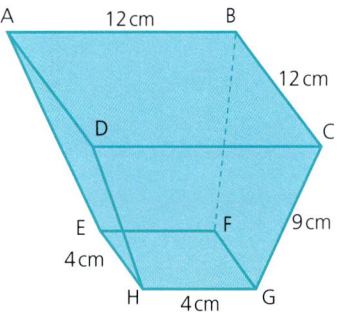

# 51 Trigonometric graphs, properties and transformations

The graphs of the trigonometric ratios $\sin x$, $\cos x$ and $\tan x$ were introduced in Chapter 48 in this topic. They each have characteristic shapes and properties.

The graph of $y = \sin x$ has:
- a period of 360° (i.e. it repeats itself every 360°)
- a maximum value of +1
- a minimum value of –1
- an amplitude of 1
- symmetry, e.g. $\sin x = \sin(180 - x)$.

### Note

The amplitude and period of a wave can be visualised in this graph:

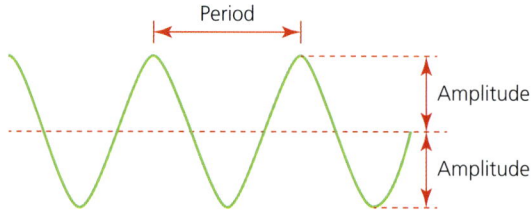

The graph of $y = \cos x$ has:
- a period of 360°
- a maximum value of +1
- a minimum value of –1
- an amplitude of 1
- symmetry, e.g. $\cos x = \cos(360 - x)$.

## Trigonometric graphs, properties and transformations

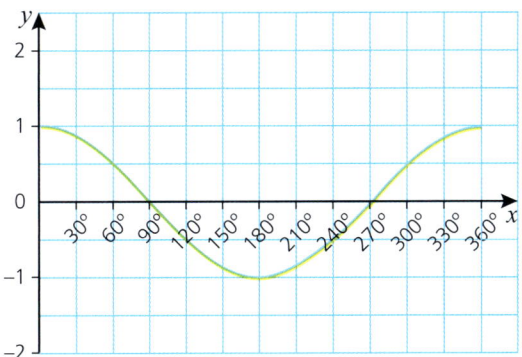

The graph of $y = \tan x$ has:
- a period of $180°$
- no maximum or minimum value
- no amplitude
- symmetry
- asymptotes at $90°$ and $270°$.

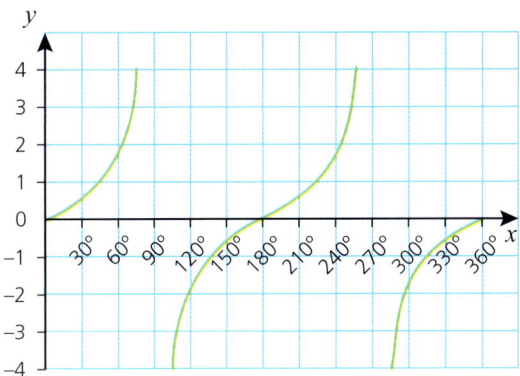

In Topic 3, various functions were transformed. These transformations can also be applied to trigonometric graphs.

Consider the function $f(x) = \sin x$. The graph of the functions $y = f(x)$ and $y = f(x) + 2$ are shown below.

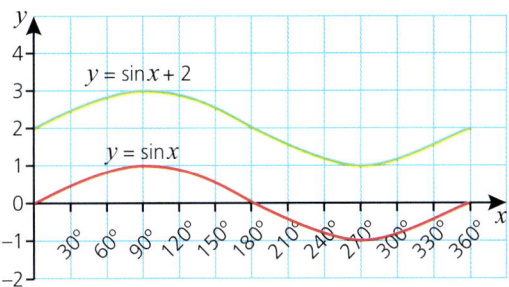

The graph of $y = \sin x$ has been translated $\begin{pmatrix} 0 \\ 2 \end{pmatrix}$.

465

# 51 TRIGONOMETRIC GRAPHS, PROPERTIES AND TRANSFORMATIONS

In general, the transformation that maps $y = f(x)$ onto $y = f(x) + a$ is the translation $\begin{pmatrix} 0 \\ a \end{pmatrix}$.

The second type of transformation also involves a translation. Consider the function $f(x) = \tan x$. The graph of the functions $y = f(x)$ and $y = f(x + 30)$ are shown below. (Note: $y = f(x + 30)$ is the same as $y = \tan(x + 30)$.

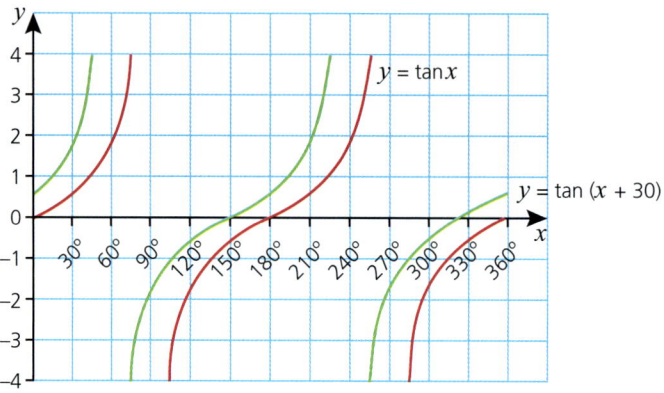

The graph of $y = \tan x$ has been translated by $\begin{pmatrix} -30 \\ 0 \end{pmatrix}$.

In general, the transformation that maps $y = f(x)$ onto $y = f(x - a)$ is the translation $\begin{pmatrix} a \\ 0 \end{pmatrix}$.

Transformations can be investigated more fully using a graphic display calculator. Instructions to graph $y = \cos x$ on a graphic display calculator, in the range for $x$ of $0 \leq x \leq 360$, are given below:

## Trigonometric graphs, properties and transformations

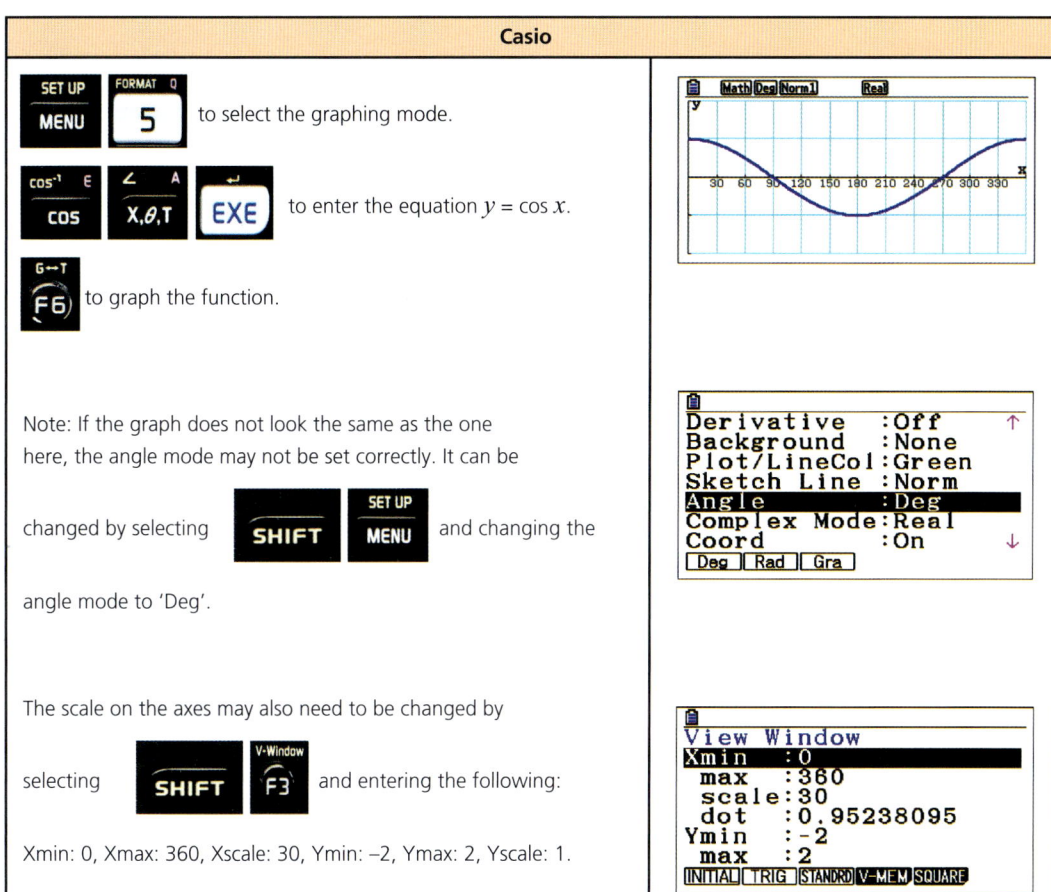

| Casio | |
|---|---|
| **SET UP / MENU** **FORMAT / 5** to select the graphing mode. | |
| **cos⁻¹ / COS** **∠ / X,θ,T** **EXE** to enter the equation $y = \cos x$. | |
| **G↔T / F6** to graph the function. | |
| Note: If the graph does not look the same as the one here, the angle mode may not be set correctly. It can be changed by selecting **SHIFT** **SET UP / MENU** and changing the angle mode to 'Deg'. | |
| The scale on the axes may also need to be changed by selecting **SHIFT** **V-Window / F3** and entering the following: Xmin: 0, Xmax: 360, Xscale: 30, Ymin: –2, Ymax: 2, Yscale: 1. | |

467

# 51 TRIGONOMETRIC GRAPHS, PROPERTIES AND TRANSFORMATIONS

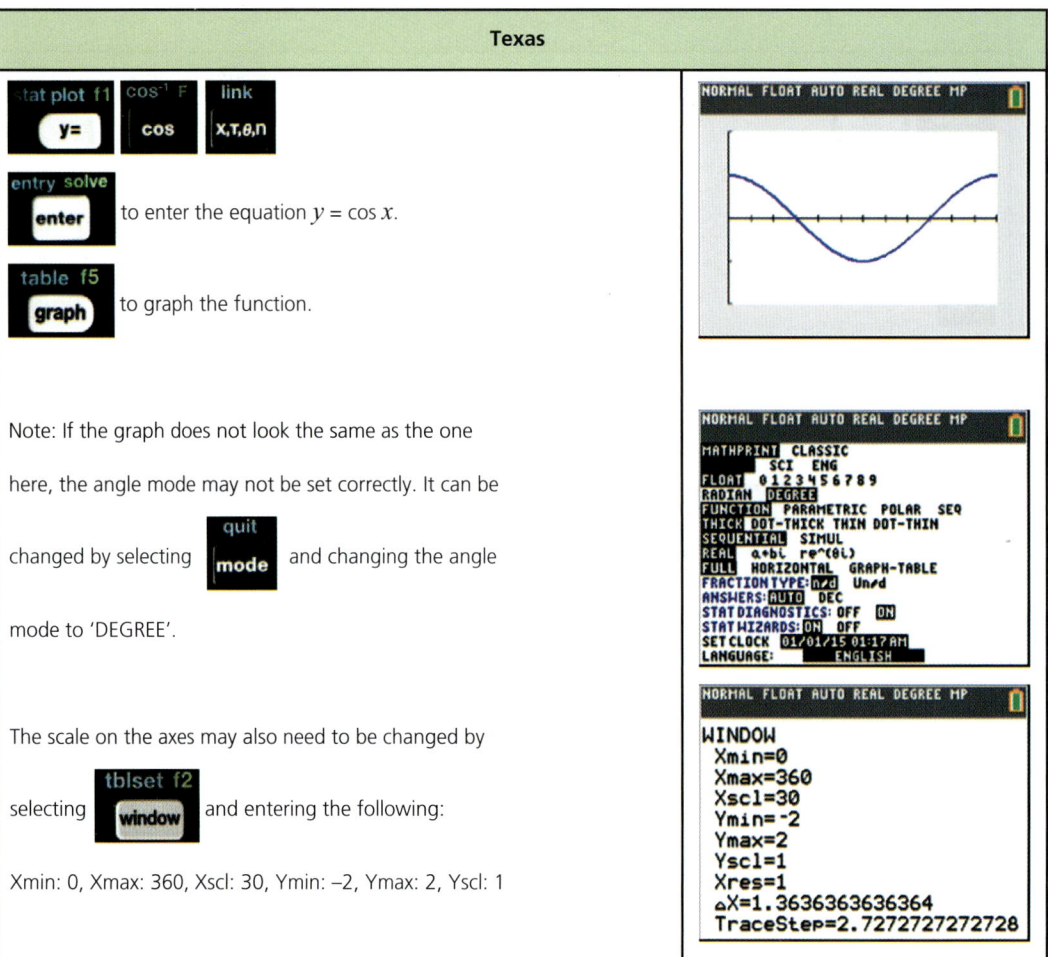

**Exercise 7.17**

1. Determine the transformation which maps $f(x) = \sin x$ onto each of the following functions:
   a. $y = \sin x - 3$
   b. $y = \sin(x + 60)$
   c. $y = \cos x$

2. a. Sketch a graph of $f(x) = \tan x$ for angles $0 \leqslant x \leqslant 360°$.
   b. On the same axes, sketch the graph of $y = \tan x + 2$.

*Further transformations of trigonometric graphs*

3 Using your graphic display calculator, produce a screen similar to the ones below. One of the functions is identified each time.

a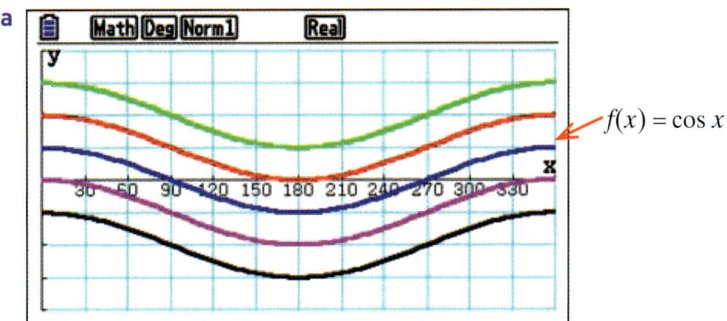

b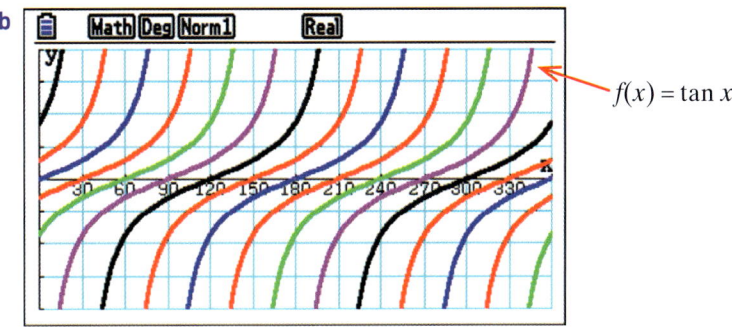

# Further transformations of trigonometric graphs

So far you have studied the effects of translations in both the $x$- and $y$-directions on the equations of a trigonometric function. However other transformations are also possible.

The graphs of $y = \sin x$ and $y = 2\sin x$ are both shown below.

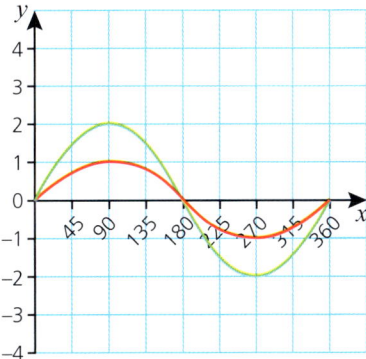

469

# 51 TRIGONOMETRIC GRAPHS, PROPERTIES AND TRANSFORMATIONS

It can be seen that, compared with the graph of $y = \sin x$, the graph of $y = 2\sin x$ has double the amplitude (so an amplitude of 2). However, the period has remained the same at 360°.

The graphs of $y = \sin x$ and $y = \sin(2x)$ are both shown below.

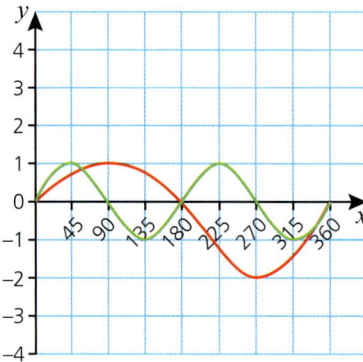

Here it can be seen that, compared with the graph $y = \sin x$, the graph of $y = \sin(2x)$ has half the period (so a period of 180°). However, the amplitude has remained the same.

**Exercise 7.18**

1 Use your graphic display calculator to graph each of the following functions and to help complete the table below.

| Function | Period | Amplitude |
|---|---|---|
| $f(x) = \sin(x)$ | 360° | 1 |
| $f(x) = \cos(x)$ | | |
| $f(x) = 2\sin(x)$ | | 2 |
| $f(x) = 3\sin(x)$ | | |
| $f(x) = 5\cos(x)$ | | |
| $f(x) = \frac{1}{2}\cos(x)$ | | |
| $f(x) = \cos(2x)$ | | |
| $f(x) = \cos(3x)$ | | |
| $f(x) = \cos(5x)$ | | |
| $f(x) = 2\cos(4x)$ | | |
| $f(x) = 3\sin(3x)$ | | |
| $f(x) = 5\sin\left(\frac{1}{4}x\right)$ | | |
| $f(x) = a\sin(bx)$ | | |
| $f(x) = a\cos(bx)$ | | |

2 Comparing the function $f(x) = \sin(x)$ with the function $f(x) = a\sin(x)$ describe the effect of the constant $a$.

3 Comparing the function $f(x) = \cos(x)$ with the function $f(x) = \cos(bx)$ describe the effect of the constant $b$.

4 i  Use your graphic display calculator to produce similar screens to the ones below.
  ii In each case write the equation you used to produce the unlabelled curve.

a
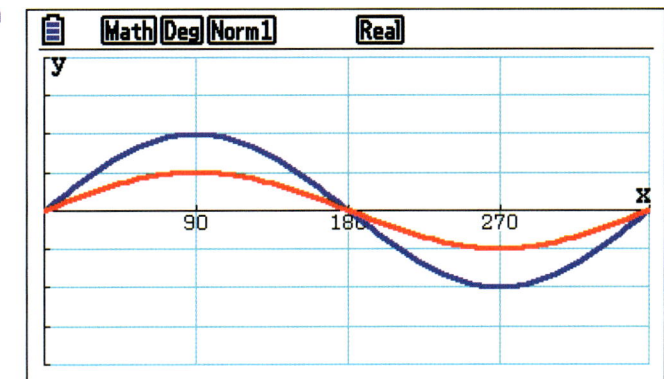

b
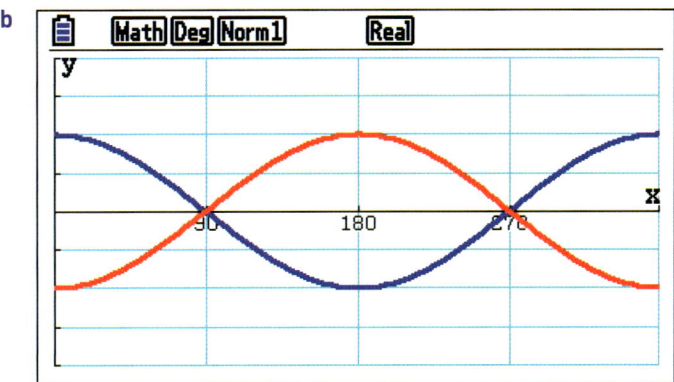

c
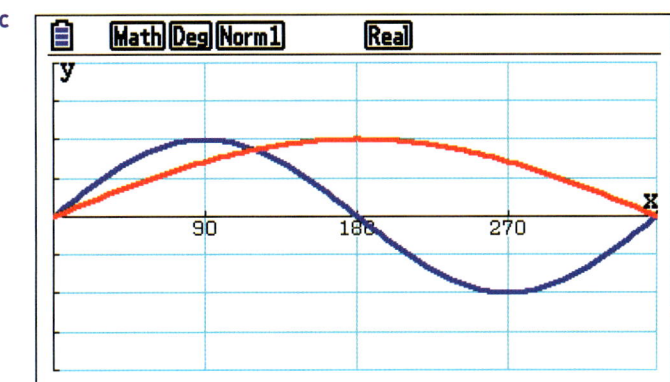

## 51 TRIGONOMETRIC GRAPHS, PROPERTIES AND TRANSFORMATIONS

**Exercise 7.18 (cont)**

d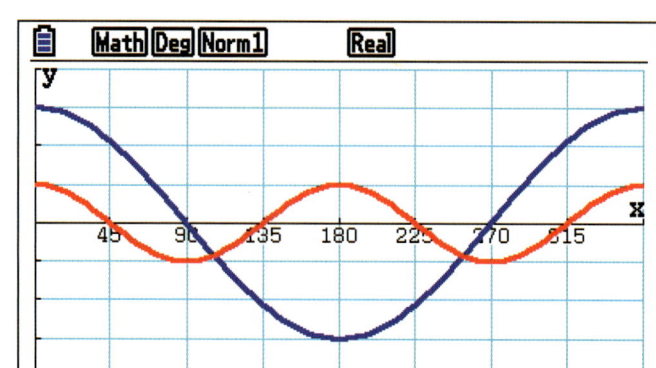

# Investigations, modelling and ICT 7

## Investigation: Pythagoras and circles

The explanation for Pythagoras' theorem usually shows a right-angled triangle with squares drawn on each of its three sides, as in the diagram below.

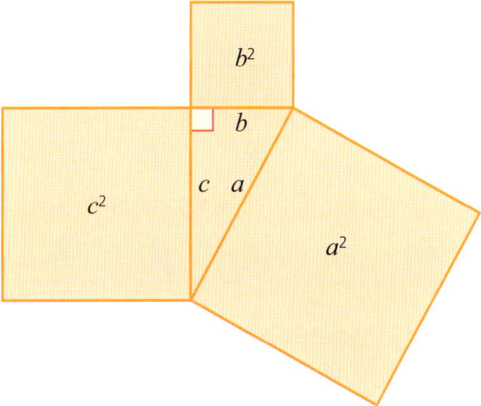

In this example, the area of the square on the hypotenuse, $a^2$, is equal to the sum of the areas of the squares on the other two sides, $b^2 + c^2$.

This gives the formula $a^2 = b^2 + c^2$.

1. Draw a right-angled triangle.
2. Using a pair of compasses, construct a semicircle off each side of the triangle. Your diagram should look similar to the one shown below.

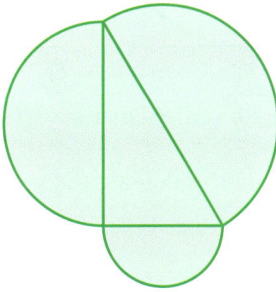

3. By measuring the diameter of each semicircle, calculate their areas.
4. Is the area of the semicircle on the hypotenuse the sum of the areas of the semicircles drawn on the other two sides? Does Pythagoras' theorem still hold for semicircles?
5. Does Pythagoras' theorem still hold if equilateral triangles are drawn on each side?
6. Investigate for other regular polygons.

# INVESTIGATIONS, MODELLING AND ICT 7

## Investigation: Numbered balls

The balls above start with the number 25 and then subsequent numbered balls are added according to a rule. The process stops when ball number 1 is added.

*Make sure your rule works all the way to the end of the sequence.*

1. Write in words the rule for generating the sequence of numbered balls.
2. What is the longest sequence of balls starting with a number less than 100?
3. Is there a strategy for generating a long sequence?
4. Use your rule to state the longest sequence of balls starting with a number less than 1000.
5. Extend the investigation by having a different term-to-term rule.

## Modelling: The ferris wheel

A circle of radius 5 cm is shown. The horizontal line XY is tangential to the circle at the point $P_0$.

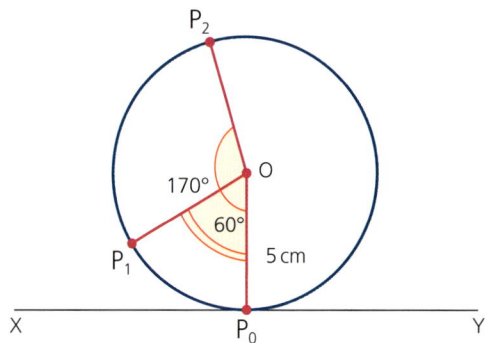

1. The angle $P_0OP_1$ is 60°; calculate the vertical distance of $P_1$ from the line XY.
2. The angle $P_0OP_2$ is 170°; calculate the vertical distance of $P_2$ from the line XY.

# Modelling: The ferris wheel

The London Eye is a large Ferris wheel in central London. It has a radius of 60 m and its lowest point is 15 m above the ground as shown.

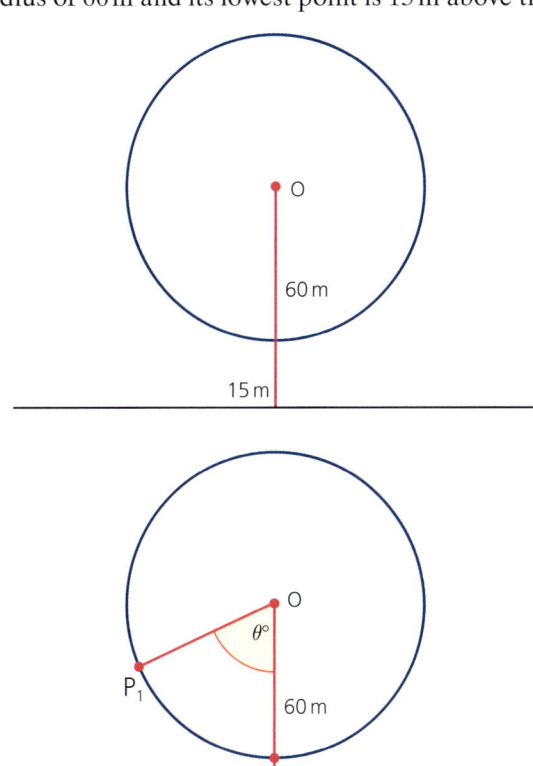

One full revolution takes 30 minutes.

A family has a ride on the wheel by entering at the lowest point ($P_0$).

**3 a** After 5 minutes, the family is at a point as shown. Calculate the angle $\theta$.
 **b** Show that the point $P_1$ is 45 m above the ground.

**4** Show that the height ($H$ metres) of the family above the ground ($t$ minutes) after entering the wheel can be modelled by the formula $H = 15(5 - 4\cos(12t)°)$.

**5** Use this formula to calculate the height of the family after:
 **a** 10 minutes
 **b** 27 minutes

**6 a** Calculate how long after entering the wheel the family will be at a height of 100 m for the first time. Give your answer to 1 d.p.
 **b** Deduce from your answer to Q.6a how long after entering the wheel the family will again be at a height of 100 m. Justify your reasoning.

# INVESTIGATIONS, MODELLING AND ICT 7

**7** Using your graphic display calculator to help, sketch a graph of the height of the family above the ground during the 30-minute ride. Label any intercepts with the axes and the coordinates of its maximum point.

## ICT Activity

In this activity, you will need to use your graphic display calculator to investigate the relationship between different trigonometric ratios.

1. **a** Using your calculator, plot the graph of
   $y = \sin x$ for $0° \leq x \leq 360°$.
   The graph should look similar to the one shown below:

   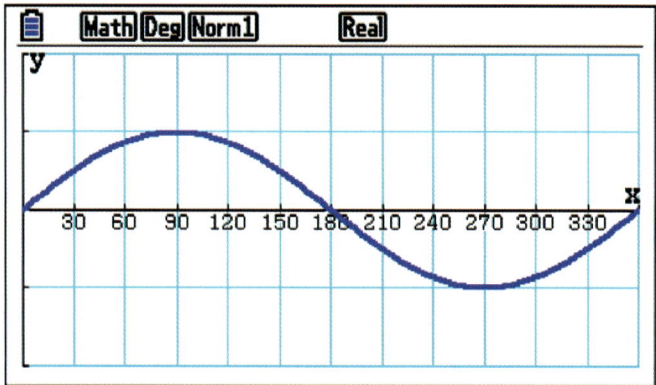

   **b** Using the graph-solving facility, Find the value of the following:
     **i** $\sin 70°$
     **ii** $\sin 125°$
     **iii** $\sin 300°$
   **c** Referring to the graph, explain why $\sin x = 0.7$ has two solutions between $0°$ and $360°$.
   **d** Use the graph to solve the equation $\sin x = 0.5$.
2. **a** On the same axes as before, plot $y = \cos x$.
   **b** How many solutions are there to the equation $\sin x = \cos x$ between $0°$ and $360°$?
   **c** What is the solution to the equation $\sin x = \cos x$ between $180°$ and $270°$?
3. By plotting appropriate graphs, solve the following for $0° \leq x \leq 360°$:
   **a** $\sin x = \tan x$
   **b** $\cos x - \tan x$

# Student assessments 7

## Student assessment 1

1. The two triangles shown below are similar.

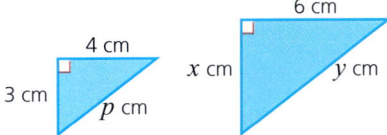

   a Using Pythagoras' theorem, calculate the value of $p$.
   b Calculate the values of $x$ and $y$.

2. A map shows three towns A, B and C. Town A is due north of C. Town B is due east of A.
   The distance AC is 75 km and the distance AB is 100 km. Calculate the distance between towns B and C.

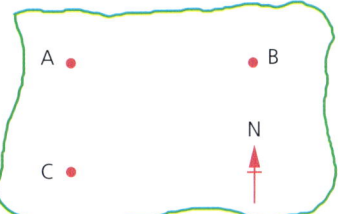

3. Calculate the distance from A to the top of each of the two trees in the diagram below.

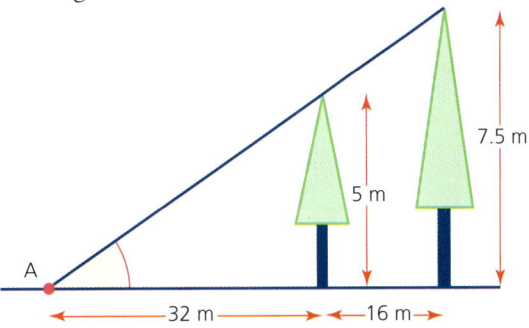

4. Two boats X and Y, sailing in a race, are shown in the diagram. Boat X is 140 m due north of a buoy B. Boat Y is due east of buoy B. Boats X and Y are 320 m apart. Calculate the distance BY.

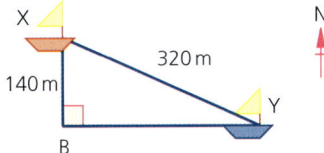

**5** If AB is the diameter of the circle, AC = 5 cm and BC = 12 cm, calculate:

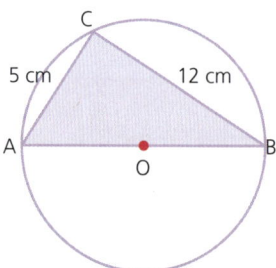

  **a** the size of angle ACB
  **b** the length of the radius of the circle.
**6** Calculate the diameter of the circle given that LM and MN are both tangents to the circle, O is its centre and OM = 18 mm.

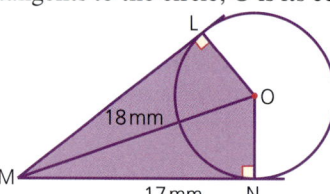

**7** In the diagram, XY and YZ are both tangents to the circle with centre O. Calculate the length OY.

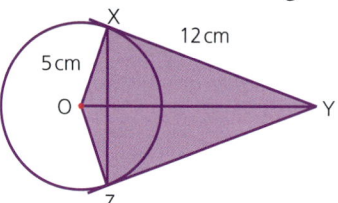

**8** In the diagram, LN and MN are both tangents to the circle centre O. NM = 40 cm and ON = 50 cm.

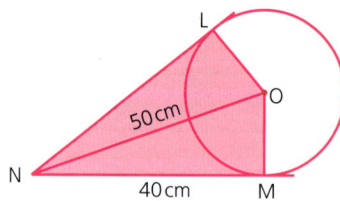

  **a** Calculate the radius of the circle.
  **b** Calculate the circumference of the circle.

*NB: Diagrams are not drawn to scale.*

## Student assessment 2

1  Calculate the length of the side marked $x$ cm in these diagrams. Give your answers correct to 3 s.f.

   a
   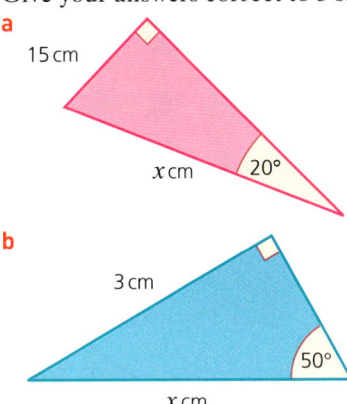

   b

2  Calculate the angle marked $\theta°$ in these diagrams. Give your answers correct to the nearest degree.

   a
   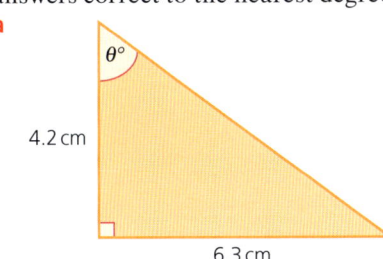

   b
   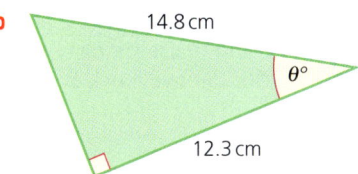

3  Calculate the length of the side marked $q$ cm in these diagrams. Give your answers correct to 3 s.f.

   a
   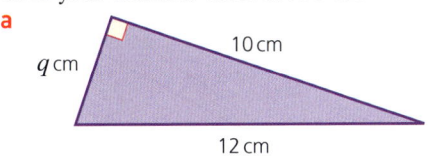

# STUDENT ASSESSMENTS 7

**b**

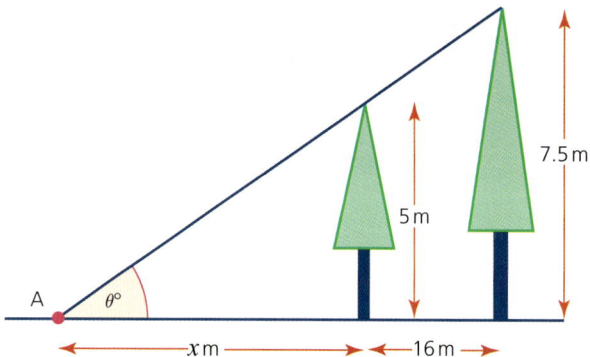

**4** Two trees stand 16 m apart. Their tops make an angle of $\theta°$ at point A on the ground.

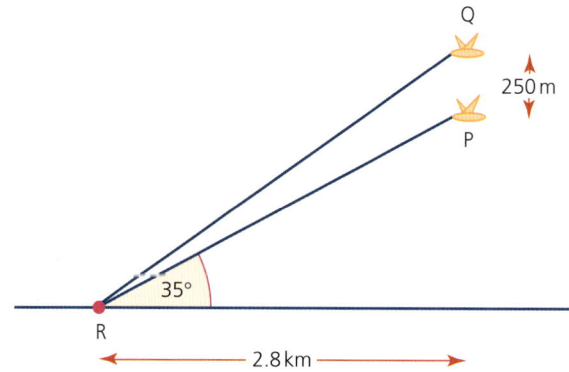

 **a** Write $\theta°$ in terms of the height of the shorter tree and its distance $x$ metres from point A.
 **b** Write $\theta°$ in terms of the height of the taller tree and its distance from A.
 **c** Form an equation in terms of $x$.
 **d** Calculate the value of $x$.
 **e** Calculate the value of $\theta$.

**5** Two hawks P and Q are flying vertically above one another. Hawk Q is 250 m above hawk P. They both spot a snake at R.

Using the information given, calculate:
 **a** the height of P above the ground
 **b** the distance between P and R
 **c** the distance between Q and R.

# Student assessment 3

1. Using the triangular prism shown calculate:

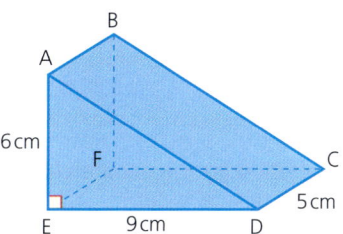

   a the length AD
   b the length AC
   c the angle AC makes with the plane CDEF
   d the angle AC makes with the plane ABFE.

2. Draw a graph of $y = \cos \theta°$, for $0° \leq \theta° \leq 360°$. Mark the angles $0°$, $90°$, $180°$, $270°$, $360°$ and also the maximum and minimum values of $y$.

3. The cosine of which other angle between 0 and 180° has the same value as:
   a $\cos 128°$
   b $-\cos 80°$?

4. For the triangle below calculate:

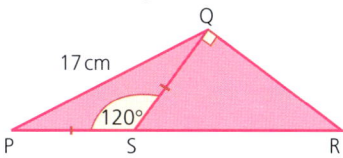

   a the length PS
   b angle QRS
   c the length SR.

5.

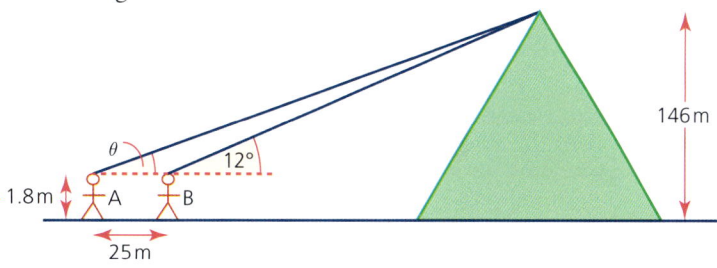

*A, B and the top of the pyramid are in the same vertical plane.*

The Great Pyramid at Giza is 146 m high. Two people A and B are looking at the top of the pyramid. The angle of elevation of the top of the pyramid from B is 12°. The distance between A and B is 25 m. If both A and B are 1.8 m tall, calculate:
   a the distance from B to the centre of the base of the pyramid
   b the angle of elevation $\theta$ of the top of the pyramid from A
   c the distance between A's head and the top of the pyramid.

# STUDENT ASSESSMENTS 7

## Student assessment 4

1. In the triangle shown opposite, $\cos\theta = \frac{5}{6}$.

   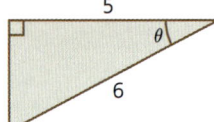

   Showing your working clearly, find the exact value of:
   
   a  $\sin\theta$
   
   b  $\cos\theta$

2. Solve the following trigonometric equations, giving all the solutions in the range $0° \leq \theta \leq 360°$.
   
   a  $\sin x = \frac{2}{5}$
   
   b  $\tan x = -\frac{\sqrt{3}}{3}$
   
   c  $\cos x = -0.1$
   
   d  $\sin x = 1$

3. Explain, with the aid of a graph, why the equation $\cos x = \frac{3}{2}$ has no solutions.

4. A circle with radius 3 cm, centre at O, is shown on the axes below.

   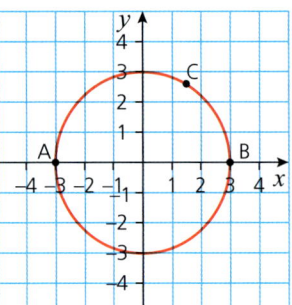

   The points A and B lie where the circumference of the circle intersects the x-axis. Point C is free to move on the circumference of the circle.
   
   a  Find, justifying your answer, the size of angle ACB.
   
   b  If BC = 3 cm, calculate the possible coordinates of point C, giving your answer in exact form.

# Student assessment 5

1. Two hot air balloons A and B are travelling in the same horizontal direction as shown in the diagram below. A is travelling at 2 m/s and B at 3 m/s. Their heights above the ground are 1.6 km and 1 km, respectively.

   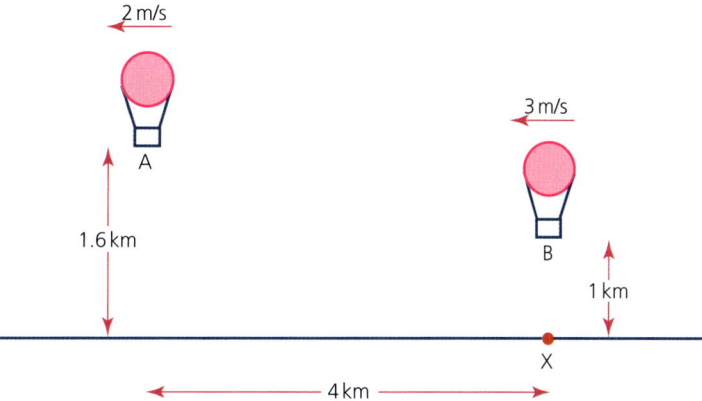

   At midday, their horizontal distance apart is 4 km and balloon B is directly above a point X on the ground.
   Calculate:
   a  the angle of elevation of A from X at midday
   b  the angle of depression of B from A at midday
   c  their horizontal distance apart at 12.30 p.m.
   d  the angle of elevation of B from X at 12.30 p.m.
   e  the angle of elevation of A from B at 12.30 p.m.
   f  how much closer A and B are at 12.30 p.m. compared with midday.

2. a  Plot the graphs of $y = \sin \theta°$ and $y = \cos \theta°$, for $0° \leqslant \theta° \leqslant 360°$, on the same axes.
   b  Use your graph to find the angles for which $\sin \theta° = \cos \theta°$.

3. This cuboid has one of its corners removed to leave a flat triangle BDC.
   Calculate:
   a  the length DC
   b  the length BC
   c  the length DB
   d  angle CBD
   e  the area of △BDC
   f  the angle AC makes with the plane AEHD.

   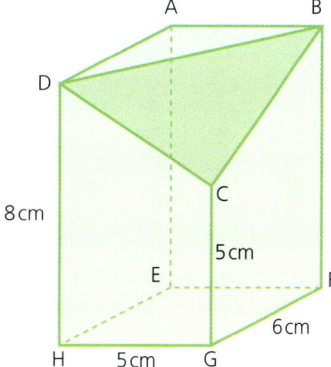

4. Describe the transformation that maps $f(x) = \cos x$ onto each of the following graphs:
   a  $y = \cos x - 5$
   b  $y = \cos(x + 120)$
   c  $y = \sin x$

# TOPIC 8

## Vectors and transformations

**Contents**

Chapter 52 Simple vectors (E8.2)
Chapter 53 Magnitude of a vector (E8.3)
Chapter 54 Transformations (C8.1, E8.1)
Chapter 55 Further transformations (C8.1, E8.1)

### Learning objectives

**C8.1**
Recognise, describe and draw the following transformations:
1. reflection of a shape in a vertical or horizontal line
2. rotation of a shape about the origin, vertices or midpoints of edges of the shapes, through multiples of 90°
3. enlargement of a shape from a given centre by a given scale factor
4. translation of a shape by a given vector $\begin{pmatrix} x \\ y \end{pmatrix}$

**E8.1**
Recognise, describe and draw the following transformations:
1. **reflection of a shape in a straight line**
2. **rotation of a shape about a given centre through multiples of 90°**
3. enlargement of a shape from a given centre by a given scale factor
4. **translation of a shape by a given vector**

**E8.2**
1. **Describe a translation using a vector represented by $\begin{pmatrix} x \\ y \end{pmatrix}$, $\overrightarrow{AB}$ or a**
2. **Add and subtract vectors**
3. **Multiply a vector by a scalar**

**E8.3**
**Calculate the magnitude of a vector $\begin{pmatrix} x \\ y \end{pmatrix}$ as $\sqrt{x^2 + y^2}$**

Elements in purple refer to the Extended curriculum only.

## The Italians

Leonardo Pisano (known today as Fibonacci) introduced new methods of arithmetic to Europe from the Hindus, Persians and Arabs. He discovered the sequence 1, 1, 2, 3, 5, 8, 13, … , which is now called the Fibonacci sequence, and some of its occurrences in nature. He also brought the decimal system, algebra and the 'lattice' method of multiplication to Europe. Fibonacci has been called the 'most talented mathematician of the middle ages'. Many books say that he brought Islamic mathematics to Europe, but in Fibonacci's own introduction to *Liber Abaci,* he credits the Hindus.

The Renaissance began in Italy. Art, architecture, music and the sciences flourished. However, the Roman Catholic Church was both powerful and resistant to change.

*Fibonacci (1170–1250)*

Girolamo Cardano (1501–1576) wrote his great mathematical book *Ars Magna* (Great Art) in which he showed, among much algebra that was new, calculations involving the solutions to cubic equations. He published this book, the first algebra book in Latin, to great acclaim. He was charged with heresy in 1570 because the church did not approve of his work on astrology. Although he was found innocent and continued to study mathematics, no other work of his was ever published.

# 52 Simple vectors

A **translation** (a sliding movement) can be described using column vectors. A column vector describes the movement of the object in both the *x* direction and the *y* direction.

### → Worked example

i Describe the translation from A to B in the diagram in terms of a column vector.

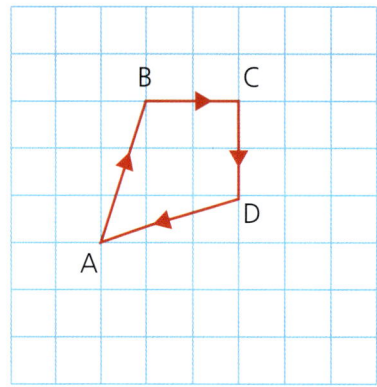

$\overrightarrow{AB} = \begin{pmatrix} 1 \\ 3 \end{pmatrix}$

i.e. 1 unit in the *x* direction, 3 units in the *y* direction

ii Describe $\overrightarrow{BC}$ in terms of a column vector.

$\overrightarrow{BC} = \begin{pmatrix} 2 \\ 0 \end{pmatrix}$

iii Describe $\overrightarrow{CD}$ in terms of a column vector.

$\overrightarrow{CD} = \begin{pmatrix} 0 \\ -2 \end{pmatrix}$

iv Describe $\overrightarrow{DA}$ in terms of a column vector.

$\overrightarrow{DA} = \begin{pmatrix} -3 \\ -1 \end{pmatrix}$

*Note that the arrow above AB shows the direction of the translation i.e. going from A to B.*

Translations can also be named by a single letter. The direction of the arrow indicates the direction of the translation.

## Simple vectors

###  Worked example

Define **a** and **b** in the diagram using column vectors.

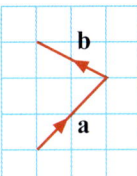

$$\mathbf{a} = \begin{pmatrix} 2 \\ 2 \end{pmatrix} \qquad \mathbf{b} = \begin{pmatrix} -2 \\ 1 \end{pmatrix}$$

Note: When you represent vectors by single letters, for example **a**, in handwritten work, you should write them as <u>a</u>.

If $\mathbf{a} = \begin{pmatrix} 2 \\ 5 \end{pmatrix}$ and $\mathbf{b} = \begin{pmatrix} -3 \\ -2 \end{pmatrix}$, they can be represented diagrammatically as shown.

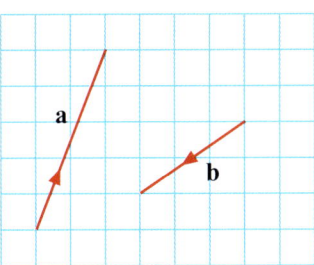

The diagrammatic representation of −**a** and −**b** is shown below.

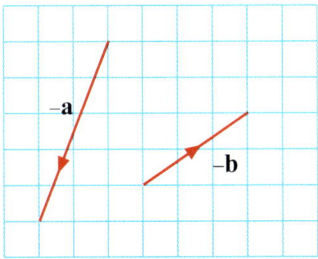

It can be seen from the diagram above that $-\mathbf{a} = \begin{pmatrix} -2 \\ -5 \end{pmatrix}$ and $-\mathbf{b} = \begin{pmatrix} 3 \\ 2 \end{pmatrix}$.

# 52 SIMPLE VECTORS

**Exercise 8.1** In Q.1 and Q.2, describe each translation using a column vector.

1. a $\vec{AB}$  b $\vec{BC}$
   c $\vec{CD}$  d $\vec{DE}$
   e $\vec{EA}$  f $\vec{AE}$
   g $\vec{DA}$  h $\vec{CA}$
   i $\vec{DB}$

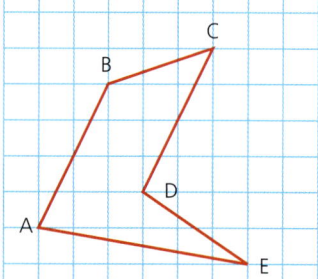

2. a **a**   b **b**   c **c**
   d **d**   e **e**   f **−b**
   g **−c**  h **−d**  i **−a**

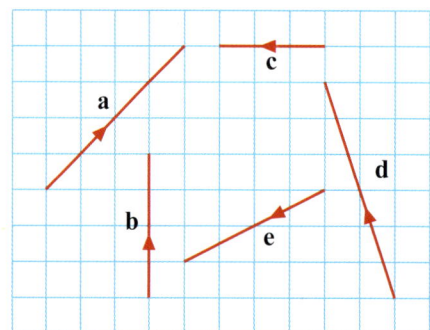

3. Draw and label the following vectors on a square grid:

   a $\mathbf{a} = \begin{pmatrix} 2 \\ 4 \end{pmatrix}$   b $\mathbf{b} = \begin{pmatrix} -3 \\ 6 \end{pmatrix}$   c $\mathbf{c} = \begin{pmatrix} 3 \\ -5 \end{pmatrix}$

   d $\mathbf{d} = \begin{pmatrix} -4 \\ -3 \end{pmatrix}$   e $\mathbf{e} = \begin{pmatrix} 0 \\ -6 \end{pmatrix}$   f $\mathbf{f} = \begin{pmatrix} -5 \\ 0 \end{pmatrix}$

   g **−c**   h **−b**   i **−f**

# Addition and subtraction of vectors

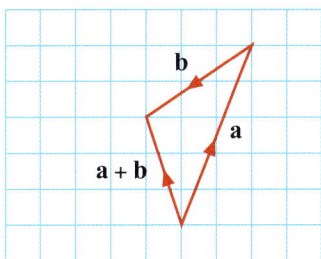

Vectors can be added together and represented diagrammatically as shown.

The translation represented by **a** followed by **b** can be written as a single transformation **a** + **b**:

i.e. $\begin{pmatrix} 2 \\ 5 \end{pmatrix} + \begin{pmatrix} -3 \\ -2 \end{pmatrix} = \begin{pmatrix} -1 \\ 3 \end{pmatrix}$

> ### Worked example

$\mathbf{a} = \begin{pmatrix} 2 \\ 5 \end{pmatrix} \quad \mathbf{b} = \begin{pmatrix} -3 \\ -2 \end{pmatrix}$

i  Draw a diagram to represent **a** − **b**, where **a** − **b** = (**a**) + (−**b**).

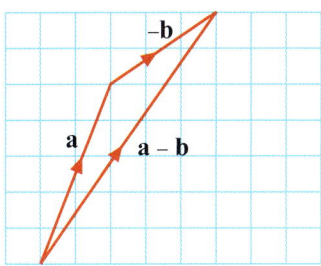

ii  Calculate the vector represented by **a** − **b**.

$\begin{pmatrix} 2 \\ 5 \end{pmatrix} - \begin{pmatrix} -3 \\ -2 \end{pmatrix} = \begin{pmatrix} 5 \\ 7 \end{pmatrix}$

## Exercise 8.2

In the following questions,

$\mathbf{a} = \begin{pmatrix} 3 \\ 4 \end{pmatrix} \quad \mathbf{b} = \begin{pmatrix} -2 \\ 1 \end{pmatrix} \quad \mathbf{c} = \begin{pmatrix} -4 \\ -3 \end{pmatrix} \quad \mathbf{d} = \begin{pmatrix} 3 \\ -2 \end{pmatrix}$

1  Draw vector diagrams to represent the following:
   a  **a** + **b**             b  **b** + **a**             c  **a** + **d**
   d  **d** + **a**             e  **b** + **c**             f  **c** + **b**

# 52 SIMPLE VECTORS

**Exercise 8.2 (cont)**

2. What conclusions can you draw from your answers to Q.1?
3. Draw vector diagrams to represent the following:
   a  b − c
   b  d − a
   c  −a − c
   d  a + c − b
   e  d − c − b
   f  −c + b + d
4. Represent each of the vectors in Q.3 by a single column vector.

## Multiplying a vector by a scalar

Look at the two vectors in the diagram.

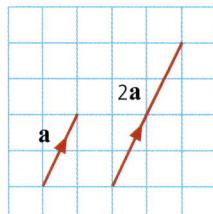

$$\mathbf{a} = \begin{pmatrix} 1 \\ 2 \end{pmatrix} \qquad 2\mathbf{a} = 2\begin{pmatrix} 1 \\ 2 \end{pmatrix} = \begin{pmatrix} 2 \\ 4 \end{pmatrix}$$

### ➔ Worked example

If $\mathbf{a} = \begin{pmatrix} 2 \\ -4 \end{pmatrix}$, write the vectors **b**, **c**, **d** and **e** in terms of **a**.

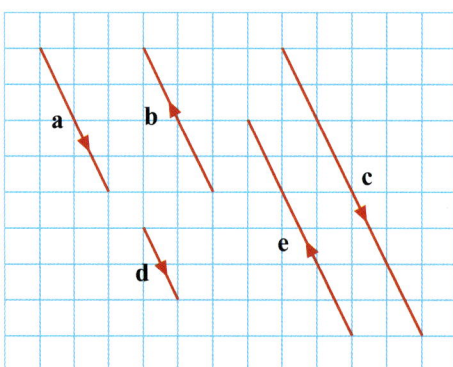

$\mathbf{b} = -\mathbf{a}$  $\qquad$  $\mathbf{c} = 2\mathbf{a}$  $\qquad$  $\mathbf{d} = \frac{1}{2}\mathbf{a}$  $\qquad$  $\mathbf{e} = -\frac{3}{2}\mathbf{a}$

## Exercise 8.3

1  a $= \begin{pmatrix} 1 \\ 4 \end{pmatrix}$   b $= \begin{pmatrix} -4 \\ -2 \end{pmatrix}$   c $= \begin{pmatrix} -4 \\ 6 \end{pmatrix}$

Write the following vectors in terms of either **a**, **b** or **c**.

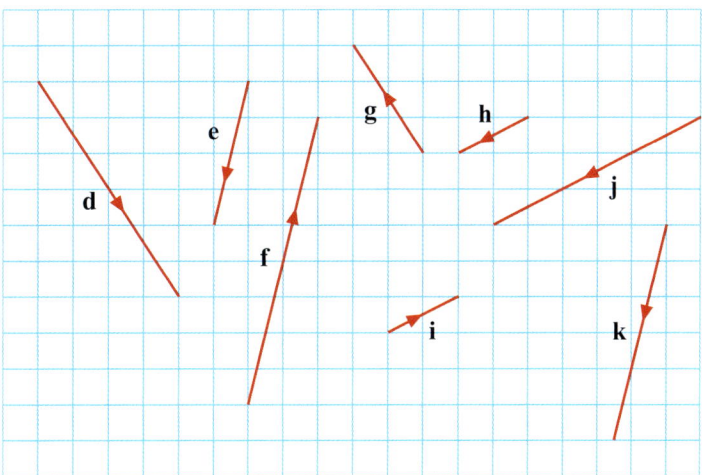

2  a $= \begin{pmatrix} 2 \\ 3 \end{pmatrix}$   b $= \begin{pmatrix} -4 \\ -1 \end{pmatrix}$   c $= \begin{pmatrix} -2 \\ 4 \end{pmatrix}$

Represent each of the following as a single column vector:
a  2**a**          b  3**b**          c  −**c**
d  **a** + **b**   e  **b** − **c**   f  3**c** − **a**
g  2**b** − **a**  h  $\frac{1}{2}$(**a** − **b**)   i  2**a** − 3**c**

3  a $= \begin{pmatrix} -2 \\ 3 \end{pmatrix}$   b $= \begin{pmatrix} 0 \\ -3 \end{pmatrix}$   c $= \begin{pmatrix} 4 \\ -1 \end{pmatrix}$

Write each of the following vectors in terms of **a**, **b** and **c**:

a  $\begin{pmatrix} -4 \\ 6 \end{pmatrix}$   b  $\begin{pmatrix} 0 \\ 3 \end{pmatrix}$   c  $\begin{pmatrix} 4 \\ -4 \end{pmatrix}$

d  $\begin{pmatrix} -2 \\ 6 \end{pmatrix}$   e  $\begin{pmatrix} 8 \\ -2 \end{pmatrix}$   f  $\begin{pmatrix} 10 \\ -5 \end{pmatrix}$

# 53 Magnitude of a vector

The **magnitude** or size of a vector is represented by its length, i.e. the longer the length, the greater the magnitude.

The magnitude of a vector **a** or $\overrightarrow{AB}$ is denoted by $|\mathbf{a}|$ or $|\overrightarrow{AB}|$ respectively and is calculated using Pythagoras' theorem.

## ➡ Worked examples

$$\mathbf{a} = \begin{pmatrix} 3 \\ 4 \end{pmatrix} \qquad \overrightarrow{BC} = \begin{pmatrix} -6 \\ 8 \end{pmatrix}$$

**a** Represent both of the above vectors diagrammatically.

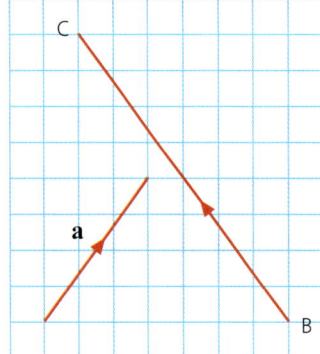

**b** **i** Calculate $|\mathbf{a}|$.

$$|\mathbf{a}| = \sqrt{(3^2 + 4^2)}$$
$$= \sqrt{25}$$
$$= 5$$

**ii** Calculate $|\overrightarrow{BC}|$.

$$\overrightarrow{BC} = \sqrt{(-6)^2 + 8^2}$$
$$= \sqrt{100}$$
$$= 10$$

## Exercise 8.4

1. Calculate the magnitude of the vectors shown below. Give your answers correct to 1 d.p.

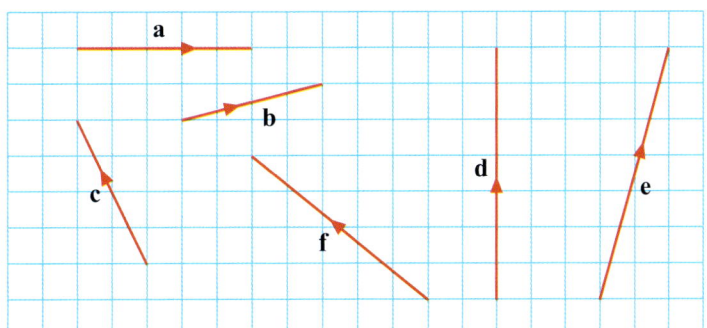

2. Calculate the magnitude of the following vectors, giving your answers to 1 d.p.

   a  $\overrightarrow{AB} = \begin{pmatrix} 0 \\ 4 \end{pmatrix}$
   b  $\overrightarrow{BC} = \begin{pmatrix} 2 \\ 5 \end{pmatrix}$
   c  $\overrightarrow{CD} = \begin{pmatrix} -4 \\ -6 \end{pmatrix}$

   d  $\overrightarrow{DE} = \begin{pmatrix} -5 \\ 12 \end{pmatrix}$
   e  $2\overrightarrow{AB}$
   f  $-\overrightarrow{CD}$

3. $\mathbf{a} = \begin{pmatrix} 4 \\ -3 \end{pmatrix}$  $\mathbf{b} = \begin{pmatrix} -5 \\ 7 \end{pmatrix}$  $\mathbf{c} = \begin{pmatrix} -1 \\ -8 \end{pmatrix}$

   Calculate the magnitude of the following, giving your answers to 1 d.p.

   a  $\mathbf{a} + \mathbf{b}$
   b  $2\mathbf{a} - \mathbf{b}$
   c  $\mathbf{b} - \mathbf{c}$
   d  $2\mathbf{c} + 3\mathbf{b}$
   e  $2\mathbf{b} - 3\mathbf{a}$
   f  $\mathbf{a} + 2\mathbf{b} - \mathbf{c}$

## Vector geometry

In general, vectors are not fixed in position. If a vector **a** has a specific magnitude and direction, then any other vector with the same magnitude and direction as **a** can also be labelled **a**.

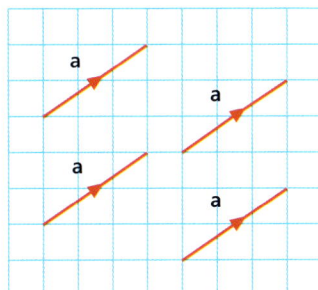

If $\mathbf{a} = \begin{pmatrix} 3 \\ 2 \end{pmatrix}$ then all the vectors shown in this diagram can also be labelled **a**, as they all have the same magnitude and direction.

This property of vectors can be used to solve problems in vector geometry.

# 53 MAGNITUDE OF A VECTOR

> **Worked examples**
>
> a  Name a vector equal to $\vec{AD}$.
>    $\vec{BC} = \vec{AD}$
>
>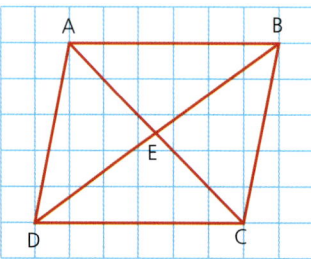
>
> b  Write $\vec{BD}$ in terms of $\vec{BE}$.
>    $\vec{BD} = 2\vec{BE}$
>
> c  Write $\vec{CD}$ in terms of $\vec{AB}$.
>    $\vec{CD} = \vec{BA} = -\vec{AB}$

**Exercise 8.5**   1  If $\vec{AG} = \mathbf{a}$ and $\vec{AE} = \mathbf{b}$, write the following in terms of **a** and **b**:

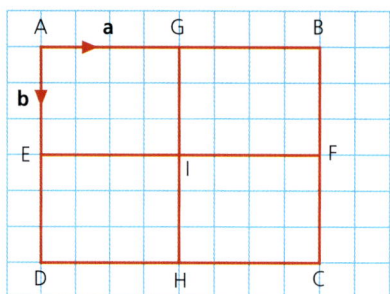

a  $\vec{EI}$   b  $\vec{HC}$   c  $\vec{FC}$
d  $\vec{DE}$   e  $\vec{GH}$   f  $\vec{CD}$
g  $\vec{AI}$   h  $\vec{GE}$   i  $\vec{FD}$

2  If $\vec{LP} = \mathbf{a}$ and $\vec{LR} = \mathbf{b}$, write the following in terms of **a** and **b**:

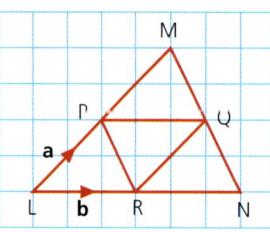

a  $\vec{LM}$   b  $\vec{PQ}$   c  $\vec{PR}$
d  $\vec{MQ}$   e  $\vec{MP}$   f  $\vec{NP}$

# Vector geometry

**3** *ABCDEF* is a regular hexagon.

If $\vec{GA} = \mathbf{a}$ and $\vec{GB} = \mathbf{b}$, write the following in terms of **a** and **b**:

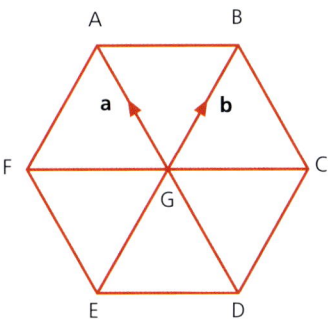

- **a** $\vec{AD}$
- **b** $\vec{FE}$
- **c** $\vec{DC}$
- **d** $\vec{AB}$
- **e** $\vec{FC}$
- **f** $\vec{EC}$
- **g** $\vec{BF}$
- **h** $\vec{FD}$
- **i** $\vec{AE}$

**4** If $\vec{AB} = \mathbf{a}$ and $\vec{AG} = \mathbf{b}$, write the following in terms of **a** and **b**:

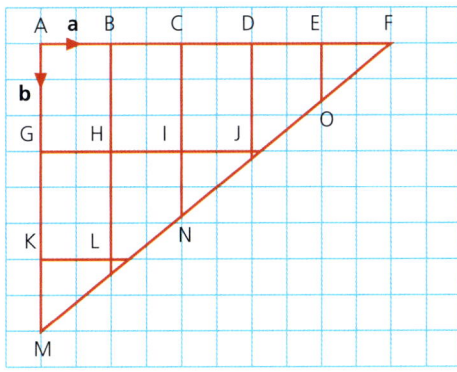

- **a** $\vec{AF}$
- **b** $\vec{AM}$
- **c** $\vec{FM}$
- **d** $\vec{FO}$
- **e** $\vec{EI}$
- **f** $\vec{KF}$
- **g** $\vec{CN}$
- **h** $\vec{AN}$
- **i** $\vec{DN}$

## Exercise 8.6

**1** In the diagram, T is the midpoint of the line PS and R divides the line QS in the ratio 1:3.

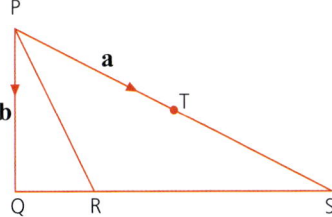

$\vec{PT} = \mathbf{a}$ and $\vec{PQ} = \mathbf{b}$.

# 53 MAGNITUDE OF A VECTOR

**Exercise 8.6 (cont)**

a Write each of the following in terms of **a** and **b**:
  i $\vec{PS}$
  ii $\vec{QS}$
  iii $\vec{PR}$

b Show that $\vec{RT} = \frac{1}{4}(2\mathbf{a} - 3\mathbf{b})$.

2 $\vec{PM} = 3\vec{LP}$ and $\vec{QN} = 3\vec{LQ}$

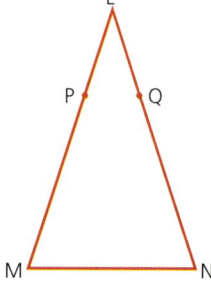

Prove that:
a the line PQ is parallel to the line MN
b the line MN is four times the length of the line PQ.

3 PQRS is a parallelogram. The point T divides the line PQ in the ratio 1:3, and U, V and W are the midpoints of SR, PS and QR respectively.

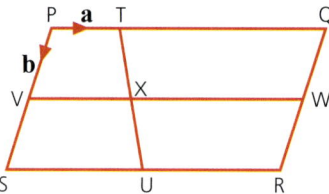

$\vec{PT} = \mathbf{a}$ and $\vec{PV} = \mathbf{b}$.

a Write each of the following in terms of **a** and **b**:
  i $\vec{PQ}$
  ii $\vec{SU}$
  iii $\vec{PU}$
  iv $\vec{VX}$

b Show that $\vec{XR} = \frac{1}{2}(5\mathbf{a} + 2\mathbf{b})$.

4 ABC is an isosceles triangle. L is the midpoint of BC. M divides the line LA in the ratio 1:5, and N divides AC in the ratio 2:5.

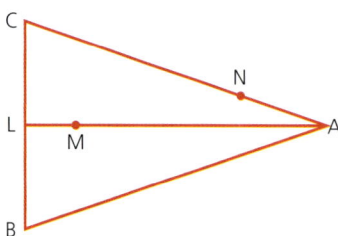

a $\vec{BC} = \mathbf{p}$ and $\vec{BA} = \mathbf{q}$. Write the following in terms of **p** and **q**:
  i $\vec{LA}$
  ii $\vec{AN}$

b Show that $\vec{MN} = \frac{1}{84}(46\mathbf{q} - 11\mathbf{p})$.

# 54 Transformations

An object undergoing a transformation changes either in position or shape. In its simplest form, this change can occur as a result of either a **reflection**, **rotation**, **translation** or **enlargement**. If an object undergoes a transformation, then its new position or shape is known as the **image**. The transformation that maps the image back onto the original object is known as an **inverse transformation**.

## Reflection

If an object is reflected, it undergoes a 'flip' movement about a dashed (broken) line known as the **mirror line**, as shown in the diagram:

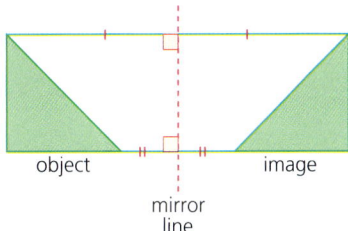

A point on the object and its equivalent point on the image are **equidistant** from the mirror line. This distance is measured at right angles to the mirror line. The line joining the point to its image is perpendicular to the mirror line.

**Exercise 8.7**

Copy the following objects and mirror lines and, in each case, draw in the position of the object under reflection in the dashed mirror line(s).

1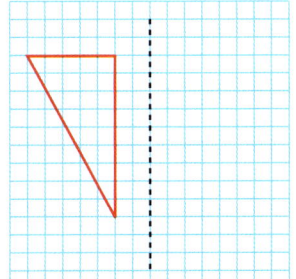

2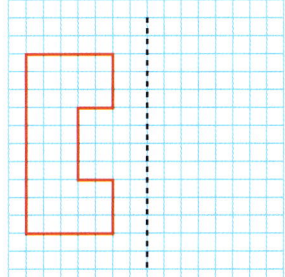

497

## 54 TRANSFORMATIONS

**Exercise 8.7 (cont)**

3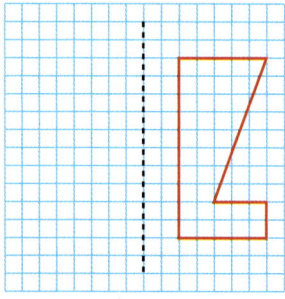
4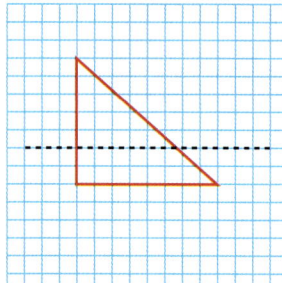
5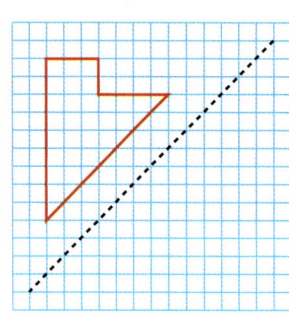
6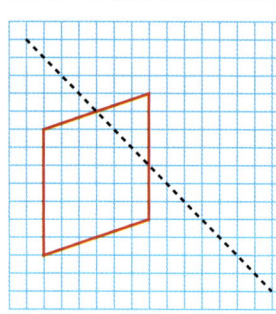
7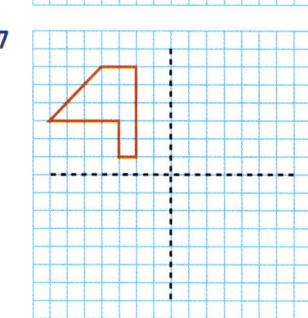
8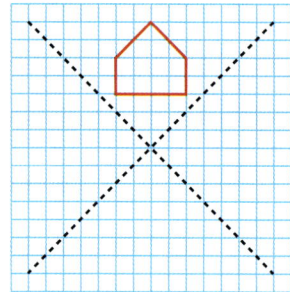

**Exercise 8.8**  Copy the following objects and images and, in each case, draw in the position of the mirror line(s).

1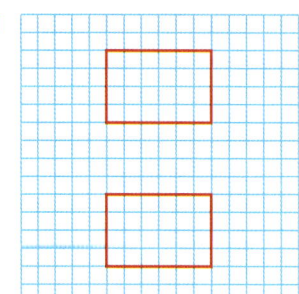
2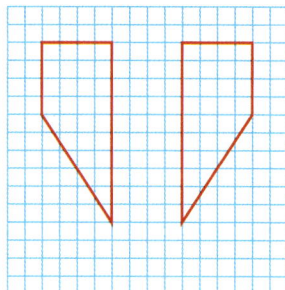

*Rotation*

**3**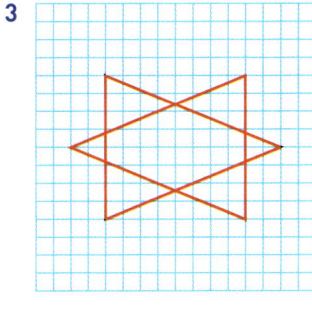

**4**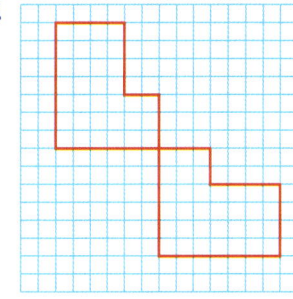

**5**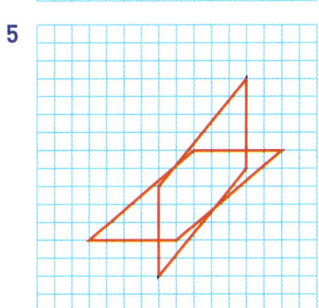

**6**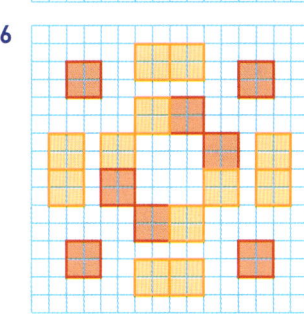

# Rotation

If an object is rotated, it undergoes a 'turning' movement about a specific point known as the **centre of rotation**. When describing a rotation, it is necessary to identify not only the position of the centre of rotation, but also the angle and direction of the turn, as shown in the diagram:

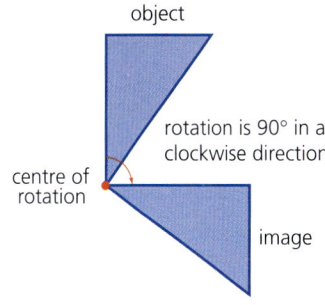

**Exercise 8.9** In the following, the object and centre of rotation have both been given. Copy each diagram and draw the object's image under the stated rotation about the marked point.

**1**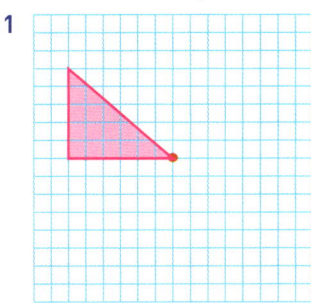

rotation 180°

**2**

rotation 90° clockwise

## 54 TRANSFORMATIONS

### Exercise 8.9 (cont)

> **Note**
>
> When shapes are rotated on a coordinate grid, Core students are expected to rotate a shape about one of the vertices of the shape or the origin.

3
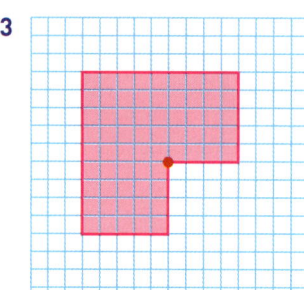
rotation 180°

4
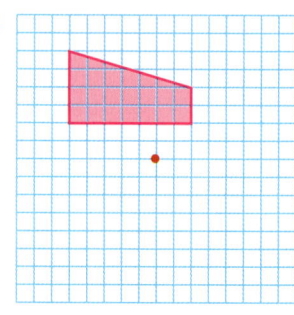
rotation 90° clockwise

5
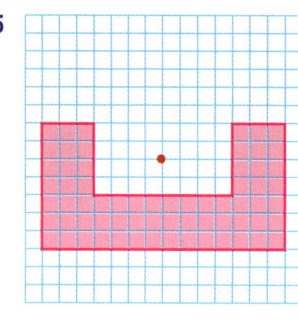
rotation 90° anti-clockwise

6
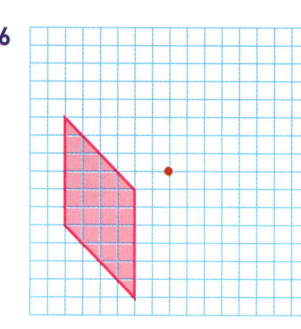
rotation 90° clockwise

### Exercise 8.10

Copy the diagrams in Q.1–6. In each case, the object (unshaded) and image (shaded) have been drawn.
a  Mark the centre of rotation.
b  Calculate the angle and direction of rotation.

*Using some tracing paper will help with these questions.*

1
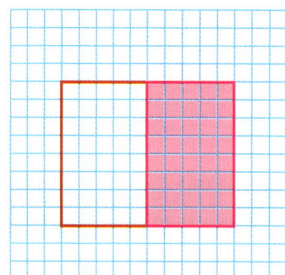

2
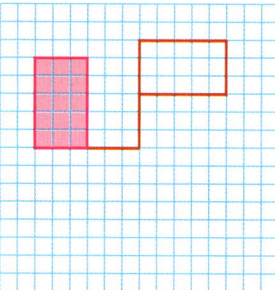

# Translation

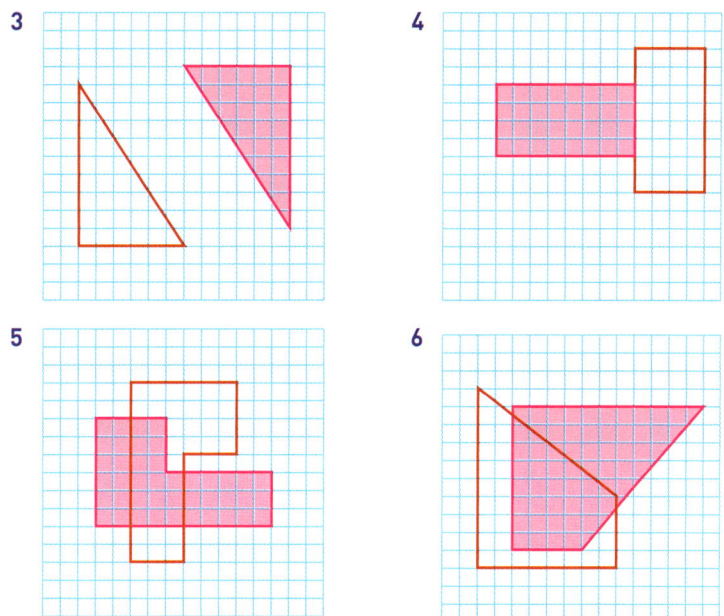

**7** For each of the rotations above, give the inverse transformation.

# Translation

If an object is translated, it undergoes a 'straight sliding' movement. When describing a translation, it is necessary to give the translation vector. As no rotation is involved, each point on the object moves in the same way to its corresponding point on the image. For example:

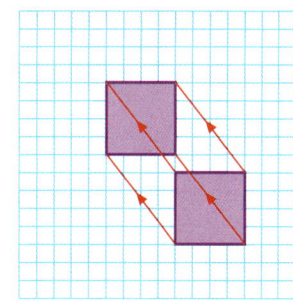

Vector = $\begin{pmatrix} 6 \\ 3 \end{pmatrix}$     Vector = $\begin{pmatrix} -4 \\ 5 \end{pmatrix}$

# 54 TRANSFORMATIONS

**Exercise 8.11**  In the following diagrams, object A has been translated to both of images B and C. Give the translation vectors.

1
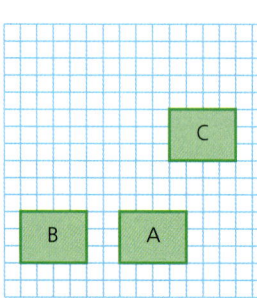

2
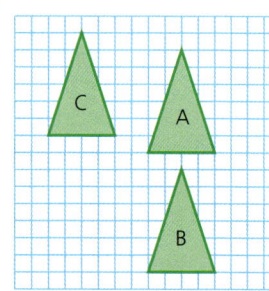

3
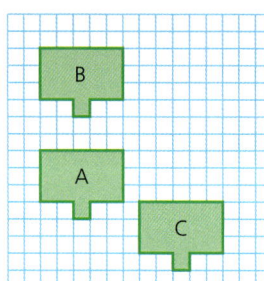

4
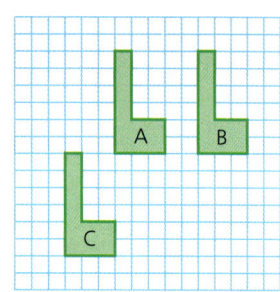

**Exercise 8.12**  Copy each of the following diagrams. Translate the object by the vector given in each case and draw the image in its position. (Note that a bigger grid than the one shown may be needed.)

1
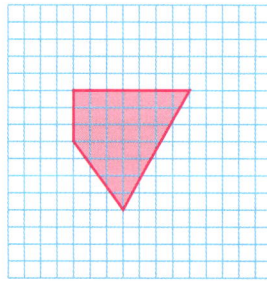

Vector = $\begin{pmatrix} 3 \\ 5 \end{pmatrix}$

2
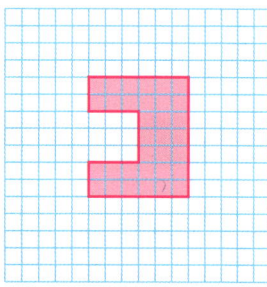

Vector = $\begin{pmatrix} 5 \\ -4 \end{pmatrix}$

3
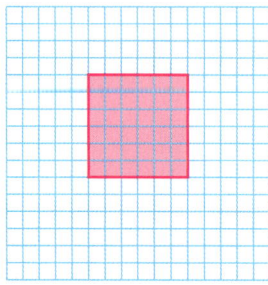

Vector = $\begin{pmatrix} -4 \\ 6 \end{pmatrix}$

4
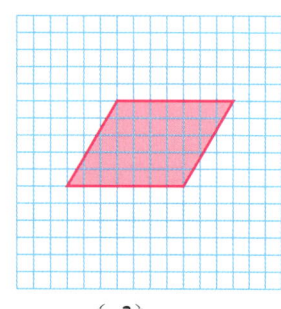

Vector = $\begin{pmatrix} -2 \\ -5 \end{pmatrix}$

*Enlargement*

5
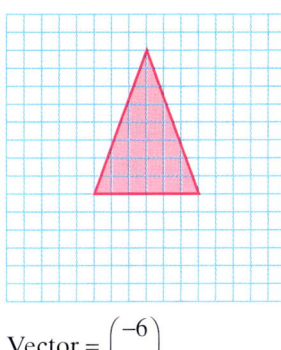
Vector = $\begin{pmatrix} -6 \\ 0 \end{pmatrix}$

6
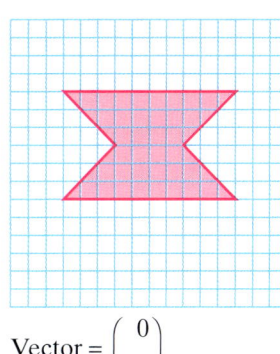
Vector = $\begin{pmatrix} 0 \\ -1 \end{pmatrix}$

7   For each of the diagrams in Q.1–6, give the vector that would map the image back on to the original object.

# Enlargement

If an object is enlarged, the result is an image which is mathematically similar to the object but of a different size. The image can be either larger or smaller than the original object. When describing an enlargement, two pieces of information need to be given, the position of the **centre of enlargement** and the **scale factor of enlargement**.

## ➡ Worked examples

a   In the diagram below, triangle ABC is enlarged to form triangle A'B'C'.

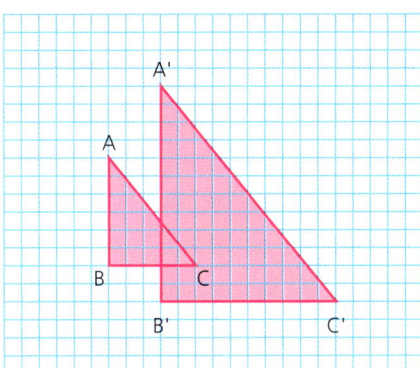

i   Find the centre of enlargement.

The centre of enlargement is found by joining corresponding points on the object and image with a straight line. These lines are then extended until they meet. The point at which they meet is the centre of enlargement, O.

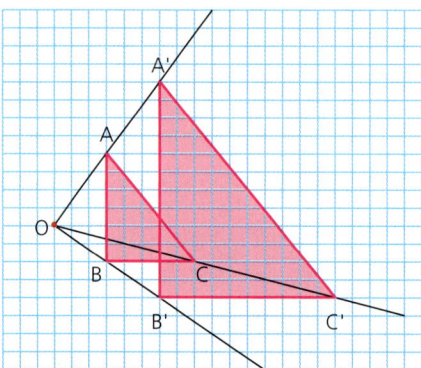

ii Calculate the scale factor of enlargement.

The scale factor of enlargement can be calculated in one of two ways. From the diagram above, it can be seen that the distance OA' is twice the distance OA. Similarly, OC' and OB' are both twice OC and OB respectively; hence the scale factor of enlargement is 2.

Alternatively, the scale factor can be found by considering the ratio of the length of a side on the image to the length of the corresponding side on the object, i.e.

$$\frac{A'B'}{AB} = \frac{12}{6} = 2$$

Hence the scale factor of enlargement is 2.

b In the diagram below, the rectangle ABCD undergoes a transformation to form rectangle A'B'C'D'.

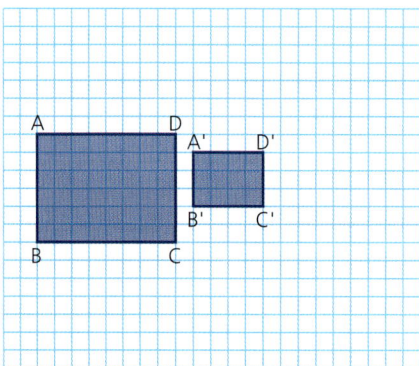

i Find the centre of enlargement.

By joining corresponding points on both the object and the image, the centre of enlargement is found at O.

# Enlargement

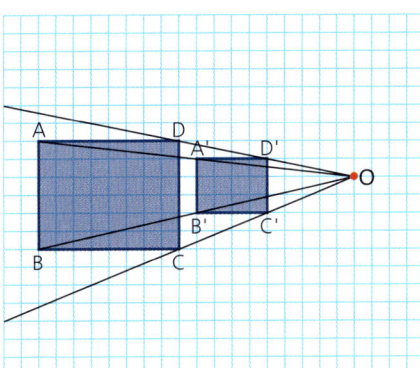

**ii** Calculate the scale factor of enlargement.

The scale factor of enlargement $\frac{A'B'}{AB} = \frac{3}{6} = \frac{1}{2}$

Note: If the scale factor of enlargement is greater than 1, then the image is larger than the object. If the scale factor lies between 0 and 1, then the resulting image is smaller than the object. In these cases, although the image is smaller than the object, the transformation mathematically is still known as an enlargement.

**Exercise 8.13**  Copy the following diagrams and find:
a the centre of enlargement          b the scale factor of enlargement.

**1**

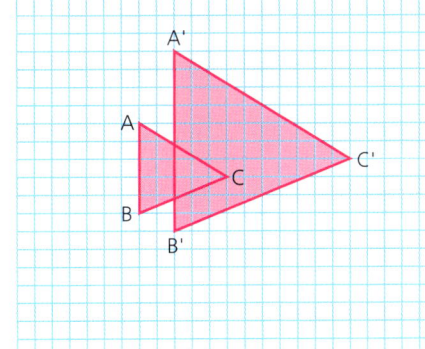

**2**

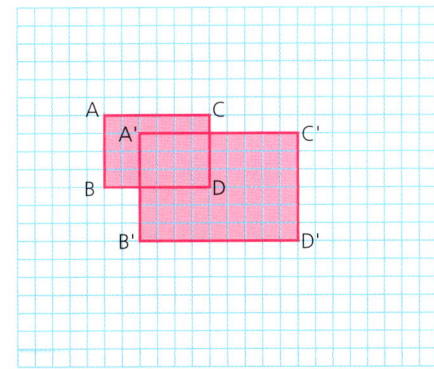

## 54 TRANSFORMATIONS

**Exercise 8.13 (cont)**

3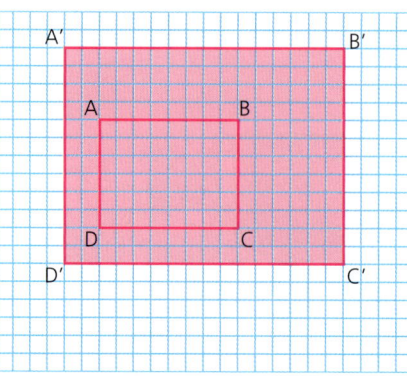

4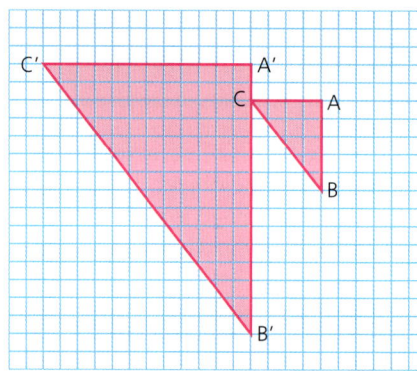

5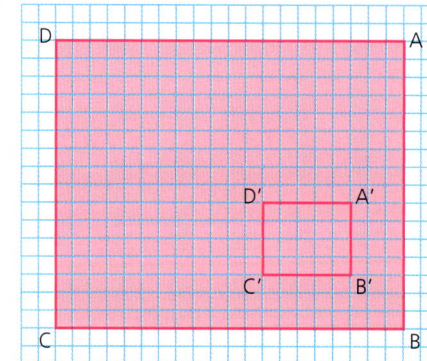

6 For each of Q.1–5, give the enlargement that would map the image back onto the original object.

# Enlargement

**Exercise 8.14** Copy each of the following diagrams. Enlarge the objects by the scale factor given and from the centre of enlargement shown. (Note that a bigger grid than the one shown may be needed.)

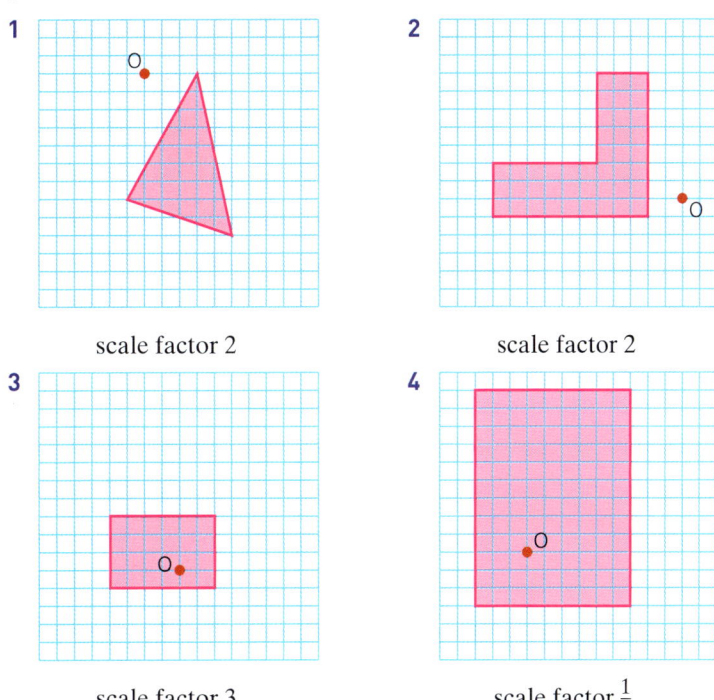

1. scale factor 2
2. scale factor 2
3. scale factor 3
4. scale factor $\frac{1}{3}$

The diagram below shows an example of **negative enlargement**.

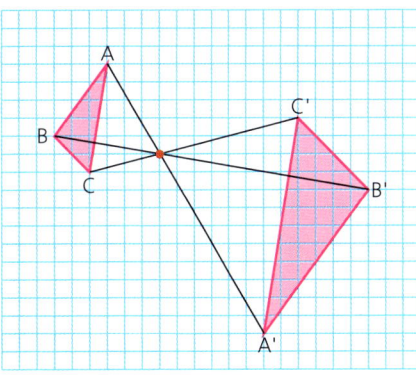

scale factor of enlargement is $-2$

With negative enlargement, each point and its image are on opposite sides of the centre of enlargement. The scale factor of enlargement is calculated in the same way, remembering, however, to write a '−' sign before the number.

# 54 TRANSFORMATIONS

**Exercise 8.15**

1 Copy the following diagram, calculate the scale factor of enlargement and show the position of the centre of enlargement.

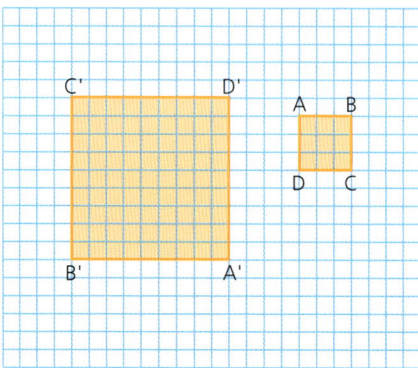

2 Copy the diagram and enlarge the object by the given scale factor and from the centre of enlargement shown.

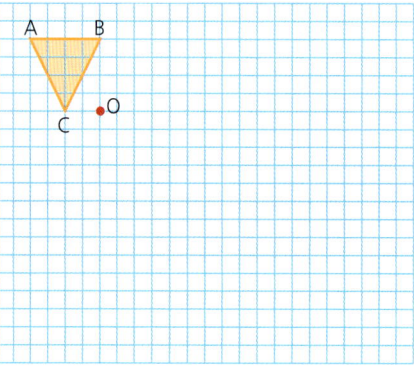

scale factor of enlargement is −2.5

3 Using the scale factor and centre of enlargement given, copy and complete the diagram.

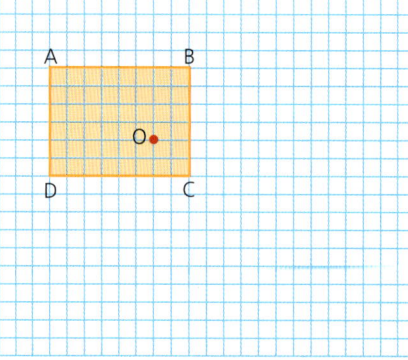

scale factor of enlargement is −2

4 Copy the diagram. Find the scale factor of enlargement and mark the position of the centre of enlargement.

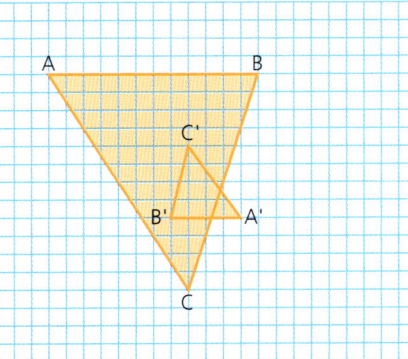

5 An object and part of its image under enlargement are given in the diagram below.

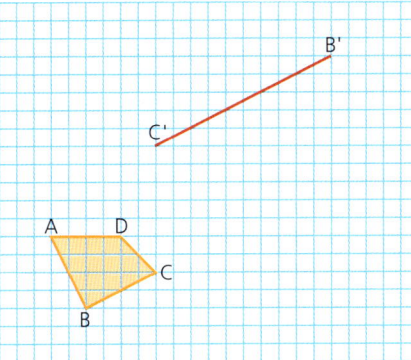

a Copy the diagram and complete the image.
b Mark the centre of enlargement and calculate the scale factor of enlargement.

6 In the diagram below, part of an object in the shape of a quadrilateral and its image under enlargement are drawn.

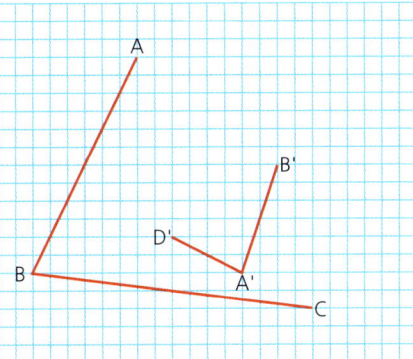

a Copy and complete the diagram.
b Mark the centre of enlargement and calculate the scale factor of enlargement.

# 55 Further transformations

Chapter 54 introduced basic aspects of transformations. However, as with most branches of mathematics, a basic principle can be extended.

## Reflection

The position of the mirror line is essential when describing a reflection. At times its equation, as well as its position, will be required.

### ➜ Worked examples

**a** Find the equation of the mirror line in the reflection given in the diagram.

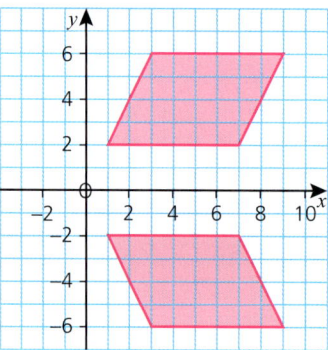

Here the mirror line is the *x*-axis. The equation of the mirror line is therefore $y = 0$.

**b** A reflection is shown in the diagram below.

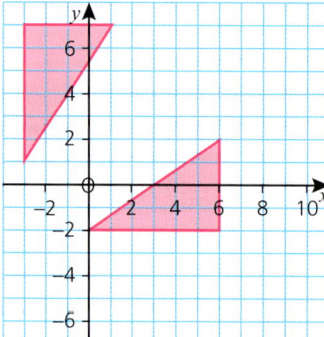

# Reflection

**i** Draw the position of the mirror line.

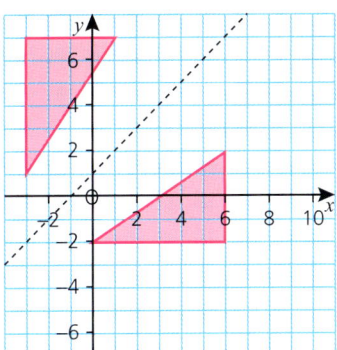

**ii** Give the equation of the mirror line.
$y = x + 1$.

**Exercise 8.16** Copy each of the following diagrams, then:

**a** draw the position of the mirror line(s)
**b** give the equation of the mirror line(s).

**1**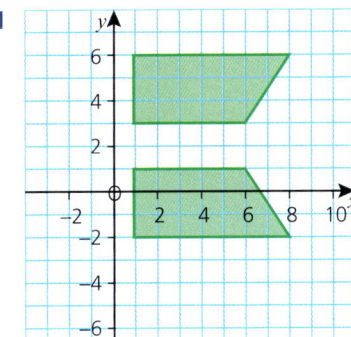

**2**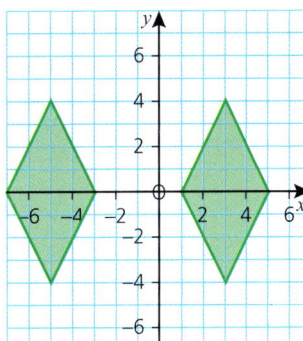

**3**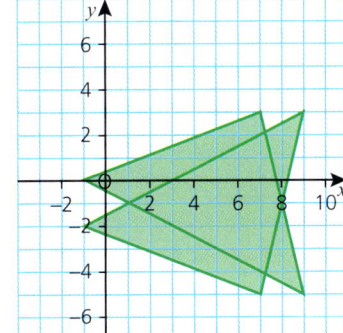

**4**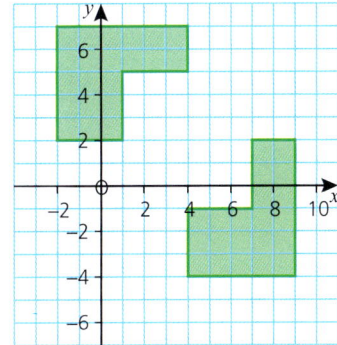

## 55 FURTHER TRANSFORMATIONS

**Exercise 8.16 (cont)**

5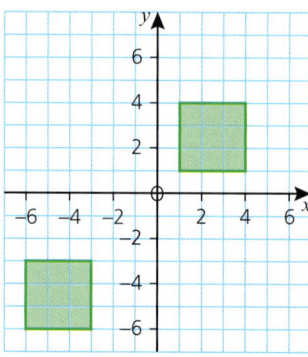

6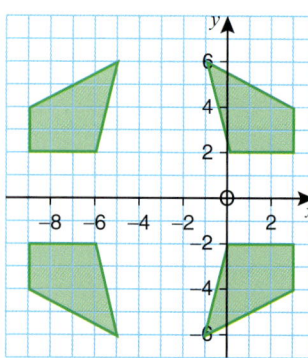

7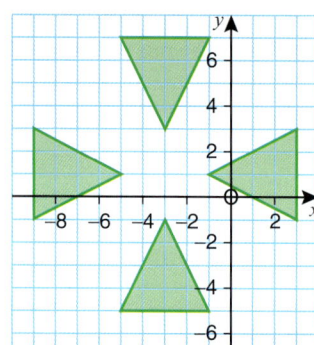

8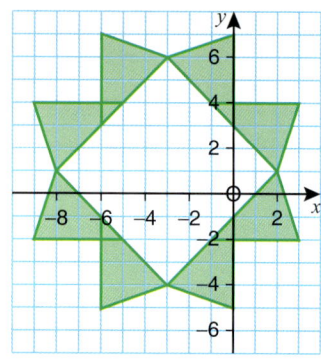

**Exercise 8.17** In Q.1 and Q.2, copy each diagram four times and reflect the object in each of the lines given.

1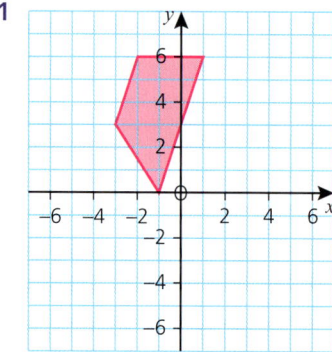

a  $x = 2$     b  $y = 0$
c  $y = x$     d  $y = -x$

2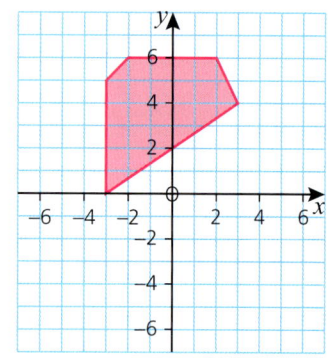

a  $x = -1$     b  $y = x + 2$
c  $y = -x - 1$  d  $x = 0$

# Combinations of transformations

3   Copy the diagram, and reflect the triangles in the lines $x = 1$ and $y = -3$.

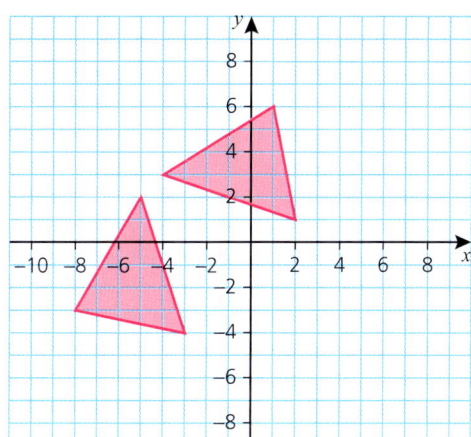

## Combinations of transformations

An object need not be subjected to just one type of transformation. It can undergo a succession of different transformations.

### → Worked examples

A triangle ABC maps onto A'B'C' after an enlargement of scale factor 3 from the centre of enlargement (0, 7). A'B'C' is then mapped onto A"B"C" by a reflection in the line $x = 1$.

i   Draw and label the image A'B'C'.

ii  Draw and label the image A"B"C".

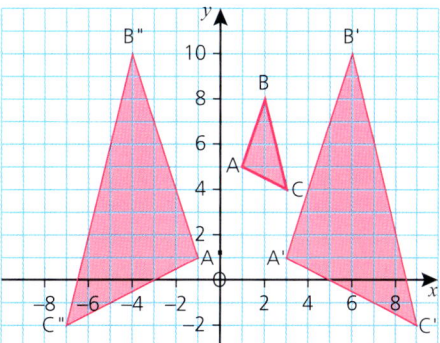

## 55 FURTHER TRANSFORMATIONS

**Exercise 8.18** In each of the following questions, copy the diagram. After both transformations, draw the images on the same grid and label them clearly.

1 The square ABCD is mapped onto A'B'C'D' by a reflection in the line $y = 3$. A'B'C'D' then maps onto A"B"C"D" as a result of a 90° rotation in a clockwise direction about the point $(-2, 5)$.

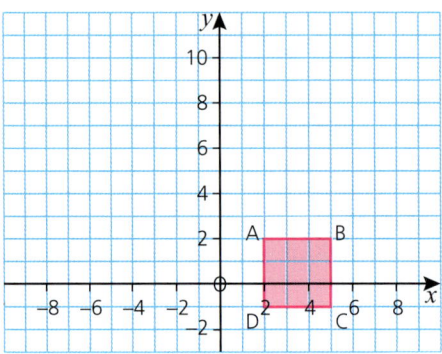

2 The rectangle ABCD is mapped onto A'B'C'D' by an enlargement of scale factor $-2$ with its centre at $(0, 5)$. A'B'C'D' then maps onto A"B"C"D" as a result of a reflection in the line $y = -x + 7$.

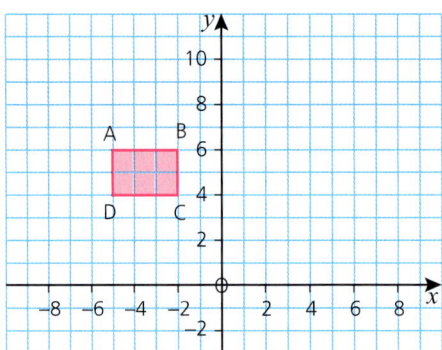

# Investigations, modelling and ICT 8

## Investigation: A painted cube

A $3 \times 3 \times 3$ cm cube is painted on the outside as shown in the left-hand diagram below:

 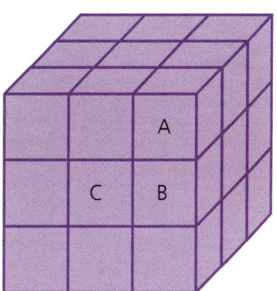

The large cube is then cut up into 27 smaller cubes, each $1\,\text{cm} \times 1\,\text{cm} \times 1\,\text{cm}$ as shown on the right.

$1 \times 1 \times 1$ cm cubes with 3 painted faces are labelled type A.

$1 \times 1 \times 1$ cm cubes with 2 painted faces are labelled type B.

$1 \times 1 \times 1$ cm cubes with 1 face painted are labelled type C.

$1 \times 1 \times 1$ cm cubes with no faces painted are labelled type D.

1  a  How many of the 27 cubes are type A?
   b  How many of the 27 cubes are type B?
   c  How many of the 27 cubes are type C?
   d  How many of the 27 cubes are type D?
2  Consider a $4 \times 4 \times 4$ cm cube cut into $1 \times 1 \times 1$ cm cubes. How many of the cubes are type A, B, C and D?
3  How many type A, B, C and D cubes are there when a $10 \times 10 \times 10$ cm cube is cut into $1 \times 1 \times 1$ cm cubes?
4  Generalise for the number of type A, B, C and D cubes in an $n \times n \times n$ cube.
5  Generalise for the number of type A, B, C and D cubes in a cuboid $l$ cm long, $w$ cm wide and $h$ cm high.

## Investigation: Triangle count

The diagram shows an isosceles triangle with a vertical line drawn from its apex to its base.

There is a total of 3 triangles in this diagram.

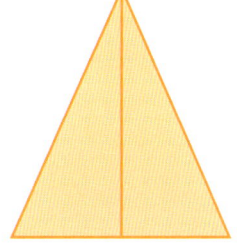

# INVESTIGATIONS, MODELLING AND ICT 8

If a horizontal line is drawn across the triangle, it will look as shown:

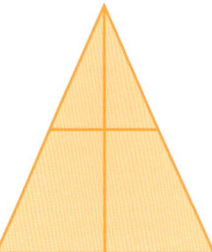

There is a total of 6 triangles in this diagram.

When one more horizontal line is added, the number of triangles increases further:

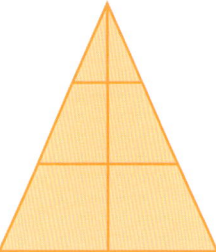

1. Calculate the total number of triangles in the diagram above with the two inner horizontal lines.
2. Investigate the relationship between the total number of triangles ($t$) and the number of inner horizontal lines ($h$). Enter your results in an ordered table.
3. Write an algebraic rule linking the total number of triangles and the number of inner horizontal lines.

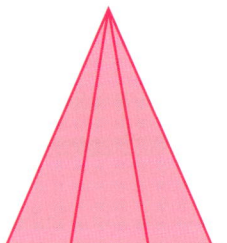

The triangle (left) has two lines drawn from the apex to the base.

There is a total of 6 triangles in this diagram.

If a horizontal line is drawn through this triangle, the number of triangles increases as shown:

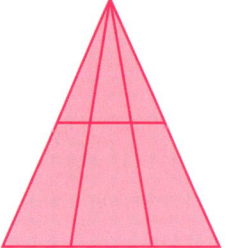

4 Calculate the total number of triangles in the diagram above with two lines from the vertex and one inner horizontal line.
5 Investigate the relationship between the total number of triangles ($t$) and the number of inner horizontal lines ($h$) when two lines are drawn from the apex. Enter your results in an ordered table.
6 Write an algebraic rule linking the total number of triangles and the number of inner horizontal lines.

# ICT Activity

An athlete (A) is training for a race that involves both swimming and running.

He can swim at a constant speed of 4 m/s and run at a constant speed of 7 m/s.

At a particular point near the end of the swimming stage, he is 40 m from the river bank and 100 m from the clothes changing station (C) as shown:

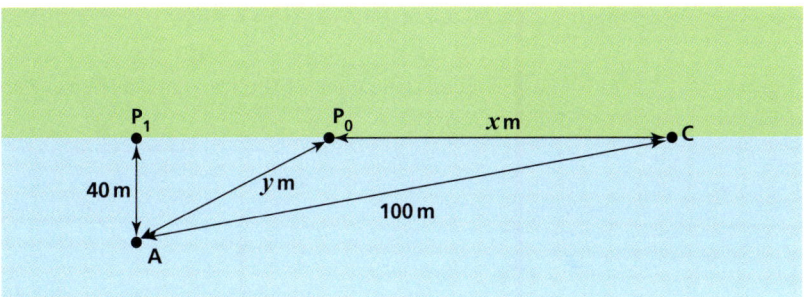

1 Calculate the time he would take to swim the distance AC.
2 Show that the distance $P_1C$ is 91.7 m correct to 3 significant figures.
3 Calculate the time he would take to swim the distance $AP_1$ and then run $P_1C$.

The position $P_0$ represents the optimum distance from C; the position where the total time taken to travel from A to C is a minimum. $P_0$ is a distance $y$ m from A and a distance $x$ m from C as shown.

4 Show that $x$ can be modelled by the formula $x = 20\sqrt{21} - \sqrt{y^2 - 1600}$.
5 Give the range of possible values of y for which the above model is valid.
6 Calculate, by trial and improvement or by using a spreadsheet, the value of $x$ m. Give your answer correct to 1 d.p.
7 Calculate the shortest time possible for the athlete to travel from A to C. Give your answer correct to 1 d.p.

# Student assessments 8

## Student assessment 1

1  Using the diagram, describe the following translations with column vectors.

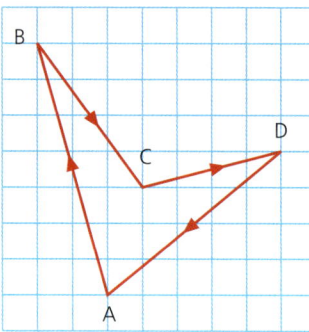

  a  $\overrightarrow{AB}$   b  $\overrightarrow{DA}$   c  $\overrightarrow{CA}$

2  Describe each of the translations below using column vectors.

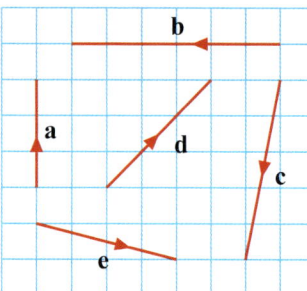

3  Using the vectors drawn in Q.2, draw diagrams to represent:
  a  $a + e$   b  $c - d$   c  $-c - e$   d  $-b + 2a$

4  $a = \begin{pmatrix} 3 \\ -5 \end{pmatrix}$  $b = \begin{pmatrix} 0 \\ 4 \end{pmatrix}$  $c = \begin{pmatrix} -4 \\ 6 \end{pmatrix}$

  Calculate:
  a  $a - c$   b  $b - a$   c  $2a + b$   d  $3c - 2a$

# Student assessment 2

**1**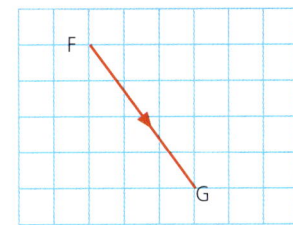

a Calculate the magnitude of the vector $\overrightarrow{FG}$ in the diagram.
b Calculate the magnitude of each of the following vectors:

$\mathbf{a} = \begin{pmatrix} 1 \\ 6 \end{pmatrix}$  $\mathbf{b} = \begin{pmatrix} 12 \\ -3 \end{pmatrix}$  $\mathbf{c} = \begin{pmatrix} -10 \\ 10 \end{pmatrix}$

**2** $\mathbf{p} = \begin{pmatrix} -3 \\ 5 \end{pmatrix}$  $\mathbf{q} = \begin{pmatrix} -4 \\ -4 \end{pmatrix}$  $\mathbf{r} = \begin{pmatrix} 8 \\ -2 \end{pmatrix}$

Calculate the magnitude of:

a $4\mathbf{p} - \mathbf{r}$  b $\frac{3}{2}\mathbf{q} - \mathbf{p}$

Give your answer to 1 d.p.

**3** If $\overrightarrow{SW} = \mathbf{a}$ and $\overrightarrow{SV} = \mathbf{b}$ in the diagram, write each of the following in terms of **a** and **b**:

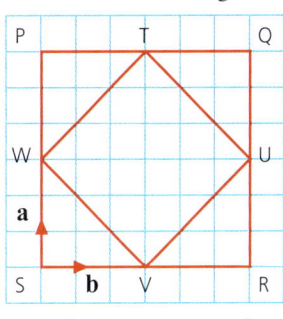

a $\overrightarrow{SP}$  b $\overrightarrow{QT}$  c $\overrightarrow{TU}$

# Student assessment 3

**1** In the triangle ABC below, $\overrightarrow{AB} = \mathbf{a}$ and $\overrightarrow{AD} = \mathbf{b}$. D divides the side AC in the ratio 1:4 and E is the midpoint of BC.

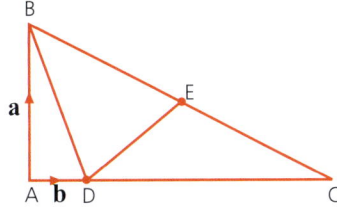

Write the following in terms of **a** and **b**:
a $\overrightarrow{AC}$  b $\overrightarrow{BC}$  c $\overrightarrow{DE}$

**2** In the square PQRS, T is the midpoint of the side PQ and U is the midpoint of the side SR. $\overrightarrow{PQ} = \mathbf{a}$ and $\overrightarrow{PS} = \mathbf{b}$.

a Write the following in terms of **a** and **b**:
 i $\overrightarrow{PT}$  ii $\overrightarrow{QS}$
b Calculate the ratio $\overrightarrow{PV} : \overrightarrow{PU}$.

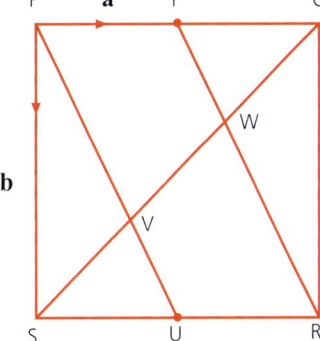

**3** ABCDEFGH is a regular octagon. $AB = \mathbf{a}$ and $AH = \mathbf{b}$. Write the following vectors in terms of **a** and **b**:

a $\overrightarrow{FE}$  b $\overrightarrow{ED}$  c $\overrightarrow{BG}$

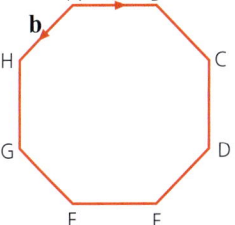

# STUDENT ASSESSMENTS 8

##  Student assessment 4

1 Reflect the object below in the mirror line shown.

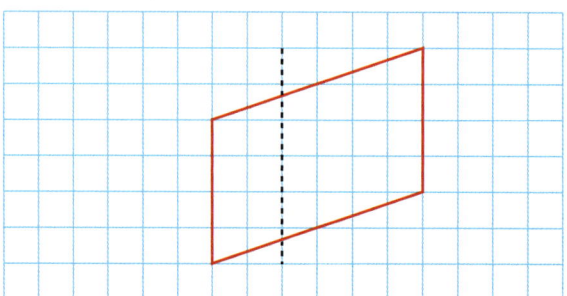

2 Rotate the object below 180° about the centre of rotation O.

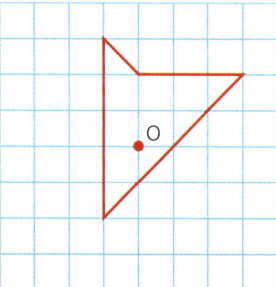

3 Write down the column vector of the translation which maps:
   a   triangle A to triangle B    b   triangle B to triangle C.

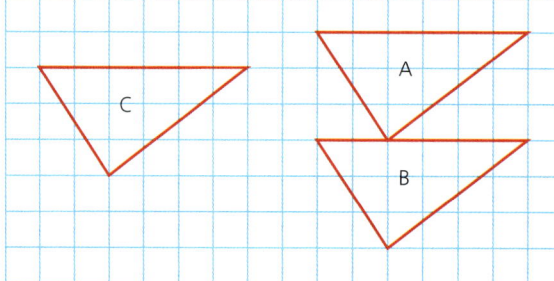

## Student assessments 8

**4** Enlarge the rectangle below by a scale factor 1.5 and from the centre of enlargement O.

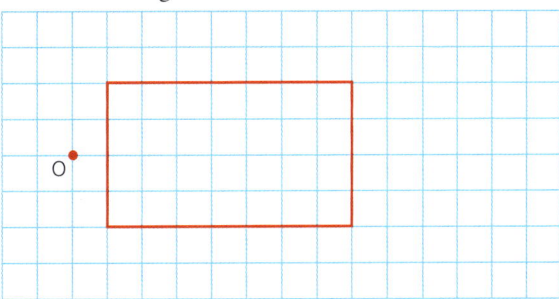

**5** An object WXYZ and its image W'X'Y'Z' are shown below.

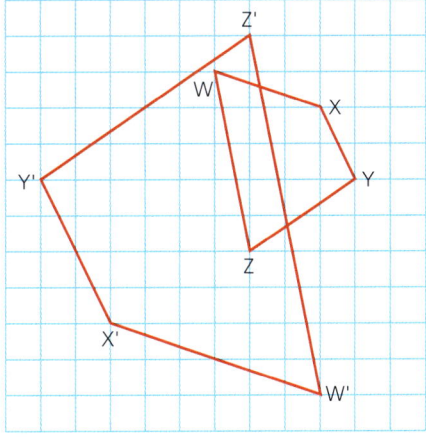

**a** Find the position of the centre of enlargement.
**b** Calculate the scale factor of enlargement.
**c** Determine the inverse transformation that maps the image back on to the original object.

**6** Square ABCD is mapped onto square A'B'C'D'. Square A'B'C'D' is then mapped onto square A"B"C"D".

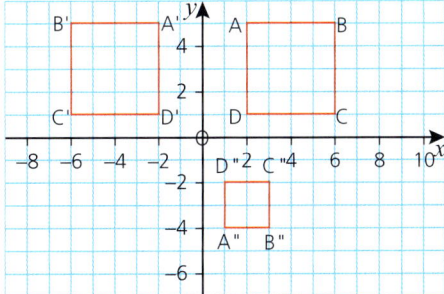

**a** Describe fully the transformation which maps ABCD onto A'B'C'D'.
**b** Describe fully the transformation which maps A'B'C'D' onto A"B"C"D".

521

# TOPIC 9

## Probability

### Contents

Chapter 56  Theoretical probability (C9.1, C9.3, E9.1, E9.3)
Chapter 57  Tree diagrams (C9.3, E9.3)
Chapter 58  Use of Venn diagrams in probability (C9.3, E9.3)
Chapter 59  Laws of probability (E9.3)
Chapter 60  Relative frequency (C9.2, E9.2)

### Learning objectives

**C9.1      E9.1**
1. Understand and use the probability scale from 0 to 1
2. **Understand and use probability notation**
3. Calculate the probability of a single event
4. Understand that the probability of an event not occurring = 1 – the probability of the event occurring

**C9.2      E9.2**
1. Understand relative frequency as an estimate of probability
2. Calculate expected frequencies

**C9.3      E9.3**
Calculate the probability of combined events using, where appropriate:
- sample space diagrams
- tree diagrams
- Venn diagrams

Elements in purple refer to the Extended curriculum only.

## Order and chaos

Blaise Pascal and Pierre de Fermat (known for his last theorem) corresponded about problems connected to games of chance.

Although Newton and Galileo had had some thoughts on the subject, this is accepted as the beginning of the study of what is now called probability. Later, in 1657, Christiaan Huygens wrote the first book on the subject, entitled *The Value of all Chances in Games of Fortune*.

In 1821, Carl Friedrich Gauss (1777–1855) worked on the normal distribution.

At the start of the nineteenth century, the French mathematician Pierre-Simon Laplace was convinced of the existence of a Newtonian universe. In other words, if you knew the position and velocities of all the particles in the universe, you would be able to predict the future because their movement would be predetermined by scientific laws. However, quantum mechanics has since shown that this is not true. Chaos theory is at the centre of understanding these limits.

Blaise Pascal (1623-1662)

# 56 Theoretical probability

## Probability of an event

Probability is the study of chance, or the likelihood of an event happening.

In this chapter you will be looking at **theoretical probability**. But, because probability is based on chance, what theory predicts does not necessarily happen in practice.

A **favourable outcome** refers to the event in question actually happening. The total number of **possible outcomes** refers to all the different types of outcome one can get in a particular situation. In general:

$$\text{Probability of an event} = \frac{\text{number of favourable outcomes}}{\text{total number of equally likely possible outcomes}}$$

Therefore:

if the probability = 0, it implies the event is impossible,

if the probability = 1, it implies the event is certain to happen.

### ➡ Worked examples

**a** An ordinary, fair dice is rolled.

  **i** Calculate the probability of getting a 6.
  
  Number of favourable outcomes = 1 (i.e. getting a 6)
  
  Total number of possible outcomes = 6 (i.e. getting a 1, 2, 3, 4, 5 or 6)
  
  Probability of getting a 6, $P(6) = \frac{1}{6}$

  **ii** Calculate the probability of not getting a 6.
  
  Number of favourable outcomes = 5 (i.e. getting a 1, 2, 3, 4, 5)
  
  Total number of possible outcomes = 6 (i.e. getting a 1, 2, 3, 4, 5 or 6)
  
  Note: the notation $P(6')$ means the probability of not getting a 6
  
  Probability of not getting a six, $P(6') = \frac{5}{6}$

From this it can be seen that the probability of not getting a 6 is equal to 1 minus the probability of getting a 6.

i.e. $P(6) = 1 - P(6')$

These are known as **complementary events**.

In general, for an event A, $P(A) = 1 - P(A')$.

**b** An ordinary fair dice is rolled. Calculate the probability of getting an even number.

Number of favourable outcomes = 3 (i.e. getting a 2, 4 or 6)

Total number of possible outcomes = 6 (i.e. getting a 1, 2, 3, 4, 5 or 6)

Probability of getting an even number is $\frac{3}{6} = \frac{1}{2}$

> **Note**
> 
> The notation for the probability of not $A$, $P(A')$, is relevant to the Extended syllabus only. Students following the Core syllabus are expected to know that the probability of an event not occurring = 1 − the probability of the event occurring.

c Thirty students are asked to choose their favourite subject out of maths, English and art. The results are shown in the table.

|  | Maths | English | Art |
|---|---|---|---|
| Girls | 7 | 4 | 5 |
| Boys | 5 | 3 | 6 |

A student is chosen at random.

i What is the probability that this student is a girl?

Total number of girls = 16

Probability of choosing a girl is $\frac{16}{30} = \frac{8}{15}$

ii What is the probability that this student is a boy whose favourite subject is art?

Number of boys whose favourite subject is art is 6

Probability is therefore $\frac{6}{30} = \frac{1}{5}$

iii What is the probability of **not** choosing a girl whose favourite subject is English?

There are two ways of approaching this:

**Method 1:**

Total number of students who are not girls whose favourite subject is English is $7 + 5 + 5 + 3 + 6 = 26$

Therefore the probability is $\frac{26}{30} = \frac{13}{15}$

**Method 2:**

Total number of girls whose favourite subject is English is 4

Probability of choosing a girl whose favourite subject is English is $\frac{4}{30}$

Therefore the probability of **not** choosing a girl whose favourite subject is English is

$$1 - \frac{4}{30} = \frac{26}{30} = \frac{13}{15}$$

---

The likelihood of an event such as 'you will play sport tomorrow' will vary from person to person. Therefore the probability of an event is unknown and can be estimated using relative frequency. However, in contrast, the probability of some events, such as the result of rolling a dice can be calculated by considering the number of possible outcomes.

A probability scale goes from 0 to 1:

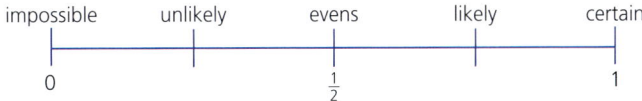

# 56 THEORETICAL PROBABILITY

**Exercise 9.1**

*You can express probabilities as fractions, decimals or percentages.*

1  Copy the probability scale above.
   Mark on the probability scale the probability that:
   a  a day chosen at random is a Saturday
   b  a coin will show tails when spun
   c  the Sun will rise tomorrow
   d  a woman will run a marathon in two hours
   e  the next car you see will be silver.
2  Express your answers to Q.1 as fractions, decimals and percentages.

 **Exercise 9.2**

1  Calculate the probability, when rolling an ordinary fair dice, of getting each of the following:
   a  a score of 1              b  a score of 5
   c  an odd number             d  a score less than 6
   e  a score of 7              f  a score less than 7.
2  a  Calculate the probability of:
      i  being born on a Wednesday
      ii not being born on a Wednesday.
   b  Explain the result of adding the answers to Q.2ai and ii together.
3  250 tickets are sold for a raffle. What is the probability of winning if you buy:
      a  1 ticket     b  5 tickets     c  250 tickets     d  0 tickets?
4  In a class there are 25 girls and 15 boys. The teacher collects all of their workbooks in a random order. Calculate the probability that the teacher will:
   a  mark a book belonging to a girl first
   b  mark a book belonging to a boy first.
5  Twenty-six tiles, each lettered with one different letter of the alphabet, are put into a bag. If one tile is drawn out at random, calculate the probability that it is:
   a  an A or P                 b  a vowel
   c  a consonant               d  an X, Y or Z
   e  a letter in your first name.
6  a  3 red, 10 white, 5 blue and 2 green counters are put into a bag. If one is picked at random, calculate the probability that it is:
      i  a green counter        ii  a blue counter.
   b  If the first counter taken out is green and it is not put back into the bag, calculate the probability that the second counter picked is:
      i  a green counter        ii  a red counter.
7  A spinner has the numbers 0 to 36 equally spaced around its edge. Assuming that it is unbiased, calculate the probability of it stopping on:
   a  the number 5              b  an even number
   c  an odd number             d  zero
   e  a number greater than 15  f  a multiple of 3
   g  a multiple of 3 or 5      h  a prime number.

*Combined events*

8  The letters R, C and A can be combined in several different ways.
   a  Write the letters in as many different combinations as possible.
   b  If a computer writes these three letters at random, calculate the probability that:
      i    the letters will be written in alphabetical order
      ii   the letter R is written before both the letters A and C
      iii  the letter C is written after the letter A
      iv   the computer will spell the word CART if the letter T is added.

9  A normal pack of playing cards contains 52 cards. These are made up of four suits (hearts, diamonds, clubs and spades). Each suit consists of 13 cards. These are labelled ace, 2, 3, 4, 5, 6, 7, 8, 9, 10, Jack, Queen and King. The hearts and diamonds are red; the clubs and spades are black. If a card is picked at random from a normal pack of cards, calculate the probability of picking:
   a  a heart              b  a black card
   c  a four               d  a red King
   e  a Jack, Queen or King   f  the ace of spades
   g  an even numbered card   h  a seven or a club.

# Combined events

You will now look at the probability of two or more events happening: combined events. If only two events are involved, then sample space diagrams can be used to show the outcomes.

## Worked examples

i   Two coins are tossed. Show all the possible outcomes in a sample space diagram.

ii  Calculate the probability of getting two heads, P(HH).

    All four outcomes are equally likely, therefore P(HH) = $\frac{1}{4}$.

iii Calculate the probability of getting a head and a tail in any order.

    The probability of getting a head and a tail in any order, i.e. HT or TH, is $\frac{2}{4} = \frac{1}{2}$.

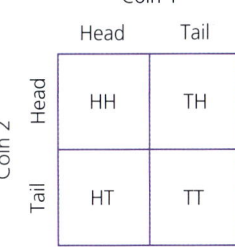

 **Exercise 9.3**

1  a  Two fair tetrahedral dice are rolled. If each is numbered 1–4, draw a sample space diagram to show all the possible outcomes.
   b  What is the probability that both dice show the same number?
   c  What is the probability that the number on one dice is double the number on the other?
   d  What is the probability that the sum of the two numbers is prime?

# 56 THEORETICAL PROBABILITY

**Exercise 9.3 (cont)**

2  Two fair dice are rolled. Copy and complete the sample space diagram to show all the possible combinations.

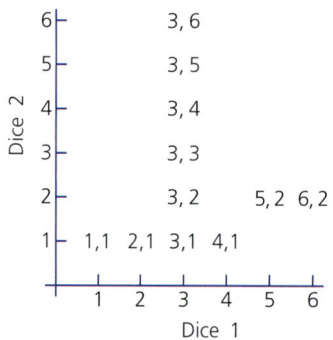

What is the probability of getting:
a  a double 3
b  any double
c  a total score of 11
d  a total score of 7
e  an even number on both dice
f  an even number on at least one dice
g  a 6 or a double
h  scores which differ by 3
i  a total which is a multiple of either 2 or 5?

# 57 Tree diagrams

When more than two combined events are being considered, sample space diagrams cannot be used and therefore another method of representing information diagrammatically is needed. Tree diagrams are a good way of doing this.

## ➜ Worked examples

i  If a coin is spun three times, show all the possible outcomes on a tree diagram, writing each of the probabilities at the side of the branches.

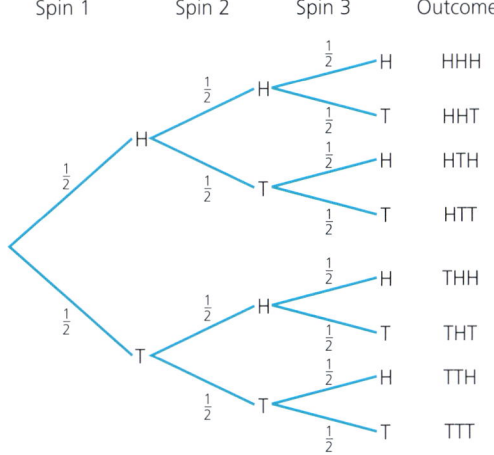

*Note how outcomes are written at the end of branches and probabilities by the side of the branches.*

ii  What is the probability of getting three heads?

To calculate the probability of getting three heads, multiply along the branches:

$P(HHH) = \frac{1}{2} \times \frac{1}{2} \times \frac{1}{2} = \frac{1}{8}$.

iii  What is the probability of getting two heads and one tail in any order?

The successful outcomes are HHT, HTH, THH.

$P(HHT) + P(HTH) + P(THH) = \left(\frac{1}{2} \times \frac{1}{2} \times \frac{1}{2}\right) + \left(\frac{1}{2} \times \frac{1}{2} \times \frac{1}{2}\right) + \left(\frac{1}{2} \times \frac{1}{2} \times \frac{1}{2}\right) = \frac{3}{8}$

Therefore the probability is $\frac{3}{8}$.

iv  What is the probability of getting at least one head?

This refers to any outcome with any of one, two or three heads, i.e. all outcomes except TTT.

P (at least one head) $= 1 - P(TTT) = 1 - \frac{1}{8} = \frac{7}{8}$

Therefore the probability is $\frac{7}{8}$.

v  What is the probability of getting no heads?

The only successful outcome for this event is TTT.

Therefore the probability is $\frac{1}{8}$.

# 57 TREE DIAGRAMS

 **Exercise 9.4**

1.  **a** A computer uses the numbers 1, 2 or 3 at random to make three-digit numbers. Assuming that a number can be repeated, show on a tree diagram all the possible combinations that the computer can print.
    **b** Calculate the probability of getting:
    **i** the number 131
    **ii** an even number
    **iii** a multiple of 11
    **iv** a multiple of 3
    **v** a multiple of 2 or 3
    **vi** a palindromic number (the same forwards as backwards e.g. 323).

2.  **a** A family has four children. Draw a tree diagram to show all the possible combinations of boys and girls. Assume P (girl) = P (boy).
    **b** Assuming the probability of having a boy or girl is equal, calculate the probability of getting:
    **i** all girls
    **ii** two girls and two boys
    **iii** at least one girl
    **iv** more girls than boys.

3.  **a** A netball team plays three matches. In each match the team is equally likely to win, lose or draw. Draw a tree diagram to show all the possible outcomes over the three matches.
    **b** Calculate the probability that the team:
    **i** wins all three matches
    **ii** wins more times than it loses
    **iii** loses at least one match
    **iv** either draws or loses all the three matches.
    **c** Explain why it is not very realistic to assume that the outcomes are equally likely in this case.

4.  A spinner is split into quarters.
    **a** If it is spun twice, draw a probability tree showing all the possible outcomes.
    **b** Calculate the probability of getting:
    **i** two greens
    **ii** a green and a blue in any order
    **iii** no whites.

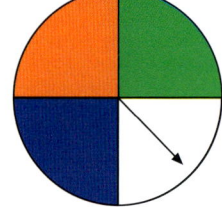

## Tree diagrams for unequal probabilities

In each of the cases considered so far, all of the outcomes have been assumed to be equally likely. However, this need not be the case.

## Tree diagrams for unequal probabilities

### → Worked examples

In winter, the probability that it rains on any one day is $\frac{5}{7}$.

i Using a tree diagram, show all the possible combinations for two consecutive days. Write each of the probabilities by the sides of the branches.

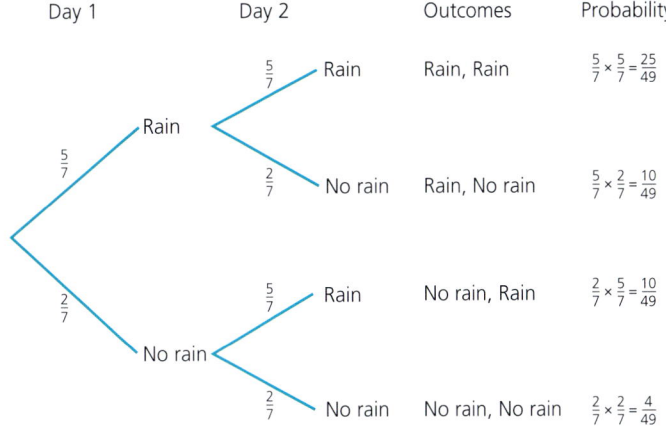

Note how the probability of each outcome is found by multiplying the probabilities for each of the branches. This is because each outcome is the result of calculating the fraction of a fraction.

ii Calculate the probability that it will rain on both days.

This is an outcome that is $\frac{5}{7}$ of $\frac{5}{7}$.

$P(R, R) = \frac{5}{7} \times \frac{5}{7} = \frac{25}{49}$

iii Calculate the probability that it will rain on the first day but not the second day.

$P(R, R') = \frac{5}{7} \times \frac{2}{7} = \frac{10}{49}$

iv Calculate the probability that it will rain on at least one day.

The outcomes which satisfy this event are (R, R) (R, R') and (R', R).

Therefore the probability $\frac{25}{49} + \frac{10}{49} + \frac{10}{49} = \frac{45}{49}$.

 **Exercise 9.5**

1 A particular board game involves players rolling a dice. However, before a player can start, they need to roll a 6.

  a Copy and complete the tree diagram below showing all the possible combinations for the first two rolls of the dice.

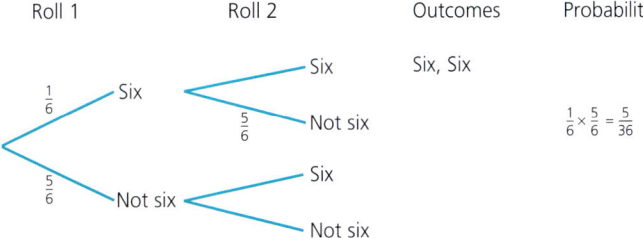

531

## 57 TREE DIAGRAMS

**Exercise 9.5 (cont)**

    **b** Calculate the probability of the following:
       **i** getting a six on the first throw
       **ii** starting within the first two throws
       **iii** starting on the second throw
       **iv** not starting within the first three throws
       **v** starting within the first three throws.
    **c** If you add the answers to Q.1biv and v, what do you notice? Explain your answer.

**2** In Italy, $\frac{3}{5}$ of the cars are made abroad. By drawing a tree diagram and writing the probabilities next to each of the branches, calculate the following probabilities:
    **a** the next two cars to pass a particular spot are both Italian
    **b** two of the next three cars are foreign
    **c** at least one of the next three cars is Italian.

**3** The probability that a morning bus arrives on time is 65%.
    **a** Draw a tree diagram showing all the possible outcomes for three consecutive mornings.
    **b** Label your tree diagram and use it to calculate the probability that:
       **i** the bus is on time on all three mornings
       **ii** the bus is late the first two mornings
       **iii** the bus is on time two out of the three mornings
       **iv** the bus is on time at least twice.

**4** Light bulbs are packaged in cartons of three. 10% of the bulbs are found to be faulty. Calculate the probability of finding two faulty bulbs in a single carton.

**5** A cricket team has a 0.25 chance of losing a game. Calculate the probability of the team achieving:
    **a** two consecutive wins
    **b** three consecutive wins
    **c** ten consecutive wins.

# Tree diagrams for probability problems without replacement

In the examples considered so far, the probability for each outcome remained the same throughout the problem. However, this need not always be the case.

### Worked examples

    **a** A bag contains three red balls and seven black balls. If the balls are put back after being picked, what is the probability of picking:
       **i** two red balls
       **ii** a red ball and a black ball in any order.

*Tree diagrams for probability problems without replacement*

This is selection with replacement. Draw a tree diagram to help visualise the problem:

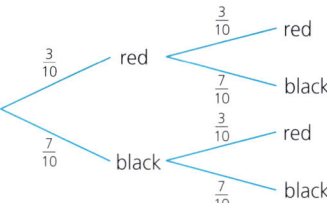

i   The probability of a red followed by a red, $P(RR) = \frac{3}{10} \times \frac{3}{10} = \frac{9}{100}$.

ii  The probability of a red followed by a black or a black followed by a red is:

$$P(RB) + P(BR) = \left(\frac{3}{10} \times \frac{7}{10}\right) + \left(\frac{7}{10} \times \frac{3}{10}\right) = \frac{21}{100} + \frac{21}{100} = \frac{42}{100}.$$

**b** Repeat the previous question, but this time each ball that is picked is not put back in the bag.

This is selection without replacement. The tree diagram is now as shown:

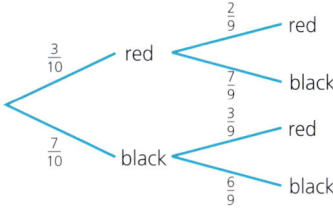

i   $P(RR) = \frac{3}{10} \times \frac{2}{9} = \frac{6}{90}$

ii  $P(RB) + P(BR) = \left(\frac{3}{10} \times \frac{7}{9}\right) + \left(\frac{7}{10} \times \frac{3}{9}\right) = \frac{21}{90} + \frac{21}{90} = \frac{42}{90}.$

## Exercise 9.6

1  A bag contains five red balls and four black balls. If a ball is picked out at random, its colour recorded and it is then put back in the bag, what is the probability of choosing:
   **a**  two red balls
   **b**  two black balls
   **c**  a red ball and a black ball in this order
   **d**  a red ball and a black ball in any order?

2  Repeat Q.1 but, in this case, after a ball is picked at random, it is not put back in the bag.

3  A bag contains two black, three white and five red balls. If a ball is picked, its colour recorded and it is then put back in the bag, what is the probability of picking:
   **a**  two black balls
   **b**  a red and a white ball in any order?

4  Repeat Q.3 but, in this case, after a ball is picked at random, it is not put back in the bag.

# 57 TREE DIAGRAMS

**Exercise 9.6 (cont)**

5  You buy five tickets for a raffle. 100 tickets are sold altogether. Tickets are picked at random. You have not won a prize after the first three tickets have been drawn.
   a  What is the probability that you win a prize with either of the next two draws?
   b  What is the probability that you do not win a prize with either of the next two draws?

6  A normal pack of 52 cards is shuffled and three cards are picked at random. Draw a tree diagram to help calculate the probability of picking:
   a  two clubs first
   b  three clubs
   c  no clubs
   d  at least one club.

7  A bowl of fruit contains one mango, one banana, two oranges and two papayas. Two pieces of fruit are chosen at random and eaten.
   a  Draw a tree diagram showing all the possible combinations of the two pieces of fruit.
   b  Use your tree diagram to calculate the probability that:
      i    both the pieces of fruit eaten are oranges
      ii   a mango and a banana are eaten
      iii  at least one papaya is eaten.

8  A class has $n$ number of girls and $n$ number of boys.
   Two students are chosen at random.
   a  Draw a tree diagram to show all the possible outcomes, labelling the probability of each branch in terms of $n$, where appropriate.
   b  Show that the probability of two girls being chosen is $\dfrac{n-1}{2(2n-1)}$

9  A bag of sweets contains $n$ red sweets and $n+3$ yellow sweets.
   A child takes two sweets from the bag at random.
   a  Draw a tree diagram to show all the possible outcomes and label the probability of each branch in terms of $n$.
   b  Calculate the probability that the child picks two yellow sweets.

# 58 Use of Venn diagrams in probability

You saw in Topic 1 how Venn diagrams can be used to represent sets. They can also be used to solve problems involving probability.

## ➡ Worked examples

a  In a survey carried out in a college, students were asked which was their favourite subject.

15 chose English

8 chose Science

12 chose Mathematics

5 chose Art

What is the probability that a student chosen at random will like Science the best?

This can be represented on a Venn diagram as:

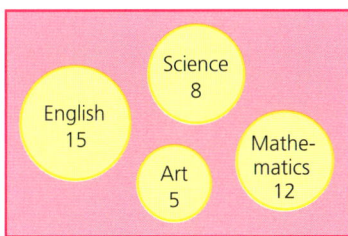

There are 40 students, so the probability is $\frac{8}{40} = \frac{1}{5}$.

b  A group of 21 friends decide to go out for the day to the local town. 9 of them decide to see a film at the cinema and 15 of them go for lunch.

  i  Draw a Venn diagram to show this information if set $A$ represents those who see a film and set $B$ represents those who have lunch.

     $9 + 15 = 24$; as there are only 21 people, this implies that 3 people see the film and have lunch. This means that $9 - 3 = 6$ only went to see a film and $15 - 3 = 12$ only had lunch.

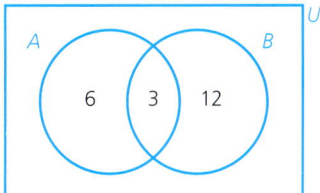

  ii  Determine the probability that a person picked at random only went to the cinema.

      The number who only went to the cinema is 6, therefore the probability is $\frac{6}{21} = \frac{2}{7}$.

# 58 USE OF VENN DIAGRAMS IN PROBABILITY

 **Exercise 9.7**

1. In a class of 30 students, 20 study French, 18 study Spanish and 5 study neither.
   a Draw a Venn diagram to show this information.
   b What is the probability that a student chosen at random studies both French and Spanish?

2. In a group of 35 students, 19 take Physics, 18 take Chemistry and 3 take neither. What is the probability that a student chosen at random takes:
   a both Physics and Chemistry
   b Physics only
   c Chemistry only?

3. 108 people visited an art gallery. 60 liked the pictures, 53 liked the sculpture, 10 liked neither.
   What is the probability that a person chosen at random liked the pictures but not the sculpture?

4. In a series of examinations in a school:
   37 students took English
   48 students took French
   45 students took Spanish
   15 students took English and French
   13 students took French and Spanish
   7 students took English and Spanish
   5 students took all three.
   a Draw a Venn diagram to represent this information.
   b What is the probability that a student picked at random took:
      i all three languages
      ii English only
      iii French only?

# 59 Laws of probability

## Mutually exclusive events

Events that cannot happen at the same time are known as mutually exclusive events. For example, a sweet bag contains 12 red sweets and 8 yellow sweets. Let picking a red sweet be event $A$ and picking a yellow sweet be event $B$. If only one sweet is picked, it is not possible to pick a sweet which is both red and yellow. Therefore these events are **mutually exclusive**.

This can be shown in a Venn diagram:

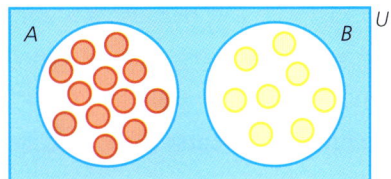

$P(A) = \frac{12}{20}$ while $P(B) = \frac{8}{20}$.

As there is no overlap, $P(A \cup B) = P(A) + P(B) = \frac{12}{20} + \frac{8}{20} = \frac{20}{20} = 1$,

i.e. the probability of mutually exclusive events $A$ or $B$ happening is equal to the sum of the probabilities of event $A$ and event $B$; the sum of the probabilities of all possible mutually exclusive events is 1.

### → Worked example

In a 50 m swim, the world record holder has a probability of 0.72 of winning. The probability of her finishing second is 0.25.

What is the probability that she either wins or comes second?

Since she cannot finish both 1st and 2nd, the events are mutually exclusive.
Therefore $P(1st \cup 2nd) = 0.72 + 0.25 = 0.97$.

## Combined events

If events are not mutually exclusive then they may occur at the same time.

These are known as **combined events**.

For example, a pack of 52 cards contains four suits: clubs (♣), spades (♠), hearts (♥) and diamonds (♦). Clubs and spades are black; hearts and diamonds are red. Each suit contains 13 cards. These are ace, 2, 3, 4, 5, 6, 7, 8, 9, 10, Jack, Queen and King.

A card is picked at random. Event $A$ represents picking a black card; event $B$ represents picking a King.

537

## 59 LAWS OF PROBABILITY

In a Venn diagram this can be shown as:

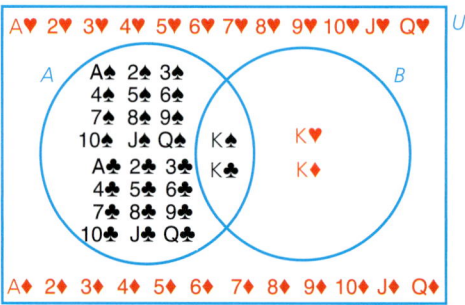

$P(A) = \frac{26}{52} = \frac{1}{2}$ and $P(B) = \frac{4}{52} = \frac{1}{13}$

However, $P(A \cup B) \neq \frac{26}{52} + \frac{4}{52}$ because K♠ and K♣ belong to both events $A$ and $B$ and have therefore been counted twice. This is shown in the overlap of the Venn diagram and needs to be taken into account.

Therefore, for combined events

*This formula is not in the syllabus.* → $P(A \cup B) = P(A) + P(B) - P(A \cap B)$

i.e. the probability of event $A$ or event $B$ is equal to the sum of the probabilities of $A$ and $B$ minus the probability of ($A$ and $B$).

$P(A \cup B) = \frac{26}{52} + \frac{4}{52} - \frac{2}{52} = \frac{28}{52} = \frac{7}{13}$

However the answer can also be obtained by looking directly at the Venn diagram where

$P(A \cup B) = \frac{24 + 2 + 2}{52} = \frac{28}{52} = \frac{7}{13}$

### → Worked examples

In a holiday survey of 100 people:
72 people have had a beach holiday
16 have had a sightseeing holiday
12 have had both.
What is the probability that one person chosen at random from the survey has had either a beach holiday ($B$) or a sightseeing holiday ($S$)?

$P(B) = \frac{72}{100}$    $P(S) = \frac{16}{100}$    $P(B \cap S) = \frac{12}{100}$

Therefore $P(B \cup S) = \frac{72}{100} + \frac{16}{100} - \frac{12}{100} = \frac{76}{100}$

The use of a Venn diagram also helps work out the solution as shown below:

Therefore $P(B \cup S) = \frac{60 + 12 + 4}{100} = \frac{76}{100}$

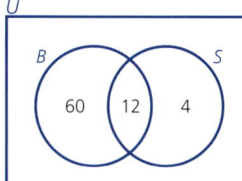

## Independent events

A student may be born on 1 June, another student in his class may also be born on 1 June. These events are independent of each other (assuming they are not twins).

If a dice is thrown and a coin spun, the outcomes of each are also independent, i.e. the outcome of one does not affect the outcome of another.

For independent events, the probability of both occurring is the product of each occurring separately, i.e. $P(A \cap B) = P(A) \times P(B)$.

### ➡ Worked examples

**a** You spin a coin and roll a dice.

  **i** What is the probability of getting a head on the coin and a five on the dice?
  $P(H) = \frac{1}{2}$  $P(5) = \frac{1}{6}$
  The events are independent and therefore
  $P(H \cap 5) = P(H) \times P(5) = \frac{1}{2} \times \frac{1}{6} = \frac{1}{12}$

  **ii** What is the probability of getting either a head on the coin or a five on the dice, but not both?

  $P(H \cup 5)$ is the probability of getting a head, a five or both.

  Therefore $P(H \cup 5) - P(H \cap 5)$ removes the probability of both events occurring. The solution is

  $P(H \cup 5) - P(H \cap 5) = P(H) + P(5) - P(H \cap 5)$
  $= \frac{1}{2} + \frac{1}{6} - \frac{1}{12} = \frac{7}{12}$

  An alternative would be to consider the information presented in a Venn diagram with the probability of getting a head (H) and the probability of getting a 5 (5) shown as follows:

  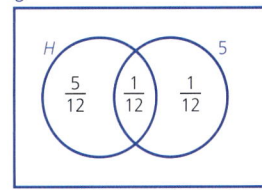

  Therefore $P(H \cup 5) = \frac{5+1+1}{12} = \frac{7}{12}$

**b** The probabilities of two events $X$ and $Y$ are given by:
$P(X) = 0.5$, $P(Y) = 0.4$, and $P(X \cap Y) = 0.2$.

  **i** Are events $X$ and $Y$ mutually exclusive?

  No: if the events were mutually exclusive, then $P(X \cap Y)$ would be 0 as the events could not occur at the same time.

  **ii** Calculate $P(X \cup Y)$.

  $P(X \cup Y) = P(X) + P(Y) - P(X \cap Y)$
  $= 0.5 + 0.4 - 0.2 = 0.7$

  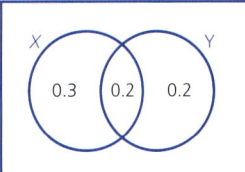

  Once again, using a Venn diagram produces the following approach:
  Therefore $P(X \cup Y) = 0.3 + 0.2 + 0.2 = 0.7$

  **iii** What kind of events are $X$ and $Y$?

  Since $P(X \cap Y) = P(X) \times P(Y)$, i.e. $0.2 = 0.5 \times 0.4$, events $X$ and $Y$ must be independent.

# 59 LAWS OF PROBABILITY

## Probability from contingency tables

A contingency table is a two-way table containing frequency data. The data allows the probabilities of events to be calculated.

> ### Worked example

An ice cream seller keeps a record of the number of different types of ice creams he sells to adults and children. The results are displayed in the contingency table below:

|  | Adults | Children |
|---|---|---|
| Vanilla | 24 | 8 |
| Strawberry | 15 | 14 |
| Chocolate | 6 | 28 |

A person was seen buying one of the ice creams. Calculate the probability of each of the following:

i  It was a child.

Total number of ice creams sold: $24 + 15 + 6 + 8 + 14 + 28 = 95$

Total number of children served: $8 + 14 + 28 = 50$

Therefore $P(\text{Child}) = \frac{50}{95} = \frac{10}{19}$

ii  The ice cream being bought was chocolate.

Total number of chocolate ice creams bought: $28 + 6 = 34$

Therefore $P(\text{Chocolate}) = \frac{34}{95}$

iii  It was an adult buying a vanilla ice cream.

Number of adults buying vanilla ice creams: 24

Therefore $P(\text{Adult buying vanilla}) = \frac{24}{95}$

### Exercise 9.8

1  In a 50 m swim, the record holder has a probability of 0.68 of winning and a 0.25 probability of finishing second. What is her probability of finishing in the first two?

2  The Jamaican 100 m women's relay team has a 0.5 chance of coming first in the final, 0.25 chance of coming second and 0.05 chance of coming third. What is the team's chance of a medal?

3  You spin a coin and throw a dice. What is the probability of getting:
  a  a head and a factor of 3
  b  a head or a factor of 3
  c  a head or a factor of 3, but not both?

4  What is the probability that two people, picked at random, both have a birthday in June?

5  Amelia takes two buses to work. On a particular day, the probability of her catching the first bus is 0.7 and the probability of catching the second bus is 0.5.
  The probability of her catching neither is 0.1. If $A$ represents catching the first bus and $B$ the second:
  a  State $P(A \cup B)'$.
  b  Find $P(A \cup B)$.
  c  Given that $P(A \cup B) = P(A) + P(B) - P(A \cap B)$, calculate $P(A \cap B)$.

540

## Probability from contingency tables

**6** The probability of Marco having breakfast is 0.75. The probability that he gets a lift to work is 0.9 if he has had breakfast and 0.8 if he has not.
  **a** What is the probability of Marco having breakfast then getting a lift?
  **b** What is the probability of Marco not having breakfast then getting a lift?
  **c** What is the probability that Marco gets a lift?

**7** The numbers and types of books on a student's bookshelf are given in the contingency table below:

|  | Hardback | Paperback |
|---|---|---|
| **Fiction** | 8 | 46 |
| **Non-fiction** | 22 | 14 |

Calculate the probability that a book picked from the shelf at random is:
  **a** a paperback book
  **b** a fiction book
  **c** a non-fiction hardback.

**8** One morning, there are three exams, Maths, French and Philosophy, occurring at the same time. The numbers of boys and girls sitting each exam are given in the contingency table below:

|  | Maths | French | Philosophy |
|---|---|---|---|
| **Boys** | 48 | 6 | 12 |
| **Girls** | 52 | 16 | 8 |

Calculate the probability that a student picked at random is:
  **a** a girl
  **b** sitting the French exam
  **c** a boy sitting the Philosophy exam.

**9** A shirt manufacturer produces a style of shirt in three different colours (white, blue and grey) and in four different sizes (XL, L, M and S).
A shop receives a large box with a delivery of these shirts. The contents of the box are summarised in the contingency table.

|  | White | Blue | Grey |
|---|---|---|---|
| **XL** | 6 | 4 | 8 |
| **L** | 10 | 20 | 6 |
| **M** | 12 | 13 | 8 |
| **S** | 4 | 7 | 2 |

  **a** If the shirts are unpacked in a random order, calculate the probability that the first shirt unpacked is:
   **i** white
   **ii** large
   **iii** large and white.
  **b** Calculate the probability that the first two shirts unpacked (without replacement) are:
   **i** blue
   **ii** the same colour
   **iii** not the same colour.

# 59 LAWS OF PROBABILITY

**Exercise 9.8 (cont)**

10 a How many students are in your class?
   b How likely do you think it is that two people in your class will share the same birthday? Very likely? Likely? About 50–50? Unlikely? Very unlikely?
   c Write down everybody's birthday. Did two people have the same birthday?
   
   Below is a way of calculating the probability that two people have the same birthday depending how many people there are. To study this, it is easiest to look at the probability of birthdays being *different*. When this probability is less than 50%, then the probability that two people will have the same birthday is greater than 50%.
   
   When the first person asks the second person, the probability of them *not* having the same birthday is $\frac{364}{365}$
   
   (i.e. it is $\frac{1}{365}$ that they have the same birthday).
   
   When the next person is asked, as the events are independent, the probability of all three having different birthdays is:
   
   $$\left(\frac{364}{365}\right) \times \left(\frac{363}{365}\right) = 99.2\%$$
   
   When the next person is asked, the probability of all four having different birthdays is:
   
   $$\left(\frac{364}{365}\right) \times \left(\frac{363}{365}\right) \times \left(\frac{362}{365}\right) = 98.4\%$$
   
   and so on …
   
   d Copy and complete the table below until the probability is 50%.

*The probabilities in the second column can be efficiently calculated using a spreadsheet to help.*

| Number of people | Probability of them *not* having the same birthday |
|---|---|
| 2 | $\left(\frac{364}{365}\right) = 99.7\%$ |
| 3 | $\left(\frac{364}{365}\right) \times \left(\frac{363}{365}\right) = 99.2\%$ |
| 4 | $\left(\frac{364}{365}\right) \times \left(\frac{363}{365}\right) \times \left(\frac{362}{365}\right) = 98.4\%$ |
| 5 | |
| 10 | |
| 15 | |
| 20 | |
| etc. | |

   e Explain in words what your solution to Q.10d means.

# 60 Relative frequency

So far the work covered has dealt with theoretical probability. However, there are many occasions when the probability of an outcome is not initially known, so experiments are carried out in order to make predictions. This is known as **experimental probability**.

For example, a six-sided dice is known to be biased, i.e. not all numbers are equally likely.

The dice is rolled 60 times and the results recorded in the table below:

| Number | 1 | 2 | 3 | 4 | 5 | 6 |
|---|---|---|---|---|---|---|
| Frequency | 3 | 7 | 12 | 8 | 20 | 10 |

In order to calculate the experimental probability of each number, its **relative frequency** is calculated. The relative frequency is the fraction representing the number of times an event occurs as a proportion of the total.

Therefore the relative frequency of each number is:

| Number | 1 | 2 | 3 | 4 | 5 | 6 |
|---|---|---|---|---|---|---|
| Frequency | 3 | 7 | 12 | 8 | 20 | 10 |
| Relative frequency | $\frac{3}{60}$ | $\frac{7}{60}$ | $\frac{12}{60}$ | $\frac{8}{60}$ | $\frac{20}{60}$ | $\frac{10}{60}$ |

Predictions can be made about future outcomes based on these results.

## ➔ Worked examples

Using the results above, calculate the following:

**a** If the dice is rolled 100 times, approximately how many times would you expect to get a three?

$P(3) = \frac{12}{60} = \frac{1}{5}$

Therefore the number of threes would be approximately $\frac{1}{5} \times 100 = 20$.

**b** The dice was rolled many times. The number two was rolled 38 times. Work out how many times the dice was rolled.

$P(2) = \frac{7}{60}$

Let $x$ be the number of times the dice was rolled:

$x \times \frac{7}{60} = 38$

$\Rightarrow \quad x = \frac{38 \times 60}{7}$

$\Rightarrow \quad x = 325.71$ (2 d.p.)

Therefore the dice was rolled approximately 326 times.

# 60 RELATIVE FREQUENCY

Accuracy is improved the more times the experiment is carried out, as any rogue results have a relatively smaller effect. So to improve the accuracy of the results, simply increase the number of trials.

### Exercise 9.9

1  A dice is rolled 100 times and the results recorded in the table below:

| Number | 1 | 2 | 3 | 4 | 5 | 6 |
|---|---|---|---|---|---|---|
| Frequency | 18 | 15 | 16 | 17 | 16 | 18 |

   a  Explain, giving reasons, whether you think the dice is fair or biased.
   Another dice is rolled 100 times and the results recorded below:

| Number | 1 | 2 | 3 | 4 | 5 | 6 |
|---|---|---|---|---|---|---|
| Frequency | 10 | 25 | 32 | 5 | 20 | 8 |

   b  Explain, giving reasons, whether you think this dice is fair or biased.
   c  Calculate the relative frequency of each number on the second dice.
   d  If the second dice was rolled 450 times, how many times would you expect to get a six?
   e  The second dice was rolled $x$ times. The number 4 was obtained 23 times. Estimate the value of $x$.
   f  Both dice were rolled 350 times. How many **more** sixes would you expect to get with the first dice compared with the second?

2  A bird spotter wants to find out the probability of different types of birds landing in his garden so that he can put out appropriate feed. He conducts a survey over a period of five hours. The results are shown below:

| Type of bird | Sparrow | Starling | Crow | Wren | Other |
|---|---|---|---|---|---|
| Frequency | 46 | 32 | 9 | 16 | 27 |

   a  Assuming conditions are similar the following day, estimate the number of crows he is likely to spot in a three-hour period.
   b  The day after, he conducts a similar survey and counts 50 starlings. Approximately how long was he recording results on this occasion?
   c  Six months later, he decides to estimate the number of sparrows visiting his garden in a two hour period. Explain, giving reasons, whether he should use the original data for his estimate.

3  I go to work by bus each day. If the bus is on time, I get to work on time. Over a 20-day period, I record whether I arrive at work on time or late. If I arrive late, I also record how late I am. The results are shown below:

| Arrival time | On time | 5 min late | 10 min late | 15 min late | 20 min late |
|---|---|---|---|---|---|
| Frequency | 12 | 2 | 1 | 4 | 1 |

   a  I work 230 days in a working year.
      i   Estimate how many times I would arrive on time.
      ii  What assumptions have you made in estimating the answer to **i**?
   b  Estimate the number of times I would arrive 20 minutes late in a working year.
   c  Estimate the total amount of time in hours that I arrive late in a working year.

**d** In the same city, there are 250 000 people who use the buses to get to work each day.
  **i** Estimate the total amount of time lost in the city per day because of late buses.
  **ii** What assumptions have you made in estimating the above answer?
**4** Check the bias of a dice in your classroom by conducting an experiment.
Explain your methods and display your results clearly. Refer to your results when deciding whether or not the dice is biased.
**5** Drawing pins, when dropped, can either land point up or point down as shown below:

  **a** Carry out an experiment to determine the probability of a drawing pin landing point up.
  **b** What factors are likely to influence whether a drawing pin lands point up or point down?
  **c** By considering one of the factors you stated in Q.5b, carry out a further experiment to determine whether it does influence how a drawing pin lands.

# Investigations, modelling and ICT 9

## Investigation: Probability drop

A game involves dropping a red marble down a chute. On hitting a triangle divider, the marble can either bounce left or right. On completing the drop, the marble lands in one of the trays along the bottom. The trays are numbered from left to right. Different sizes of game exist; the four smallest versions are shown below:

Game 1

Game 2

Game 3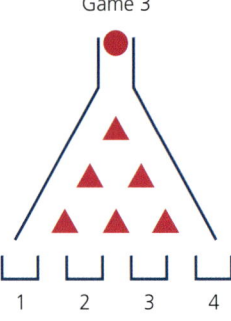

Game 4

To land in tray 2 in the second game above, the ball can travel in one of two ways. These are: Left – Right or Right – Left.

This can be abbreviated to LR or RL.

1. State the different routes the marble can take to land in each of the trays in the third game.
2. State the different routes the marble can take to land in each of the trays in the fourth game.
3. State, giving reasons, the probability of a marble landing in tray 1 of the fourth game.
4. State, giving reasons, the probability of a marble landing in each of the other trays in the fourth game.
5. Investigate the probability of the marble landing in each of the different trays of larger games.
6. Using your findings from your investigation, predict the probability of a marble landing in tray 7 of the tenth game (11 trays at the bottom).
7. Investigate the links between this game and the sequence of numbers generated in Pascal's triangle.

## Investigation: Dice sum

Two ordinary dice are rolled and their scores added together.

Below is an incomplete table showing the possible outcomes:

|  |  | Dice 1 | | | | | |
|---|---|---|---|---|---|---|---|
|  |  | 1 | 2 | 3 | 4 | 5 | 6 |
| Dice 2 | 1 | 2 |   |   | 5 |   |   |
|  | 2 |   |   |   |   |   |   |
|  | 3 |   |   |   | 7 |   |   |
|  | 4 |   |   |   | 8 |   |   |
|  | 5 |   |   |   | 9 | 10 | 11 |
|  | 6 |   |   |   |   |   | 12 |

1  Copy and complete the table to show all possible outcomes.
2  How many possible outcomes are there?
3  What is the most likely total when two dice are rolled?
4  What is the probability of getting a total score of 4?
5  What is the probability of getting the most likely total?
6  How many times more likely is a total score of 5 compared with a total score of 2?

Now consider rolling two four-sided dice each numbered 1–4. Their scores are also added together.

7  Draw a table to show all the possible outcomes when the two four-sided dice are rolled and their scores added together.
8  How many possible outcomes are there?
9  What is the most likely total?
10 What is the probability of getting the most likely total?
11 Investigate the number of possible outcomes, the most likely total and its probability when two identical dice are rolled together and their scores are added, i.e. consider 8-sided dice, 10-sided dice, etc.
12 Consider two $m$-sided dice rolled together and their scores added.
   a  What is the total number of outcomes in terms of $m$?
   b  What is the most likely total, in terms of $m$?
   c  What, in terms of $m$, is the probability of the most likely total?
13 Consider an $m$-sided and an $n$-sided dice rolled together, where $m > n$.
   a  In terms of $m$ and $n$, find the total number of outcomes.
   b  In terms of $m$ and/or $n$, find the most likely total(s).
   c  In terms of $m$ and/or $n$, find the probability of getting the most likely total.

# INVESTIGATIONS, MODELLING AND ICT 9

## Modelling: Infection screening

Medical organisations often run large scale medical screening on populations for certain types of diseases. As a result of this screening people will receive either a positive test result (implying that they are infected) or a negative test result (implying that they are not infected). However, no test is 100% accurate. Occasionally people will receive a **false positive** result. This means getting a positive result, when in fact they **are not** infected. Similarly, some people receive a **false negative** result. This means getting a negative result, when in fact they **are** infected.

### Scenario 1

Infection rates of a certain disease are known to be running at 30% of a population. 10 000 people are tested for the disease using a screening method that is 95% accurate.

**1 a** Copy and complete the diagram below:

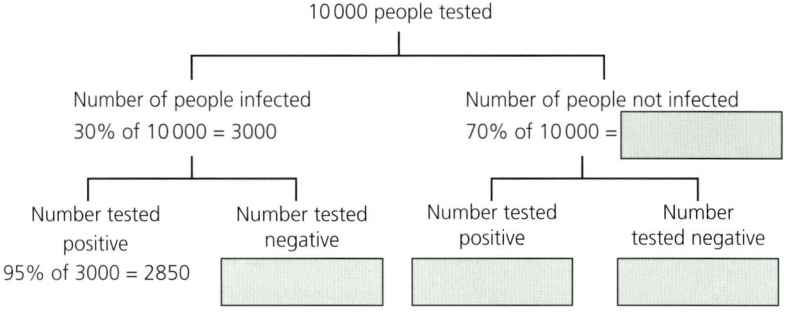

  **b** How many people received a false positive?
  **c** What percentage of positive results were false positives?

### Scenario 2

Infection rates for the same disease are known to be running at 2% in another population. 10 000 people are tested for the disease also using the screening method that is 95% accurate.

2  a  Copy and complete the diagram below:

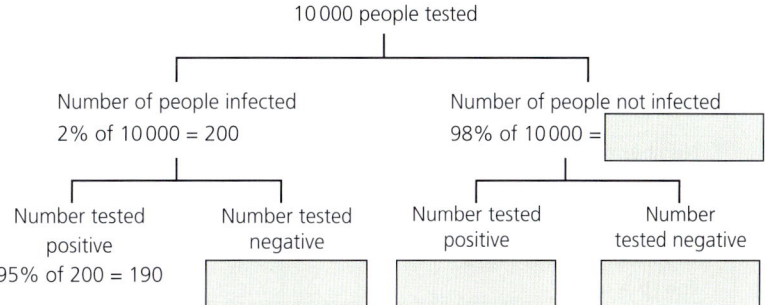

   b  How many people received a false positive?
      What percentage of positive results were false positives?
3  What can you conclude from your answers to Q.1 and 2 above?
4  A government wants to mass screen 1 000 000 people for a disease that has an infection rate of 4% of the population. It wants the percentage of false positives out of the total number of positive results to be less than 5%. Calculate the minimum percentage accuracy that the test must have, for the screening to be conducted. Give your answer correct to 1 d.p.

# ICT Activity: Buffon's needle experiment

You will need to use a spreadsheet for this activity.

The French count Le Comte de Buffon devised the following probability experiment.
1  Measure the length of a match (with the head cut off) as accurately as possible.
2  On a sheet of paper, draw a series of straight lines parallel to each other. The distance between each line should be the same as the length of the match.
3  Take ten identical matches and drop them randomly on the paper. Count the number of matches that cross or touch any of the lines. For example, in the diagram below, the number of matches crossing or touching lines is six.

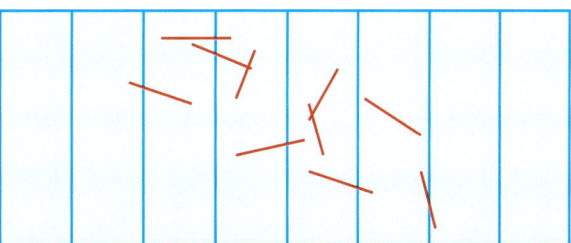

## INVESTIGATIONS, MODELLING AND ICT 9

4 Repeat the experiment a further nine times, making a note of your results, so that altogether you have dropped 100 matches.
5 Set up a spreadsheet similar to the one shown below and enter your results in cell B2.

|   | A | B | C | D | E | F | G | H | I | J | K |
|---|---|---|---|---|---|---|---|---|---|---|---|
| 1 | Number of drops (N) | 100 | 200 | 300 | 400 | 500 | 600 | 700 | 800 | 900 | 1000 |
| 2 | Number of matches crossing/touching lines (n) | | | | | | | | | | |
| 3 | Probability of crossing a line ($p = n/N$) | | | | | | | | | | |
| 4 | $2/p$ | | | | | | | | | | |

6 Repeat 100 match drops again, making a total of 200 drops, and enter cumulative results in cell C2.
7 By collating the results of your fellow students, enter the cumulative results of dropping a match 300–1000 times in cells D2–K2 respectively.
8 Using an appropriate formula, get the spreadsheet to complete the calculations in Rows 3 and 4.
9 Use the spreadsheet to plot a line graph of N against $\frac{2}{p}$.
10 What value does $\frac{2}{p}$ appear to get closer to?

# Student assessments 9

## Student assessment 1

1. Calculate the probability of:
   a being born on a Saturday
   b being born on the 5th of a month in a non-leap year
   c being born on 20 June in a non-leap year
   d being born on 29 February.

2. When rolling an ordinary fair dice, calculate the probability of getting:
   a a 2
   b an even number
   c a 3 or more
   d less than 1.

3. A bag contains 12 white counters, 7 black counters and 1 red counter.
   a If, when a counter is taken out, it is not replaced, calculate the probability that:
      i the first counter is white
      ii the second counter removed is red, given that the first was black.
   b If, when a counter is picked, it is then put back in the bag, how many attempts will be needed before it is mathematically certain that a red counter will have been picked out?

4. A coin is tossed and an ordinary fair dice is rolled.
   a Draw a sample space diagram showing all the possible combinations.
   b Calculate the probability of getting:
      i a head and a six
      ii a tail and an odd number
      iii a head and a prime number.

5. Two spinners A and B are split into quarters and coloured as shown. Both spinners are spun.

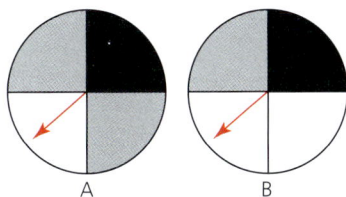

   a Draw a fully labelled tree diagram showing all the possible combinations on the two spinners. Write beside each branch the probability of each outcome.
   b Use your tree diagram to calculate the probability of getting:
      i two blacks
      ii two greys
      iii a grey on spinner A and a white on spinner B.

**STUDENT ASSESSMENTS 9**

6  A coin is tossed three times.
   a  Draw a tree diagram to show all the possible outcomes.
   b  Use your tree diagram to calculate the probability of getting:
      i   three tails
      ii  two heads
      iii no tails
      iv  at least one tail.
7  A goalkeeper expects to save one penalty out of every three. Calculate the probability that he:
   a  saves one penalty out of the next three
   b  fails to save any of the next three penalties
   c  saves two out of the next three penalties.
8  A board game uses a fair dice in the shape of a tetrahedron. The sides of the dice are numbered 1, 2, 3 and 4. Calculate the probability of:
   a  not throwing a 4 in two throws
   b  throwing two consecutive 1s
   c  throwing a total of 5 in two throws.
9  A normal pack of 52 cards is shuffled and three cards are picked at random. Calculate the probability that all three cards are picture cards.

## Student assessment 2

1  A card is drawn from a standard pack of cards.
   a  Draw a Venn diagram to show the following:
      $A$ is the set of aces
      $B$ is the set of picture cards
      $C$ is the set of clubs
   b  From your Venn diagram, find the following probabilities:
      i    P(ace or picture card)
      ii   P(not an ace or picture card)
      iii  P(club or ace)
      iv   P(club and ace)
      v    P(ace and picture card)
2  Students in a school can choose to study one or more science subjects from Physics, Chemistry and Biology.
   In a year group of 120 students, 60 took Physics, 60 took Biology and 72 took Chemistry; 34 took Physics and Chemistry, 32 took Chemistry and Biology and 24 took Physics and Biology; 18 took all three.
   a  Draw a Venn diagram to represent this information.
   b  If a student is chosen at random, what is the probability that:
      i   the student chose to study only one subject
      ii  the student chose Physics or Chemistry and did not choose Biology.

**3** A class took an English test and a Mathematics test. 40% passed both tests and 75% passed the English test.
What percentage of those who passed the English test also passed the Mathematics test?

**4** A jar contains blue and red counters.
Two counters are chosen without replacement. The probability of choosing a blue then a red counter is 0.44. The probability of choosing a blue counter on the first draw is 0.5.
What is the probability of choosing a red counter on the second draw if the first counter chosen was blue?

**5** In a group of children, the probability that a child has black hair is 0.7. The probability that a child has brown eyes is 0.55. The probability that a child has either black hair or brown eyes is 0.85.
What is the probability that a child chosen at random has both black hair and brown eyes?

**6** It is not known whether a six-sided dice is biased. It is rolled 80 times and the results recorded in the table below:

| Number | 1 | 2 | 3 | 4 | 5 | 6 |
|---|---|---|---|---|---|---|
| Frequency | 11 | 12 | 6 | 18 | 16 | 17 |

**a** Explain, giving reasons, whether you think the dice is fair or biased.
**b** Calculate the relative frequency of each number.
**c** If the dice was rolled 220 times, estimate the number of times a 3 would be rolled.
**d** The dice was rolled $x$ times. The number 6 was rolled 48 times. Estimate the value of $x$.

# TOPIC 10

## Statistics

### Contents

Chapter 61   Basic graphs and charts (C10.1, C10.2, C10.3, C10.7, E10.1, E10.2, E10.3, E10.7)
Chapter 62   Stem-and-leaf diagrams (C10.2, C10.7, E10.2, E10.7)
Chapter 63   Averages and ranges (C10.4, C10.6, E10.4, E10.5, E10.6)
Chapter 64   Cumulative frequency (E10.8)
Chapter 65   Scatter diagrams, correlation and lines of best fit (C10.9, E10.9)

## Learning objectives

**C10.1   E10.1**
Classify and tabulate statistical data

**C10.2   E10.2**
Read, interpret and draw inferences from tables and statistical diagrams

Compare sets of data using tables, graphs and statistical measures

Appreciate restrictions on drawing conclusions from given data

**C10.3   E10.3**
Distinguish between discrete and continuous data

**C10.4   E10.4**
Calculate the mean, median, mode, quartiles, range and interquartile range for discrete data and distinguish between the purposes for which these are used

**E10.5**
Calculate an estimate of the mean for grouped discrete or continuous data

Identify the modal class from a grouped frequency distribution

**C10.6   E10.6**
Use a graphic display calculator to calculate:
1. mean, median and quartiles for discrete data
2. mean for grouped data

**C10.7   E10.7**
Draw and interpret bar charts, pie charts, pictograms, stem-and-leaf diagrams, frequency distributions and frequency diagrams

**E10.8**
Draw and interpret cumulative frequency tables and diagrams

Estimate and interpret the median, percentiles, quartiles and interquartile range from cumulative frequency diagrams

**C10.9   E10.9**
Draw and interpret scatter diagrams

Understand what is meant by and describe positive, negative and zero correlation

Draw by eye, interpret and use a straight line of best fit

Use a graphic display calculator to find and use the equation of linear regression

Elements in purple refer to the Extended curriculum only.

## Statistics in history

The earliest writing on statistics was found in a ninth-century book entitled *Manuscript on Deciphering Cryptographic Messages*, written by the Arab philosopher Al-Kindi (801–873), who lived in Baghdad. In his book, he gave a detailed description of how to use statistics to unlock coded messages.

The *Nuova Cronica*, a fourteenth-century history of Florence by the Italian banker Giovanni Villani, includes much statistical information on population, commerce, trade and education.

Florence Nightingale (1820–1910) was a famous British nurse who treated casualties in the Crimean War (1853–1856). By using statistics she realised that most of the deaths that occurred were not as a result of battle injuries but from preventable illnesses afterwards, such as cholera and typhoid. By understanding these statistics, Florence Nightingale was able to improve the sanitary conditions of the injured soldiers and therefore reduce their mortality rates.

Early statistics served the needs of states, state-istics. By the early nineteenth century, statistics included the collection and analysis of data in general. Today, statistics are widely employed in government, business, and natural and social sciences. The use of modern computers has enabled large-scale statistical computation and has also made possible new methods that are impractical to perform manually.

# 61 Basic graphs and charts

## Discrete and continuous data

**Discrete data** can only take specific values; for example, the number of tickets sold for a concert can only be a positive integer value.

**Continuous data**, on the other hand, can take any value within a certain range; for example, the time taken to run 100 m will typically fall in the range 10–20 seconds. Within that range, however, the time stated will depend on the accuracy required. So a time stated as 13.8 s could have been 13.76 s, 13.764 s or 13.7644 s, etc.

**Exercise 10.1**

State whether the data below is discrete or continuous.
1. Your shoe size
2. Your height
3. Your house number
4. Your weight
5. The total score when two dice are thrown
6. A Mathematics exam mark
7. The distance from the Earth to the Moon
8. The number of students in your school
9. The speed of a train
10. The density of lead

## Displaying simple discrete data

Data can be displayed in many different ways. It is therefore important to choose the method that displays the data most clearly and effectively.

### Tally tables and bar charts

The figures in the list below are the numbers of chocolate buttons in each of twenty packets of buttons.

35  36  38  37  35  36  38  36  37  35

36  36  38  36  35  38  37  38  36  38

*Displaying simple discrete data*

The figures can be shown on a tally table:

| Number | Tally | Frequency |
|---|---|---|
| 35 | IIII | 4 |
| 36 | ЖII | 7 |
| 37 | III | 3 |
| 38 | ЖI | 6 |

When the tallies are added up to get the frequency, the table is usually called a **frequency table**. The information can then be displayed in a variety of ways.

## Pictograms

● = 4 packets, ◕ = 3 packets, ◑ = 2 packets, ◔ = 1 packet

| Buttons per packet | |
|---|---|
| 35 | ● |
| 36 | ● ◕ |
| 37 | ◕ |
| 38 | ● ◑ |

## Bar charts

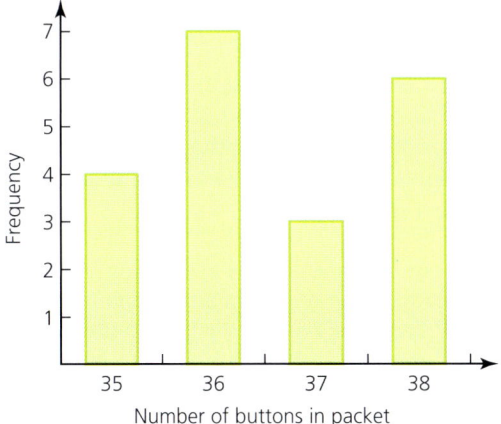

Number of buttons in packet

A bar chart has certain properties. These are:
- The height of the bar represents the frequency.
- The bars do not touch.

# 61 BASIC GRAPHS AND CHARTS

## Frequency histogram

The frequency table shows the British shoe sizes of 20 students in a class.

| Shoe size | 6 | $6\frac{1}{2}$ | 7 | $7\frac{1}{2}$ | 8 | $8\frac{1}{2}$ | 9 |
|---|---|---|---|---|---|---|---|
| Frequency | 2 | 3 | 3 | 6 | 4 | 1 | 1 |

This can be displayed as a **frequency histogram**.

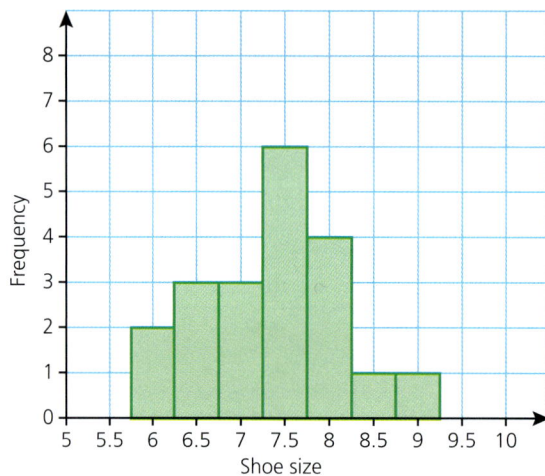

Shoe sizes are an example of discrete data as the data can only take certain values. As a result, the frequency histogram has certain properties.
- Each bar is of equal width and its height represents the frequency.
- The bars touch (this is not the case with a bar chart).
- The value is written at the mid-width of each bar. This is because students with a foot size in the range 6.75–7.25, for example, would have a shoe size of 7.

**Exercise 10.2**

1 Record the shoe sizes of everybody in your class.
 a Present the results in a tally and frequency table.
 b Present the results as a pictogram.
 c Present the data as a frequency histogram.
 d What conclusions can you draw from your results?

## Grouped discrete or continuous data

If there is a big range in the data, it is sometimes easier and more useful to group the data in a grouped frequency table.

The discrete data below shows the scores for the first round of a golf competition.

```
 71  75  82  96  83  75  76  82 103  85  79  77  83  85  88
104  76  77  79  83  84  86  88 102  95  96  99 102  75  72
```

One possible way of grouping this data in a grouped frequency table is shown.

Note: The groups are arranged so that no score can appear in two groups.

| Score | Frequency |
|---|---|
| 71–75 | 5 |
| 76–80 | 6 |
| 81–85 | 8 |
| 86–90 | 3 |
| 91–95 | 1 |
| 96–100 | 3 |
| 101–105 | 4 |

With grouped continuous data, the groups are presented in a different way.

### Exercise 10.3

1. The following data gives the percentage scores obtained by students from two classes, 11X and 11Y, in a Mathematics exam.
   11X
   42 73 93 85 68 58 33 70 71 85 90 99 41 70 65
   80 73 89 88 93 49 50 57 64 78 79 94 80 50 76 99
   11Y
   70 65 50 89 96 45 32 64 55 39 45 58 50 82 84
   91 92 88 71 52 33 44 45 53 74 91 46 48 59 57 95
   a Using group sizes of 10, draw a grouped tally and frequency table for each of the classes.
   b Comment on any similarities or differences between the results.

2. The number of apples collected from 50 trees is recorded below:
   35   78   15   65   69   32   12    9    89  110  112  148   98
   67   45   25   18   23   56   71   62   46  128    7  133   96
   24   38   73   82  142   15   98    6  123   49   85   63   19
   111   52   84   63   78   12   55  138  102   53   80
   Choose suitable groups for this data and represent it in a grouped frequency table.

## 61 BASIC GRAPHS AND CHARTS

The results below are the times given (in h:min:s) for the first 50 people completing a marathon.

```
2:07:11  2:08:15  2:09:36  2:09:45  2:10:45
2:10:46  2:11:42  2:11:57  2:12:02  2:12:11
2:13:12  2:13:26  2:14:26  2:15:34  2:15:43
2:16:25  2:16:27  2:17:09  2:18:29  2:19:26
2:19:27  2:19:31  2:20:00  2:20:23  2:20:29
2:21:47  2:21:52  2:22:32  2:22:48  2:23:08
2:23:17  2:23:28  2:23:46  2:23:48  2:23:57
2:24:04  2:24:12  2:24:15  2:24:24  2:24:29
2:24:45  2:25:18  2:25:34  2:25:56  2:26:10
2:26:22  2:26:51  2:27:14  2:27:23  2:27:37
```

The data can be arranged into a grouped frequency table as follows:

| Group | Frequency |
|---|---|
| $2:05:00 < t \leq 2:10:00$ | 4 |
| $2:10:00 < t \leq 2:15:00$ | 9 |
| $2:15:00 < t \leq 2:20:00$ | 10 |
| $2:20:00 < t \leq 2:25:00$ | 18 |
| $2:25:00 < t \leq 2:30:00$ | 9 |

Note that, as with discrete data, the groups do not overlap. However, as the data is continuous, the groups are written using inequalities. The first group includes all times from, but not including, 2 h 5 min, up to and including 2 h 10 min.

With continuous data, the upper and lower bound of each group are the numbers written as the limits of the group. In the example above, for the group $2:05:00 < t \leq 2:10:00$, the upper bound is 2:10:00 whilst the lower bound is 2:05:00 despite it not being included in the inequality.

## Pie charts

Pie charts are a popular way of displaying data clearly. A pie chart consists of a circle divided into sectors where the angle of each sector is proportional to the relative size of the quantity it represents.

# Composite bar chart

The table below shows the number of goals scored by a football team (A) over the 40 games it played during one season.

| Number of goals | 0 | 1 | 2 | 3 | 4 |
|---|---|---|---|---|---|
| Frequency | 22 | 8 | 6 | 3 | 1 |

The table shows that Team A scored no goals in 22 of their matches, 1 goal in 8 of their matches, and so on.

As the angle at the centre of a circle is 360°, each frequency must be converted to a fraction out of 360 as shown:

| Number of goals | Frequency | Relative frequency | Angle (degrees) |
|---|---|---|---|
| 0 | 22 | $\frac{22}{40}$ | $\frac{22}{40} \times 360 = 198$ |
| 1 | 8 | $\frac{8}{40}$ | $\frac{8}{40} \times 360 = 72$ |
| 2 | 6 | $\frac{6}{40}$ | $\frac{6}{40} \times 360 = 54$ |
| 3 | 3 | $\frac{3}{40}$ | $\frac{3}{40} \times 360 = 27$ |
| 4 | 1 | $\frac{1}{40}$ | $\frac{1}{40} \times 360 = 9$ |

The pie chart can now be drawn using a protractor.

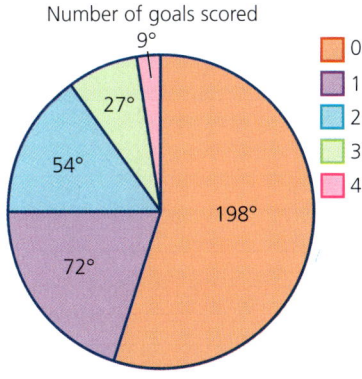

## Composite bar chart

Composite bar charts are useful for comparing different sets of data. Each bar includes more than one set of data stacked on top of each other.

*Another name for a **composite** bar chart is a **stacked** bar chart.*

# 61 BASIC GRAPHS AND CHARTS

*Notice how there is only one bar and the frequencies are stacked on top of each other.*

Consider the data for the number of goals scored by Team A in the example above. The data can be graphed as a composite bar chart:

If, however, two sets of data with different totals need to be compared, then the above composite bar chart is of limited use. A percentage composite bar chart is more useful.

Using the previous example of Team A's goals, the relative frequency table can be adapted to percentages as shown:

| Number of goals | Frequency | Relative frequency | Percentage |
| --- | --- | --- | --- |
| 0 | 22 | $\frac{22}{40}$ | $\frac{22}{40} \times 100 = 55$ |
| 1 | 8 | $\frac{8}{40}$ | $\frac{8}{40} \times 100 = 20$ |
| 2 | 6 | $\frac{6}{40}$ | $\frac{6}{40} \times 100 = 15$ |
| 3 | 3 | $\frac{3}{40}$ | $\frac{3}{40} \times 100 = 7.5$ |
| 4 | 1 | $\frac{1}{40}$ | $\frac{1}{40} \times 100 = 2.5$ |

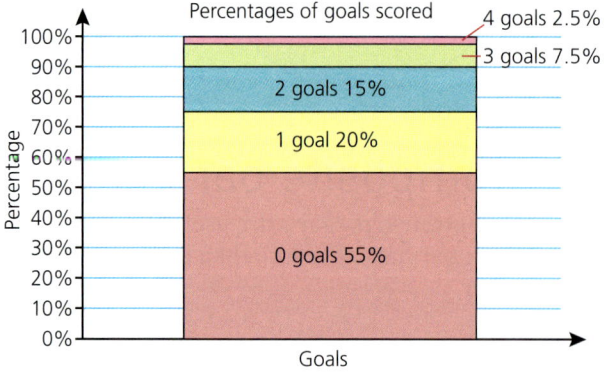

Composite bar charts are particularly useful when comparisons need to be made between different sets of data.

## Composite bar chart

### Worked example

Another football team (B) in the same league as Team A has the following goal-scoring record over the same season.

| Number of goals | 0 | 1 | 2 | 3 | 4 | 5 |
|---|---|---|---|---|---|---|
| Frequency | 8 | 12 | 10 | 5 | 3 | 2 |

**a** Draw a composite bar chart for Team B on the same axes as the one for Team A.

A relative frequency and percentage table can be calculated to give:

| Number of goals | Frequency | Relative frequency | Percentage |
|---|---|---|---|
| 0 | 8 | $\frac{8}{40}$ | $\frac{8}{40} \times 100 = 20$ |
| 1 | 12 | $\frac{12}{40}$ | $\frac{12}{40} \times 100 = 30$ |
| 2 | 10 | $\frac{10}{40}$ | $\frac{10}{40} \times 100 = 25$ |
| 3 | 5 | $\frac{5}{40}$ | $\frac{5}{40} \times 100 = 12.5$ |
| 4 | 3 | $\frac{3}{40}$ | $\frac{3}{40} \times 100 = 7.5$ |
| 5 | 2 | $\frac{2}{40}$ | $\frac{2}{40} \times 100 = 5$ |

Drawing both composite charts on the same axes means that the results of both teams can be compared more easily.

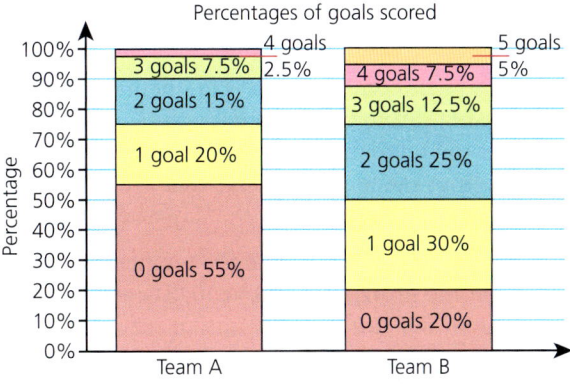

**b** From the charts, decide which of the two teams is likely to be doing better in the league. Explain your answer.

563

# 61 BASIC GRAPHS AND CHARTS

The bar charts suggest that Team B is the more successful team as it scores no goals less often than Team A and also has a higher proportion of goals scored than Team A.

c   Explain why your conclusion to part b may be incorrect.

The chart only displays the number of goals scored and does not show the number of goals conceded. Team B may be better than Team A at scoring goals, but it may also let in more goals.

## Line graphs

A line graph is a good way of analysing data over a period of time. Data is collected and plotted as coordinates on a graph and a line is then drawn passing through each of the points.

The table shows the temperature taken at four-hourly intervals during one day in New York.

| Time | 00:00 | 04:00 | 08:00 | 12:00 | 16:00 | 20:00 | 24:00 |
|---|---|---|---|---|---|---|---|
| Temperature (°C) | 15 | 9 | 16 | 25 | 31 | 20 | 16 |

Plotting a line graph shows the changes in temperature during the day:

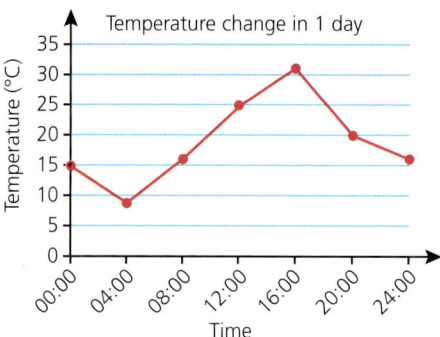

As time is continuous, it is mathematically acceptable to draw a line passing through each of the points. It is a valid assumption to make that the temperature changes at a constant rate between each of the readings taken. This means that predictions can be made about the temperature at times of the day when readings were not actually taken.

For example, to estimate the temperature at 10:00, a line is drawn up from the horizontal axis at 10:00 until it meets the graph. A line is then drawn horizontally until it reaches the vertical axis, where the temperature can be read.

*Line graphs*

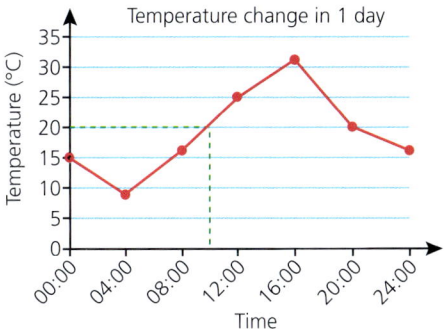

The temperature at 10:00 was approximately 20°C.

**Exercise 10.4**

1  A teenager decides to keep a record of her spending habits over a period of a month. Her results are displayed in the table below:

| Category | Amount spent ($) |
| --- | --- |
| Clothes | 110 |
| Entertainment | 65 |
| Food | 40 |
| Transport | 20 |
| Gifts | 20 |
| Other | 15 |

  **a**  Calculate the angle represented by each category for the purpose of drawing a pie chart.
  **b**  Draw a pie chart representing the teenager's spending habits.

2  The brother of the teenager in Q.1 also keeps a record of his spending over the same period of time. His results are shown below:

| Category | Amount spent ($) |
| --- | --- |
| Clothes | 10 |
| Entertainment | 150 |
| Food | 10 |
| Transport | 20 |
| Gifts | 5 |
| Other | 50 |

  **a**  Find each of these amounts as a percentage of the total.
  **b**  Draw a stacked bar chart of his data.

# 61 BASIC GRAPHS AND CHARTS

**Exercise 10.4 (cont)**

*Another name for a dual bar chart is a side-by-side bar chart.*

3  The heights of boys and girls in one class are plotted on the dual bar chart shown below.

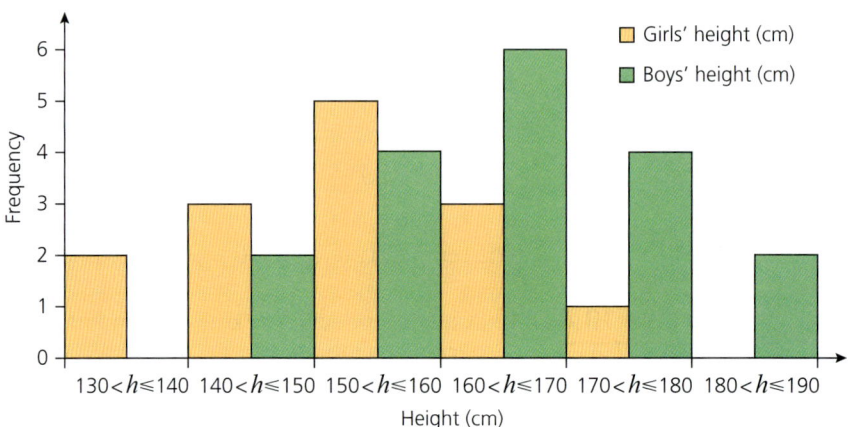

a  How many girls are there in the class?
b  How many more boys than girls are there in the height range $160 < h \leqslant 170$ cm?
c  Describe the differences in heights between boys and girls in the class.
d  Construct a dual bar chart for the heights of boys and girls in your own class.

4  Two fishing boats return to port and the mass of two types of fish caught by each boat is recorded.
The quantities are shown in the composite (stacked) bar chart below.

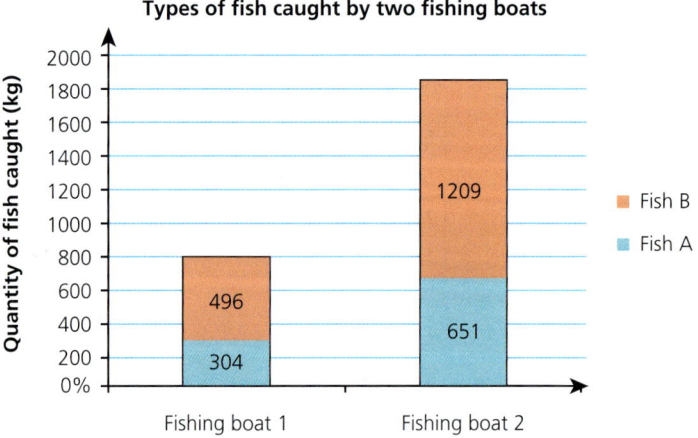

a  Which boat caught the most of fish type A?
b  Assuming only the two types of fish were caught, which boat's catch had a higher percentage of fish type A? Show your working clearly.

The above composite bar chart shows the quantity of fish in kg on the vertical axis.

c  Construct a composite bar chart comparing the catches of both boats, but with percentages on the vertical axis.

# 62 Stem-and-leaf diagrams

**Stem-and-leaf diagrams** (or stem-and-leaf plots) are a special type of bar chart in which the bars are made from the data itself. This has the advantage that the original data can be recovered easily from the diagram. The stem is the first digit of the numbers in a group, so if the numbers are 63, 65, 67, 68, 69, the stem is 6. The leaves are the remaining numbers written in order.

## Worked example

The ages of people on a coach transferring them from an airport to a ski resort are as follows:

| 22 | 24 | 25 | 31 | 33 | 23 | 24 | 26 | 37 | 42 |
| 40 | 36 | 33 | 24 | 25 | 18 | 20 | 27 | 25 | 33 |
| 28 | 33 | 35 | 39 | 40 | 48 | 27 | 25 | 24 | 29 |

Display the data on a stem-and-leaf diagram.

```
1 | 8
2 | 0 2 3 4 4 4 4 5 5 5 5 6 7 7 8 9
3 | 1 3 3 3 3 5 6 7 9
4 | 0 0 2 8
```

Key 2 | 5 means 25

Note that the key states what the stem means. If the data was, for example, 1.8, 2.7, 3.2, the key could state that 2 | 7 means 2.7.

### Exercise 10.5

1  A test in Mathematics is marked out of 40. The scores for the class of 32 students are shown below:

| 24 | 27 | 30 | 33 | 26 | 27 | 28 | 39 |
| 21 | 18 | 16 | 33 | 22 | 38 | 33 | 21 |
| 16 | 11 | 14 | 23 | 37 | 36 | 38 | 22 |
| 28 | 15 | 9  | 17 | 28 | 33 | 36 | 34 |

Display the data on a stem-and-leaf diagram.

2  A basketball team played 24 matches in the 2016 season. Their scores are shown below:

| 62 | 48 | 85 | 74 | 63 | 67 | 71 | 83 |
| 46 | 52 | 63 | 65 | 72 | 76 | 68 | 58 |
| 54 | 46 | 88 | 55 | 46 | 52 | 58 | 54 |

Display the scores on a stem-and-leaf diagram.

## 62 STEM-AND-LEAF DIAGRAMS

**Exercise 10.5 (cont)**

**3** A class of 27 students was asked to draw a line 8 cm long with a straight edge rather than with a ruler. The lines were then measured and their lengths to the nearest millimetre were recorded.

| 8.8 | 6.2 | 8.3 | 7.9 | 8.0 | 5.9 | 6.2 | 10.0 | 9.7 |
|-----|-----|-----|-----|-----|-----|-----|------|-----|
| 7.9 | 5.4 | 6.8 | 7.3 | 7.7 | 8.9 | 10.4 | 5.9 | 8.3 |
| 6.1 | 7.2 | 8.3 | 9.4 | 6.5 | 5.8 | 8.8 | 8.0 | 7.3 |

Present this data using a stem-and-leaf diagram.

## Back-to-back diagrams

Stem-and-leaf diagrams are often used as an easy way to compare two sets of data. The leaves are usually put 'back-to-back' on either side of the stem.

### ➡ Worked example

The stem-and-leaf diagram for the ages of people on a coach to a ski resort (as in the previous worked example) is shown below. The data is easily accessible.

```
1 | 8
2 | 0 2 3 4 4 4 5 5 5 5 6 7 7 8 9
3 | 1 3 3 3 3 5 6 7 9
4 | 0 0 2 8
```
Key 2 | 5 means 25

A second coach from the airport is taking people to a golfing holiday. The ages of the people are shown below:

| 43 | 46 | 52 | 61 | 65 | 38 | 36 | 28 | 37 | 45 |
| 69 | 72 | 63 | 55 | 46 | 34 | 35 | 37 | 43 | 48 |
| 54 | 53 | 47 | 36 | 58 | 63 | 70 | 55 | 63 | 64 |

Display the two sets of data on a back-to-back stem-and-leaf diagram.

```
              Golf          |   |          Skiing
                            | 1 | 8
                          8 | 2 | 0 2 3 4 4 4 5 5 5 5 6 7 7 8 9
            8 7 7 6 6 5 4 | 3 | 1 3 3 3 3 5 6 7 9
            8 7 6 6 5 3 3 | 4 | 0 0 2 8
                8 5 5 4 3 2 | 5 |
              9 5 4 3 3 3 1 | 6 |
                        2 0 | 7 |
```
Key 6 | 4 means 46    Key 3 | 5 means 35

## Exercise 10.6

1. Write three sentences commenting on the back-to-back stem-and-leaf diagram in the worked example above.

2. The basketball team in Q.2 of Exercise 10.5 replaced their team coach at the end of the 2016 season. Their scores for the 24 matches played in the 2017 season are shown below:

| 82 | 32 | 88 | 24 | 105 | 63 | 86 | 42 |
| 35 | 88 | 78 | 106 | 64 | 72 | 88 | 26 |
| 35 | 41 | 100 | 48 | 54 | 36 | 28 | 33 |

Display the scores from both seasons on a back-to-back stem-and-leaf diagram and comment on it.

3. The Mathematics test results shown in Q.1 of Exercise 10.5 were for test B. Test A had already been set and marked and the teacher had gone over some of the questions with the class. The marks out of 40 for test A are shown below:

| 22 | 18 | 9 | 11 | 38 | 33 | 21 | 14 |
| 16 | 8 | 12 | 37 | 39 | 25 | 23 | 18 |
| 34 | 36 | 23 | 16 | 14 | 12 | 22 | 29 |
| 33 | 35 | 12 | 17 | 22 | 28 | 32 | 39 |

Construct a back-to-back stem-and-leaf diagram for scores from both tests and comment on it.

# 63 Averages and ranges

## Averages

'Average' is a word which, in general use, is taken to mean somewhere in the middle. For example, a woman may describe herself as being of average height. A student may think that he or she is of average ability in Science. Mathematics is more precise and uses three main methods to measure average.

1. The **mode** is the value occurring most often.
2. The **median** is the middle value when all the data is arranged in order of size.
3. The **mean** is found by adding together all the values of the data and then dividing the total by the number of data values.

### ➡ Worked example

The numbers below represent the number of goals scored by a hockey team in the first 15 matches of the season. Find the mean, median and mode of the goals.

$$1\ 0\ 2\ 4\ 1\ 2\ 1\ 1\ 2\ 5\ 5\ 0\ 1\ 2\ 3$$

Mean = $\dfrac{1+0+2+4+1+2+1+1+2+5+5+0+1+2+3}{15} = 2$

Arranging all the data in order and then picking out the middle number gives the median:

0 0 1 1 1 1 ②  2 2 2 3 4 5 5

The mode is the number that appears most often. Therefore the mode is 1.

Note: If there is an even number of data values, then there will not be one middle number, but a middle pair. The median is calculated by working out the mean of the middle pair.

## Quartiles and range

Just as the median takes the middle value by splitting the data into two halves, quartiles split the data into quarters.

Taking the example above with the data still arranged in order:

0 0 1 1 1 1 2 2 2 2 3 4 5 5

Splitting the data into quarters produces the following:

0 0 1 | 1 1 1 | 2 2 2 2 | 3 4 5 5

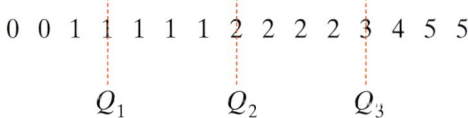
$Q_1 \qquad Q_2 \qquad Q_3$

## Quartiles and range

$Q_1$ is known as the **lower quartile**, $Q_2$, as already described, is the median and $Q_3$ is known as the **upper quartile**.

The position of the quartiles can be calculated using simple formulae. For $n$ data values, $Q_1 = \frac{1}{4}(n+1)$ and $Q_3 = \frac{3}{4}(n+1)$.

As with the median, if the position of a quartile falls midway between two data values, then its value is the mean of the two.

### ➜ Worked example

Calculate the lower and upper quartiles of this set of numbers:

$$7\ \ 7\ \ 8\ \ 12\ \ 12\ \ 12\ \ 15\ \ 16\ \ 21$$

Position of $Q_1 = \frac{n+1}{4} = \frac{10}{4} = 2.5$

Position of $Q_3 = \frac{3(n+1)}{4} = \frac{30}{4} = 7.5$

The data set can therefore be split as shown below:

$$7\ \ 7\ \vdots\ 8\ \ 12\ \ 12\ \ 12\ \ 15\ \vdots\ 16\ \ 21$$
$$\phantom{7\ \ 7\ \ }Q_1 \phantom{\ 8\ \ 12\ \ 12\ \ 12\ \ } Q_3$$

Therefore $Q_1 = 7.5$ and $Q_3 = 15.5$.

The **interquartile range** is the spread of the middle 50% of the data and can be calculated as the difference between the upper and lower quartiles, i.e. interquartile range = upper quartile − lower quartile.

To find the **range** of a data set, simply subtract the smallest data value from the largest data value.

Your graphic display calculator is also capable of calculating the mean, median and quartiles of a set of discrete data. To calculate these for the data set at the start of this chapter, follow the instructions below:

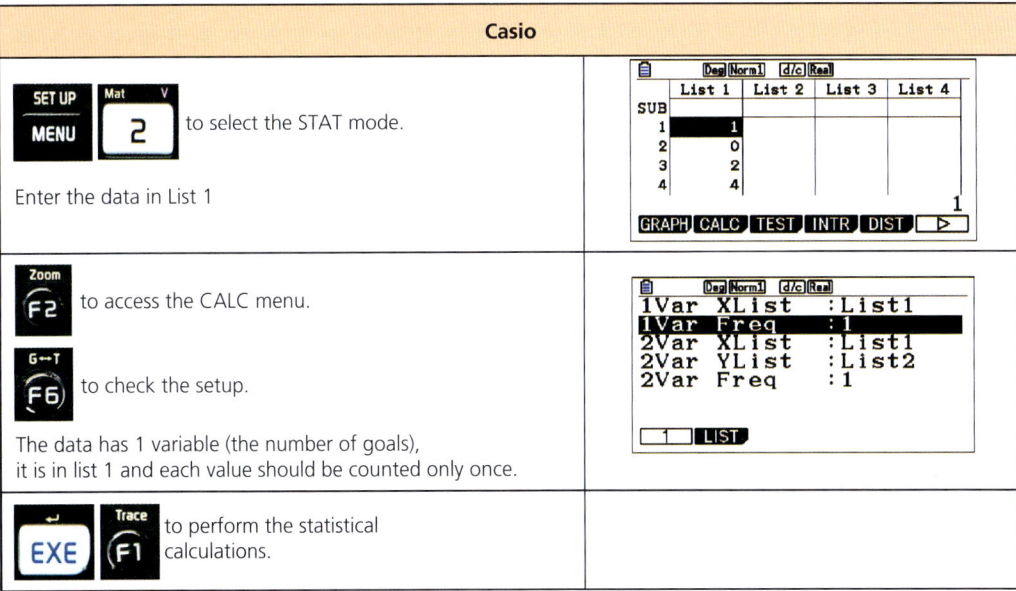

## 63 AVERAGES AND RANGES

| | |
|---|---|
| The following screen summarises the results of many calculations. | ```
1-Variable
x̄     =2
Σx    =30
Σx²   =96
σx    =1.54919333
sx    =1.60356745
n     =15
``` |
| The screen can be scrolled to reveal further results. | ```
1-Variable
minX  =0
Q1    =1
Med   =2
Q3    =3
maxX  =5
Mod   =1
``` |

Note: the mean is given by $\bar{x}$ and the median by 'Med'.
The range can be calculated by subtracting 'minX' from 'maxX' and the interquartile range by subtracting Q1 from Q3.

### Texas

| | |
|---|---|
| **stat** **1** to enter the data into lists. Enter the data in List 1. | (screen showing L1 list with data) |
| **stat** ▶ select the 'CALC' menu. **1** to select the 1-variable option (1-Var Stats). | ```
EDIT CALC TESTS
1:1-Var Stats
2:2-Var Stats
3:Med-Med
4:LinReg(ax+b)
5:QuadReg
6:CubicReg
7:QuartReg
8:LinReg(a+bx)
9↓LnReg
``` |
| Ensure the list selected is 'L1' and the 'FreqList' is left blank. This will result in each value in L1 being counted only once. | ```
1-Var Stats
List:L1
FreqList:
Calculate
``` |
| **enter** to carry out the statistical calculations on the 1-variable data in List 1. The following screen summarises the results of many calculations. | ```
1-Var Stats
x̄=2
Σx=30
Σx²=96
Sx=1.603567451
σx=1.549193338
n=15
minX=0
↓Q₁=1
``` |
| The screen can be scrolled to reveal further results. | ```
1-Var Stats
↑Sx=1.603567451
σx=1.549193338
n=15
minX=0
Q₁=1
Med=2
Q₃=3
maxX=5
``` |

Note: the mean is given by $\bar{x}$ and the median by 'Med'. The range can be calculated by subtracting 'minX' from 'maxX' and the interquartile range by subtracting Q1 from Q3.

*Large amounts of data*

**Exercise 10.7**

1. Find the mean, median, mode, quartiles and range for each set of data.
   a. The number of goals scored by a water polo team in each of 15 matches:
      1 0 2 4 0 1 1 1 2 5 3 0 1 2 2
   b. The total scores when two dice are rolled:
      7 4 5 7 3 2 8 6 8 7 6 5 11 9 7 3 8 7 6 5
   c. The number of students present in a class over a three-week period:
      28 24 25 28 23 28 27 26 27 25 28 28 28 26 25
   d. An athlete's training times (in seconds) for the 100 m race:
      14.0 14.3 14.1 14.3 14.2 14.0 13.9 13.8
      13.9 13.8 13.8 13.7 13.8 13.8 13.8

2. The mean mass of the 11 players in a football team is 80.3 kg. The mean mass of the team plus a substitute is 81.2 kg. Calculate the mass of the substitute.

3. After eight matches, a basketball player had scored a mean of 27 points. After three more matches his mean was 29. Calculate the total number of points he scored in the last three games.

# Large amounts of data

When there are only three values in a set of data, the median value is given by the second value,

i.e.  1  ②  3.

When there are four values in a set of data, the median value is given by the mean of the second and third values,

i.e.  1  ②  ③  4.

When there are five values in a set of data, the median value is given by the third value.

If this pattern is continued, it can be deduced that for $n$ values in a set of data, the median value is given by the value at $\frac{n+1}{2}$. This is useful when finding the median of large sets of data.

## ➡ Worked example

The British shoe sizes of 49 people are recorded in the table below. Calculate the median, mean and modal shoe size.

| Shoe size | 3 | $3\frac{1}{2}$ | 4 | $4\frac{1}{2}$ | 5 | $5\frac{1}{2}$ | 6 | $6\frac{1}{2}$ | 7 |
|---|---|---|---|---|---|---|---|---|---|
| Frequency | 2 | 4 | 5 | 9 | 8 | 6 | 6 | 5 | 4 |

As there are 49 data values, the median value is the 25th value. This occurs within shoe size 5, as 2 + 4 + 5 + 9 = 20 but 2 + 4 + 5 + 9 + 8 = 28.

So the median shoe size is 5.

## 63 AVERAGES AND RANGES

To calculate the mean shoe size:

$$\frac{(3\times 2)+\left(3\tfrac{1}{2}\times 4\right)+(4\times 5)+\left(4\tfrac{1}{2}\times 9\right)+(5\times 8)+\left(5\tfrac{1}{2}\times 6\right)+(6\times 6)+\left(6\tfrac{1}{2}\times 5\right)+(7\times 4)}{49}=\frac{250}{49}$$

So the mean shoe size is 5.10 (correct to three significant figures).

Note: The mean value is not necessarily a data value which appears in the set or a real shoe size.

The modal shoe size is $4\tfrac{1}{2}$

The range of the data is, as before, the smallest data value subtracted from the largest data value. In this case the largest recorded shoe size is 7 and the smallest 3. Therefore the range is 4.

These calculations can also be carried out on your graphic display calculator by entering the frequency tables:

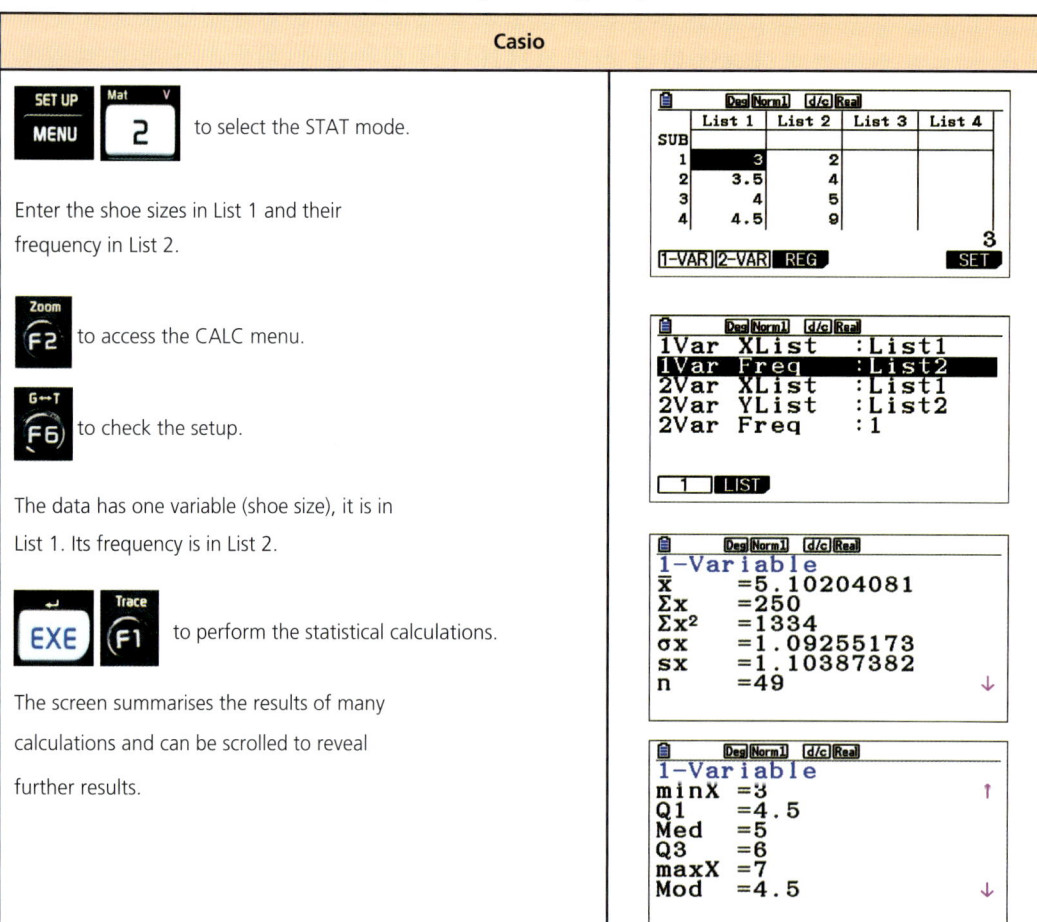

*Large amounts of data*

| Texas |
|---|

 to enter the data into lists.

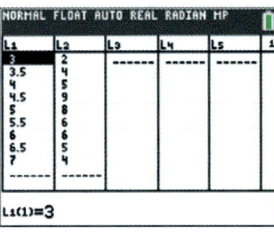

Enter the shoe size in List 1 and the frequency in List 2.

 to select the 'CALC' menu.

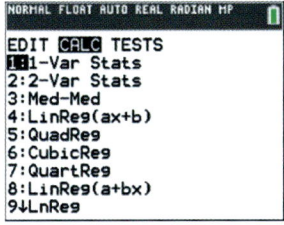

 to select the 1-variable option (1-Var Stats).

Ensure the List selected is $L_1$

  to enter List 2 as the frequency.

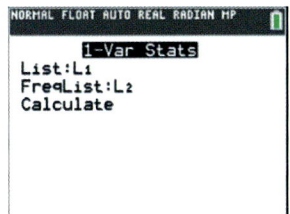

 to perform statistical calculations on the

1-variable data in List 1 with frequency in List 2.

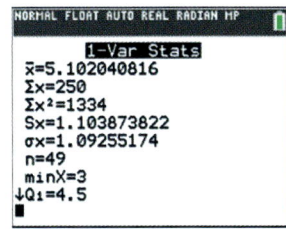

The screen summarises the results of many calculations and can be scrolled to reveal further results.

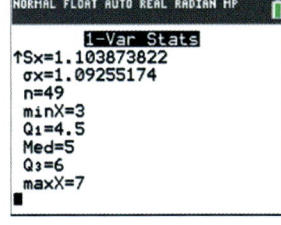

## 63 AVERAGES AND RANGES

**Exercise 10.8**

1  An ordinary dice was rolled 60 times. The results are shown in the table below. Calculate the mean, median and mode of the scores.

| Score | 1 | 2 | 3 | 4 | 5 | 6 |
|---|---|---|---|---|---|---|
| Frequency | 12 | 11 | 8 | 12 | 7 | 10 |

2  Two dice were rolled 100 times. Each time their combined score was recorded. Below is a table of the results. Calculate the mean, median and mode of the scores.

| Score | 2 | 3 | 4 | 5 | 6 | 7 | 8 | 9 | 10 | 11 | 12 |
|---|---|---|---|---|---|---|---|---|---|---|---|
| Frequency | 5 | 6 | 7 | 9 | 14 | 16 | 13 | 11 | 9 | 7 | 3 |

3  Sixty flowering bushes are planted. At their flowering peak, the number of flowers per bush is counted and recorded. The results are shown in the table below:

| Flowers per bush | 0 | 1 | 2 | 3 | 4 | 5 | 6 | 7 | 8 |
|---|---|---|---|---|---|---|---|---|---|
| Frequency | 0 | 0 | 0 | 6 | 4 | 6 | 10 | 16 | 18 |

  a  Calculate the mean, median and mode of the number of flowers per bush.
  b  Which of the mean, median and mode would be most useful when advertising the bush to potential buyers?

# Mean and mode for grouped data

As has already been described, sometimes it is more useful to group data, particularly if the range of values is very large. However, by grouping data, some accuracy is lost.

The results below are the distances (to the nearest metre) run by twenty students in one minute.

  256  271  271  274  275  276  276  277  279  280
  281  282  284  286  287  288  296  300  303  308

Table 1: Class interval of 5

| Group | 250–254 | 255–259 | 260–264 | 265–269 | 270–274 | 275–279 | 280–284 | 285–289 | 290–294 | 295–299 | 300–304 | 305–309 |
|---|---|---|---|---|---|---|---|---|---|---|---|---|
| Frequency | 0 | 1 | 0 | 0 | 3 | 5 | 4 | 3 | 0 | 1 | 2 | 1 |

Table 2: Class interval of 10

| Group | 250–259 | 260–269 | 270–279 | 280–289 | 290–299 | 300–309 |
|---|---|---|---|---|---|---|
| Frequency | 1 | 0 | 8 | 7 | 1 | 3 |

## Mean and mode for grouped data

Table 3: Class interval of 20

| Group | 250–269 | 270–289 | 290–309 |
|---|---|---|---|
| Frequency | 1 | 15 | 4 |

The three tables above highlight the effects of different group sizes. Table 1 is perhaps too detailed, while in Table 3 the group sizes are too big and therefore most of the results fall into one group. Table 2 is the most useful in that the spread of the results is still clear. In the 270–279 group, you can see that there are eight students, but some detail is lost; without the raw data, you would not know where in the group they lie.

To find the mean of grouped data, you assume that all the data within a group takes the **mid-interval value**. For example, using Table 2 above,

| Group | 250–259 | 260–269 | 270–279 | 280–289 | 290–299 | 300–309 |
|---|---|---|---|---|---|---|
| Mid-interval value | 254.5 | 264.5 | 274.5 | 284.5 | 294.5 | 304.5 |
| Frequency | 1 | 0 | 8 | 7 | 1 | 3 |

*As it is assumed that all values take the mid-interval value, the result is only an estimate of the mean.*

Estimated mean = 
$$\frac{(254.5 \times 1) + (264.5 \times 0) + (274.5 \times 8) + (284.5 \times 7) + (294.5 \times 1) + (304.5 \times 3)}{20} = 282.5$$

The **estimate** of mean distance run is 282.5 metres.

The **modal class** is 270–279.

Note: In the example above, the distance data is rounded to the nearest natural number. This has the effect of presenting continuous data as discrete data. If the data had been truly continuous, the groupings would need to be presented differently as shown below.

| Group | $250 < d \leq 260$ | $260 < d \leq 270$ | $270 < d \leq 280$ | $280 < d \leq 290$ | $290 < d \leq 300$ | $300 < d \leq 310$ |
|---|---|---|---|---|---|---|
| Mid-interval value | 255 | 265 | 275 | 285 | 295 | 305 |
| Frequency | 1 | 0 | 8 | 7 | 1 | 3 |

The group $250 < d \leq 260$ in the table above means any result that falls in the group from 250 up to, but not including, 260. It therefore has a group width of 10 rather than 9 as before. The mid-interval values are therefore affected as is the estimate for the mean.

Estimated mean = 
$$\frac{(255 \times 1) + (265 \times 0) + (275 \times 8) + (285 \times 7) + (295 \times 1) + (305 \times 3)}{20} = 283 \text{ m}$$

The graphic display calculator can work out the mean and median of grouped data. The mid-interval value should be entered in List 1 and the frequency in List 2. Then proceed as before.

## 63 AVERAGES AND RANGES

**Exercise 10.9**

1. A pet shop has 100 tanks containing fish. The number of fish in each tank is recorded in the table below.

| No. of fish | 0–9 | 10–19 | 20–29 | 30–39 | 40–49 |
|---|---|---|---|---|---|
| Frequency | 7 | 12 | 24 | 42 | 15 |

   a Calculate an estimate for the mean number of fish in each tank.
   b Give the modal class size.

2. A school has 148 Year 11 students studying Mathematics. Their percentage scores in their Mathematics mock examination are recorded in the grouped frequency table.

| % Score | 0–9 | 10–19 | 20–29 | 30–39 | 40–49 | 50–59 | 60–69 | 70–79 | 80–89 | 90–99 |
|---|---|---|---|---|---|---|---|---|---|---|
| Frequency | 3 | 2 | 4 | 6 | 8 | 36 | 47 | 28 | 10 | 4 |

   a Calculate the mean percentage score for the mock examination.
   b What was the modal class score?

3. A stationmaster records how many minutes late each train is. The table of results is shown below:

| No. of minutes late | $0 < t \leqslant 5$ | $5 < t \leqslant 10$ | $10 < t \leqslant 15$ | $15 < t \leqslant 20$ | $20 < t \leqslant 25$ | $25 < t \leqslant 30$ |
|---|---|---|---|---|---|---|
| Frequency | 16 | 9 | 3 | 1 | 0 | 1 |

   a Calculate an estimate for the mean number of minutes late a train is.
   b What is the modal number of minutes late?
   c The stationmaster's report concludes: 'Trains are, on average, less than five minutes late'. Comment on this conclusion.

# 64 Cumulative frequency

Calculating the cumulative frequency is done by adding up the frequencies as you go along. A cumulative frequency graph is particularly useful when trying to calculate the median of a large set of data, grouped data or continuous data, or when trying to establish how consistent the results in a set of data are.

## ➔ Worked example

The duration of two different brands of battery, A and B, is tested. Fifty batteries of each type are randomly selected and tested in the same way. The duration of each battery is then recorded. The results of the tests are shown in the tables below.

| Type A: duration (h) | Frequency | Cumulative frequency |
|---|---|---|
| $0 < t \leq 5$ | 3 | 3 |
| $5 < t \leq 10$ | 5 | 8 |
| $10 < t \leq 15$ | 8 | 16 |
| $15 < t \leq 20$ | 10 | 26 |
| $20 < t \leq 25$ | 12 | 38 |
| $25 < t \leq 30$ | 7 | 45 |
| $30 < t \leq 35$ | 5 | 50 |

| Type B: duration (h) | Frequency | Cumulative frequency |
|---|---|---|
| $0 < t \leq 5$ | 1 | 1 |
| $5 < t \leq 10$ | 1 | 2 |
| $10 < t \leq 15$ | 10 | 12 |
| $15 < t \leq 20$ | 23 | 35 |
| $20 < t \leq 25$ | 9 | 44 |
| $25 < t \leq 30$ | 4 | 48 |
| $30 < t \leq 35$ | 2 | 50 |

# 64 CUMULATIVE FREQUENCY

i Plot a cumulative frequency curve for each brand of battery.

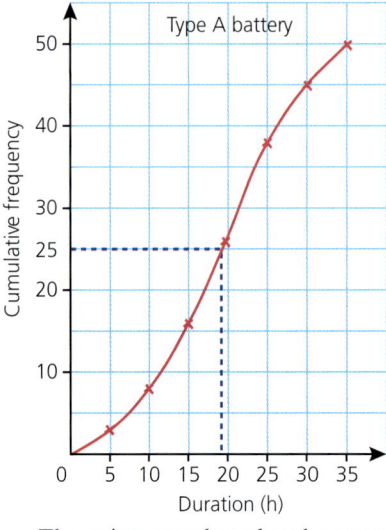
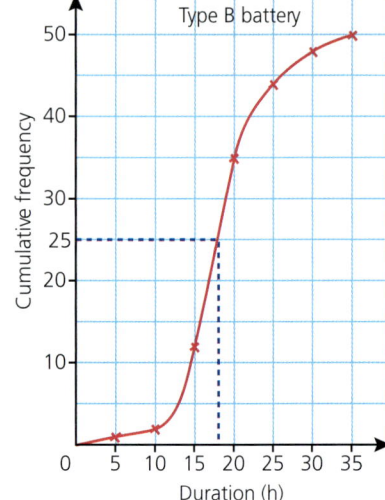

The points are plotted at the upper boundary of each class interval rather than at the middle of the interval. So, for Type A, points are plotted at (5, 3), (10, 8), etc. The points are joined with a smooth curve which is extended to include (0, 0).

ii Estimate the median duration for each brand.

The median value is the value which occurs halfway up the cumulative frequency axis. This is shown with broken lines on the graph. Therefore:

Median for type A batteries ≈ 19 h

Median for type B batteries ≈ 18 h

This tells you that, on average, batteries of type A last longer (19 hours) than batteries of type B (18 hours).

## Exercise 10.10

1 Sixty athletes enter a cross-country race. Their finishing times are recorded and are shown in the table below:

| Finishing time (h) | Frequency | Cumulative freq. |
|---|---|---|
| $0 < h \leq 0.5$ | 0 | |
| $0.5 < h \leq 1.0$ | 0 | |
| $1.0 < h \leq 1.5$ | 6 | |
| $1.5 < h \leq 2.0$ | 34 | |
| $2.0 < h \leq 2.5$ | 16 | |
| $2.5 < h \leq 3.0$ | 3 | |
| $3.0 < h \leq 3.5$ | 1 | |

a Copy the table and calculate the values for the cumulative frequency.
b Draw a cumulative frequency curve of the results.
c Show how your graph could be used to find the approximate median finishing time.
d What does the median value tell you?

# Quartiles and the interquartile range

2  Three Mathematics classes take the same test in preparation for their final examination. Their raw scores are shown below:
   **Class A**  12, 21, 24, 30, 33, 36, 42, 45, 53, 53, 57, 59, 61, 62, 74, 88, 92, 93
   **Class B**  48, 53, 54, 59, 61, 62, 67, 78, 85, 96, 98, 99
   **Class C**  10, 22, 36, 42, 44, 68, 72, 74, 75, 83, 86, 89, 93, 96, 97, 99, 99
   a  Using the class intervals $0 < x \leqslant 20$, $20 < x \leqslant 40$ etc., draw up a grouped frequency table and cumulative frequency table for each class.
   b  Draw a cumulative frequency curve for each class.
   c  Show how your graph could be used to find the median score for each class.
   d  What does the median value tell you?

3  The table below shows the heights of students in a class over a three-year period.

| Height (cm) | Frequency 2008 | Frequency 2009 | Frequency 2010 |
|---|---|---|---|
| $150 < h \leqslant 155$ | 6 | 2 | 2 |
| $155 < h \leqslant 160$ | 8 | 9 | 6 |
| $160 < h \leqslant 165$ | 11 | 10 | 9 |
| $165 < h \leqslant 170$ | 4 | 4 | 8 |
| $170 < h \leqslant 175$ | 1 | 3 | 2 |
| $175 < h \leqslant 180$ | 0 | 2 | 2 |
| $180 < h \leqslant 185$ | 0 | 0 | 1 |

   a  Construct a cumulative frequency table for each year.
   b  Draw the cumulative frequency curve for each year.
   c  Show how your graph could be used to find the median height for each year.
   d  What conclusions can be made from the median values?

## Quartiles and the interquartile range

The cumulative frequency axis can also be represented in terms of **percentiles**. A percentile scale divides the cumulative frequency scale into hundredths. The maximum value of cumulative frequency is found at the 100th percentile. Similarly, the median, being the middle value, is called the 50th percentile. The 25th percentile is known as the lower quartile, and the 75th percentile is called the upper quartile, as introduced in Chapter 63.

75% of 120 is 90, therefore 90 is the 75th percentile.

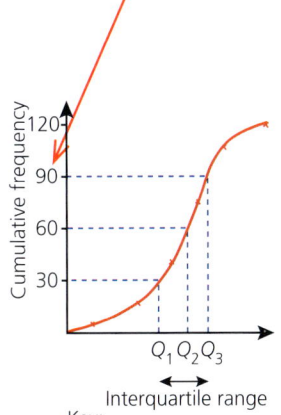
Key:
$Q_1$ Lower quartile
$Q_2$ Median
$Q_3$ Upper quartile

The range of a distribution is found by subtracting the lowest value from the highest value. Sometimes this will give a useful result, but often it will not. A better measure of spread is given by looking at the spread of the middle half of the results, that is, the difference between the upper and lower quartiles. This is known as the **interquartile range**.

## Worked example

Consider again the two types of batteries A and B discussed earlier.

**a** Using the graphs, estimate the upper and lower quartiles for each battery.

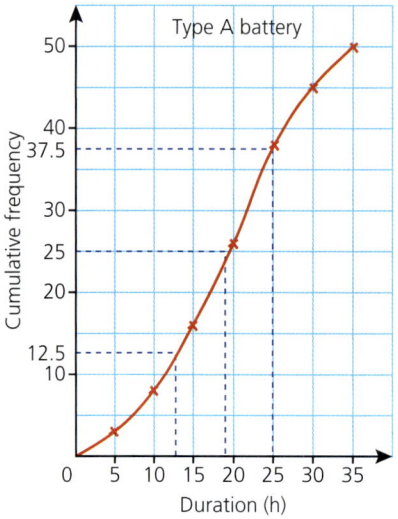
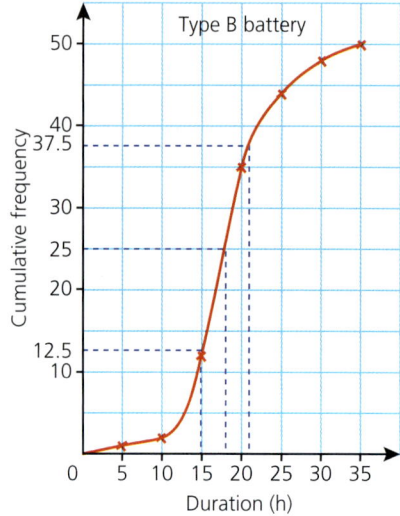

Lower quartile of Type A ≈ 13 h
Upper quartile of Type A ≈ 25 h
Lower quartile of Type B ≈ 15 h
Upper quartile of Type B ≈ 21 h

**b** Calculate the interquartile range for each type of battery.

Interquartile range of type A ≈ 12 h
Interquartile range of type B ≈ 6 h

**c** Based on these results, how might the manufacturers advertise the two types of batteries?

Type A: on 'average' the longer-lasting battery
Type B: the more reliable battery

## Quartiles and the interquartile range

The interquartile range can be calculated using a graphic display calculator:

| Casio |
|---|

The screen opposite is the result of performing statistical calculations on the Battery A data above.

The interquartile range can be calculated as a result:

IQR = $Q_3 - Q_1$ = 22.5 − 12.5 = 10

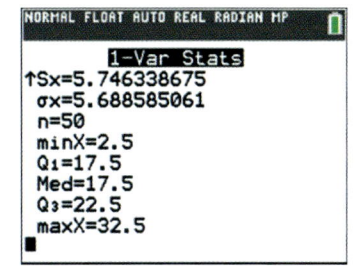

| Texas |
|---|

The screen opposite is the result of performing statistical calculations on the Battery B data above.

The interquartile range can be calculated as a result:

IQR = $Q_3 - Q_1$ = 22.5 − 17.5 = 5.

### Exercise 10.11

1. Using the results obtained from Q.2 of Exercise 10.10:
   a. find the interquartile range of each of the classes taking the Mathematics test
   b. analyse your results and write a short summary comparing the three classes.
2. Using the results obtained from Q.3 of Exercise 10.10:
   a. find the interquartile range of the students' heights each year
   b. analyse your results and write a short summary comparing the three years.
3. Forty boys enter a school javelin competition. The distances (d m) thrown are recorded below:

| Distance thrown (m) | Frequency |
|---|---|
| $0 < d \leqslant 20$ | 4 |
| $20 < d \leqslant 40$ | 9 |
| $40 < d \leqslant 60$ | 15 |
| $60 < d \leqslant 80$ | 10 |
| $80 < d \leqslant 100$ | 2 |

a. Construct a cumulative frequency table for the above results.
b. Draw a cumulative frequency curve.
c. If the top 20% of boys are considered for the final, estimate (using the graph) the qualifying distance.
d. Calculate the interquartile range of the throws.
e. Calculate the median distance thrown.

583

**Exercise 10.11 (cont)**

4 The masses ($m$ g) of two different types of oranges are compared. Eighty oranges are randomly selected from each type and weighed. The results are shown below:

| Type A | |
|---|---|
| Mass (g) | Frequency |
| $75 < m \leqslant 100$ | 4 |
| $100 < m \leqslant 125$ | 7 |
| $125 < m \leqslant 150$ | 15 |
| $150 < m \leqslant 175$ | 32 |
| $175 < m \leqslant 200$ | 14 |
| $200 < m \leqslant 225$ | 6 |
| $225 < m \leqslant 250$ | 2 |

| Type B | |
|---|---|
| Mass (g) | Frequency |
| $75 < m \leqslant 100$ | 0 |
| $100 < m \leqslant 125$ | 16 |
| $125 < m \leqslant 150$ | 43 |
| $150 < m \leqslant 175$ | 10 |
| $175 < m \leqslant 200$ | 7 |
| $200 < m \leqslant 225$ | 4 |
| $225 < m \leqslant 250$ | 0 |

a Construct a cumulative frequency table for each type of orange.
b Draw a cumulative frequency graph for each type of orange.
c Calculate the median mass for each type of orange.
d Using your graphs, estimate for each type of orange:
 i the lower quartile
 ii the upper quartile
 iii the interquartile range.
e Write a brief report comparing the two types of oranges.

5 Two competing brands of batteries are compared. One hundred batteries of each brand are tested and the duration of each is recorded. The results of the tests are shown in the cumulative frequency graphs below.

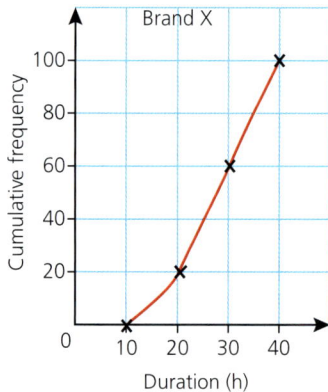
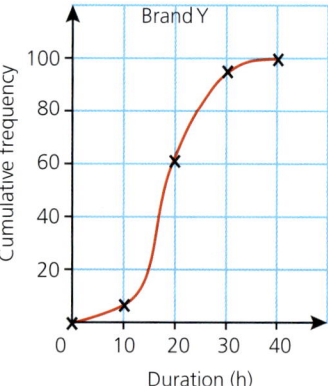

a The manufacturers of brand X claim that on average their batteries will last at least 40% longer than those of brand Y. Showing your method clearly, decide whether this claim is true.
b The manufacturers of brand X also claim that their batteries are more reliable than those of brand Y. Is this claim true? Show your working clearly.

# 65 Scatter diagrams, correlation and lines of best fit

When information about two different aspects (**variables**) of a data item is recorded, such as the height and mass of children, you are collecting **bivariate data**. You can use the values of the two variables as the coordinates of a point to represent it on a **scatter diagram** (or scatter graph).

Scatter diagrams are particularly useful if you wish to see if there is a **correlation** (relationship) between the two variables. How the points lie when plotted indicates the type of relationship between the two sets of data.

## ➔ Worked example

The heights and weights (masses) of 20 children under the age of five were recorded. The heights were recorded in centimetres and the masses in kilograms. The data is shown below with the heights written in red and the masses in blue.

| Height | 32 | 34 | 45 | 46 | 52 | 59 | 63 | 64 | 71 | 73 |
|---|---|---|---|---|---|---|---|---|---|---|
| Mass | 5.834 | 3.792 | 9.037 | 4.225 | 10.149 | 6.188 | 9.891 | 16.010 | 15.806 | 9.929 |
| Height | 86 | 87 | 95 | 96 | 96 | 101 | 108 | 109 | 117 | 121 |
| Mass | 11.132 | 16.443 | 20.895 | 16.181 | 14.000 | 19.459 | 15.928 | 12.047 | 19.423 | 14.331 |

i   Plot a scatter diagram for this data.

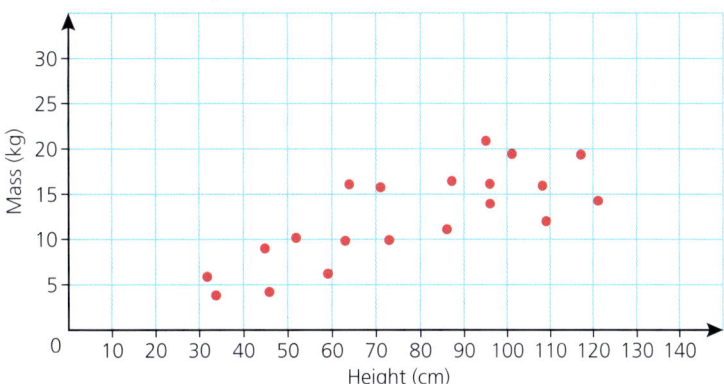

ii  Comment on any relationship that you see.

The points tend to lie in a diagonal direction from bottom left to top right. This suggests that as height increases then, in general, mass increases too. Therefore there is a positive correlation between height and mass.

## 65 SCATTER DIAGRAMS, CORRELATION AND LINES OF BEST FIT

**iii** Estimate the mass of another child of height of 80 cm.

You have to assume that this child will follow the trend set by the other 20 children. To find an approximate value for the mass, you draw a line of best fit. This is a solid straight line which best passes through the points. It also passes through the point $(\bar{x}, \bar{y})$ as shown below.

Note: $\bar{x}$ and $\bar{y}$ are the means of the $x$ and $y$ values respectively, in this example (77.75, 12.535).

A line of best fit need not pass through the origin.

*Your line of best fit can be drawn by eye, but make sure it:*

- *extends across the full data set, but take care if extending it beyond the data*

- *does not need to coincide exactly with any of the points, but there should be approximately the same number of points either side of the line over its entire length.*

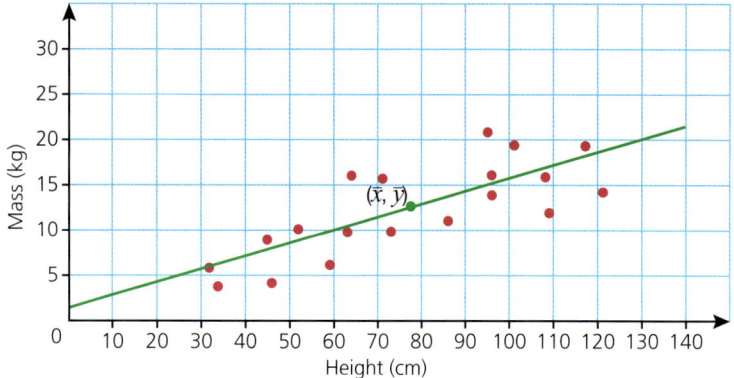

The line of best fit can now be used to give an approximate solution to the question. If a child has a height of 80 cm, you would expect their mass, by reading from the graph below, to be in the region of 13 kg.

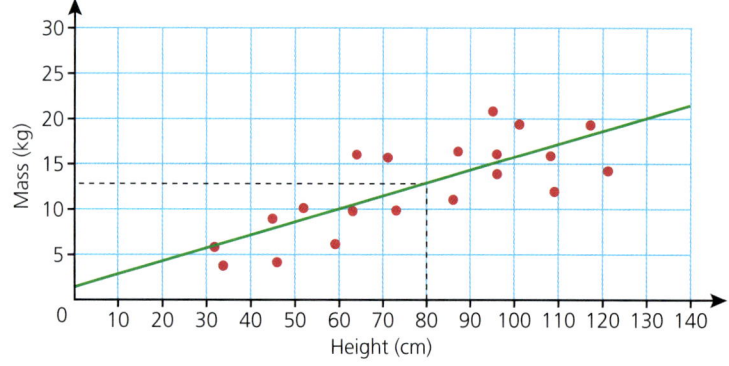

Your graphic display calculator will plot scatter diagrams and analyse them. For example, plot the following data for an ice cream vendor's sales on a scatter diagram and, if appropriate, draw a line of best fit.

| Temperature (°C) | 15 | 24 | 18 | 24 | 19 | 26 | 22 | 24 | 27 | 28 | 30 | 25 | 22 | 17 |
|---|---|---|---|---|---|---|---|---|---|---|---|---|---|---|
| Number of ice creams sold | 8 | 34 | 20 | 38 | 28 | 37 | 32 | 29 | 33 | 35 | 44 | 28 | 30 | 25 |

## Scatter diagrams, correlation and lines of best fit

| Casio |
|---|

  to select the STAT mode.

Enter the temperature data in List 1 and the number of ice creams sold in List 2.

 to access the statistical graphing menu.

 to check the setup.

The graph type is 'Scatter' with the $x$-values from List 1 and the $y$-values from List 2. Each data value is to be counted once.

 to plot the scatter diagram.

 to select the graph 'CALC' menu.

 as the line of best fit required is linear.

 if the linear information is wanted in the format $y = ax + b$.

The following screen summarises the properties of the line of best fit in the form $y = ax + b$.

 to plot the line of best fit.

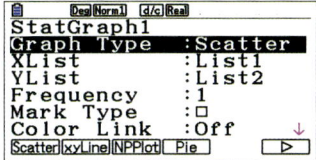
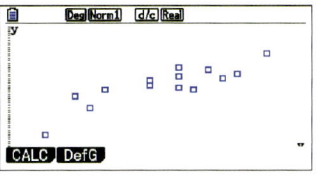
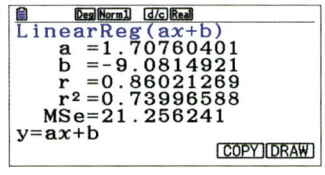
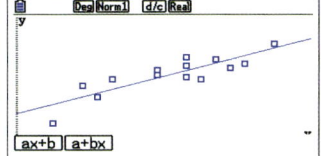

*Another term for describing a line of best fit is a **linear regression**.*

Note: The screen which gives the properties of the line of best fit also gives the value of $r$. This is an indicator of how tight the data is to the line of best fit. It is, however, beyond the scope of this syllabus.

587

# 65 SCATTER DIAGRAMS, CORRELATION AND LINES OF BEST FIT

| Texas |
|---|

 to enter the data into lists.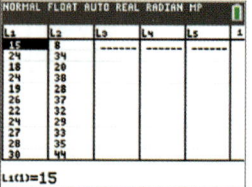

Enter the temperature data in List 1 and the number of ice creams sold in List 2.

   to enter the statistical plot setup.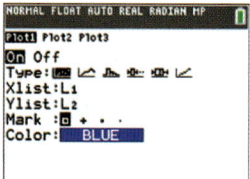

Turn 'Plot 1' to 'On'. Choose the scatter graph and ensure the $x$-values are from List 1 and the $y$-values from List 2.

 to set scale for each axis.

 to plot the scatter diagram.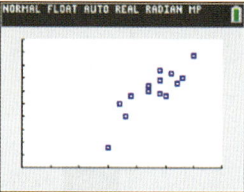

  to select the 'CALC' menu.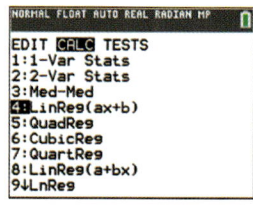

Select   to find the linear equation of the line of best fit through the points.

Ensure that the $x$-values are from L1, the $y$-values from L2. The 'FreqList' should be left blank as this implies that each data value will be counted only once.

With the cursor on 'Store RegEq'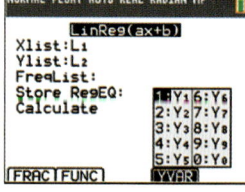

588

*Types of correlation*

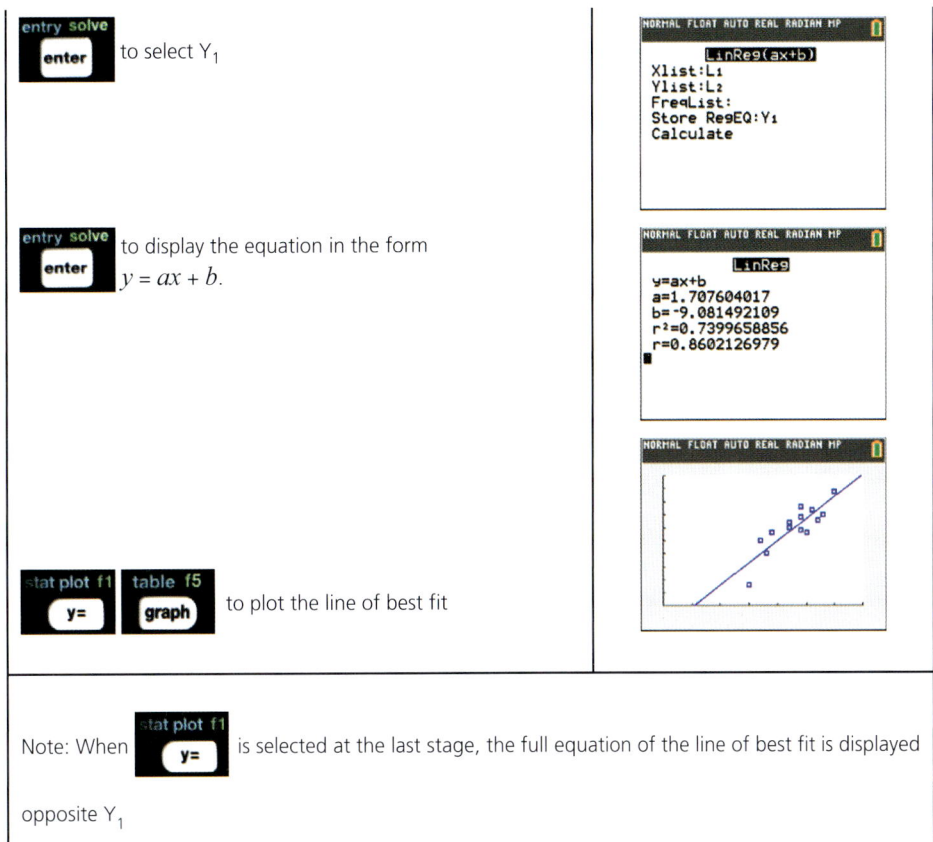

| | |
|---|---|
| **enter** to select Y₁ | |
| **enter** to display the equation in the form $y = ax + b$. | |
| **y=** **graph** to plot the line of best fit | |

Note: When **stat plot f1** **y=** is selected at the last stage, the full equation of the line of best fit is displayed opposite Y₁

## Types of correlation

There are several types of correlation depending on the arrangement of the points plotted on the scatter diagram. These are described below.

### A **strong positive correlation**

The points lie tightly around the line of best fit. As $x$ increases, so does $y$.

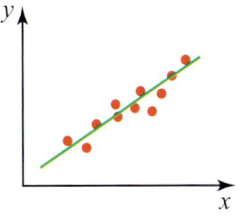

### A **weak positive correlation**

Although there is direction to the way the points are lying, they are not tightly packed around the line of best fit. As $x$ increases, $y$ tends to increase too.

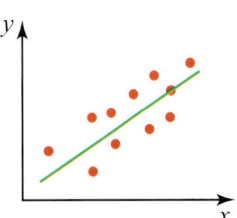

# 65 SCATTER DIAGRAMS, CORRELATION AND LINES OF BEST FIT

**No correlation**

There is no pattern to the way in which the points are lying; i.e. there is no correlation between the variables $x$ and $y$. As a result, there can be no line of best fit.

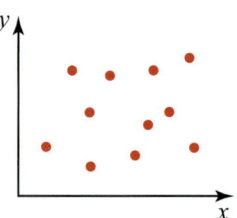

**A strong negative correlation**

The points lie tightly around the line of best fit. As $x$ increases, $y$ decreases.

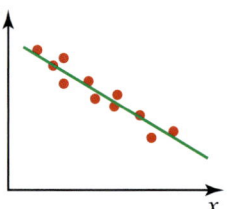

**A weak negative correlation**

The points are not tightly packed around the line of best fit. As $x$ increases, $y$ tends to decrease.

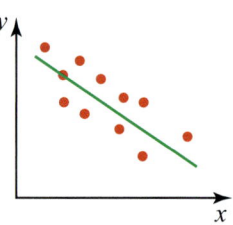

**Exercise 10.12**

1 State what type of correlation you might expect, if any, if the following data was collected and plotted on a scatter diagram. Give reasons for your answer.
   a A student's score in a Mathematics exam and their score in a Science exam
   b A student's hair colour and the distance they have to travel to school
   c The outdoor temperature and the number of cold drinks sold by a shop
   d The age of a motorcycle and its second-hand selling price
   e The number of people living in a house and the number of rooms the house has
   f A person's height and their age
   g A car's engine size and its fuel consumption

2 The table shows the readings for the number of hours of sunshine and the amount of rainfall in millimetres over a 24 hr period for several cities and towns in Europe.

| Place | Hours of sunshine | Rainfall (mm) |
|---|---|---|
| Athens | 12 | 6 |
| Belgrade | 10 | 61 |
| Copenhagen | 8 | 71 |
| Dubrovnik | 12 | 26 |
| Edinburgh | 5 | 83 |
| Frankfurt | 7 | 70 |
| Geneva | 10 | 64 |
| Helsinki | 9 | 68 |
| Innsbruck | 7 | 134 |
| Krakow | 7 | 111 |
| Lisbon | 12 | 3 |
| Marseilles | 11 | 11 |
| Naples | 10 | 19 |
| Oslo | 7 | 82 |
| Plovdiv | 11 | 37 |
| Reykjavik | 6 | 50 |
| Sofia | 10 | 68 |
| Tallinn | 10 | 68 |
| Valletta | 12 | 0 |
| York | 6 | 62 |
| Zurich | 8 | 136 |

a  Using your graphic display calculator, plot a scatter diagram of hours of sunshine against amount of rainfall.
b  What type of correlation, if any, is there between the two variables? Comment on whether this is what you would expect.

## 65 SCATTER DIAGRAMS, CORRELATION AND LINES OF BEST FIT

**Exercise 10.12 (cont)**

**3** The United Nations keeps an up-to-date database of statistical information on its member countries. The table below shows some of the information available.

| Country | Life expectancy at birth (years, 2005–10) | | Adult illiteracy rate (%, 2007) Total | Infant mortality rate (per 1000 births, 2005–10) |
|---|---|---|---|---|
| | Female | Male | | |
| Australia | 84 | 79 | 0 | 5 |
| Barbados | 80 | 74 | 2 | 10 |
| Brazil | 76 | 69 | 10 | 23 |
| Chad | 50 | 47 | 68 | 130 |
| China | 75 | 71 | 7 | 23 |
| Colombia | 77 | 69 | 7 | 19 |
| Congo | 55 | 53 | 26 | 79 |
| Cuba | 81 | 77 | 0 | 5 |
| Egypt | 72 | 68 | 34 | 35 |
| France | 85 | 78 | 0 | 4 |
| Germany | 82 | 77 | 0 | 4 |
| India | 65 | 62 | 34 | 55 |
| Iraq | 72 | 63 | 26 | 33 |
| Israel | 83 | 79 | 4 | 5 |
| Japan | 86 | 79 | 0 | 3 |
| Kenya | 55 | 54 | 26 | 64 |
| Mexico | 79 | 74 | 7 | 17 |
| Nepal | 67 | 66 | 43 | 42 |
| Portugal | 82 | 75 | 5 | 4 |
| Russian Federation | 73 | 60 | 0 | 12 |
| Saudi Arabia | 75 | 71 | 15 | 19 |
| United Kingdom | 82 | 77 | 0 | 5 |
| United States of America | 81 | 77 | 0 | 5 |

**a** Using your graphic display calculator, plot a scatter diagram and decide if there is a correlation between the adult illiteracy rate and the infant mortality rate.
**b** Are your findings in part a what you expected? Explain your answer.
**c** Without plotting a scatter diagram, decide if you think there is likely to be a correlation between male and female life expectancy at birth. Explain your reasons.
**d** Use your graphic display calculator to plot a scatter diagram to test if your predictions in part c were correct.

## Types of correlation

**4** The table below gives the average time taken for 30 students in a class to get to school each morning and the distance they live from the school.

| Distance (km) | 2 | 10 | 18 | 15 | 3 | 4 | 6 | 2 | 25 | 23 | 3 | 5 | 7 | 8 | 2 |
|---|---|---|---|---|---|---|---|---|---|---|---|---|---|---|---|
| Time (min) | 5 | 17 | 32 | 38 | 8 | 14 | 15 | 7 | 31 | 37 | 5 | 18 | 13 | 15 | 8 |
| Distance (km) | 19 | 15 | 11 | 9 | 2 | 3 | 4 | 3 | 14 | 14 | 4 | 12 | 12 | 7 | 1 |
| Time (min) | 27 | 40 | 23 | 30 | 10 | 10 | 8 | 9 | 15 | 23 | 9 | 20 | 27 | 18 | 4 |

    **a** Using your graphic display calculator, plot a scatter diagram of distance travelled against time taken.
    **b** Describe the correlation between the two variables.
    **c** Explain why some students who live further away may get to school quicker than some of those who live nearer.
    **d** Draw a line of best fit on your scatter diagram.
    **e** A new student joins the class. Use your line of best fit to estimate how far away she might live if she takes, on average, 19 minutes to get to school each morning.

**5** A department store decides to investigate if there is a correlation between the number of pairs of gloves it sells and the outside temperature. Over a one-year period, it records, every two weeks, how many pairs of gloves are sold and the mean daytime temperature during the same period. The results are given in the table below:

| Mean temp (°C) | 3 | 6 | 8 | 10 | 10 | 11 | 12 | 14 | 16 | 16 | 17 | 18 | 18 |
|---|---|---|---|---|---|---|---|---|---|---|---|---|---|
| Number of pairs of gloves | 61 | 52 | 49 | 54 | 52 | 48 | 44 | 40 | 51 | 39 | 31 | 43 | 35 |
| Mean temp (°C) | 19 | 19 | 20 | 21 | 22 | 22 | 24 | 25 | 25 | 26 | 26 | 27 | 28 |
| Number of pairs of gloves | 26 | 17 | 36 | 26 | 46 | 40 | 30 | 25 | 11 | 7 | 3 | 2 | 0 |

    **a** Using your graphic display calculator, plot a scatter diagram of mean temperature against number of pairs of gloves.
    **b** What type of correlation is there between the two variables?
    **c** How might this information be useful for the department store in the future?

# Investigations, modelling and ICT 10

## Investigation: Heights and percentiles

The graphs below show the height charts for males and females from the age of 2 to 20 years in the United States.

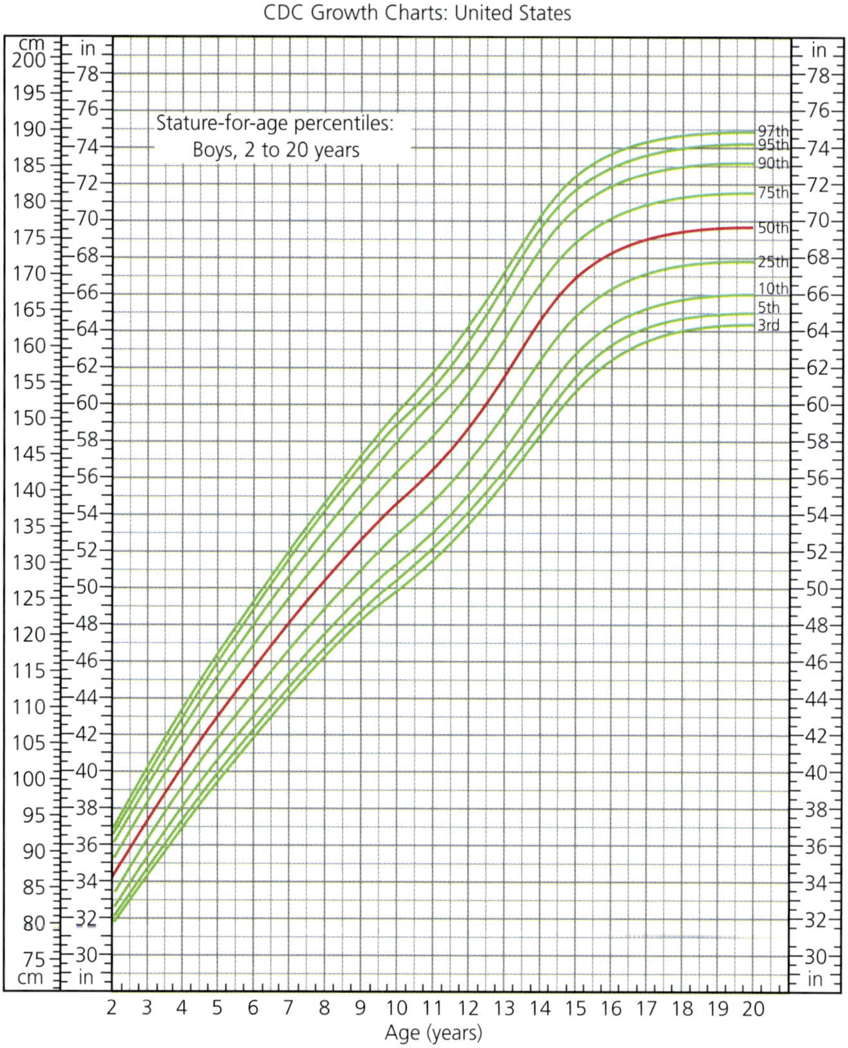

CDC Growth Charts: United States

Stature-for-age percentiles: Boys, 2 to 20 years

Note: Heights have been given in both centimetres and inches.

## Investigation: Heights and percentiles

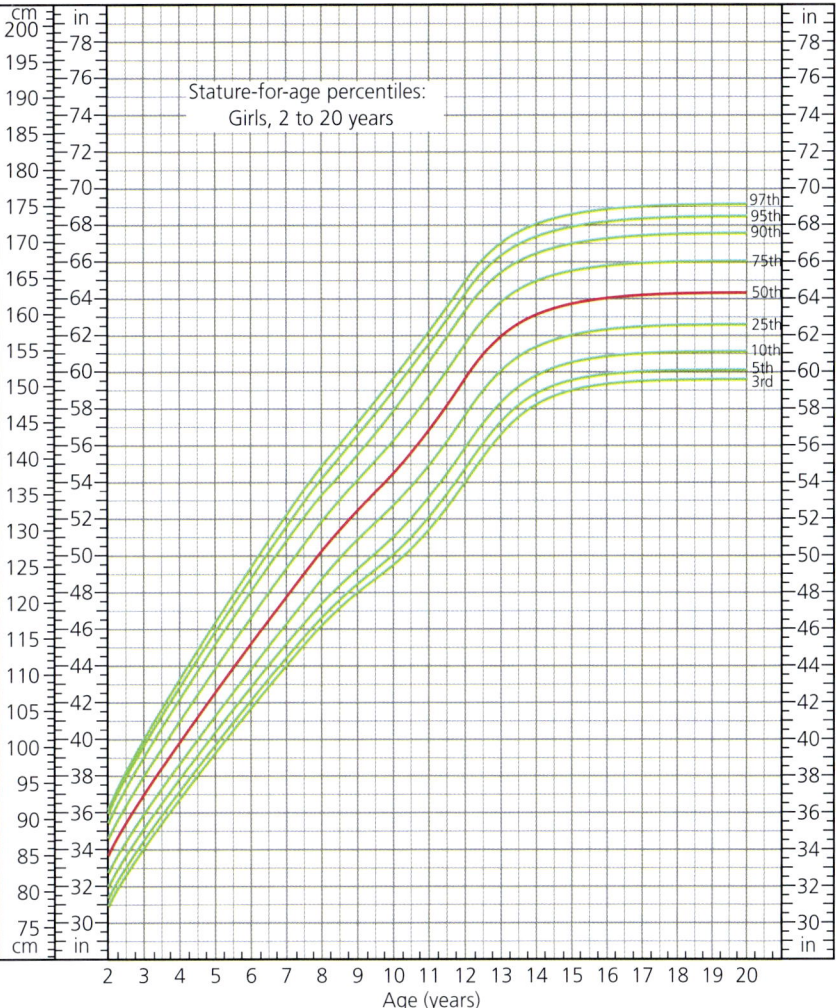

CDC Growth Charts: United States
Stature-for-age percentiles: Girls, 2 to 20 years

1. From the graph, find the height corresponding to the 75th percentile for 16-year-old girls.
2. Find the height which 75% of 16-year-old boys are taller than.
3. What is the median height for 12-year-old girls?
4. Measure the heights of students in your class. By carrying out appropriate statistical calculations, write a report comparing your data to that shown in the graphs.
5. Would all cultures use the same height charts? Explain your answer.

# INVESTIGATIONS, MODELLING AND ICT 10

## Modelling: Reading ages

Depending on their target audience, newspapers, magazines and books have different levels of readability. Some are easy to read and others more difficult.

1. Decide on some factors that you think would affect the readability of a text.
2. Write down the names of two newspapers which you think would have different reading ages. Give reasons for your answer.

There are established models for calculating the reading age of different texts in English.
One of these is the Gunning Fog Index. It calculates the reading age as follows:

Reading age = $\frac{2}{5}\left(\frac{A}{n}+\frac{100L}{A}\right)$ where

$A$ = number of words

$n$ = number of sentences

$L$ = number of words with three or more syllables

3. Choose one article from each of the two newspapers you chose in Q.2. Use the Gunning Fog Index to calculate the reading ages for the articles. Do the results support your predictions?
4. Write down some factors which you think may affect the reliability of your results.

## ICT Activity

In this activity you will be collecting the height data of all the students in your class and plotting a cumulative frequency graph of the results.

1. Measure the heights of all the students in your class.
2. Group your data appropriately.
3. Enter your data into graphing software.
4. Produce a cumulative frequency graph of the results.
5. From your graph, find:
   a. the median height of the students in your class
   b. the interquartile range of the heights.
6. Compare the cumulative frequency graph from your class with one produced from data collected from another class in a different year group. Comment on any differences or similarities between the two.

# Student assessments 10

## Student assessment 1

1. Find the mean, median and mode of the following sets of data:
   a  63  72  72  84  86
   b  6  6  6  12  18  24
   c  5  10  5  15  5  20  5  25  15  10
2. The mean mass of the 15 players in a rugby team is 85 kg. The mean mass of the team plus a substitute is 83.5 kg. Calculate the mass of the substitute.
3. Thirty families live in a street. The number of children in each family is given in the table below:

| Number of children | 0 | 1 | 2 | 3 | 4 | 5 | 6 |
|---|---|---|---|---|---|---|---|
| Frequency | 3 | 5 | 8 | 9 | 3 | 0 | 2 |

   a  Calculate the mean number of children per family.
   b  Calculate the median number of children per family.
   c  Calculate the modal number of children per family.
4. The numbers of people attending thirty screenings of a film at a local cinema are given below:
   21  30  66  71  10  37  24  21  62  50  27  31  65  12  38
   34  53  34  19  43  70  34  27  28  52  57  45  25  30  39
   a  Using groups 10–19, 20–29, 30–39, etc., present the above data in a grouped frequency table.
   b  Using your grouped data, calculate an estimate for the mean number of people attending each screening.
5. Identify which of the following types of data are discrete and which are continuous:
   a  The number of cars passing the school gate each hour
   b  The time taken to travel to school each morning
   c  The speed at which students run in a race
   d  The wingspan of butterflies
   e  The height of buildings
6. A businesswoman travels to work in her car each morning in one of two ways: using the small country roads or using the motorway. She records the time taken to travel to work each day. The results are shown in the table below:

| Time (min) | $10 < t \leqslant 15$ | $15 < t \leqslant 20$ | $20 < t \leqslant 25$ | $25 < t \leqslant 30$ | $30 < t \leqslant 35$ | $35 < t \leqslant 40$ | $40 < t \leqslant 45$ |
|---|---|---|---|---|---|---|---|
| Motorway frequency | 3 | 5 | 7 | 2 | 1 | 1 | 1 |
| Country roads frequency | 0 | 0 | 9 | 10 | 1 | 0 | 0 |

   a  Complete a cumulative frequency table for each of the sets of results shown above.
   b  Using your cumulative frequency tables, plot two cumulative frequency curves – one for the time taken to travel to work using the motorway, the other for the time taken to travel to work using country lanes.

**STUDENT ASSESSMENTS 10**

   **c** Use your graphs to work out the following for each method of travel:
      **i** the median travelling time
      **ii** the upper and lower quartile travelling times
      **iii** the interquartile range for the travelling times.
   **d** With reference to your graphs or calculations, explain which is the most reliable way of getting to work.
   **e** If she had to get to work one morning within 25 minutes of leaving home, which way would you recommend she take? Explain your answer fully.

**7** Twenty students take three long jumps. The best result for each student (in metres) is recorded below:

   4.3   5.4   4.3   4.0   3.8   5.1   3.6   5.5   6.2   4.7
   5.2   3.8   2.4   4.7   3.9   5.6   5.8   4.7   3.3   2.9

The students were then coached in long jump technique and given three further jumps. Their individual best results are recorded below:

   4.7   5.9   4.8   4.6   4.5   5.3   5.2   5.5   6.3   4.9
   5.2   4.9   5.6   5.3   6.8   5.4   5.8   5.4   4.3   5.5

Draw a back-to-back stem-and-leaf diagram of their long jumps before and after coaching. Comment on your diagram.

**8** The popularity of a group of professional football players and their yearly salary is given in the table below:

| Popularity | 1 | 2 | 3 | 4 | 5 | 6 | 7 | 8 | 9 | 10 |
|---|---|---|---|---|---|---|---|---|---|---|
| Salary ($ million) | 4.8 | 3.6 | 4.5 | 3.1 | 7.7 | 6.3 | 2.9 | 3.1 | 4.1 | 1.8 |
| Popularity | 11 | 12 | 13 | 14 | 15 | 16 | 17 | 18 | 19 | 20 |
| Salary ($ million) | 4.5 | 3.1 | 2.7 | 3.9 | 6.2 | 5.8 | 4.1 | 5.3 | 7.2 | 6.5 |

  **a** Using your graphic display calculator, find the equation of the line of best fit.
  **b** This statement is made in a newspaper: 'Big money footballers are not popular with fans.' Comment on this statement in the light of your answer to part a.

# Student assessment 2

**1** Find the mean, median and mode of the following sets of data:
  **a** 4 5 5 6 7
  **b** 3 8 12 18 18 24
  **c** 4 9 3 8 7 11 3 5 3 8
**2** The mean mass of the 11 players in a football team is 76 kg. The mean mass of the team plus a substitute is 76.2 kg. Calculate the mass of the substitute.

3   Thirty children were asked about the number of pets they had. The results are shown in the table below.

| Number of pets | 0 | 1 | 2 | 3 | 4 | 5 | 6 |
|---|---|---|---|---|---|---|---|
| Frequency | 5 | 5 | 3 | 7 | 3 | 1 | 6 |

   a   Calculate the mean number of pets per child.
   b   Calculate the median number of pets per child.
   c   Calculate the modal number of pets.

4   The numbers of people attending discos at a club's over-30s evenings are:
   89 94 32 45 57 68 127 138 23 77 99 47 44 100 106
   132 28 56 59 49 96 103 90 84 136 38 72 47 58 110
   a   Using groups 0–19, 20–39, 40–59 etc., present the above data in a grouped frequency table.
   b   Using your grouped data, calculate an estimate for the mean number of people going to the disco each night.

5   Identify which of the following types of data are discrete and which are continuous:
   a   The number of goals scored in a hockey match
   b   Dress sizes
   c   The time taken to fly from Hong Kong to Beijing
   d   The price of a kilogram of carrots
   e   The speed of a police car

6   Four hundred students sit their IGCSE Mathematics exam. Their marks (as percentages) are shown in the table below:

| Mark (%) | Frequency | Cumulative frequency |
|---|---|---|
| 31–40 | 21 | |
| 41–50 | 55 | |
| 51–60 | 125 | |
| 61–70 | 74 | |
| 71–80 | 52 | |
| 81–90 | 45 | |
| 91–100 | 28 | |

   a   Copy and complete the above table by calculating the cumulative frequency.
   b   Draw a cumulative frequency curve of the results.
   c   Using the graph, estimate a value for:
      i   the median exam mark
      ii  the upper and lower quartiles
      iii the interquartile range.

# STUDENT ASSESSMENTS 10

**7** Eight hundred students sit an exam. Their marks (as percentages) are shown in the table below:

| Mark (%) | Frequency | Cumulative frequency |
|---|---|---|
| 1–10 | 10 | |
| 11–20 | 30 | |
| 21–30 | 40 | |
| 31–40 | 50 | |
| 41–50 | 70 | |
| 51–60 | 100 | |
| 61–70 | 240 | |
| 71–80 | 160 | |
| 81–90 | 70 | |
| 91–100 | 30 | |

**a** Copy and complete the above table by calculating the cumulative frequency.
**b** Draw a cumulative frequency curve of the results.
**c** A grade 9 is awarded to any student achieving at or above the upper quartile. Using your graph, identify the minimum mark required for a grade 9.
**d** Any student below the lower quartile is considered to have failed the exam. Using your graph, identify the minimum mark needed so as not to fail the exam.
**e** How many students failed the exam?
**f** How many students achieved a grade 9?

# Glossary

$=$ $=$ means is equal to. For example, $3 + 4 = 7$.

$\neq$ $\neq$ means is not equal to. For example, $3 + 4 \neq 8$.

$>$ $>$ means is greater than. For example, $8 > 3 + 4$.

$<$ $<$ means is less than. For example, $3 + 4 < 8$.

$\geq$ $\geq$ means is greater than or equal to. For example, $x \geq 5$ means $x$ is any number greater than or equal to 5.

$\leq$ $\leq$ means is less than or equal to. For example, $x \leq 5$ means $x$ is any number less than or equal to 5.

$\in$ $\in$ means is an element of $e \in S$. So means the element e belongs to the set S.

$\notin$ $\notin$ means is NOT an element of. So $e \notin S$ means the element e does not belong to the set S.

$\subseteq$ $\subseteq$ means is a subset of. So $X \subseteq Y$ means X is a subset of Y.

$\nsubseteq$ $\nsubseteq$ means is NOT a subset of. So $X \nsubseteq Y$ means X is not a subset of Y.

$\subset$ $\subset$ means is a proper subset of. So $X \subset Y$ means X is a proper subset of Y.

$\not\subset$ $\not\subset$ means is NOT a proper subset of. So $X \not\subset Y$ means X is not a proper subset of Y.

$A \cap B$ $A \cap B$ means all the elements that belong to BOTH set $A$ and set $B$. $A \cap B$ denotes the elements that are in the intersection of $A$ and $B$ on a Venn diagram.

$A \cup B$ The union of sets $A$ and $B$, $A \cup B$, is all the elements that belong to EITHER set $A$ OR set $B$ OR both sets.

**n(A)** The number of elements in set A.

**A'** The complement of set A.

$A \subseteq B$ A is a subset of B.

$A \subset B$ A is a proper subset of B.

$A \nsubseteq B$ A is not a subset of B.

$A \not\subset B$ A is not a proper subset of B.

$\cup$ The universal set, $\cup$, for any particular problem is the set which contains all the possible elements for that problem.

**∅ or { }** The empty set is a set with no elements. It is written as ∅ or { }.

**12-hour clock** 12-hour clock is when the day is split into two halves 'am' and 'pm. The times before 12 noon are written using am and times after 12 noon are written as pm.

**24-hour clock** 24-hour clock is when the time is given as the number of hours that have passed since midnight. The hours part of the time is given two digits. For example, 01 30 is 1.30 am and 13 30 is 1.30 pm.

**accuracy** The accuracy of a measurement tells you how close the measurement is to the true value. For example, if you measure a pencil correct to the nearest centimetre, your measurement will be within 0.5 cm of the true measurement.

**acute angle** An acute angle lies between 0° and 90°.

**acute-angled triangle** In an acute-angled triangle, all three angles are less than 90°.

**addition** Addition is one of the four operations: addition, subtraction, multiplication and division. It means to find the total or sum of two or more numbers or quantities.

**adjacent** In a right-angled triangle, the adjacent is the side which is next to the angle.

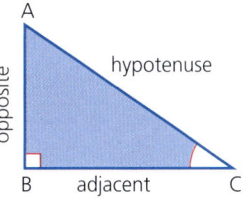

**algebraic fraction** An algebraic fraction is a fraction with a denominator that is an algebraic expression. For example, $\frac{3}{x}$ or $\frac{1}{x+3}$.

**alternate angles** Alternate angles are formed when a line crosses two other lines. Alternate angles are equal only if the lines are parallel. Look for a Z shape.

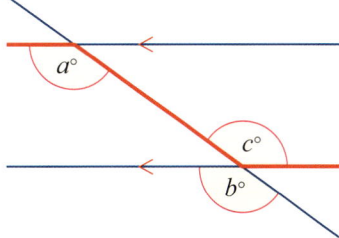

**altitude** The altitude of a triangle is the perpendicular height.

**angle of depression** The angle of depression is the angle below the horizontal through which a line of view is lowered.

**angle of elevation** The angle of elevation is the angle above the horizontal through which a line of view is raised.

601

**angles at a point** The angles at a point add up to 360°.

**angles on a straight line** The angles on a straight line add up to 180°.

**apex** The apex of a pyramid is the point where the triangular faces of the pyramid meet.

**arc** An arc is part of the circumference of a circle between two radii. When the angle between the two radii of length $r$ is $\emptyset$, then: arc length = $\frac{\emptyset}{360} \times 2\pi r$

**area** The area of a shape is the amount of surface that it covers. Area is measured in mm², cm², m², km², etc.

**area factor** When shape A is an enlargement by scale factor k of shape B, then the area factor is k2.

**area of a circle** The area, $A$, of a circle of radius r is: $A = \pi r^2$

**area of a parallelogram** The area, $A$, of a parallelogram of base length b and perpendicular height h is: $A = bh$

**area of a rectangle** The area, $A$, of a rectangle of length $l$ and breadth $b$ is: $A = lb$

**area of a trapezium** The area, $A$, of a trapezium is: $A = \frac{1}{2}h(a+b)$

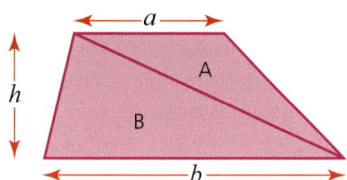

**area of a sector** The area of a sector is given by: $\frac{\emptyset}{360} \times \pi r^2$

**area of a triangle** The area, $A$, of a triangle of base $b$ and perpendicular height $h$ is: $A = \frac{1}{2}bh$

**area of any triangle** The area of any triangle is given by: area = $\frac{1}{2}ab\sin C$

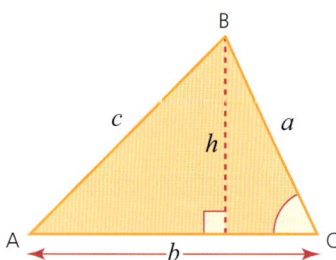

**asymptote** An asymptote is a line that a graph tends towards but does not meet. Here the x-axis and y-axis are both asymptotes:

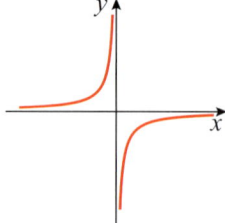

**average** An average is a measure of the typical value in a data set.

**average speed** average speed = $\frac{\text{total distance}}{\text{total time}}$

**back bearing** If the bearing of B from A is given, then the back bearing is the bearing of A from B. It is the bearing that takes you from B back to A. The back bearing is in the reverse direction to the original bearing – it represents the direction of the return journey.

**bar chart** A bar chart is a chart that uses rectangular bars to display data. The height of each bar represents the frequency.

**base** The base of a triangle is one of its sides. Any side can be the base, but the height must be measured perpendicular to the chosen base.

**basic pay** Basic pay is the fixed pay that an employee is given for working a certain number of hours.

**basic week** A basic week is the fixed number of hours that an employee is expected to work each week.

**bearing** A bearing is a direction. It is the angle measured clockwise from North. Bearings are given as 3 figures so, for example, for an angle of 45° the three-figure bearing is 045°.

**bisect** Bisect means to divide in half.

**bonus** A bonus is an extra payment that is sometimes added to an employee's basic pay.

**breadth** The breadth of a rectangle is the measure of its shortest side.

**capacity** The capacity of a container is the space inside that container.

**centre of enlargement** The centre of enlargement is a specific point about which an object is enlarged.

**centre of rotation** The centre of rotation is a specific 'pivot' point about which an object is rotated.

**changing the subject of a formula** This means rearranging that formula.

**chord** A chord joins two points on the circumference of a circle.

**circumference** The circumference is the perimeter of a circle.

**circumference of a circle** The circumference, C, of a circle of radius r is: $C = 2\pi r$

**co-interior angles** Co-interior angles face each other when two parallel lines are crossed by another diagonal line. Co-interior angles add up to 180°.

**column vector** A column vector describes the movement of an object in both the x direction and the y direction.

**common difference** The common difference, d, is the difference between one term and the next in an arithmetic sequence.

**common ratio** The common ratio, r, is the ratio between one term and the next in a geometric sequence.

**complement** The complement of set A is the set of elements which are in $\mathscr{E}$ but not in A. The complement of A is written as A'.

**composite bar chart** A composite bar chart shows the bars stacked on top of each other.

**composite function** A composite function is when one function is applied to the results of another function. fg(x) means apply function g first, and then apply function f to the result.

**compound interest** Compound interest is interest that is paid not only on the principal amount, but also on the interest itself. So the amount of interest earned each year increases.

**compound measure** A compound measure is one made up of two or more other measures.

**conditional probability** Conditional probability is when the probability of event A happening is changed by event B happening.

**cone** A cone is a like a pyramid, but with a circular base.

**congruent** Congruent shapes are exactly the same shape and size – they are identical.

**constant of proportionality** When two quantities, x and y, are in direct proportion, $\frac{y}{x} = k$ (or $y = kx$), where k is the constant of proportionality.

**construction** A construction is an accurate drawing made using a ruler and a pair of compasses.

**continuous data** Continuous data is numerical data that can take on any value in a certain range. For example, height and weight (mass) are continuous data.

**conversion graph** A conversion graph is a straight-line graph used to convert one set of units to another.

**correlation** Correlation is the relationship between two sets of data.

**corresponding angles** Corresponding angles are formed when a line crosses a pair of parallel lines. Corresponding angles are equal. Look for an F shape.

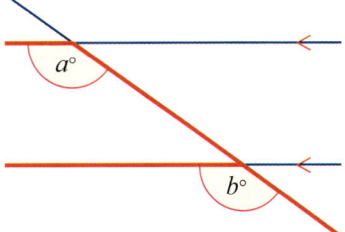

**cosine** The cosine of an angle, cos θ, in a right-angled triangle is the ratio of the adjacent side to the hypotenuse: $\cos\theta = \frac{\text{length of adjacent side}}{\text{length of hypotenuse}}$

**cosine rule** The cosine rule is: $a^2 = b^2 + c^2 - 2bc\cos A$ or $\cos A = \frac{b^2 + c^2 - a^2}{2bc}$.

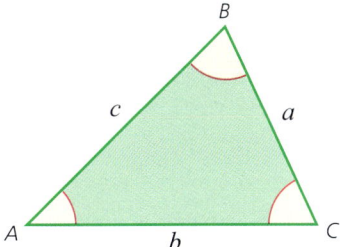

**cost price** The cost price is the total amount of money that it costs to produce a good or service, before any profit is made.

**cube number** A cube number is the result when an integer is multiplied by itself twice. The cube numbers are 1, 8, 27, 64, 125, …

**cube root** The cube root of a number is the number which when multiplied by itself twice gives the original number. The inverse of cubing is cube rooting. For example, the cube root of 27 is 3 (as $3 \times 3 \times 3 = 27$). The symbol $\sqrt[3]{\phantom{x}}$ is used for the cube root of a number, so $\sqrt[3]{27} = 3$.

**cubic** A cubic function has the form $f(x) = ax^3$

**cubic curve** A cubic curve is the graph of a cubic function.

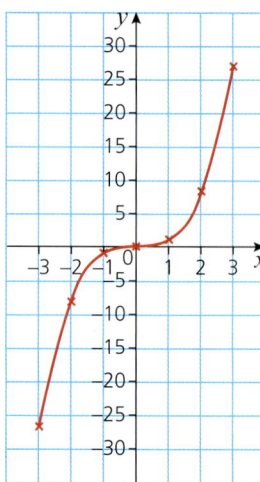

**cuboid** A cuboid is a prism with a rectangular cross-section.

**cumulative frequency** The cumulative frequency is the running total of the frequencies in a data set.

**cyclic quadrilateral** A cyclic quadrilateral is a quadrilateral whose vertices all lie on the circumference of a circle.

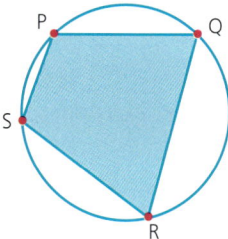

**cylinder** A cylinder is three-dimensional shape with a constant circular cross-section.

**decagon** A decagon is a 10-sided polygon.

**decimal** A decimal is a number with digits after the decimal point. It is a number which is not an integer.

**decimal place** The decimal place is the number of digits after the decimal point. For example, 3.2 has 1 decimal place and 5.678 has 3 decimal places.

**denominator** The denominator is the bottom line of a fraction; it tells you how many equal parts the whole is divided into. For example, $\frac{3}{8}$ has a denominator of 8, so the 'whole' has been divided into 8 equal parts.

**deposit** A deposit can have different meanings depending on the context. It can refer to money being put into a bank account. It can also mean paying a sum of money in advance in order to secure a payment.

**depreciate** When the value of something decreases over a period of time, it is said to depreciate.

**diameter** A diameter is a line which passes through the centre of a circle and joins two points on the circumference.

**difference of two squares** The difference of two squares is an expression in the form $x^2 - y^2$. Note, $x^2 - y^2$ factorises to give $(x + y)(x - y)$.

**direct proportion** Two quantities, $x$ and $y$, are in direct proportion when the ratio $\frac{y}{x}$ is a constant. So $y = kx$ and the graph of $y$ against $x$ is a straight line passing through the origin. An increase in one quantity causes an increase in the other. For example, when the amount of ingredients is doubled for some cakes, the number of cakes made also doubles.

**directed number** A directed number is a number that is positive or negative. A number has size (magnitude) and its sign (+ or −) tells you which *direction* to move along a number line from 0 in order to reach that number.

**discount** An item sold at 10% discount is 10% cheaper than the full selling price.

**discrete data** Discrete data is numerical data that can only take on certain values, usually whole numbers. For example, the number of peas in a pod is discrete data.

**distance between two points** The distance, $d$, between two points $(x_1, y_1)$ and $(x_2, y_2)$ is:
$$d = \sqrt{(x_1 - x_2)^2 + (y_1 - y_2)^2}$$

**division** Division is one of the four operations: addition, subtraction, multiplication and division. To divide one number by another means to find how many times one number goes into another number.

**dodecagon** A dodecagon is a 12-sided polygon.

**double time** Overtime is often paid at a higher rate. When overtime is paid at twice the basic pay, it is called double time.

**dual bar chart** A dual bar chart shows the bars side by side.

**element** An object or symbol in a set is called an element.

**elevation** An elevation is a two-dimensional view of a three-dimensional object. A side elevation is the view from one side of the object and the front elevation is the view from the front.

**elimination method** The elimination method is a method for solving simultaneous equations. One of the unknowns is eliminated by either adding or subtracting the pair of equations.

**empty set** If a set is empty, it is called the empty set. It is represented by the symbol ∅.

**enlargement** An enlargement changes the size of an object. When a shape is enlarged, the image is mathematically similar to the object but is a different size. Note: the image may be larger or smaller than the original object.

**equation** An equation says that one expression is equal to another. For example, $6 + 4 = 16 - 6$. When an expression contains an unknown, it can be solved. For example, the solution to the equation $x + 4 = 16 - x$ is $x = 6$.

**equation of a straight line** The equation of a straight line can be written in the form $y = mx + c$, where $m$ is the gradient of the line and the $y$-intercept is at $(0, c)$.

**equilateral triangle** An equilateral triangle has three equal angles (all 60°) and three sides of equal length.

**equivalent fraction** Equivalent fractions have the same decimal value. For example, $\frac{3}{5} = 0.6$ and $\frac{9}{15} = 0.6$, so $\frac{3}{5}$ and $\frac{9}{15}$ are equivalent.

**estimate / estimation** Estimation is a way of working out the approximate answer or estimate to a calculation. The numbers in the calculation are rounded (usually to 1 significant figure) so that the calculation is easier to work out without a calculator.

**evaluate** Evaluate means to work out the value of something.

**expand** Expand means to multiply out or remove the brackets.

**exponential** An exponential expression is in the form $a^x$ where $a$ is a positive constant and $x$ is a variable.

**exponential equation** An exponential equation has a variable (unknown) as the index. For example, $y = 2^x$ is an exponential equation.

**exponential function** An exponential function is in the form $f(x) = a^x$, where $a$ is a positive constant.

**exponential sequence** An exponential sequence is a sequence where there is a common ratio ($r$) between successive terms. The $n$th term can be written as $T_n = ar^{n-1}$, where $a$ is the first term.

**expression** An expression is an algebraic statement. It does not have an equal sign (unlike an equation or a formula).

**exterior angle** The exterior angle of an $n$-sided convex regular polygon = $\frac{360°}{n}$.

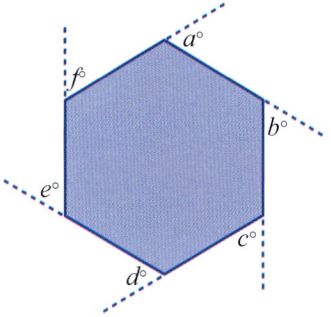

**factor** A factor of a number divides into that number exactly. For example, the factors of 18 are 1, 2, 3, 6, 9 and 18.

**factorise** Factorise means to factorise fully, removing common factors and writing an equivalent expression using brackets. For example, $2x - 6$ factorises to give $2(x - 3)$.

**favourable outcome** A favourable outcome refers to the event in question (for example, getting a 6 when a dice is thrown) actually happening.

**fraction** A fraction represents a part of a whole.

**frequency** Frequency is the number of times a particular outcome happens.

**frequency table** A frequency table shows the frequency of each data value in a data set.

**frustum** A frustum is the base part of a cone or pyramid when the top of the cone or pyramid is removed.

**gradient** Gradient is a measure of how steep a line is. The gradient of the line joining the points $(x_1, y_1)$ and $(x_2, y_2)$ is given by: gradient $= \frac{y_2 - y_1}{x_2 - x_1}$

**gradient-intercept form** The gradient-intercept form of the equation of a straight line is the form $y = mx + c$, where $m$ is the gradient and $c$ is the $y$-intercept.

**gross earnings** Gross earnings are the total earnings *before* all the deductions such as tax, insurance and pension contributions are made.

**grouped frequency table** A grouped frequency table is a method of displaying a large data set so that it is easier to handle.

**height** The height of a triangle is the perpendicular distance from its base to its third vertex.

**hemisphere** A hemisphere is made when a sphere is cut into two congruent halves. It is a half a sphere.

**hexagon** A hexagon is a 6-sided polygon.

**highest common factor** The highest common factor (HCF) of two numbers is the greatest integer that divides exactly into both numbers. For example, the highest common factor of 6 and 15 is 3.

**histogram** A histogram is a chart used to display grouped continuous data as bars. Both axes have continuous scales, and the vertical axis shows the frequency density of each bar.

**hypotenuse** The hypotenuse is the longest side of a right-angled triangle.

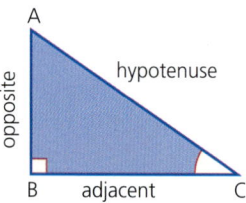

**image** When an object undergoes a transformation, the resulting position or shape is the image.

**improper fraction (or vulgar fraction)** In an improper fraction, the numerator is more than the denominator. For example, is an improper fraction.

**index** The index is the power to which a number is raised. For example, in $4^3$ the power (or index) is 3 and so $4^3 = 4 \times 4 \times 4$.

**inequality** An inequality compares the size of two values, showing if one is less than, or greater than, the other. For example, $x + 2 < 8$ or $7 > 6$

**integer** An integer is a positive or negative whole number (including zero). The set of integers is given the symbol $\mathbb{Z}$, where $\mathbb{Z} = \{..., -3, -2, -1, 0, 1, 2, 3, ...\}$.

**interest** Interest is the money added by a bank to a sum deposited by a customer. Interest is also the money charged by a bank for a loan to a customer.

**interior angle** The sum of the interior angles of an $n$-sided polygon is $180(n-2)°$.

**interquartile range** interquartile range = upper quartile – lower quartile

**intersection** The intersection of two sets is the elements that are common to both sets. It is represented by the symbol $\cap$.

**inverse of a function** The inverse of a function is its reverse, i.e. it 'undoes' the function's effects. The inverse of the function $f(x)$ is written $f^{-1}(x)$.

**inverse proportion** Two quantities, $x$ and $y$, are in inverse proportion when the product of the two quantities $xy$ is constant, i.e. when an increase in one quantity causes a decrease in the second quantity.

**irrational number** An irrational number is any number (positive or negative) that cannot be written as a fraction. Any decimal which neither terminates nor recurs is irrational. The square root of any number other than square numbers is also irrational. Some examples of irrational numbers are $\pi$, $\sqrt{2}$ and $\sqrt{10}$.

**irregular polygon** An irregular polygon is a polygon that does not have equal sides or equal angles.

**isosceles trapezium** An isosceles trapezium is a quadrilateral with the following properties:
- one pair of parallel sides
- the other pair of sides are equal in length
- two pairs of equal base angles
- opposite base angles add up to 180°.

**isosceles triangle** An isosceles triangle has two equal angles and two sides of equal length.

**kite** A kite is a quadrilateral with the following properties:
- two pairs of equal sides
- one pair of equal angles
- diagonals which cross at right angles.

**laws of indices** The laws of indices are:
- $a^m \times a^n = a^{m+n}$
- $a^m \times a^n$ or $\dfrac{a^m}{a^n} = a^{m-n}$
- $(a^m)^n = a^{mn}$
- $a^1 = a$
- $a^0 = 1$
- $a^{-m} = \dfrac{1}{a^m}$
- $a^{\frac{1}{n}} = \sqrt[n]{a}$
- $a^{\frac{m}{n}} = \sqrt[n]{(a^m)}$ or $\left(\sqrt[n]{(a)}\right)^m$

**length** The length of a rectangle is the measure of its longest side.

**line** A line is a one-dimensional object with length but no width. It has infinite length.

**line of best fit** A line of best fit is a straight line that passes through the points on a scatter diagram as closely as possible.

**line of symmetry** A line of symmetry divides a two-dimensional shape into two congruent (identical) shapes.

**line segment** A line segment is part of a line.

**line symmetry** A shape has line symmetry if it has one or more lines or planes of symmetry

**linear** A linear function is a function whose graph is a straight line. It is of the form $f(x) = ax$.

**linear equation** The graph of a linear equation is a straight line. The highest power of the variable is 1.

**linear function** The graph of a linear function is a straight line.

**linear inequality** A linear inequality is an inequality which involves a linear function. In a linear inequality, the highest power of the variable is 1.

**linear sequence** A linear sequence is a sequence where the difference between any two terms is a constant.

**local maximum** A local maximum of the function f is a stationary point where f(x) reaches a maximum within a given range.

**local minimum** A local minimum of the function f is a stationary point where f(x) reaches a minimum within a given range.

**logarithm** A logarithm is the power to which a base number must be raised to make a given number.

**loss** When an item is sold for less than it cost to make, it is sold at a loss: loss = cost price − selling price

**lower quartile** The lower quartile is the 25th percentile.

**lowest common multiple** The lowest common multiple (LCM) of two numbers is the lowest integer that is a multiple of both numbers. For example, the lowest common multiple of 6 and 15 is 30.

**lowest terms (or simplest form)** A fraction is in its lowest terms when the highest common factor of the numerator and denominator is 1. In other words, the fraction cannot be cancelled down any further. For example, $\frac{30}{45} = \frac{6}{9} = \frac{2}{3}$, so $\frac{2}{3}$ is a fraction in its lowest terms.

**magnitude** Magnitude means size.

**mean** The mean is found by adding together all of the data values and then dividing this total by the number of data values. The mean is one of the three main ways to measure an average.

**median** The median is the middle value when the data set is organised in order of size. The median is one of the three main ways to measure an average.

**metric units of capacity** 1 litre (l) = 1000 millilitres (ml) 1 ml = 1 cm$^3$

**metric units of length** 1 kilometre (km) = 1000 metres (m) 1 metre (m) = 100 centimetres (cm) 1 centimetre (cm) = 10 millimetres (mm)

**metric units of mass** 1 tonne (t) = 1000 kilograms (kg) 1 kilogram (kg) = 1000 grams (g) 1 gram (g) = 1000 milligrams (mg)

**midpoint** The midpoint of a line segment AB, where $A(x_1, y_1)$ and $B(x_2, y_2)$ is: $\left( \frac{x_1 + x_2}{2}, \frac{y_1 + y_2}{2} \right)$

**mirror line** The mirror line is the line about which an object is reflected.

**mixed number** A mixed number is made up of a whole number and a proper fraction. For example, $2\frac{3}{8}$ is a mixed number.

**modal class** The modal class is the class or group in a grouped frequency table with highest frequency.

**mode** The mode is the value occurring most often in a data set. The mode is one of the three main ways to measure an average.

**multiple** The multiple of a number is the result when you multiply that number by a positive integer. For example, the multiples of 6 are 6, 12, 18, 24, 30, …

**multiplication** Multiplication is one of the four operations: addition, subtraction, multiplication and division. Multiplication is repeated addition, so 3 multiplied by 4 means 3 + 3 + 3 + 3.

**natural number** A natural number is a whole number (integer) that is used in counting. Natural numbers start at zero and continue 0, 1, 2, 3, … The set of natural numbers is given the symbol $\mathbb{N} = \{…, -3, -2, -1, 0, 1, 2, 3, …\}$.

**negative correlation** Two quantities have negative correlation if, in general, one decreases as the other increases.

**negative enlargement** In a negative enlargement, the object and image are on opposite sides of the centre of enlargement. The scale factor of enlargement is negative.

**negative number** A negative number is any number less than 0.

**net** A net is a two-dimensional shape which can be folded up to form a three-dimensional shape.

**net pay** Net pay is sometimes called 'take-home' pay. It is the money left *after* all the deductions such as tax, insurance and pension contributions are made.

**no correlation** Correlation is the relationship between two sets of data. If there is no correlation, then there is no relationship between the two data sets. In a scatter graph showing no correlation, there is no pattern in the plotted points.

**numerator** The numerator is the top line of a fraction. It represents the number of equal parts of the whole. For example, $\frac{3}{8}$ has a numerator of 3, so there are 3 equal parts and each part is equal to $\frac{1}{8}$ of the 'whole'.

**obtuse angle** An obtuse angle lies between 90° and 180°.

**obtuse-angled triangle** In an obtuse-angled triangle, one angle is greater than 90°.

**octagon** An octagon is an 8-sided polygon.

**opposite** In a right-angled triangle, the opposite side is the one which is opposite the angle.

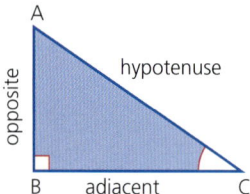

**order of operations** When a calculation contains a mixture of brackets and/or the operations (×, ÷, + and −), the order that the operations should be carried out in is:

First work out any … Brackets
… then carry out any … Multiplication and Division
… finally Addition and Subtraction

When a calculation contains operations of equal priority (e.g. + and −, or × and ÷), work from left to right. For example, $10 - 7 + 2 = 3 + 2 = 5$.

**order of rotational symmetry** The order of rotational symmetry is the number of times a shape, when rotated about a central point, fits its outline during a complete revolution.

**origin** The origin is the point at which the $x$-axis and the $y$-axis meet.

**overtime** Overtime is any hours worked in excess of the basic week.

**parallel** A pair of parallel lines can be continued to infinity in either direction without meeting. Parallel lines have the same gradient.

**parallelogram** A parallelogram is a quadrilateral with the following properties:
- two pairs of parallel sides
- opposite sides are equal
- opposite angles are equal.

**pentagon** A pentagon is a 5-sided polygon.

**per cent (%)** Per cent means parts per 100.

**percentage** A percentage is the number of parts per 100.

**percentage interest (or interest rate)** Interest is earned on a fixed percentage of the principal. The interest rate gives the percentage interest earned.

**percentage loss** percentage loss = $\frac{\text{loss}}{\text{cost price}} \times 100\%$

**percentage profit**

percentage profit = $\frac{\text{profit}}{\text{cost price}} \times 100\%$

**percentile** The cumulative frequency can be divided into percentiles. The maximum value of the cumulative frequency is the 100th percentile.

**perfect square** A quadratic equation is called a perfect square if it is in the form $y = x^2 + 2ax + a^2$, where $a$ is a constant. This factorises to give $y = (x + a)^2$.

**perimeter** The perimeter of a shape is the distance around the outside edge of the shape. Perimeter is measured in mm, cm, m, km, etc.

**perimeter of a rectangle** The perimeter of a rectangle of length $l$ and breadth $b$ is: $2l + 2b$

**perpendicular** Two lines are perpendicular if they meet at right angles. The product of the gradients of two perpendicular lines is −1. So if the gradient of a line is $m_1$, then the gradient of a line perpendicular to it is: $m_2 = -\frac{1}{m_1}$

**perpendicular bisector** The perpendicular bisector of a line $AB$ is another line which meets $AB$ at right angles and cuts $AB$ exactly in half.

**pictogram** A pictogram is a chart that uses pictures or symbols to display data.

**pie chart** A pie chart is a circular chart divided into sectors that is used to display data. The area of each sector is proportional to the frequency.

**piece work** Piece work is when an employee is paid for the number of articles made (rather than the time spent working).

**plan** A plan of an object is a scale diagram of the view from above the object, looking directly down on the object.

**plane of symmetry** A plane of symmetry divides a three-dimensional shape into two congruent (identical) three-dimensional shapes.

**point** A point is an exact location or position.

**polygon** A polygon is a closed two-dimensional shape made up of straight lines.

**polynomials** A polynomial function is a function such as a quadratic, cubic, etc. that includes only non-negative powers of $x$.

**positive correlation** Two quantities have positive correlation if, in general, one increases as the other increases.

**positive number** A positive number is any number greater than 0.

**power** For example, in $4^3$ the power is 3 and so $4^3 = 4 \times 4 \times 4$.

**prime factor** A prime factor of a number is any factor of that number that is also a prime. For example, the prime factors of 60 are 2, 3 and 5.

**prime number** A prime number is a number with exactly two factors: one and itself. The prime numbers are 2, 3, 5, 7, 11, … Note: 1 is not a prime number as it only has one factor.

**principal** The principal is the amount of money deposited by a customer in a bank account.

**prism** A prism is a three-dimensional object with a constant cross-sectional area.

**probability** Probability is the study of chance. The probability of an event happening is a measure of how likely that event is to happen. Probability is given on a scale of 0 (an impossible event) to 1 (a certain event): probability

$$= \frac{\text{number of favourable outcomes}}{\text{total number of equally likely outcomes}}$$

**probability scale** A probability scale is a scale that indicates how likely an event is, ranging from impossible to certain.

**product (e.g. of prime factors)** The product of two numbers is the answer when they are multiplied together.

**profit** When an item is sold for more than it cost to make, it is sold at a profit:
profit = selling price − cost price

**proper fraction** In a proper fraction, the numerator is less than its denominator. For example, $\frac{3}{8}$ is a proper fraction.

**pyramid** A pyramid is a three-dimensional shape. It has a polygon for a base and the other faces are triangles which meet at a common vertex, called the apex.

**Pythagoras' theorem** Pythagoras' theorem states the relationship between the lengths of the three sides of a right-angled triangle. Pythagoras' theorem is: $a^2 = b^2 + c^2$

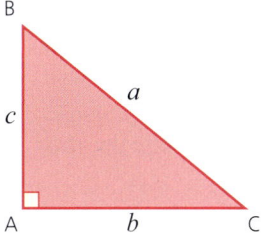

**quadratic equation** A quadratic equation can be written in the form $ax^2 + bx + c = 0$, where $a$, $b$ and $c$ are constants.

**quadratic expression** In a quadratic expression, the highest power of any of the terms is 2. A quadratic expression can be written in the form $ax^2 + bx + c$, where $a$, $b$ and $c$ are constants.

**quadratic formula** The quadratic formula is used to solve a quadratic equation in the form $ax^2 + bx + c = 0$. The formula is: $x = \frac{-b \pm \sqrt{b^2 - 4ac}}{2a}$

**quadratic function** A quadratic function is in the form $y = ax^2 + bx + c$, where $a$, $b$ and $c$ are constants.

**quadrilateral** A quadrilateral is a 4-sided polygon.

**radius** A radius is a line which joins the centre of a circle to a point on the circumference.

**range** Range is a measure of the spread of a data set. The range is the difference between the largest and smallest data values.

**rate** Rate is a ratio of two measurements, usually the second measurement is time. For example, water flows through a pipe at a rate of 1 litre per second or a computer programmer types at a rate of 30 words per minute.

**ratio** A ratio is the comparison of one quantity with another.

**ratio method** The ratio method is used to solve problems involving direct proportion by comparing the ratios. For example, a bottling machine fills 500 bottles in 15 minutes. How many bottles will it fill in 90 minutes?

$$\frac{x}{90} = \frac{500}{15} \text{ so } x = \frac{500 \times 90}{15} = 3000$$

3000 bottles are filled in 90 minutes.

**rational number** A rational number is any number (positive or negative) that can be written as a fraction. All integers and all terminating and recurring decimals are rational numbers. The set of rational numbers is given the symbol $\mathbb{Q}$.

**real number** The real numbers are the all the rational and irrational numbers. The set of real numbers is given the symbol R. So any integer, fraction or decimal is a real number.

**reciprocal** The reciprocal of a number is 1 divided by that number. So the reciprocal of 4 is $1 \div 4 = \frac{1}{4} = 0.25$ and the reciprocal of $\frac{1}{5}$ is $1 \div \frac{1}{5} = 5$.

**reciprocal function** A reciprocal function is in the form $y = \frac{k}{x}$, where $k$ is a constant.

**rectangle** A rectangle is a quadrilateral with the following properties:

- two pairs of parallel sides
- opposite sides are equal
- four equal angles (each 90°).

**recurring decimal** A recurring decimal has digits that repeat forever. For example, $\frac{2}{9} = 0.2222\ldots = 0.\dot{2}$ and $\frac{415}{999} = 0.415415415\ldots = 0.\dot{4}1\dot{5}$

**reflection** A reflection is a 'flip' movement about a mirror line. The mirror line is the line of symmetry between the object and its image.

**reflex angle** A reflex angle lies between 180° and 360°.

**region** A region is a part of a graph or shape.

**regular polygon** A regular polygon has all sides of equal length and all angles of equal size.

**relative frequency**
$$\text{relative frequency} = \frac{\text{number of successful trials}}{\text{total number of trials}}$$

**rhombus** A rhombus is a quadrilateral with the following properties:
- two pairs of parallel sides
- four equal sides
- opposite angles are equal
- diagonals which cross at right angles.

**right angle** A right angle is 90°.

**right-angled triangle** In a right-angled triangle, one angle is 90°.

**rotation** A rotation is a 'turning' movement about a specific point known as the centre of rotation.

**rotational symmetry** A shape has rotational symmetry if, when rotated about a central point, it fits its outline.

**round (or rounding)** Rounding is a way of rewriting a number so it is simpler than the original number. A rounded number should be approximately equal to the unrounded (exact) number and be of the same order of magnitude (size). Rounded numbers are often given to 2 decimal places (2 d.p.) or 3 significant figures (3 s.f.), for example.

**sample space diagram** A sample space diagram shows all the possible outcomes of an experiment.

**scalar** A scalar is a quantity with magnitude (size) only.

**scale** A scale on a drawing shows the ratio of a length on the drawing to the length on the actual object.

**scale factor of enlargement** The scale factor of enlargement is the ratio between corresponding sides on an object and its image.

**scalene triangle** In a scalene triangle, none of the angles are of equal size and none of the sides are of equal length.

**scatter diagram** A scatter diagram is a graph of plotted points which shows the relationship between two data sets.

**sector** A sector is the region of a circle enclosed by two radii and an arc.

**segment** A segment is an area of a circle formed by a line (chord) and an arc.

**selling price** The selling price is the total amount of money that an item is sold for.

**semicircle** A semicircle is made when a circle is cut into two congruent halves. A semicircle is half a circle.

**sequence** A sequence is a collection of terms arranged in a specific order, where each term is obtained according to a rule.

**set** A set is a well-defined group of objects or symbols.

**significant figures** The first significant figure of a number is the first non-zero digit in the number. The second significant figure is the next digit in the number, and so on. For example, in the numbers 78 046 and 0.0078 046 the first significant figure is 7, the second significant figure is 8 and the third significant figure is 0.

**similar** Two shapes are similar if the corresponding angles are equal and the corresponding sides are in proportion to each other.

**simple interest** Simple interest is calculated only on the principal (initial) amount deposited in an account. When simple interest is earned, the amount of interest paid is the same each year.

$$\text{simple interest} = \frac{\text{principal} \times \text{time in years} \times \text{rate per cent}}{100}$$

**simplest form (or lowest terms)** A fraction is in its simplest form when the highest common factor of the numerator and denominator is 1. In other words, the fraction cannot be cancelled down any further. For example, $\frac{30}{45} = \frac{6}{9} = \frac{2}{3}$, so $\frac{2}{3}$ is a fraction in its simplest form.

**simultaneous equations** Simultaneous equations are a pair of equations involving two unknowns.

**sine** The sine of an angle, $\sin \theta$, in a right-angled triangle is the ratio of the side opposite the angle and the hypotenuse.

$$\sin \theta = \frac{\text{length of opposite side}}{\text{length of hypotenuse}}$$

**sine rule** The sine rule is: $\frac{a}{\sin A} = \frac{b}{\sin B} = \frac{c}{\sin C}$ or: $\frac{\sin A}{a} = \frac{\sin B}{b} = \frac{\sin C}{c}$

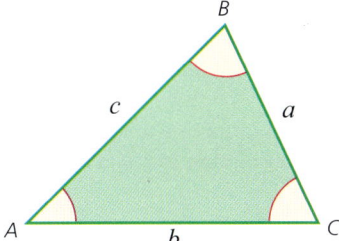

**speed** speed = $\frac{\text{distance}}{\text{time}}$ When the speed is not constant: average speed = $\frac{\text{total distance}}{\text{total time}}$

**sphere** A sphere is a three-dimensional shape which is a ball.

**square** A square is a quadrilateral with the following properties:
- two pairs of parallel sides
- four equal sides
- four equal angles (90°)
- diagonals which cross at right angles.

**square number** A square number is the result when an integer is multiplied by itself. The square numbers are 1, 4, 9, 16, 25, …

**square root** The square root of a number is the number which when multiplied by itself gives the original number. The inverse of squaring is square rooting. Every number has two square roots, for example, the square root of 9 is 3 (as $3 \times 3 = 9$) and $-3$ (as $-3 \times -3 = 9$). The symbol $\sqrt{\phantom{x}}$ is used for the positive square root of a number, so $\sqrt{9} = 3$.

**standard form** Standard form is a way of writing very large or very small numbers. A number in standard form is written as $A \times 10^n$, where $1 \leq A < 10$ and $n$ is a positive or negative integer. Examples of numbers in standard from are $5 \times 10^3$ and $2.7 \times 10^{-18}$.

**stem and leaf diagram** A stem and leaf diagram is a diagram where each date value is split into two parts – the 'stem' and the 'leaf' (usually the last digit). The data is then grouped so that data values with the same stem appear on the same line.

**strong negative correlation** In a scatter graph with strong negative correlation:
- as variable $x$ increases, variable $y$ decreases
- all points lie closely to the line of best fit.

**strong positive correlation** In a scatter graph with strong positive correlation:
- as variable $x$ increases, so does variable $y$
- all points lie closely to the line of best fit.

**subject** The subject of a formula is the single variable (often on the left-hand side of a formula) that the rest of the formula is equal to. For example, in $C = 2\pi r$, $C$ is the subject and in $a^2 + b^2 = c^2$, $c^2$ is the subject.

**subset** When all the elements of set X are also elements of set Y, then X is a subset of Y. Every set has itself and the empty set as subsets.

**substitute / substitution** Substitution is replacing the variables (letter symbols) in an expression or formula with numbers.

**substitution method** The substitution method is a method for solving simultaneous equations, where one unknown is made the subject of one of the equations, and then this expression is substituted into the second equation.

**subtraction** Subtraction is one of the four operations: addition, subtraction, multiplication and division. It means to take one number away from another.

**supplementary angle** Two angles that add together to total 180° are called supplementary angles.

**surd** A surd is a square root or cube root of a number which cannot be simplified by removing the root. A surd is an irrational number. For example, $\sqrt{4}$ is not a surd as $\sqrt{4} = 2$. $\sqrt{5} = 2.2360679…$ is a surd.

**surface area of a cuboid** The surface area of a cuboid of length $l$, width $w$ and height $h$ is: surface area = $2(wl + lh + wh)$

**surface area of a cylinder** The surface area of a cylinder of radius $r$ and height $h$ is: surface area = $2\pi r(r + h)$

**surface area of a sphere** The surface area of a sphere is $4\pi r^2$.

**tally table** A tally table is table where the frequencies of each outcome are recorded using marks like ||| for 3 or ||||| for 5.

**tangent** The tangent to a curve at a point is a straight line that just touches the curve at that point. The gradient of the tangent is the same as the gradient of the curve at that point.

**tangent ($\theta$)** The tangent of an angle, $\tan \theta$, in a right-angled triangle is the ratio of the sides opposite and adjacent to the angle.

$\tan \theta = \frac{\text{length of opposite side}}{\text{length of adjacent side}}$

**term** Each number in a sequence is a called a term.

**terminating decimal** A terminating decimal has digits after the decimal point that do not continue forever. For example, 0.123 and 0.987654321.

**term-to-term rule** A term-to-term rule describes how to use one term in a sequence to find the next term.

**three-figure bearing** A three-figure bearing is a measure of the direction in which an object is travelling. North is 000° and South is 180°.

**time and a half** Overtime is often paid at a higher rate. When overtime is paid at 1.5 × basic pay, it is called time and a half.

**total number of possible outcomes** The total number of possible outcomes refers to all the different types of outcomes one can get in a particular situation.

**transformation** A transformation changes either the position or shape of an object.

**translation** A translation is a sliding movement. Each point on the object moves in the same way to its corresponding point on the image, as described by its translation vector.

**translation vector** A translation vector describes a translation in terms of its horizontal and vertical movement. For example, the translation vector $\begin{pmatrix} 2 \\ -3 \end{pmatrix}$ describes a translation of 2 units right and 3 units down.

**trapezium** A trapezium is a quadrilateral with one pair of parallel sides.

**travel graph** A travel graph is a diagram showing the journey of one or more objects on the same pair of axes. The vertical axis is distance and the horizontal axis is time.

**triangle** A triangle is a 3-sided polygon.

**triangle numbers** Triangle numbers form the sequence 1, 3, 6, 10, 15... The difference between consecutive numbers increases by one each time.

**turning point** The turning point of a quadratic graph is its highest or lowest point. If the $x^2$ term is positive, the graph will have a lowest point. If the $x^2$ term is negative, it will have a highest point.

**two-way table** A two-way table is a way of displaying data from two categories.

**union** The union of two sets is everything that belongs to EITHER or BOTH sets. It is represented by the symbol ∪.

**universal set** The universal set for any particular problem is the set which contains all the possible elements for that problem. It is represented by the symbol ∪.

**upper quartile** The upper quartile is the 75th percentile.

**vector** A vector is a quantity with both magnitude (size) and direction. A vector can be used to describe the position of one point in space relative to another.

**Venn diagram** A Venn diagram is a diagram comprising of overlapping circles, which is used to display sets.

**vertex (plural: vertices)** A vertex of a shape is a point where two sides meet.

**vertically opposite angles** Vertically opposite angles are formed when two lines cross. Vertically opposite angles are equal.

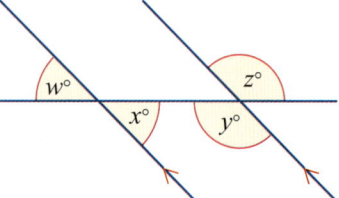

**volume** The volume of a 3D solid is the amount of space the solid fills.

**volume factor** When shape A is an enlargement by scale factor k of shape B, the volume factor is $K^3$.

**volume of a cylinder** The volume of a cylinder of radius $r$ and height $h$ is given by: $volume = \pi r^2 h$

**volume of a cone** The volume of a cone with height $h$ and a base of radius $r$ is given by: $volume = \frac{1}{3}\pi r^2 h$

**volume of a prism** The volume of a prism is given by: volume = area of cross-section × length

**volume of a pyramid** The volume of pyramid is given by: volume = $\frac{1}{3}$ × area of base × perpendicular height

**volume of a sphere** The volume of a sphere is given by: $volume = \frac{4}{3}\pi r^3$

**x-axis** The x-axis is the horizontal axis on a graph.

**y-axis** The y-axis is the vertical axis on a graph.

**zero (of an equation)** The zero(s) of an equation are the value(s) of $x$ when $y = 0$. On a graph, these are the values of $x$ where the curve crosses the $x$-axis. These are also sometimes known as roots of an equation.

# Index

## A

abacuses 3
acceleration 137
accuracy 544, 548–9
   appropriate 65
   continuous data 556
   grouping data 576
addition
   of algebraic fractions 129–31
   of fractions 40
   and order of operations 31
   of vectors 489
al-Karaji, Muhammad 115
al-Khwarizmi 115
al-Kindi 555
algebra 24, 103–5, 114–205, 207
   algebraic fractions 128–31
   algebraic representation and manipulation 116–31
   changing the subject of formulas 119–20, 126–7
   complex equations 147–49
   difference of two squares 123
   direct variation 189–93
   evaluation 123–4
   expansion of brackets 116–17, 121–2
   factorisation 117–18, 122–6
   formula construction 138–40
   indices 174–6
   inequalities 162–73, 197
   inverse variation 191–4
   linear equations 132–3
   quadratic equations 124–6, 150–6
   sequences 177–88
   simultaneous equations 141–6
   substitution 118–19
alternate segment theorem 347–9
altitude 304
amplitude 464–5, 470
angles 304
   acute 304
   alternate 319, 324
   alternate segment theorem 347–49

arc length 377
   at a point 320–2
   between a line and a plane 460–3
   of a circle 341–9, 377, 379
   co-interior 325
   and construction of simple equations 133–7
   corresponding 324
   of depression 451–5
   of elevation 451–5
   exterior 327–8, 332
   interior 327–8, 331–2
   measurement and drawing of 316–8
   obtuse 304
   opposite 346–7
   of pie charts 560–1
   and polygons 326–33
   properties 320–33
   reflex 304
   right 304–5, 347–8, 416–22, 423–31, 441–2, 444–5, 449, 460
   of rotation 499–500
   special 432–40
   on a straight line 320–22
   supplementary 325, 347
   vertically opposite 323
   within parallel lines 322–6
   see also trigonometry
apex 393
Apollonius of Perga 303
approximation 65–9
Archimedes 303
arcs 306, 341, 344–5
   arc length 377–9
   major/minor 306, 341, 377
   and sector area 379
area
   of a circle 373–6
   conversions 368–9
   formula 139
   of a parallelogram 382–4
   of a sector 379–81
   of similar shapes 338–40, 370–2
   of a trapezium 382–4
   of a triangle 449–50
   see also surface area
area factor 334, 338

Aryabatta 23
asymptotes 218–9, 436, 465
averages 17–21, 570–4, 5760–8, 580–1, 583–4
axes
   asymptotes 218
   rotational symmetry 313
   x-axis 276–7, 284
   y-axis 228–9, 276–7, 284

## B

back-to-back diagrams 568–9
bar charts 556, 557
   composite (stacked) 561–3, 566
   dual (side-by-side) 565–6
   stem-and-leaf diagrams 567–9
basic week 92
bearings 318–9
Bernoulli family 415
Bhascara 23
bias 543–5
binomial theorem 115
bonuses 92
brackets
   expansion 116–17, 121–2
   order of operations 31
Brahmagupta 23
British mathematicians 365

## C

calculus 365
capacity, unit conversions 368
Cardano, Girolamo 485
Chang Tshang 207
chaos 523
charts 556–66
Chinese mathematicians 207
chords 306, 341, 347
   equal 349–50
circles 306
   angles 341–9, 377, 379
   arc length 377–9
   area 373–6
   circumference 306, 341–6, 351, 373–7
   and the construction of simple equations 135
   diameter 306, 341, 373
   properties 341–4

and Pythagoras's theorem 422, 473
sector area 379–81
and trigonometry 434–6
circumference 306, 341–6, 351, 373–7
class intervals 576–7, 580–1
common ratio 184
completing the square 153–4
composite functions 253–6
compound interest 263–4
concyclic points 347
cones
similar 340
surface area 402–3
symmetry 314–15
and trigonometry in three dimensions 458
volume 397–401
constant of proportionality/variation 189
contingency tables, probability from 540–2
conversion, unit measures 367–9
coordinate geometry 274–301
coordinates 276–80
equation of a straight line 285–95
line segments 280–4
correlation 585, 589–93
cosine (cos) 423, 428–35, 437–8, 440–1
graphs 435, 437, 464, 467–9, 471–2
cosine rule 445–8
cube roots 29–30, 115
cubes (numbers) 25, 188
cubes (shapes)
nets 309
painted cube 515
symmetry 314–15
and trigonometry in three dimensions 456–7
and the volume of a pyramid 394–5
cubic expressions, factorisation 124–6
cubic functions 216–17, 247, 249
cubic rules, sequences with 181–4
cuboids
similar 338–9
surface area 384–7
symmetry 313, 314
and trigonometry in three dimensions 457, 460–2

cumulative frequency 579–84
currency conversion 91
curve stitching 267–8
cyclic quadrilaterals, opposite angles 346
cylinders
similar 339
surface area 384–7
symmetry 314–15

### D

data
averages 570, 576–9
bivariate 585
contingency tables 540
continuous 556, 559–60, 577, 579
discrete 556–60, 571, 577
display 556–61, 564, 567–61, 586–90, 590
grouped 559–60, 576–9, 596
large amounts 573–66
lists 17–21
mean/mode for grouped 576–8
quartiles 571–2
data sets 571, 586
decagons 308, 331
decimal places, approximation 65–6, 67
decimals 23, 36–7, 115, 207
changing from/to a fraction 44–6
percentage equivalent 47
denominators 34, 39–41, 80–1, 129–30
depreciation 94
Descartes, René 275
diameter 306, 341, 373
difference of two squares 123
discounts 93
distance 85–90
division
of fractions 43–4
order of operations 31
of a quantity in a given ratio 61–3
dodecagons 308, 331
domain 208–10
double time 92

### E

earnings 92–3
elimination 141
enlargements 497, 503–9, 513–14

centre of 503–9, 513–14
negative 507
scale factor 503–9, 513–14
equations
calculators 6, 155–61
of common functions 247–52
complex 147–9
exponential 175–6, 197
functions 208
linear 132–40, 155–58
root of 13
simultaneous 23, 141–8
of a straight line 285–95
trigonometric 439–40
see also quadratic equations
estimation 67–9, 577, 578
Euclid of Alexandria 303
Eudoxus of Asia Minor 303
Euler, Leonhard 415
evaluation (algebra) 123–4
events
combined 527–8, 537–8
complementary 524
independent 539
mutually exclusive 537
see also outcomes
exponential equations 175–6, 197, 262–5
exponential functions 218–20, 251, 259
inverse 259
see also logarithmic functions
exponential (geometric) sequences 184–7
exponential growth 56

### F

factorisation 117–18, 122–3
quadratic equations 150–2
quadratic/cubic expressions 124–6
factorised form 234–7
factors 26
common 27
highest common 27, 117
prime 26–7
false negatives/positives 548–9
Fermat, Pierre de 275, 523
Fermat's Last Theorem 275
Fibonacci sequence 485
finance 50–1, 91–94
formulas 103–4
changing the subject 119–20, 126–7
construction 138–40

for the terms of a linear
sequence 177–80
fractional indices 72–3
fractions 34–6
   addition 40
   algebraic 128–31
   of an amount 34–5
   changing from/to a decimal
      38–39, 44–6
   division 43–4
   equivalent 38–44
   improper 34, 35–6, 40
   multiplication 41–2
   percentage equivalent 47
   proper 34
   subtraction 40
French mathematicians 275
frequency histograms 558
frequency tables 19–21, 557,
     561–3
   grouped 559–60
frustums 397
functions 206–73
   calculators 226–33
   composite 253–6
   cubic 216–17, 247, 249
   equation of common 247–52
   exponential 218–20, 251, 259
   graphs of common 211–20
   inverse 257–8
   linear 211–13, 247–49
   logarithmic 259–65
   many-to-one 208
   as mappings 208–10
   notation 208–10
   one-to-one 208
   quadratic 214–15, 234–46
   range calculation from
      domain 209–10
   reciprocal 217–8, 247
   transformation of graphs
      221–5
   zeros (roots) 227–8

## G

Galileo 523
Gauss, Carl Friedrich 523
geometry 23, 207, 302–63
   angle measurement/drawing
      316–8
   angle properties 320–33
   bearing measurement/
      drawing 318–9
   circle properties 341–4
   geometrical vocabulary
      304–9

similarity 334–40
symmetry 310–15
vector geometry 493–7
*see also* coordinate geometry
gradients
   direct variation 189
   equation of a line through
      two points 292
   equation of a straight line
      290–1
   of a horizontal line 283
   linear functions 247
   negative 283
   positive 283
   of a straight line 282–4
   of a vertical line 283
graphic display calculators
     3–21, 300
   appropriate accuracy 67
   cubic functions 217
   equations 6, 155–61
   exponential functions 220
   functions 220, 226–33
   graphs 6, 9–14, 300
   history 4
   inequalities 165–6, 171–3,
      300
   interquartile range 583
   intersections 12–14
   linear functions 212–13
   lists 17–21
   logarithmic functions
      259–60
   order of operations 31–3
   quadratic equations 150
   quadratic functions 215
   scatter diagrams 586–9
   sequences 189–90
   simultaneous equations
      142–4
   speed, distance and time
      86–7
   standard form 75
   statistics 5, 571–2, 574–5,
      583
   substitution 118–19
   surds 79
   tables 6, 15–17
   trigonometric graphs 466–9,
      471–2
graphs 556–66
   calculators 6, 9–14, 300
   of common functions
      211–20
   of cumulative frequency 580
   of direct variation 189–91

of equations of common
     functions 247–52
of equations of a straight line
     285–95
and inequalities 164–73, 300
of inverse variation 191
linear 294–5, 564
and linear equations 155–58
of mass against extension 299
of quadratic equations
     158–61
of quadratic functions
     234–42, 244–6
of simultaneous equations
     143–4
of speed, distance and time
     88–90
of square roots 29
straight line 294–5
transformation 221–5
trigonometric 434–7, 439,
     442, 464–72
Greek mathematicians 303

## H

height 304, 371
hemispheres 393
heptagons 308, 331
hexagons 308
   angle properties 331, 332
   regular 331, 332
   similar 337
highest common factor (HCF)
     27, 117
Hindu mathematics 23
histograms, frequency 558
Huygens, Christiaan 523
hyperbolas, rectangular 268
hypotenuse 423, 426–9, 473

## I

images 497
indices (index) 25, 115
   algebra 174–6
   exponential functions 259,
     260
   fractional 72–3
   laws of 70–3
   negative 72, 77
   positive 70–1, 74–6
   rules of 174
   zero index 71
inequalities 162–73, 197
   linear 162–9, 300
   practical problems 168–9
   quadratic 169–73

integers, negative/positive 24
interest
    compound 54–7, 185–7, 263–4
    percentage 52–4
    simple 52–4
interquartile range 571, 581–4
intersections 12–14, 267–8, 284
    sets 98–101
inverse functions 257–8
inverse variation 191–4
Italian mathematicians 485

# K

Khayyam, Omar 115
kites, properties 307–8

# L

Laghada 23
Laplace, Pierre-Simon 523
latitude 276
length 367, 371
line graphs 564
line segments 280–4
line symmetry 310–11
linear equations 132–40, 155–58
linear functions 211–13, 247–49
linear inequalities 300
linear regression 587
    see also line of best fit
linear sequences 177–80
lines 304
    angle between a line and a plane 460–3
    angles formed within parallel lines 322–6
    angles on a straight line 320
    equation of a straight line 285–95
    gradient of a straight line 282–4
    mirror 497–8, 510–11
    parallel 283, 304, 322–6
    perpendicular 284, 304, 349–50, 497
    straight line graphs 294–5
lines of best fit 586–90
lists 17–21
logarithmic functions 259–65
longitude 276
loss 93–4
lowest common multiple (LCM) 28, 129

# M

mappings, functions as 208–10
mass, unit conversions 367–8
maxima, local 229–30
mean 17–21, 570–4, 576–80
measures 366–9
median 570–3, 576–7, 580–1, 583–4
    calculator list facility 17–21
    grouped data 577
mensuration 364–413
    arc length 377–9
    area of plane shapes 370–2, 382–7
    area of a sector 379–81
    circle circumference/area 373–6
    measures 366–9
    perimeter of simple plane shapes 370–2
    surface area of solids 392–3, 402–3
    volume of prisms 387–9
    volume of solids 390–401
metric units 366
mid-internal value 577
midpoints, of line segments 281–2
minima, local 229–30
mirror lines 497–8, 510–11
mixed numbers 34–6, 40–1, 43
mode 17–21, 570, 573–4, 576–8
money 50–1, 91–94
multiples 28
multiplication
    and bracket expansion 116–17, 121–2
    of fractions 41–2
    order of operations 31
multipliers 55–6, 185

# N

nets 309
Newton, Isaac 275, 365, 415, 523
Nightingale, Florence 555
nonagons 308, 331
notation
    function 208–10
    for sets 95–102
    numbers 22–113
        approximation 65–9
        cube 25, 188
        cube roots 29–30, 115
        decimals 36–7, 44–7
        factors 26–7
        fractions 34–6
        integers 24
        irrational 24–5, 78
        laws of indices 70–73
        mixed 34–6, 40–1, 43
        money/finance 50–1, 91–94
        multiples 28
        natural 24
        order of operations 31–3
        percentages 38–9, 47–58
        prime 25, 78, 105
        prime factors 26–7
        proportion 59–61, 63–4
        ratio 61–3
        rational 24, 78
        reciprocals 30, 43
        rounding 65–9
        set notation 95–102
        sexagesimal 86
        square 25, 105, 119, 123, 153–4, 188
        square roots 28–30, 115
        standard form 74–6
        surds 78–81
        time 82–90
        triangle 26, 188
        Venn diagrams 97–102
        vocabulary for sets of 24–30
numerators 34, 39, 41

# O

octagons 308, 331
optimum solutions 168
order 523
order of operations 31–3
origin 276
outcomes 547
    combined events 527
    experimental probability 543
    favourable 524
    independent events 539
    possible 524–5
    tree diagrams 529–32, 534
    see also events
overtime 92

# P

parabolas 158–60
parallelograms
    area 382–4
    properties 307–8
    similar 337

Parimala, Raman   23
Pascal, Blaise   275, 523
Pascal's Theorem   275
patterns   103
pay
pentagons   308, 331–2
percentages   37–8, 47–58
   changing a fraction to   37–8
   and compound interest   54–7
   decreases   50–1
   and discounts   93
   equivalent   47
   increases   50–1
   and loss   94
   and profit   94
   of a quantity   49
   reverse   58
   simple   47–8
percentiles   581, 594–5
perimeters   147–9, 370–2
period   464–5, 470
perpendicular bisectors   349–50
perpendicular lines   284, 304, 497
Persian mathematicians   115
pi (π)   23, 25, 373–5, 415
   and arc length   377
   leaving answers in terms of   385
   and sector area   379–80
   and the surface area of a cone   402–3
   and the surface area of a sphere   392–3
   and the volume of a cone   397–9
   and the volume of a sphere   390–2
pictograms   557
pie charts   561–2, 565
piece work   92
plane, angle between a line and a   460–3
points   304
   angles at   320–2
   concyclic   347
   equation of a line through two points   292
   equations of a quadratic function given a vertex and another point   242–6
   see also midpoints
polygons
   angle properties   327–33
   irregular   308, 331

properties   308, 326–7
see also specific polygons
powers   28–30
   laws of indices   70–3
   order of operations   31
   squares   105
   of two   188
price, cost/selling   93
prime factors   26–7
prime numbers   25, 78, 105
principal (capital)   52–4
prisms
   circular   387
   rectangular   387
   symmetry   313–14
   triangular   387
   volume of   387–89
probability   522–53
   of an event   524–7
   experimental   543
   from contingency tables   540–2
   laws of   537–42
   relative frequency   543–5
   theoretical   275, 524–8, 543
   tree diagrams   529–34
   unequal   530–2
   Venn diagrams   535–9
profit   93–4
proportion   59–61, 63–4
   direct   59–61
   inverse   63–4
proportionality, constant of   189
protractors   316
pyramids
   similar   339
   symmetry   314
   and trigonometry in three dimensions   457–60, 462–3
   volume   393–7
Pythagoras of Samos   303
Pythagoras' theorem   25, 207, 280, 416–22, 432–3
   and circles   422, 473
   and the cosine rule   445
   and trigonometry in three dimensions   456–7, 461

## Q

quadratic equations, solution   23, 115, 150–4
   by calculators   158–61
   by completing the square   153–4

   by factorisation   150–2
   using the quadratic formula   152–4
   zeros (roots)   159
quadratic expressions, factorisation   124–6
quadratic formula   152–4
quadratic function   214–15, 234–46
   equation from a vertex and another point   242–6
   from key values   234–46
   from the vertex   238–42
quadratic inequalities   169–73
quadratic rules, sequences with   181–4
quadrilaterals   307–8
   angle properties   329–30, 331, 346
   cyclic   346
   regular   330
   simple equations   133–8
   see also specific quadrilaterals
quartiles   570–3, 581–4

## R

radius   306, 341–4, 347, 349, 351
   angle between a tangent and   342–4
   and arc length   377
   and circumference   373
   and sector area   379–80
   and the surface area of a cone   402–3
   and the surface area of a sphere   392–3
   and the volume of a cone   397–400
   and the volume of a sphere   390–2
range   208, 570–3, 574, 576
   calculation from the domain   209–10
   interquartile   571, 581–4
ratio   59–63
reciprocal functions   217–9, 247
reciprocals   30, 43
rectangles   307–8, 370
rectangular hyperbola   268
reflection   497–9, 510–14
regions   267–8
relative frequency   543–5
rhombus   307–8, 337
roots   28–30

of an equation  13
cube roots  29–30, 115
of functions  227–8
square roots  28–30, 115
surds  78–81
rotation  497, 499–501, 514
rotational symmetry  312–13
rounding  65–9
Run-Matrix  5

## S

scalars, vector multiplication  490–1
scale, calculator graphs  9–10
scale factors  334, 338, 503–9, 513–14
scatter diagrams  585–93
sectors  306, 341, 373–4
    area  379–81
segments  306, 341, 345–7
semicircles  306, 341–2
sequences  177–88
    combinations of  187–8
    with cubic rules  181–3
    exponential (geometric)  184–7
    linear  177–80
    with quadratic rules  181–3
    term-to-term rules  177, 187
sets
    complement  97
    elements  95–9
    empty  96
    intersection  98–101
    notation  95–102
    subsets  96–7
    union  98–102
    universal  97
    vocabulary for  24–30
shapes
    angle properties  320–33
    area  370–2, 373–6
    perimeter  370–2
    properties  305–9
    similar  334–40
    symmetry  310–15
    see also specific shapes
significant figures  66, 67
similarity  334–40
simplest form  39
simplification
    algebraic fractions  128, 131
    simplifying the expression  116

simultaneous equations  23, 141–6
sine (sin)  423, 426–8, 432–5, 437–41
    and area of a triangle  449
    graphs  434–5, 437, 439, 442, 464–5, 469–70
sine rule  441–4
slide rules  4
solids
    surface area  392–3, 402–3
    volume  390–401
speed  85–90
spheres
    surface area  392–3
    symmetry  314–15
    volume  390–2
spreadsheets  6, 107
square roots  28–30, 115
squares (numbers)  25, 105
    completing the square  153–4
    difference of two squares  123
    perfect  121
    sequences  188
squares (shapes)
    properties  307–8
    similar  338
standard form  74–7
statistics  554–600
    averages  570–4
    basic graphs and charts  556–66
    calculators  4, 571–2, 574–5, 583
    correlation  585, 589–93
    cumulative frequency  579–84
    in history  555
    lines of best fit  586–90
    quartiles  570–3
    ranges  570–3
    scatter diagrams  585–93
    stem-and-leaf diagrams  567–9
stem-and-leaf diagrams  567–9
subsets  96–7
substitution  118–19, 142–6
subtraction
    of algebraic fractions  129–31
    of fractions  40
    order of operations  31

of vectors  489
surds (radicals)  78–81
surface area
    of a cone  402–3
    of a cuboid  384–7
    of a cylinder  384–7
    of a sphere  392–3
Swiss mathematicians  415
symmetry  311–15
    line  311–12
    rotational  312–13
    in three dimensions  313–15

## T

tables
    and calculators  6, 15–17
    table of values  231–2
    tally tables  556–7
tangent (tan)  423–6, 432–6, 438–41
    angle between a radius and  342–4
    graphs  436, 439, 464–6, 469
tangents  306, 341, 347, 351
term-to-term rules  177, 187
tessellation  326–7
tetrahedrons  309
Thales of Alexandria  303
three dimensions
    and nets  309
    symmetry in  313–15
    trigonometry in  456–63
    see also solids
time  82–90
    as sexagesimal number system  86
    and velocity  138
time and a half  92
transformations  484–5, 497–514
    combinations of  513–14
    enlargements  497, 503–9, 513–14
    of graphs  221–5
    inverse  497
    reflections  497–9, 510–14
    rotations  497, 499–501, 514
    translations  464, 467, 486, 489, 497, 501–3
    on trigonometric graphs  465–72
translations  497, 501–3
    on trigonometric graphs  466, 469
    and vectors  486, 489

trapeziums
   area  382–4
   properties  307–8
   similar  336, 337
tree diagrams  529–34
   for probability problems without replacement  532–4
   for unequal probabilities  530–2
triangle numbers  26, 188
triangles  305–6, 308, 515–17
   acute-angled  305
   adjacent side  423, 428–9
   angle properties  327–31
   area of a  371–2, 449–50
   base  371
   and circle properties  341–51
   congruent  306, 342, 349–51
   equilateral  305, 432–3
   height (altitude)  371
   isosceles  305, 347, 349, 432
   obtuse-angled  305
   opposite side  423, 426–7
   and Pythagoras' theorem  416–22
   regular  331
   right-angled  305, 416–42, 444–5, 449, 460, 473
   scalene  305
   similar  306, 334–6, 338, 340
   and simple equations  133–7
   vertex  371
trigonometric equations  439–40
trigonometric graphs  464–72
trigonometric ratios  432–40
trigonometry  23, 414–83
   applications  451–63
   cosine  423, 428–35, 437–8, 440–1
   cosine rule  445–8
   Pythagoras' theorem  416–22
   sine  423, 426–8, 432–5, 437–41
   sine rule  441–4
   special angles  432–40
   tangent  342–4, 423–6, 432–6, 438–41, 464–6, 469
   in three dimensions  456–63
trinomial expressions  121
two dimensions
   and coordinates  276
   symmetry in  310–13

## U
unitary method  59–61

## V
variables  585
variation
   constant of  189
   direct  189–92
vectors  482–496
   addition  489
   column  486–93, 501–3
   geometry  493–496
   magnitude  492–496
   multiplication by a scalar  490–1
   simple  486–90
   subtraction  489
   translation  501–3
velocity  85, 138

Venn diagrams  97–102, 535–9
vertices (vertex)  229
   of cyclic quadrilaterals  346
   of polygons  331
   and quadratic functions  238–46
   of triangles  327, 329, 371
   and the volume of a pyramid  395
Villani, Giovanni  555
volume
   of a cone  397–401
   conversions  368–9
   of a prism  387–9
   of a pyramid  393–7
   of similar shapes  338–40
   of a sphere  390–2
   *see also* capacity
volume factor  338

## W
Wang Xiaotong  207

## X
$x$-axis  276–7, 284
$x$-intercept  13

## Y
$y$-axis  276–7, 284
   intersection  228–9
$y$-intercept  13–14, 229, 247, 290–1

## Z
zero  23
zero index  71